AF581895

Integrated Water Resources Research

Integrated Water Resources Research: Advancements in Understanding to Improve Future Sustainability

Editor

Jason A. Hubbart

MDPI • Basel • Beijing • Wuhan • Barcelona • Belgrade • Manchester • Tokyo • Cluj • Tianjin

Editor
Jason A. Hubbart
West Virginia University
USA

Editorial Office
MDPI
St. Alban-Anlage 66
4052 Basel, Switzerland

This is a reprint of articles from the Special Issue published online in the open access journal *Water* (ISSN 2073-4441) (available at: https://www.mdpi.com/journal/water/special_issues/Water_Future_Sustainability).

For citation purposes, cite each article independently as indicated on the article page online and as indicated below:

LastName, A.A.; LastName, B.B.; LastName, C.C. Article Title. *Journal Name* **Year**, *Volume Number*, Page Range.

ISBN 978-3-0365-0228-1 (Hbk)
ISBN 978-3-0365-0229-8 (PDF)

Contents

About the Editor

Jason A. Hubbart (Professor and Director) serves as Director of the West Virginia University (WVU) Institute of Water Security and Science (IWSS). He is Professor of Physical Hydrology and Water Quality in the Division of Plant and Soil Sciences (School of Agriculture and Food) and Division of Forestry and Natural Resources (School of Natural Resources) in the Davis College of Agriculture, Natural Resources and Design (DCANRD). Dr. Hubbart also serves as Assistant Director of the West Virginia Agricultural and Forestry Experiment Station, and is the West Virginia gubernatorial appointee to the Science and Technical Advisory Committee of the Chesapeake Bay Program (STAC). Hubbart advises postdocs and graduate students, conducts research, and has published over 115 peer-reviewed articles in the fields of physical hydrology, watershed management and water quality, and environmental biophysics in addition to directing the WVU DCANRD Interdisciplinary Hydrology Laboratory (IHL).

Editorial

Integrated Water Resources Research: Advancements in Understanding to Improve Future Sustainability

Jason A. Hubbart [1,2,3]

1 Institute of Water Security and Science, West Virginia University, 3109 Agricultural Sciences Building, Morgantown, WV 26506, USA; jason.hubbart@mail.wvu.edu; Tel.: +1-30-4293-2472
2 Division of Plant and Soil Sciences, Davis College of Agriculture, Natural Resources and Design, West Virginia University, Agricultural Sciences Building, Morgantown, WV 26506, USA
3 Division of Forestry and Natural Resources, Davis College of Agriculture, Natural Resources and Design, West Virginia University, Agricultural Sciences Building, Morgantown, WV 26506, USA

Received: 1 August 2020; Accepted: 4 August 2020; Published: 6 August 2020

Abstract: Anthropogenic and natural disturbances to freshwater quantity and quality is a greater issue for society than ever before. To successfully restore water resources in impaired watersheds requires understanding the interactions between hydrology, climate, land use, water quality, ecology, social and economic pressures. Current understanding of these interactions is limited primarily by a lack of innovation, investment, and interdisciplinary collaboration. This Special Issue of Water includes 18 articles broadly addressing investigative areas related to experimental study designs and modeling ($n = 8$), freshwater pollutants of concern ($n = 7$), and human dimensions of water use and management ($n = 3$). Results demonstrate the immense, globally transferable value of the experimental watershed approach, the relevance and critical importance of current integrated studies of pollutants of concern, and the imperative to include human sociological and economic processes in water resources investigations. Study results encourage cooperation, trust and innovation, between watershed stakeholders to reach common goals to improve and sustain the resource. The publications in this Special Issue are substantial; however, managers remain insufficiently informed to make best water resource decisions amidst combined influences of land use change, rapid ongoing human population growth, and changing environmental conditions. There is thus, a persistent need for further advancements in integrated and interdisciplinary research to improve scientific understanding, management and future sustainability of water resources.

Keywords: watershed management; water quality; water resources; human dimensions of water; watershed modeling; hydrological modeling; water pollutants

1. Introduction

The requirements to understand, synthesize, and resolve water resource challenges are more complex today than any time in human history. The scope of contemporary water resource challenges requires an unprecedented amount of information that spans a continuum of physical, natural and socioeconomic sciences [1]. For the first time in history, a freshwater crisis has grown global in proportion. This is at least in part attributable to economic globalization and continuing human-induced perturbation of natural resource commodities including the water cycle. This crisis is manifested in many sectors including (but not limited to) politics, public health, agriculture, development and the environment [2]. In the United States alone, unwavering aggressive use of freshwater resources, leading to scarcity and quality problems, requires innovative, interdisciplinary and integrated scientific, technological, and training solutions. This perspective is important given that cumulative anthropogenic drivers (e.g., land use, population growth, climate change) confound the uncertainties of decision-making, and many critical information gaps remain. Such cumulative alterations impact water quality (e.g., chemical composition, pathogen presence and persistence) and quantity (e.g., low flows, peak flows, flooding) regimes [3–6], and can ultimately result in further freshwater ecosystem degradation. Land managers are thus often inadequately informed to make correct management decisions in regions where sources of impairment are simultaneously shifting due to combined influences of land use change, rapid ongoing human population growth, and changing environmental conditions. Given the inherent complexity of these dynamic systems, it is not surprising that effective management is dynamic, characterized by ongoing scientific advancements and policy refinements [7,8]. The intent of this Special Issue of Water was to provide some of the latest integrated and multidisciplinary water resources research that advances the understanding, sustainability and therefore management of water resources.

2. Special Issue Overview

This Special Issue of Water entitled, "Integrated Water Resources Research: Advancements in Understanding to Improve Future Sustainability" includes 18 articles broadly addressing investigative areas related to experimental study designs and modeling (n = 8), fresh water pollutants of concern (n = 7), and human dimensions of water use and management (n = 3), presented, in brief, in the following text.

2.1. Experimental Study Designs and Modeling

A case study was presented by Hubbart et al. [9] using an experimental watershed study design and collaborative adaptive management (CAM) framework [1] to illustrate how these approaches can be used in mixed-land-use (including municipal) watersheds to provide quantitative information to identify and address past, present, and future sources of impairment, and thus better utilize limited taxpayer funds towards mitigation. Results identified challenges associated with CAM, and how the experimental watershed approach can help to objectively elucidate causal factors, target critical source areas, and provide the science-based information needed to make informed management decisions. Results further demonstrate the immense, globally transferrable value of the experimental watershed approach for municipal watersheds. Mixed land use influences on stream physical habitat was investigated by Zeiger and Hubbart [10] who directly measured channel geomorphology, and stream substrate composition every 100 m over 56 km (n = 561). Results showed that agricultural and urban land use explained nearly all the variance in average river width to depth ratios (R^2 = 0.960; p = 0.020; n = 5), and maximum bank angle (R^2 = 0.896; p = 0.052; n = 5). Streambed substrate samples indicated significantly ($p < 0.001$) increased embeddedness at agricultural and urban sites. Results demonstrate how hydrogeomorphological assessments can help guide regional stream restoration efforts. Cao et al. [11] used a three-stage data envelopment analysis (DEA) model and Chinese provincial panel data to analyze input efficiency of water-saving irrigation. Results showed that overall efficiency of water-saving irrigation practices are increasing nationally (China). Authors noted that efficiency of water-saving irrigation input will significantly increase investment in areas such as Hainan, Chongqing, Guizhou, Tibet, and Qinghai. Kutta and Hubbart [12] investigated land cover alteration (e.g., reforestation) feedbacks on climate with respect to implications for ecology, water resources, and watershed management. Results indicated an increasingly wet and temperate climate for the Northeast United States, and specifically the state of West Virginia, characterized by warming summertime minimum temperatures, cooling maximum temperatures, and increased annual precipitation that accelerated during the second half of the period of record (1959–2016). Trends were elevation dependent and may be accelerating due to local to regional ecohydrological feedbacks including increasing forest age and density, changing forest species composition, and increasing globally averaged atmospheric moisture. Importantly, results imply that excessive wetness may become the primary ecosystem stressor associated with climate change in the USA's rugged and flood prone Appalachian region and by extension, similar physiographic regions, globally. Work by Gaertner et al. [13] supported these findings using historic and future predicted climate and water balance data to quantify streamflow sensitivity and project future streamflow changes for 29 forested catchments in the Northeast United States. Results showed that streamflow is expected to increase under the low-emission and decrease under the high-emission climate change pathway. In related work, Rojano et al. [14] showed that modeled urban flow regimes were correlated to net ecosystem production (NEP), and under hypoxic conditions, local inflows were correlated with specific conductance. Results show the value of using integrated modeling approaches with observed data to resolve big-river challenges. El Hafyani et al. [15] developed a method to assess regional water balances using remote sensing techniques in the Boufakrane river watershed in Meknes Region (Morocco). Using a supervised classification procedure and combined mapping procedure, the water balance was developed considering changing management and consumption patterns. Results showed that urban areas, natural vegetation, arboriculture and cereals increased the water balance by approximately 184%, 13%, 35% and 49%, respectively, while forests and bare soils decreased the water balance by approximately 79% and 17%, respectively. Further, increased water consumption by human activities was largely mitigated by evapotranspiration savings from deforestation, a practice that can no longer be sustained. Finally, Gootman et al. [16] conducted a study validating five saturated hydraulic conductivity (K_{sat}) pedotransfer functions in a catchment of the Chesapeake Bay Watershed, Northeast United States. The study showed that dry bulk density and porosity were significantly different by location ($p < 0.05$) and four different models corroborated that spatial variability in farm-scale K_{sat} estimates was small (CV < 0.5), thereby validating the use of simple, soil-property-based models to predict K_{sat}, thereby increasing model applicability and transferability. These investigations advance understanding of land use and modeling practices on water resources and therefore predictive confidence in water resources management decisions.

2.2. Fresh Water Pollutants of Concern

Multiple articles were published that quantitatively characterize relationships between Escherichia (E) coli concentration, suspended particulate matter (SPM) particle size class, physicochemical concentrations and land use practices. These articles are some of the first (globally) that use the study design described in Hubbart et al. [1,9] to advance aquatic microbial process understanding. For example, Petersen and Hubbart [17] showed that statistically significant relationships exist between E. coli concentration, size interval ($p < 0.0001$) and suspended particulate matter (SPM) ($p = 0.05$). Results showed a predominance (90% of total) of E. coli colony forming units (CFU) in the <5 μm SPM interval. Petersen and Hubbart [18] showed that Annual average E. coli concentration increased by approximately 112% from acid mine drainage (AMD) impacted headwaters to lower watershed reaches (approximate averages of 177 CFU per 100 mL vs. 376 CFU per 100 mL, respectively). Significant Spearman's correlations ($p < 0.05$) were identified from analyses of pH and E. coli concentration data representing 77% of sample sites. Results highlight legacy effects of historic coal mining drainage on microbial CFUs in fresh water. A tipping point of 25–30% mixed development was identified as leading to significant ($p < 0.05$) negative correlations between chloride and E. coli concentrations. Agricultural land use sub-catchments were shown to have elevated E. coli concentrations (avg. 560 CFU per 100 mL) relative to proximate mixed development (avg. 330 CFU per 100 mL) and forested (avg. 206 CFU per 100 mL) sub-catchments [19]. Additionally, agricultural land use showed statistically significant relationships ($p < 0.01$) between annual E. coli and SPM concentration. Quarterly principle component analysis (PCA) biplots indicated temporal variability in land use impacts on E. coli and SPM concentrations, with agricultural land use being closely correlated with both pollutants during spring and summer quarters but not fall and winter quarters. Finally, Petersen and Hubbart [20] provided an overview of factors known to impact the survival of E. coli in the environment. Findings indicated (1) large knowledge gaps regarding environmental factors influencing E. coli's survival in the environment, and (2) a lack of implemented management strategies assessed at larger field scales, thus leaving their actual impact(s) largely unknown. Kessler et al. [21] used an in silico ontological pathway analysis to identify the genes affected by the most commonly detected endocrine-disrupting chemicals

(EDC) in large river water supplies, grouped by organismal injuries, cell death, cancer, and behavior. Results highlighted the critical need of additional investigations with a potential emphasis on the effects linked to increased DNA impacts. Spatio-temporal variability in total dissolved solids (TDS) was investigated using a modeling approach within a large river basin of the Northeastern United States to assess the extent and drivers of vulnerability to TDS exceedance [22]. It was shown that consistently low TDS from contributing surface waters to receiving water reduced vulnerability to elevated TDS. Authors identified that management should include efforts to increase assimilative capacity and manage upstream reservoirs. Finally, Horne and Hubbart [23] used the design explicated by Hubbart et al. [1,9] and Petersen and Hubbart [17–19] to investigate stream water temperature (Tw) and land use practices. Using data from 21 stream temperature monitoring sites, results showed that forested land use was negatively correlated ($p = 0.05$) with mean and maximum Tw and agricultural land use was significantly positively correlated ($p = 0.05$) to maximum Tw. Mixed development and Tw were also shown to be significantly correlated ($p = 0.05$) depending on time of year. Correlation trends in some reaches were reversed between the winter and summer seasons, contradicting previous research. Independently, and collectively, these studies advance understanding of land use impacts on many water quality constituents of concern, and aid in the decision making of effective water quality management practices and policies.

2.3. Human Dimensions of Water Use and Management

Under the research theme of resilience by means of service and rapid recovery after disasters it was shown that sufficient technology and good water quality are not enough for achieving resilient water services, but education and institutional management are essential components of that process that can be achieved by a deliberate education system, capacity building, and good governance [24]. Spatial patterns of water quality perceptions were gathered in a survey of southwest West Virginia (WV), United States residents to identify significant differences across counties labeled as socioeconomically transitional, at-risk, and distressed, relative to water quality perceptions, education level, and income level [25]. Findings highlighted the importance of location on water quality perceptions and presented an analytical framework that could be applied to future research. Finally, a study performed in Tanzania showed quantitative relationships between increasing population food insecurity and climate change relationships related to small farm landowners ($n = 701$) standards of living [26]. A bivariate logistic regression model was developed to relate application of water conservation techniques (WCT) to household socio-economic, and farmer perception related variables. Results, suggest that policies must encourage conservation behavior, emphasize the economic and food security-related benefits of adopting WCTs, include strategies that make adoption of WCTs attractive, attempt to reach greater number of farmers via social networks and provide better access to public funds.

3. Conclusions

Published articles from this Special Issue of Water address many aspects of integrated and multidisciplinary water resources research. Article contributions include advancements in effective ways of conducting integrated water research and communicating results to promote deliberate advancements in management, human well-being, and resource sustainability. Assuming human-induced environmental changes continue as anticipated, there is a need for highly organized efforts to continuously monitor, model and improve best management practice decisions to mitigate anthropogenic and natural pressures on water resources. This pursuit is critical because, in the absence of advances in integrated and interdisciplinary observed data and modeling, sources of impairment may remain unrecognized and unaddressed.

Author Contributions: Conceptualization: J.A.H.; formal analysis: J.A.H.; investigation: J.A.H.; data curation: J.A.H.; writing-original draft preparation: J.A.H.; writing—review and editing: J.A.H.; supervision: J.A.H.;

project administration: J.A.H.; funding acquisition: J.A.H. All authors have read and agreed to the published version of the manuscript.

Funding: Funding was provided by the National Science Foundation under Award Number OIA-1458952, the USDA National Institute of Food and Agriculture, Hatch project 1011536, and the West Virginia Agricultural and Forestry Experiment Station. Results presented may not reflect the views of the sponsors, and no official endorsement should be inferred.

Conflicts of Interest: The authors declare no conflict of interest.

References

1. Hubbart, J.A. Water Resources Education: Preparing the Next Generation of Water Resource Professionals. Special edition. *Water Resour. Impact* **2010**, *12*, 13–15.
2. Kirshen, P.; Vogel, R.M.; Rogers, B.L. Challenges in Graduate Education in Integrated Water Resources Management. *J. Water Resour. Plan. Manag.* **2004**, *130*, 185–186. [CrossRef]
3. Yates, M.V.; Gerba, C.P.; Kelley, L.M. Virus persistence in groundwater. *Appl. Environ. Microbiol.* **1985**, *49*, 778–781. [CrossRef] [PubMed]
4. Poff, N.L.; Allan, J.D.; Bain, M.B.; Karr, J.R.; Prestegaard, K.L.; Richter, B.D.; Sparks, R.E.; Stromberg, J.C. The Natural Flow Regime. *Bioscience* **1997**, *47*, 769–784. [CrossRef]
5. Kellner, E.; Hubbart, J.; Ikem, A. A comparison of forest and agricultural shallow groundwater chemical status a century after land use change. *Sci. Total. Environ.* **2015**, *529*, 82–90. [CrossRef]
6. Zell, C.; Kellner, E.; Hubbart, J. Forested and agricultural land use impacts on subsurface floodplain storage capacity using coupled vadose zone-saturated zone modeling. *Environ. Earth Sci.* **2015**, *74*, 7215–7228. [CrossRef]
7. Allan, C.; Curtis, A.; Stankey, G.; Shindler, B. Adaptive Management and Watersheds: A Social Science Perspective1. *JAWRA* **2008**, *44*, 166–174. [CrossRef]
8. Prato, T. Multiple-attribute evaluation of ecosystem management for the Missouri River system. *Ecol. Econ.* **2003**, *45*, 297–309. [CrossRef]
9. Hubbart, J.; Kellner, E.; Zeiger, S. A Case-Study Application of the Experimental Watershed Study Design to Advance Adaptive Management of Contemporary Watersheds. *Water* **2019**, *11*, 2355. [CrossRef]
10. Zeiger, S.J.; Hubbart, J.A. Characterizing Land Use Impacts on Channel Geomorphology and Streambed Sedimentological Characteristics. *Water* **2019**, *11*, 1088. [CrossRef]
11. Cao, Y.; Zhang, W.; Ren, J. Efficiency Analysis of the Input for Water-Saving Agriculture in China. *Water* **2020**, *12*, 207. [CrossRef]
12. Kutta, E.; Hubbart, J. Climatic Trends of West Virginia: A Representative Appalachian Microcosm. *Water* **2019**, *11*, 1117. [CrossRef]
13. Gaertner, B.; Fernandez, R.; Zegre, N. Twenty-First Century Streamflow and Climate Change in Forest Catchments of the Central Appalachian Mountains Region, US. *Water* **2020**, *12*, 453. [CrossRef]
14. Rojano, F.; Huber, D.H.; Ugwuanyi, I.R.; Noundou, V.L.; Kemajou-Tchamba, A.L.; Chavarria-Palma, J.E. Net Ecosystem Production of a River Relying on Hydrology, Hydrodynamics and Water Quality Monitoring Stations. *Water* **2020**, *12*, 783. [CrossRef]
15. El Hafyani, M.; Essahlaoui, A.; Van Rompaey, A.; Mohajane, M.; El Hmaidi, A.; El Ouali, A.; Moudden, F.; Serrhini, N.-E. Assessing Regional Scale Water Balances through Remote Sensing Techniques: A Case Study of Boufakrane River Watershed, Meknes Region, Morocco. *Water* **2020**, *12*, 320. [CrossRef]
16. Gootman, K.S.; Kellner, E.; Hubbart, J.A. A Comparison and Validation of Saturated Hydraulic Conductivity Models. *Water* **2020**, *12*, 2040. [CrossRef]
17. Petersen, F.; Hubbart, J.A. Quantifying Escherichia coli and Suspended Particulate Matter Concentrations in a Mixed-Land Use Appalachian Watershed. *Water* **2020**, *12*, 532. [CrossRef]
18. Petersen, F.; Hubbart, J.A. Advancing Understanding of Land Use and Physicochemical Impacts on Fecal Contamination in Mixed-Land-Use Watersheds. *Water* **2020**, *12*, 1094. [CrossRef]
19. Petersen, F.; Hubbart, J.A. Spatial and Temporal Characterization of Escherichia coli, Suspended Particulate Matter and Land Use Practice Relationships in a Mixed-Land Use Contemporary Watershed. *Water* **2020**, *12*, 1228. [CrossRef]

20. Petersen, F.; Hubbart, J.A. Physical Factors Impacting the Survival and Occurrence of Escherichia coli in Secondary Habitats. *Water* **2020**, *12*, 1796. [CrossRef]
21. Kessler, J.; Dawley, D.; Crow, D.; Garmany, R.; Georgel, P.T. Potential Health Risks Linked to Emerging Contaminants in Major Rivers and Treated Waters. *Water* **2019**, *11*, 2615. [CrossRef]
22. Merriam, E.R.; Petty, J.T.; O'Neal, M.; Ziemkiewicz, P.F. Flow-Mediated Vulnerability of Source Waters to Elevated TDS in an Appalachian River Basin. *Water* **2020**, *12*, 384. [CrossRef]
23. Horne, J.P.; Hubbart, J.A. A Spatially Distributed Investigation of Stream Water Temperature in a Contemporary Mixed-Land-Use Watershed. *Water* **2020**, *12*, 1756. [CrossRef]
24. Laitinen, J.; Kallio, J.; Katko, T.S.; Hukka, J.J.; Juuti, P. Resilient Urban Water Services for the 21th Century Society—Stakeholder Survey in Finland. *Water* **2020**, *12*, 187. [CrossRef]
25. Andrew, R.G.; Burns, R.C.; Allen, M.E. The Influence of Location on Water Quality Perceptions across a Geographic and Socioeconomic Gradient in Appalachia. *Water* **2019**, *11*, 2225. [CrossRef]
26. Jha, S.; Kaechele, H.; Sieber, S. Factors Influencing the Adoption of Water Conservation Technologies by Smallholder Farmer Households in Tanzania. *Water* **2019**, *11*, 2640. [CrossRef]

Article

A Case-Study Application of the Experimental Watershed Study Design to Advance Adaptive Management of Contemporary Watersheds

Jason A. Hubbart [1,*], Elliott Kellner [2] and Sean J. Zeiger [3]

1 West Virginia University, Institute of Water Security and Science, Davis College of Agriculture, Natural Resources and Design, Schools of Agriculture and Food, and Natural Resources, 3109 Agricultural Sciences Building, Morgantown, WV 26506, USA
2 West Virginia University, Institute of Water Security and Science, Davis College of Agriculture Natural Resources and Design, Division of Plant and Soil Sciences, 3011 Agricultural Sciences Building, Morgantown, WV 26506, USA; Elliott.Kellner@mail.wvu.edu
3 School of Natural Resources, University of Missouri, 203-T ABNR Building, Columbia, MO 65211, USA; sjzxbb@mail.missouri.edu
* Correspondence: Jason.Hubbart@mail.wvu.edu; Tel.: +1-3042932472

Received: 14 September 2019; Accepted: 6 November 2019; Published: 9 November 2019

Abstract: Land managers are often inadequately informed to make management decisions in contemporary watersheds, in which sources of impairment are simultaneously shifting due to the combined influences of land use change, rapid ongoing human population growth, and changing environmental conditions. There is, thus, a great need for effective collaborative adaptive management (CAM; or derivatives) efforts utilizing an accepted methodological approach that provides data needed to properly identify and address past, present, and future sources of impairment. The experimental watershed study design holds great promise for meeting such needs and facilitating an effective collaborative and adaptive management process. To advance understanding of natural and anthropogenic influences on sources of impairment, and to demonstrate the approach in a contemporary watershed, a nested-scale experimental watershed study design was implemented in a representative, contemporary, mixed-use watershed located in Midwestern USA. Results identify challenges associated with CAM, and how the experimental watershed approach can help to objectively elucidate causal factors, target critical source areas, and provide the science-based information needed to make informed management decisions. Results show urban/suburban development and agriculture are primary drivers of alterations to watershed hydrology, streamflow regimes, transport of multiple water quality constituents, and stream physical habitat. However, several natural processes and watershed characteristics, such as surficial geology and stream system evolution, are likely compounding observed water quality impairment and aquatic habitat degradation. Given the varied and complicated set of factors contributing to such issues in the study watershed and other contemporary watersheds, watershed restoration is likely subject to physical limitations and should be conceptualized in the context of achievable goals/objectives. Overall, results demonstrate the immense, globally transferrable value of the experimental watershed approach and coupled CAM process to address contemporary water resource management challenges.

Keywords: urban watershed management; municipal watershed; water quality impairment; collaborative adaptive management; water resources; urban watersheds

1. Challenges in Contemporary Watershed Management

1.1. Collaborative Adaptive Management

Contemporary watershed management problems are complex, consisting of multiple, conflicting, and non-linear and/or stochastic variables. Research has shown that the management of watersheds is most effective using an adaptive and integrated approach based on iterative applications of best (or better) practices guided by ecosystem process responses. Adaptive management comprises critical steps that include (but are not limited to) problem assessment, remediation design, implementation, monitoring, evaluation, and management plan adjustment [1–5]. Based on initial problem assessments, a project is often designed and implemented, and then, with regular monitoring and (re)evaluation, adjustments may be applied, and projects revised. This iterative process helps update management plans over time while incorporating additional precautions and experiences garnered from new information. In addition to complexity and uncertainty, natural resource management is interconnected through equally complex and intermingled land use needs (practices) of humans. Land and water management, therefore, benefit from a collaborative approach that includes multiple stakeholders. Collaborative Adaptive Management (CAM) facilitates the introduction of local stakeholders as a major component of sustainable decision-making. It is recognized that there may be other approximate derivatives to CAM, but for simplicity, here we reference CAM. The primary goal of CAM is to integrate knowledge and science with experience and the perspectives of scientists, stakeholders, and managers for more effective management decision-making [6–9].

Collaborative Adaptive Management has been applied broadly in landscape-level planning and management globally [10]. There have been various applications of the collaborative adaptive process in regions of the world including (but not limited to) Southeast Asia [11], Brazil [12,13], and Europe [14–20]. Most CAM applications have been in the United States and Australia [21–23]. One such application in the United States is the Chesapeake Bay Program (CBP), a globally recognized model for the collaborative management and restoration of large aquatic ecosystems [24]. The Chesapeake Bay Total Maximum Daily Load (TMDL) [25], administered by the U.S. Environmental Protection Agency (EPA), is one of the largest and longest-running pollution control programs in history. The Chesapeake Bay Watershed Agreement (2014) was drafted based on contributions from numerous federal and state agencies, citizens, stakeholders, academic institutions, local governments, and non-profit organizations [26]. The regulatory program is currently implemented in six states (Delaware, Maryland, New York, Pennsylvania, Virginia, West Virginia), and the District of Columbia, and mandates reductions of three primary constituents of concern (i.e., nitrogen, phosphorus, and suspended sediment) to improve various indicators of Bay-water quality and aquatic habitat (e.g., dissolved oxygen, turbidity, submerged aquatic vegetation). The Chesapeake Bay Program provides a model for CAM activities, including stream restoration, upland pollutant source reduction, infrastructure improvements such as urban green streets, and retrofitting existing stormwater facilities to improve water quality [27]. Another leading example of CAM in the United States is the Mississippi River Basin, Gulf Hypoxia Program coordinated by the Mississippi River/Gulf of Mexico Watershed Nutrient Task Force (Hypoxia Task Force or HTF). The HTF is a collaborative management effort by state and federal agencies established, at least in part, to reduce the size and persistence of "The Dead Zone", a large area of hypoxia (i.e., oxygen concentration less than 2 mg L^{-1}) in the Gulf of Mexico [28–32]. The CAM model has been similarly utilized to improve the management of various natural resources, including (but not limited to) wild fisheries management [33], surface and groundwater resource allocation [34], and urban water management [35] in Australia. Collectively, action plans associated with CAM programs highlight the importance of accounting for future monitoring information, changing environmental conditions, and lessons learned globally [21,31,32], thereby emphasizing the need for high-quality environmental data to inform effective management of natural resources.

1.2. Environmental Monitoring to Improve Management

Over 100 years ago, watershed managers recognized the need to better understand land use and water quality and quantity relationships, in order to improve management practices and stewardship and to sustain natural resource commodities. There was an urgency to understand how the water balance of a given watershed is controlled by climate, soil, and vegetation interactions, and how alterations of such factors may affect the water regime (i.e., timing and quantity of water), water quality, and various related natural resources [36,37]. Among the first studies in the United States to address nationwide watershed issues was the Wagon Wheel Gap Experimental Watershed Study, which started in 1909 to protect navigable streams at the watershed scale [38,39]. Other early studies focused on the effects of road building and forest harvest practices on water quantity and quality (e.g., flow and flow velocity, erosion, sediment, nutrients). Later studies included agricultural impacts. However, despite advancements generated through early studies, watershed mismanagement continues to be identified as primarily responsible for anthropogenic disturbances of waterways [40–43], and land managers remain poorly equipped to address contemporary mixed-land-use watershed issues that are set in a continuum of forested, agriculture, *and* urban land use types and are associated with aggressive human population growth. For context, a recent report published by the United Nations [44] showed approximately 30% of the global human population (751 million people) lived in metropolitan areas during 1950. By 2018 that percentage had grown to 55% (4.2 billion urban inhabitants) [44]. By 2050, it is projected that almost 70% of the global human population, approximately 6.7 billion people, will live in metropolitan areas [44]. Moreover, there are concomitant, growing, human health and quality-of-life issues related to water resources that are global in scale. There are increasing demands for management solutions and guarantees of sustainable water resources and water quality for future generations, which will depend on research, education, outreach, collaboration, adaptive management, and understanding the cultural anthropology of water [45]. Considering the scope of these complexities, is there any question that we must reconsider all that we think we know, and reimagine watershed management, given the rapid succession of intermingled impacts in recent decades alone?

1.3. Contemporary Application of the Experimental Watershed Approach

The nested-scale and/or paired experimental watershed study designs (and other derivations) have been shown to be effective approaches for quantitatively characterizing hydrologic and water quality perturbations in mixed-land-use watersheds [46–57]. Nested watershed study designs utilize a series of sub-catchments inside a larger watershed to monitor land use impacts on environmental variables of interest. A paired watershed study design includes data collection from at least two watersheds (control and treatment) with similar physiographical characteristics. Sub-catchments are delineated to isolate land use types and hydrologic characteristics. While often applied at the watershed scale, the design concept can be applied at any scale, from the reach level (scale-nested) to the basin level. Ultimately, the design enables researchers to partition and quantify the influencing processes observed at the sub-catchment scale [58], and thereby determine the influence and cumulative effect of dominant land use types on the response variable of interest. By applying a nested-scale experimental watershed approach, factors (e.g., land use, hydroclimate) contributing to a given variable of interest may be more effectively (objectively) disentangled, producing quantitative information regarding hydrologic and water quality regimes related to specific land-uses. For example, Tetzlaff et al. [57] discussed how the experimental watershed approach has yielded several benefits, including (but not limited to) (1) science-based information to answer site-specific management questions, (2) quantitative information needed for ongoing model development, and (3) ground-truthing of large-scale remote sensing data. Experimental watershed studies can elucidate unknown problems in a watershed of interest, where unique combinations of natural and anthropogenic (legacy and ongoing) conditions would otherwise confound planning efforts [59]. Additionally, experimental watershed studies can enable the discovery of previously unknown phenomena and/or processes that contribute to globally important natural resources security issues. For example, results from Hubbard Brook Experimental Forest involving the acid rain phenomenon were especially transformative, with important global implications for natural resources management [48]. A study by Felson and Pickett [60] showed that scientists and urban designer partnerships could result in a deeper understanding of urban land-use influence on ecological response across spatial scales, and rural-urban land use gradients [57]. Such information is important considering the combined influence of hydroclimate extremes and land use change that is expected to continue to degrade water resources and ecological health in future decades [41].

Despite the potential for experimental watershed studies to yield valuable information for land and water resource managers [59], the approach is rarely applied in contemporary, mixed-land-use watersheds due to seemingly daunting challenges. Felson and Pickett [60] noted challenges associated with collaborative urban planning, and science-based experimental design efforts include, (1) the need for enhanced communication between planners, scientists, and stakeholders, (2) a lack of control over experimental installation during initial urban planning and development, and (3) associated costs and financial limitations typically faced by local municipalities. However, given the increasing rate and intensity of global water resource degradation, effective methods must be utilized to overcome obstacles to implementation, regardless of the level of complexity. Therefore, while the challenges noted by previous authors are certainly affirmed, investment costs in the shorter term might be outweighed by long-term irretrievable effects of less informed management. Contemporary application of the experimental watershed approach is an effective method that can provide the detailed information necessary to improve management, conservation, and sustainability of water resources while driving down long-term costs. To date, one of the few examples of a mixed-land-use, contemporary experimental watershed study is Hinkson Creek Watershed [53]. The purpose of this article is to provide an example of the successful integration of the experimental watershed study design and collaborative adaptive management to advance policy and management practices in contemporary watersheds.

2. Case Study: Hinkson Creek Watershed

2.1. *Case Study Setting*

To provide important context for the reader, in particular, pertaining to transferability to other mixed (multi) use watersheds globally, we provide information about the case-study watershed used for the current work as follows. Hinkson Creek Watershed (HCW) is located within the Lower Missouri-Moreau River Basin (LMMRB) in central Missouri, USA (Figure 1) [53]. The main channel, Hinkson Creek, is a 3rd order stream that flows through a basin of approximately 231 km^2 [53]. At the time of this work, urban areas of HCW were primarily residential with progressive commercial expansion from the City of Columbia (population approximately 122,000) [61]. Land use in the watershed was approximately 32% forest, 37% pasture or cropland, and 29% urban (Table 1). The regional climate is dominated by continental polar air masses and maritime and continental tropical air masses during the winter and summer, respectively. The mean annual total precipitation is approximately 1096 mm, and the mean annual air temperature is approximately 13.5 °C. A wet season occurs primarily from March through June. A portion of the LMMRB was targeted as critical for controlling erosion and nonpoint source pollution in 1998 [53]. Watershed restoration efforts in the LMMRB were accelerated by mandates of the Clean Water Act (CWA) and subsequent lawsuits. Hinkson Creek Watershed is representative of the LMMRB, and many developing watersheds globally, with respect to hydrologic processes, water quality, climate, and land use. Similar to many watersheds, the impaired use for Hinkson Creek was identified as "protection of warm water aquatic life" from unknown pollutants [62].

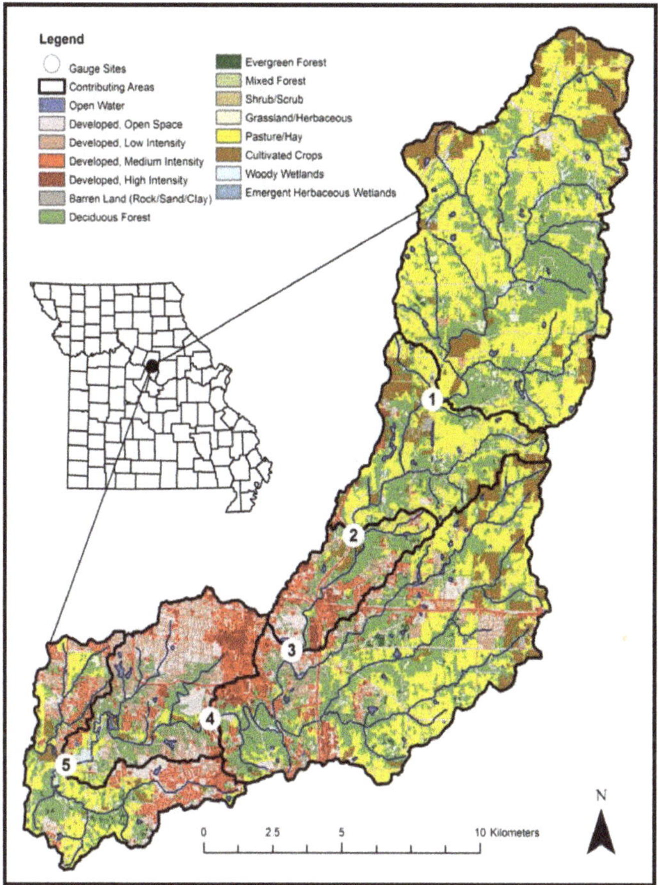

Figure 1. Locations of gauge sites (where #4 includes the USGS gauging station) and corresponding drainage area to each gauge (bold line) in the Hinkson Creek Watershed (HCW), in Central Missouri, USA. A model urban nested-scale experimental watershed study.

Table 1. Cumulative land use and land cover (LULC), drainage area, and stream length corresponding to each gauging site located in Hinkson Creek Watershed (HCW), Missouri, USA. Percent cumulative LULC is shown parenthetically.

Variable [km^2 (%)]	Site #1	Site #2	Site #3	Site #4	Site #5	HCW
Agricultural	45.0 (57.0)	56.4 (54.9)	57.6 (49.5)	78.5 (43.1)	79.7 (38.4)	85.4 (36.7)
Forested	28.4 (35.9)	37.5 (36.4)	41.1 (35.4)	62.8 (34.5)	68.6 (33.1)	74.9 (32.2)
Urban	3.7 (4.7)	6.6 (6.4)	15 (13.0)	37.1 (20.4)	54.9 (26.5)	67.6 (29.0)
Wetland	1.9 (2.4)	2.4 (2.3)	2.5 (2.1)	3.6 (2.0)	4.4 (2.0)	4.9 (2.1)
Total area	79.0	102.9	116.2	182.0	207.5	232.8
Stream length §	22.8	29.8	35.4	43.6	53.0	56.1

§ Stream length is shown in km.

Hinkson Creek's listing on the Clean Water Act (CWA) 303(d) list as impaired by unknown pollutants in 1998 [59] came about due to many issues identified by state and federal agencies and local residents, including (but not limited to), (1) larger and more frequent floods; (2) lower base flows; (3) increased soil erosion in construction and development areas with subsequent transport of the soil to streams (i.e., altered suspended sediment regimes); (4) water contamination from urban stormwater flows; (5) degradation of habitat for aquatic organisms due to the concerns listed above; and (6) degradation of aquatic habitat due to the physical alteration of stream channels and streamside (riparian) corridors [55,63–69]. In 2008, the watershed was instrumented with a nested-scale experimental watershed study design [53] to generate data that address the uncertainties of the 303(d) listing, while providing a scientific basis for developing a TMDL target. The experimental watershed program was designed to investigate the problems suspected to have led to the 1998 listing and improve understanding of contemporary land-use effects on hydrologic processes (stream response, water yield), water quality, and biological community health. Each nested monitoring site in Hinkson Creek was designed to monitor water stage and a complete suite of climate variables. Multiple additional water quality variables (e.g., suspended sediment, nitrogen, phosphorus, chloride, pH, and other constituents) were monitored at the nested sites shortly after implementation of the study. A United States Geological Survey gauging station (USGS-06910230) had collected stage data intermittently since 1966 and provided flow data for site 4 (Figure 1). Articles from the Hinkson Creek Experimental Watershed (HCEW) program were being published as early as 2010. To date, there have been over 50 publications in peer-reviewed journals and 21 graduate student theses and dissertations.

2.2. Collaborative Adaptive Management

In 2011, a Collaborative Adaptive Management (CAM) program was developed to provide direction and support for the 303(d)-delisting process (www.helpthehinkson.org) [64]. The CAM process was designed to be fundamentally science-based as doing so acknowledges uncertainties and/or unknowns about complex systems, engages scientists, decision-makers, and stakeholders, and applies continuous process improvements to reduce those uncertainties and maximize the opportunity for success [70,71]. In this manner, a science-driven CAM process can support efforts aimed at improving water quality and aquatic habitat in contemporary watersheds, because scientific results and the understanding they foster can guide informed decision-making. This approach is important because, in complex contemporary watershed systems, applying a mitigation strategy may improve one or more characteristics of the stream, but not achieve the ultimate goal. Typically, when a stream or other water body is listed as impaired, a Total Maximum Daily Load (TMDL) analysis is conducted to define the maximum pollutant load compatible with full compliance of the stream with designated uses [72]. However, this approach can be confounded when no specific pollutant has been identified. From the outset of the regulatory process, impairment of Hinkson Creek was assumed predominantly a result of urban development. Given the listing of the creek for "unknown pollutants", a volume-based flow reduction strategy was initially adopted, which was focused on urban stormwater runoff reduction as a means to reduce unknown pollutant concentrations and loadings [53]. Specifically, a target of 50% volume reduction was set for HCW in the waste load allocation (proportion of stormwater attributed to point sources) developed by the Missouri Department of Natural Resources (MDNR) [73]. The wasteload allocation was required to be met by urban and developed areas, while the load allocation (proportion of stormwater attributed to nonpoint sources) was assigned to rural areas [53]. Such volume-based approaches are encouraged by USEPA and the National Research Council [74]; therefore, the application of stormwater reduction as a surrogate for pollutants is not uncommon [53]. These details, and those that follow, chronicle the cumulative results of research conducted within the context of the HCW study and demonstrate the immeasurable value of the experimental watershed approach and the CAM program to water resource management.

2.3. Experimental Watershed Design Outcomes

The experimental watershed study design applied in HCW facilitated the identification and quantification of factors contributing to impairment of the stream and provided the information needed to target mechanistic drivers, both natural and anthropogenic, of hydrologic alteration. Detail is provided here to give the reader a sense of the scope of possible findings that can be obtained via the methodological approach. The analysis showed that annual streamflow metrics (i.e., peak flow, baseflow) had not significantly increased or decreased in Hinkson Creek from 1967 to 2010 [63]. However, more recent work indicated that significant changes in runoff volume and timing in the watershed (largely due to urbanization) have occurred in the years up to 2015 [75]. Additionally, event-based (30-min interval) rainfall-streamflow response showed increased explained variance at urban sites relative to rural sites, indicating the potential for increased streamflow response to rainfall events at urban sites [61]. Multiple (n = 12) event-based streamflow regime metrics (e.g., peak flow magnitude and timing), which were calculated from observed paired-independent storm events were correlated with urban land use [67]. A positive relationship between developed land uses (i.e., urban and suburban) and volumetric streamflow was consistently observed through various analyses [76–81], thus highlighting the importance of land use impacts on streamflow characteristics and sediment transport [56,59,77,80,82,83].

Suspended sediment levels in Hinkson Creek may be high for the region [82]. There was a disproportionately high contribution of fine sediment reported from the City of Columbia, relative to Hinkson Creek [84,85]. While the variability of spatiotemporal distributions of suspended sediment particle densities (e.g., organic material) in Hinkson Creek can confound loading estimations [86,87], work conclusively showed that average suspended sediment particle size decreased in Hinkson Creek as cumulative urban land use increased in the watershed. Moreover, a doubling of streamflow more than doubled (i.e., a non-linear relationship) fine suspended sediment concentrations in Hinkson Creek [88,89]. In addition, studies showed that nearly all (99%) of the total suspended sediment load was transported during high flows (Q_{10}) [76].

A study in 2011 showed that stream bank erosion contributed approximately 67% of suspended sediment loading over the 2011 water year, illustrating the potential contribution of in-stream vs. terrestrial suspended sediment in the watershed [90]. Kellner and Hubbart [81] showed that channel widening and incision in Hinkson Creek (e.g., erosion of streambed and banks) were spatially correlated to developed land uses, and associated streamflow characteristics, in the middle and lower watershed. Increased erosion of streambeds and banks due to urban runoff may help explain observed suspended sediment patterns and further emphasizes the importance of streamflow to sediment and pollutant transport dynamics. However, suspended sediment is only one of the set of factors influencing water quality. Alterations to multiple nutrient constituents, driven by land use practices, were observed in HCW [56,76]. Zeiger and Hubbart [56] showed total inorganic nitrogen and nitrate concentrations were relatively higher in the agricultural headwaters. Increased nitrate levels are quite common in the agricultural areas of the Midwest, particularly the Upper Mississippi River Basin, where nitrogen fertilizer applications can exceed 2.5 t km^{-2} yr^{-1}. However, total ammonia yields greater than 1.25 kg ha^{-1} yr^{-1} and total phosphorus yields exceeding 2.0 kg ha^{-1} yr^{-1} in Hinkson Creek were high for the Mississippi River Basin [56]. Total phosphorus concentrations exceeded 1.13 mg L^{-1} at suburban/urban sites.

Urban land uses also correlated with adverse physicochemical characteristics in Hinkson Creek, including toxic chloride concentrations and loadings [64], altered dissolved oxygen trends (both above and below established water quality standards [91,92]), and increased pH and total dissolved solids [83]. Hubbart et al. [64] showed chloride in Hinkson Creek reaches seasonally-mediated acute (860 mg L^{-1}) and chronic (230 mg L^{-1}) concentrations with lower concentrations persisting in floodplain shallow groundwater year-round. Collectively, the results of stream physicochemistry investigations suggest the potential for aquatic biota stress throughout the main stem of Hinkson Creek and identify land-use practices as a primary driver of water resource degradation. Results also showed that urbanization (Columbia, Missouri) has resulted in significantly ($p < 0.05$) altered stream water temperature regimes [54,93,94]. Daily maximum stream temperature exceeded a threshold of potential mortality of warm-water biota (i.e., 35.0 °C). Additionally, maximum stream temperature was 4.0 °C greater at an urban monitoring site, relative to a rural site for 10.5 h, indicating urban land use exacerbates the influence of summertime drought on thermal stream conditions. Sudden increases in stream temperature (stream temperature surges) were observed at urban sites. Stream temperature surges were significantly correlated to urban land use, downstream distance, and discharge ($p = 0.02$).

Studies identified an urban micro-climate gradient and an urban heat island (UHI) effect in the city of Columbia, and noted that strategically located urban forest patches can be used to optimize localized cooling, carbon storage and cycling [95]. Similarly, floodplain work indicated that bottomland hardwood forest soils in Hinkson Creek Watershed store larger amounts of carbon relative to non-woody floodplain sites in the urban environment [96,97]. This information was not only useful in CAM discussions, and for local restoration policies, but also in current management discussions regarding the potential for bottomland hardwood forest restoration to meet carbon sequestration targets globally. Moreover, results repeatedly and conclusively supported the reestablishment of floodplain forests, where practicable, for the conservation of both groundwater and surface water quality. Studies showed that floodplain forests reduce subsurface shallow groundwater temperature fluctuations [98], can accept and thus process significantly ($\alpha = 0.05$, approximately 120 mm yr^{-1}) more water to storage than agricultural or grassland areas [99], significantly increase soil infiltration and soil volumetric water content holding capacity [100], increase consumptive use by vegetation [58], and improve freshwater routing, water quality, aquatic ecosystem conservation, and flood mitigation in mixed-land-use watersheds [98,101–103].

Program results also highlighted the effects of agricultural practices, specifically in the upper watershed, on the hydrologic regime of Hinkson Creek. For example, Zeiger and Hubbart [56] reported high concentrations of suspended sediment in the upper watershed, related to agricultural land uses. Similarly, Kellner and Hubbart [83] found indications of poor water quality in the agricultural upper watershed, as illustrated by levels of dissolved oxygen and pH values outside the recommended range for aquatic biota [91,92]. Such results suggest water quality and aquatic habitat degradation in Hinkson Creek is not limited to the activities and spatial extent of the city of Columbia, but rather is a complex watershed-scale issue involving integrated anthropogenic and natural processes. Similarly, a physical habitat assessment (PHA) showed that Hinkson Creek is altered by agricultural and urban land uses [104,105] that have also impacted macroinvertebrate assemblages in Hinkson Creek. This information was important in CAM discussions, considering macroinvertebrates are key species indicating general aquatic ecosystem status [55,106]. Results from the PHA clearly identified agricultural and urban land use alterations to channel geomorphology [105]. Results also showed increased substrate embeddedness (e.g., 80% vertical embeddedness of pool habitats) in the agricultural headwaters and in the lower urbanized reaches of Hinkson Creek [105]. The PHA assessment also revealed an increased frequency of fine streambed sediments coupled to increased substrate embeddedness in urbanized reaches. These results are in agreement with sediment studies in HCW that showed increased suspended sediment concentrations and increased fine suspended sediment particles in urban reaches [87–89].

Long-term multi-constituent datasets collected across a rural-urban land use gradient during the study included wet, average, and dry water years, and thus provided a distinct opportunity to assess the Soil Water Assessment Tool (SWAT). Results indicated "satisfactory" (Nash-Sutcliffe efficiency (NSE) values greater than 0.5) estimates of streamflow in Hinkson Creek during successive wet years [107]. The SWAT model also produced satisfactory estimates of monthly streamflow without model calibration [108]. However, uncalibrated SWAT model estimates of monthly sediment, total phosphorus, nitrate, nitrite, ammonium, and total inorganic nitrogen were unsatisfactory with NSE values less than 0.05. Model calibration at nested gauging sites increased NSE values above the aforementioned "satisfactory" threshold. The SWAT model was also used to simulate daily stream temperature with satisfactory results in Hinkson Creek [54]. Results identified useful model applications, including forecasting future hydrologic responses to urban growth and climate change [109–111], and pre-settlement land use model assessment [68,69]. Sunde et al. [109–111] simulated potential hydrologic consequences of increased impervious surfaces and climate change in HCW. For example, Sunde et al. [109] simulated three impervious growth scenarios using the

Imperviousness Change Analysis Tool (I-CAT) in HCW, and utilized climate change modeling results from the Coupled Model Intercomparison Project—Phase 5 (CMIP5) multimodel ensemble [110]. The simulated impervious growth and climate change data were used as model forcing's in SWAT to quantify the influence of projected impervious growth and climate change on water balance components. Collectively, results highlight the potential for combined and competing influences of climate change and development to result in decreased annual streamflow (−6.1%), and increased evapotranspiration (3.9%) in HCW [111]. The SWAT model was also used to simulate pre-settlement hydrologic conditions in HCW. Results confirmed the potential for agricultural and urban land-use influences on ecologically relevant daily streamflow regime metrics (streamflow magnitude, frequency, duration, timing, and rate of change) [68], and pollutant loading [69] in HCW. Critically, results indicated restoration of historic (i.e., pre-settlement) streamflow regimes are not fiscally obtainable targets in HCW and similar watersheds, where past and present land uses have extensively altered watershed hydrology and pollutant loading processes. This information is in agreement with the current understanding of environmental flows [41]. Ultimately, modeling results emphasize the great utility of the experimental design in advancing predictive potential and improving the accuracy of management practices.

2.4. *Identified Unrecognized and "Unknown" Sources of Impairment*

While anthropogenic pressures such as land use practices can exert driving influences on hydrologic and pollutant transport regimes, natural processes and landscape characteristics can compound impacts and confound the attribution of simple causal relationships to observed effects. For example, abnormal spatiotemporal streamflow relationships alerted the program director (Dr. Jason Hubbart) to possible (previously unidentified) hydrologic sink/source behavior in the upper-watershed [112]. Subsequent research uncovered archival evidence of historical subsurface coal mining, which may provide at least a partial explanation. Additional investigation identified hydrologic processes associated with natural landscape evolution, noted by early-20th-century researchers, which, when considered in the context of recent works, provide compelling alternative explanations for water quality and flow regime observations. Despite best-intentioned management, regulatory agencies, scientists, and local decision-makers did not account for such legacy practices and processes and instead relied on recent urban development as the proximate cause of designated impairment. Therefore, it is likely that historical land-use (coal mining) and landscape processes comprise cumulative, yet often unconsidered effects that contribute systemically to the observed hydrologic regimes of contemporary developing watersheds. In this regard, findings in HCW hold important implications for contemporary watershed management and suggest rethinking the case-by-case appropriateness of federal and state water impairment listings, and the achievability of restoration requirements therein.

Similarly, an investigation of the spatiotemporal variability of suspended sediment particle size class distribution (PSD) showed that the parameter best explaining the spatial pattern of PSD was not land use, but rather the surficial geology of the watershed [59]. The spatial pattern of surficial geology in the watershed (e.g., bedrock depth/constraints) also explained observations regarding suspended sediment concentrations [77,82], and stream geomorphology [81]. Finally, evidence was found to support the observation that the natural evolution of the Hinkson Creek hydrologic system is a contributing factor to observed water quality and stream geomorphology trends [81,112]. Specifically, historic Missouri River (confluence located approximately 8 km downstream) head-cutting and back-watering processes, at least in part, explain both channel incision and suspended sediment particle size characteristics in the lower watershed [81,112]. Notably, in conceptualizing the condition, management, and potential restoration of Hinkson Creek, the contribution of natural factors has often been overlooked in favor of a focus on anthropogenic disturbance [112]. However, these studies showed that a proper accounting of all contributing factors is required for accurate descriptions of system function and effective management [112].

3. Discussion

Synthesis and Implications

Synthesized salient, emergent results (i.e., "takeaways") of the work conducted during the HCW program include, (1) anthropogenic land use in HCW, including urban/suburban development and agriculture, is a primary driver of water quality degradation in Hinkson Creek; (2) land use practices impact suspended sediment characteristics and dynamics in HCW, including the flux of fine particles, which disproportionately contribute to water quality and aquatic habitat degradation; (3) streamflow alterations due to urban/suburban development result in increased streambed and bank erosion in Hinkson Creek, which increases sediment transport and disrupts aquatic habitat; (4) considering spatiotemporal water quality trends in Hinkson Creek, including dissolved oxygen levels, chloride concentrations, pH, water temperature, and suspended sediment concentrations, it is reasonable to expect stress conditions for aquatic biota throughout the stream, not only in urbanized/developed reaches; (5) several natural processes and watershed characteristics, such as surficial geology and stream system evolution, are likely compounding observed water quality and aquatic habitat degradation; and (6) given the varied and complicated set of factors contributing to water quality and aquatic habitat degradation in HCW, restoration of Hinkson Creek is likely subject to physical limitations and should be conceptualized in the context of achievable goals/objectives.

Results of the program highlight the compounding impacts of land use practices, hydroclimatic variability, and physical watershed characteristics on the suspended sediment, streamflow, and water quality regimes of Hinkson Creek. Land-use-driven alterations to the streamflow regime (e.g., increased runoff and flow magnitude, advanced peak hydrograph) of Hinkson Creek have resulted in increased pollutant transport and loading, and disturbance of aquatic habitat (bed incision, bank erosion, elevated stream temperature) that disrupts the biological integrity of the aquatic ecosystem. Restated, anthropogenic activities in the watershed exacerbate ecosystem vulnerabilities. Due to the many investigations concluded by the program, a more detailed and comprehensive description of the system is now available to stakeholders and decision-makers, which can be subsequently used to improve the management of the watershed [59,64,78,80,81,83,112]. The program also provides valuable insights regarding potential successes and challenges faced by collaborative adaptive management programs. Three issues emerged that should be emphasized to improve future CAM applications. First is the integrated approach with multiple objectives and multiple beneficiaries. For sustainable management, environmental factors need to be equally, if not more, greatly emphasized, relative to the economic aspects of project implementation. Second, local stakeholders must be involved as much and as early as possible. Local knowledge, gained by time and experience is critical for stakeholder buy-in, and project implementation and success. Ultimately, understanding values and opinions held by local communities is of critical importance [21,113], and stakeholders should be encouraged, and provided opportunities, to volunteer as team members to engage in the process [114]. Third, there must be regular updates and improvements to the plan. Given the inherent complexity of natural ecosystems, it is not surprising that effective resource management is dynamic, characterized by ongoing updates and refinements [21,115]. Collectively, more scientific and socioeconomic information and effective involvement of stakeholders are the primary components of collaborative adaptive management that lead to improved management decision-making [116–119].

4. Conclusions

Assuming human-induced land use change and in-tandem environmental change continue as expected, there is a need for streamlined (and normalized) collaborative adaptive management efforts to continuously monitor and respond to anthropogenic and natural pressures on water resources. Results from the experimental watershed approach and CAM processes highlighted here show the value of integrating knowledge and science with experience and the perspectives of scientists, stakeholders, and managers for more effective management decision-making. This is critical because, in the absence of adequate observed data, sources of impairment are often unrecognized and/or listed as "unknown" in contemporary mixed-use watersheds. Additionally, sources of impairment can shift over time due to the combined influences of land use change, human population growth, and changing environmental conditions. Results from the case-study presented here clearly show that the experimental watershed study design can be used to provide science-based information critically needed to make informed management decisions in contemporary mixed-use watersheds. The design has the potential to be systematically applied in any watershed, thereby normalizing and standardizing study designs across watershed systems. In so doing, comparable inter- and intra-watershed information is collected, broader (multi-watershed) practices are implemented, and multi-scale costs are driven down over time.

In the Hinkson Creek Watershed in the Midwest USA, long-term monitoring of hydroclimate variables, streamflow, and multiple water quality constituents in nested sub-basins provided answers to specific questions generated during the CAM process. Additionally, hydrologic data collected and analyzed informed regional managers and advanced policy and science via generation of over 50 peer-review publications and 21 graduate student theses and dissertations. Key findings from the program showed (1) legacy effects, urban/suburban development and agriculture are primary drivers of alterations to watershed hydrology, streamflow regimes, multiple water quality constituents, and physical habitat degradation in Hinkson Creek; (2) several natural processes and watershed characteristics, such as surficial geology and stream system evolution, are likely compounding observed water quality and aquatic habitat degradation; and (3) given the varied and complicated set of factors contributing to water quality and aquatic habitat degradation, restoration of many USA CWA 303(d) listed streams and rivers like Hinkson Creek are likely subject to physical and fiscal limitations and should be conceptualized in the context of achievable goals/objectives. To this end, the nested-scale experimental watershed monitoring approach has served as a scalable model for studying natural and anthropogenic influences on water quantity, water quality, and stream physical habitat in contemporary mixed-use watersheds.

Author Contributions: conceptualization: J.A.H.; formal analysis: J.A.H., E.K., and S.J.Z.; investigation: J.A.H.; data curation: J.A.H.; writing-original draft preparation: J.A.H., E.K., and S.J.Z.; writing-review and editing: J.A.H., E.K., and S.J.Z.; supervision: J.A.H.; project administration: J.A.H.; funding acquisition: J.A.H.

Funding: The Hinkson Creek Experimental Watershed Program was founded and Directed by Jason Hubbart from 2007 to 2016. Funding was provided by the Missouri Department of Conservation and the U.S. Environmental Protection Agency Region 7 through the Missouri Department of Natural Resources (P.N: G08-NPS-17) under Section 319 of the Clean Water Act. Additional funding was provided by the partners of the Hinkson Creek Watershed Collaborative Adaptive Management program, the National Science Foundation under Award Number OIA-1458952, the USDA National Institute of Food and Agriculture, Hatch project 1011536, and the West Virginia Agricultural and Forestry Experiment Station. A glowing acknowledgment is due to John Schulz for support and guidance early in the program. Results presented may not reflect the views of the sponsors, and no official endorsement should be inferred. Finally, greatest thanks go to the many students, graduates, scientists, and contributors that devoted their work and time to the program.

Conflicts of Interest: The authors declare no conflict of interest.

References

1. Hasselman, L. Adaptive management; adaptive co-management; adaptive governance: what's the difference? *Australas. J. Environ. Manag.* **2017**, *24*, 31–46. [CrossRef]
2. Richter, B.D.; Warner, A.T.; Meyer, J.L.; Lutz, K. A collaborative and adaptive process for developing environmental flow recommendations. *River Res. Appl.* **2006**, *22*, 297–318. [CrossRef]
3. Johnson, B.L. The role of adaptive management as an operational approach for resource management agencies. *Conserv. Ecol.* **1999**, *3*, 8. [CrossRef]
4. Walters, C.J.; Holling, C.S. Large-scale management experiments and learning by doing. *Ecology* **1990**, *71*, 2060–2068. [CrossRef]
5. Walters, C. *Adaptive Management of Renewable Resources*; MacMillan: New York, NY, USA, 1986.

6. Fujitani, M.; McFall, A.; Randler, C.; Arlinghaus, R. Participatory adaptive management leads to environmental learning outcomes extending beyond the sphere of science. *Sci. Adv.* **2017**, *3*, e1602516. [CrossRef] [PubMed]
7. Wong-Parodi, G.; Strauss, B.H. Team science for science communication. *Proc. Natl. Acad. Sci. USA* **2014**, *111*, 13658–13663. [CrossRef]
8. Scheufele, D.A. Communicating science in social settings. *Proc. Natl. Acad. Sci. USA* **2013**, *110*, 14040–14047. [CrossRef]
9. Stern, P.C. Deliberative methods for understanding environmental systems. *BioScience* **2005**, *55*, 976–982. [CrossRef]
10. Varady, R.G.; Zuniga-Teran, A.A.; Garfin, G.M.; Martín, F.; Vicuña, S. Adaptive management and water security in a global context: Definitions, concepts, and examples. *Curr. Opin. Environ. Sustain.* **2016**, *21*, 70–77. [CrossRef]
11. Cookey, P.E.; Darnswasdi, R.; Ratanachai, C. Local People's Perceptions of Lake Basin Water Governance Performance in Thailand. *Ocean Coast. Manag.* **2016**, *120*, 11–28. [CrossRef]
12. Engle, N.L.; Johns, O.R.; Lemos, M.; and Nelson, D.R. Integrated and adaptive management of water resources: Tensions, legacies, and the next best thing. *Ecol. Soc.* **2011**, *16*, 19. [CrossRef]
13. Kumler, L.M.; Lemos, M.C. Managing waters of the Paraíba do Sul river basin, Brazil: A case study in institutional change and social learning. *Ecol. Soc.* **2008**, *13*, 22. [CrossRef]
14. Summers, M.F.; Holman, I.P.; Grabowski, R.C. Adaptive Management of River Flows in Europe: A Transferable Framework for Implementation. *J. Hydrol.* **2015**, *531*, 696–705. [CrossRef]
15. Contador, J.L. Adaptive management, monitoring, and the ecological sustainability of a thermal-polluted water ecosystem: A case in SW Spain. *Environ. Monit. Assess.* **2005**, *104*, 19. [CrossRef]
16. Hernandez-Mora, N.; Cabello, V.; De Stefano, L.; Del Moral, L. Networked water citizen organizations in Spain: Potential for transformation of existing power structures in water management. *Water Altern.* **2015**, *8*, 99–124.
17. Pedregal, B.; Cabello, V.; Hernandez-Mora, N.; Limones, N.; Del Moral, L. Information and knowledge for water governance in the networked society. *Water Altern.* **2015**, *8*, 1–19.
18. Ercolani, G.; Chiaradia, E.A.; Gandolfi, C.; Castelli, F.; Masseroni, D. Evaluating performances of green roofs for stormwater runoff mitigation in a high flood risk urban catchment. *J. Hydrol.* **2018**, *566*, 830–845. [CrossRef]
19. Masseroni, D.; Ercolani, G.; Chiaradia, E.A.; Gandolfi, C. A procedure for designing natural water retention measures in new development areas under hydraulic-hydrologic invariance constraints. *Hydrol. Res.* **2019**, *50*, 1293–1308. [CrossRef]
20. Masseroni, D.; Ercolani, G.; Chiaradia, E.A.; Maglionico, M.; Toscano, A.; Gandolfi, C.; Bischetti, G.B. Exploring the performances of a new integrated approach of grey, green and blue infrastructures for combined sewer overflows remediation in high-density urban areas. *J. Agric. Eng.* **2018**, *49*, 233–241. [CrossRef]
21. Allan, C.; Curtis, A.; Stankey, G.; Shindler, B. Adaptive Management and Watersheds: A Social Science Perspective. *JAWRA* **2018**, *44*, 166–174.
22. Melis, T.S.; Walters, C.J.; Korman, J. Surprise and Opportunity for Learning in Grand Canyon: The Glen Canyon Dam Adaptive Management Program. *Ecol. Soc.* **2015**, *20*, 22. [CrossRef]
23. Bennett, J.; Lawrence, P.; Johnstone, R.; Shaw, R. Adaptive management and its role in managing Great Barrier Reef water quality. *Mar. Pollut. Bull.* **2005**, *51*, 70–75. [CrossRef] [PubMed]
24. Kleinman, P.J.A.; Fanelli, R.M.; Hirsch, R.M.; Buda, A.R.; Easton, Z.M.; Wainger, L.A.; Brosch, C.; Lowenfish, M.; Collick, A.S.; Shirmohammadi, A.; et al. Phosphorus and the Chesapeake Bay: Lingering Issues and Emerging Concerns for Agriculture. *J. Environ. Qual.* **2019**, *48*, 1191–1203. [CrossRef] [PubMed]
25. USEPA. *Chesapeake Bay Total Maximum Daily Load for Nitrogen, PhosphoRus, and Sediment*; USEPA: Philadelphia, PA, USA, 2010. Available online: http://www.epa.gov/ches-apeake-bay-tmdl/chesapeake-bay-tmdl-document (accessed on 28 July 2019).
26. Chesapeake Bay Watershed Agreement. Chesapeake Bay Watershed Agreement. 2014. Available online: http://www.chesapeakebay.net/chesapeakebaywatershedagreement/page (accessed on 28 July 2019).
27. Berg, J. Stream restoration as a means of meeting Chesapeake Bay TMDL goals. *Water Resour. Impact* **2014**, *16*, 16–18.

28. Rabalais, N.N.; Diaz, R.J.; Levin, L.A.; Turner, R.E.; Gilbert, D.; Zhang, J. Dynamics and distribution of natural and human-caused hypoxia. *Biogeosciences* **2010**, *7*, 585. [CrossRef]
29. Rabalais, N.N.; Turner, R.E.; Gupta, B.S.; Boesch, D.F.; Chapman, P.; Murrell, M.C. Hypoxia in the northern Gulf of Mexico: Does the science support the plan to reduce, mitigate, and control hypoxia? *Estuar. Coasts* **2007**, *30*, 753–772. [CrossRef]
30. Turner, R.E.; Rabalais, N.N.; Justic, D. Gulf of Mexico hypoxia: Alternate states and a legacy. *Environ. Sci. Technol.* **2008**, *42*, 2323–2327. [CrossRef]
31. Mississippi River/Gulf of Mexico Watershed Nutrient Task Force. *Gulf Hypoxia Action Plan 2008: For Reducing Mitigating, and Controlling Hypoxia in the Northern Gulf of Mexico and Improving Water Quality in the Mississippi River Basin*; US Environmental Protection Agency, Office of Wetlands, Oceans, and Watersheds: Washington, DC, USA, 2008.
32. Mississippi River/Gulf of Mexico Watershed Nutrient Task Force. *Action Plan for Reducing, Mitigating, and Controlling Hypoxia in the Northern Gulf of Mexico*; Office of Wetlands, Oceans, and Watersheds, US Environmental Protection Agency: Washington, DC, USA, 2001.
33. Hauser, C.E.; Possingham, H.P. Experimental or precautionary? Adaptive management over a range of time horizons. *J. Appl. Ecol.* **2008**, *45*, 72–81. [CrossRef]
34. Brodie, R.; Sundaram, B.; Tottenham, R.; Hostetler, S.; Ransley, T. *An Adaptive Management Framework for Connected Groundwater-Surface Water Resources in Australia*; Department of Agriculture, Fisheries and Forestry: Canberra, Australia, 2007.
35. Gilmour, A.; Walkerden, G.; Scandol, J. Adaptive management of the water cycle on the urban fringe: Three Australian case studies. *Conserv. Ecol.* **1999**, *3*, 11. [CrossRef]
36. Eagleson, P.S. Climate soil, and vegetation 1. Introduction to water balance dynamics. *Water Resour. Res.* **1978**, *14*, 705–712. [CrossRef]
37. Bari, M.A.; Smettem, K.R.J.; Sivapalan, M. Understanding changes in annual runoff following land use changes: A systematic data-based approach. *Hydrol. Process.* **2005**, *19*, 2463–2479. [CrossRef]
38. Bates, C.G.; Henry, A.J. Forest and streamflow experiments at Wagon Wheel Gap, Colorado. *Mon. Weather Rev. Suppl.* **1928**, *30*, 79. [CrossRef]
39. Ice, G.; Stednick, J.D. Forest Watershed Research in the United States. *For. Hist. Today* **2004**, *17*, 16–26.
40. National Research Council (NRC). *New Strategies for America's Watersheds*; National Academy Press: Washington, DC, USA, 1999.
41. Poff, N.L. Beyond the natural flow regime? Broadening the hydro-ecological foundation to meet environmental flows challenges in a non-stationary world. *Freshw. Biol.* **2018**, *63*, 1011–1021. [CrossRef]
42. Vörösmarty, C.J.; Green, P.; Salisbury, J.; Lammers, R.B. Global water resources: Vulnerability from climate change and population growth. *Science* **2000**, *289*, 284–288. [CrossRef]
43. Kundzewicz, Z.W. Water resources for sustainable development. *Hydrol. Sci.* **1997**, *42*, 467–480. [CrossRef]
44. *World Urbanization Prospects 2018: Highlights (ST/ESA/SER.A/421)*; United Nations, Department of Economic and Social Affairs, Population Division: New York, NY, USA, 2019.
45. Hubbart, J.A. Considering the Future of Water Resources: A Call for Investigations that Include the Cultural Anthropology of Water. *Glob. J. Archaeol. Anthropol.* **2018**, *3*. [CrossRef]
46. Hewlett, J.D.; Lull, H.W.; Reinhart, K.G. In defense of experimental watersheds. *Water Resour. Res.* **1969**, *5*, 306–316. [CrossRef]
47. Leopold, L.B. Hydrologic research on instrumented watersheds, Results of research on representative and experimental basins. In Proceedings of the Wellington Symposium, IAHSIAISH-UNESCO, Wellington, NZ, USA, 1–8 December 1970; pp. 135–150.
48. Likens, G.E.; Bormann, F.H.; Pierce, R.S.; Eaton, J.S.; Johnson, N.M. *Biogeochemistry of a Forested Ecosystem*; Springer: New York, NY, USA, 1977; p. 146.
49. Bosch, J.M.; Hewlett, J.D. A review of catchment experiments to determine the effect of vegetation changes on water yield and evapotranspiration. *J. Hydrol.* **1982**, *55*, 3–23. [CrossRef]
50. Stednick, J.D. Monitoring the effects of timber harvest on annual water yield. *J. Hydrol.* **1996**, *176*, 79–95. [CrossRef]
51. Brown, A.E.; Zhang, L.; McMahon, T.A.; Western, A.W.; Vertessy, R.A. A review of paired catchment studies for determining changes in water yield resulting from alterations in vegetation. *J. Hydrol.* **2005**, *310*, 28–61. [CrossRef]

52. Hubbart, J.A.; Kavanagh, K.L.; Pangle, R.; Link, T.E.; Schotzko, A. Cold air drainage and modeled nocturnal leaf water potential in complex forested terrain. *Tree Physiol.* **2007**, *27*, 631–639. [CrossRef] [PubMed]
53. Hubbart, J.A.; Holmes, J.; Bowman, G. TMDLs: Improving stakeholder acceptance with science based allocations. *Watershed Sci. Bull.* **2010**, *1*, 19–24.
54. Zeiger, S.J.; Hubbart, J.A.; Anderson, S.H.; Stambaugh, M.L. Quantifying and modeling urban stream temperature: A central US watershed study. *Hydrol. Process.* **2015**, *30*, 503–514.
55. Nichols, J.; Hubbart, J.A. Using Macroinvertebrate Assemblages and Multiple Stressors to Infer Urban Stream System Condition: A Case Study in the Central US. *Urban Ecosyst.* **2016**, *19*, 679–704. [CrossRef]
56. Zeiger, S.J.; Hubbart, J.A. Quantifying suspended sediment flux in a mixed-land-use urbanizing watershed using a nested-scale study design. *Sci. Total Environ.* **2016**, *542*, 315–323. [CrossRef]
57. Tetzlaff, D.; Carey, S.K.; McNamara, J.P.; Laudon, H.; Soulsby, C. The essential value of long-term experimental data for hydrology and water management. *Water Resour. Res.* **2017**, *53*, 2598–2604. [CrossRef]
58. Hubbart, J.A. Urban Floodplain Management: Understanding Consumptive Water-Use Potential in Urban Forested Floodplains. *Stormwater J.* **2011**, *12*, 56–63.
59. Kellner, E.; Hubbart, J.A. Spatiotemporal variability of suspended sediment particle size in a mixed-land-use watershed. *Sci. Total Environ.* **2018**, *615*, 1164–1175. [CrossRef]
60. Felson, A.J.; Pickett, S.T. Designed experiments: New approaches to studying urban ecosystems. *Front. Ecol. Environ.* **2005**, *3*, 549–556. [CrossRef]
61. Zeiger, S.J.; Hubbart, J.A. Quantifying Land use Influences on Event-Based Flow Frequency, Timing, Magnitude, and Rate of Change in an Urbanizing Watershed of the Central USA. *Environ. Earth Sci.* **2018**, *77*, 107. [CrossRef]
62. Missouri Department of Natural Resources (MDNR). *Stream Survey Sampling Report. Phase III Hinkson Creek Stream Study, Columbia, Missouri, Boone County*; Prepared by the Missouri Department of Natural Resources, Field Services Division, Environmental Services Program, Water Quality Monitoring Section: Jefferson City, MO, USA, 2006.
63. Hubbart, J.A.; Zell, C. Considering Streamflow Trend Analyses Uncertainty in Urbanizing Watersheds: A Case Study in the Central U.S. *Earth Interact.* **2013**, *17*, 1–28. [CrossRef]
64. Hubbart, J.A.; Kellner, E.; Hooper, L.W.; and Zeiger, S.J. Quantifying loading, toxic concentrations, and systemic persistence of chloride in a contemporary mixed-land-use watershed using an experimental watershed approach. *Sci. Total Environ.* **2018**, *581–582*, 822–832. [CrossRef] [PubMed]
65. Kellner, E.; Hubbart, J.A.; Stephan, K.; Morrissey, E.; Freedman, Z.; Kutta, E.; Kelly, C. Characterization of sub-watershed-scale stream chemistry regimes in an Appalachian mixed-land-use watershed. *Environ. Monit. Assess.* **2018**, *190*, 586. [CrossRef] [PubMed]
66. Kellner, E.; Hubbart, J.A. Land use impacts on floodplain water table response to precipitation events. *Ecohydrology* **2018**, *11*, e1913. [CrossRef]
67. Zeiger, S.J.; Hubbart, J.A. Rainfall-Stream Flow Responses in a Mixed-Land-use and Municipal Watershed of the Central USA. *Water* **2018**, *77*, 438. [CrossRef]
68. Zeiger, S.J.; Hubbart, J.A. Assessing Environmental Flow Targets using Pre-Settlement Land Cover. *Water* **2018**, *10*, 791. [CrossRef]
69. Zeiger, S.J.; Hubbart, J.A. Assessing the Difference between SWAT Simulated Pre-Development and Observed Developed Loading Regimes. *Hydrology* **2018**, *5*, 29. [CrossRef]
70. Susskind, L.; Camacho, A.E.; Schenk, T. A critical assessment of collaborative adaptive management in practice. *J. Appl. Ecol.* **2012**, *49*, 47–51. [CrossRef]
71. Scarlett, L. Collaborative adaptive management: Challenges and opportunities. *Ecol. Soc.* **2013**, *18*, 3. [CrossRef]
72. National Research Council (NRC). *Assessing the TMDL Approach to Water Quality Management*; The National Academies Press: Washington, DC, USA, 2001.
73. *Total Maximum Daily Load (TMDL) for Hinkson Creek, Boone County, Missouri, Draft*; Missouri Department of Natural Resources (MDNR): Jefferson City, MO, USA, 2010.
74. National Research Council (NRC). *Urban Stormwater Management in the United States*; The National Academies Press: Washington, DC, USA, 2008.
75. Wei, L.; Hubbart, J.A.; Zhou, H. Variable Streamflow Contributions in Nested Subwatersheds of a US Midwestern Urban Watershed. *Water Resour. Manag.* **2017**, *32*, 213–228. [CrossRef]

76. Zeiger, S.J.; Hubbart, J.A. Quantifying flow interval–pollutant loading relationships in a rapidly urbanizing mixed-land-use watershed of the Central USA. *Environ. Earth Sci.* **2017**, *76*, 484. [CrossRef]
77. Kellner, E.; Hubbart, J.A. Improving understanding of mixed-land-use watershed suspended sediment regimes: Mechanistic progress through high-frequency sampling. *Sci. Total Environ.* **2017**, *598*, 228–238. [CrossRef] [PubMed]
78. Kellner, E.; Hubbart, J.A. Application of the experimental watershed approach to advance urban watershed precipitation/discharge understanding. *Urban Ecosyst.* **2017**, *20*, 799–810. [CrossRef]
79. Zeiger, S.J.; Hubbart, J.A. Quantifying relationships between watershed characteristics and hydroecological indices of Missouri streams. *Sci. Total Environ.* **2019**, *654*, 1305–1315. [CrossRef]
80. Kellner, E.; Hubbart, J.A. Flow class analyses of suspended sediment concentration and particle size in a mixed-land-use watershed. *Sci. Total Environ.* **2019**, *648*, 973–983. [CrossRef]
81. Kellner, E.; Hubbart, J.A. A method for advancing understanding of streamflow and geomorphological characteristics in mixed-land-use watersheds. *Sci. Total Environ.* **2019**, *657*, 634–643. [CrossRef]
82. Zeiger, S.J.; Hubbart, J.A. Nested-Scale Nutrient Flux in a Mixed-Land-Use Urbanizing Watershed. *Hydrol. Process.* **2016**, *30*, 1475–1490. [CrossRef]
83. Kellner, E.; Hubbart, J.A. Advancing understanding of the surface water quality regime of contemporary mixed-land-use watersheds: An application of the experimental watershed method. *Hydrology* **2017**, *4*, 31. [CrossRef]
84. Kellner, E. Quantifying Urban Stormwater Suspended Sediment Particle Size Class Distribution in the Central U.S. Master's Thesis, University of Missouri, Columbia, MO, USA, 2013.
85. Kellner, E.; Hubbart, J.A. Quantifying Urban Land-Use Impacts on Suspended Sediment Particle Size Class Distribution: A Method and Case Study. *Stormwater J.* **2014**, *15*, 40–50.
86. Freeman, G. Quantifying Suspended Sediment Loading in a Mid-Missouri Urban Watershed Using Laser Particle Diffraction. Master's Thesis, University of Missouri, Columbia, MO, USA, 2011.
87. Hubbart, J.A.; Freeman, G. Sediment Laser Diffraction: A New Approach to an Old Problem in the Central U.S. *Stormwater J.* **2010**, *11*, 36–44.
88. Hubbart, J.A.; Gebo, N.A. Quantifying the Effects of Land-Use and Erosion by Particle Size Class Analysis in the Central U.S. *Eros. Control J.* **2010**, *17*, 24–36.
89. Hubbart, J.A.; Kellner, E.; Freeman, G. A Case Study Considering the Comparability of Mass and Volumetric Suspended Sediment Data. *Environ. Earth Sci.* **2013**, *10*, 4051–4060. [CrossRef]
90. Huang, D. Quantifying Stream Bank Erosion and Deposition Rates in a Central U.S. Urban Watershed. Master's Thesis, University of Missouri, Columbia, MO, USA, 2012.
91. Missouri Department of Natural Resources (MDNR). *Water Quality*; In Rules of Department of Natural Resources; Division 20: Clean Water Commission; Missouri Department of Natural Resources: Jefferson City, MO, USA, 2014; p. 102.
92. Missouri Department of Natural Resources (MDNR). Water Chemistry. In *Volunteer Water Quality Monitoring*; Missouri Department of Natural Resources: Jefferson City, MO, USA, 2013; p. 35.
93. Zeiger, S.J. Measuring and Modeling Stream and Air Temperature Relationships in a Multi-Land Use Watershed of the Central United States. Master's Thesis, University of Missouri, Columbia, MO, USA, 2014.
94. Zeiger, S.J.; Hubbart, J.A. Quantifying urban stormwater temperature surges: A central US watershed study. *Hydrology* **2015**, *2*, 193–209. [CrossRef]
95. Hubbart, J.A.; Kellner, E.; Hooper, L.; Lupo, A.R.; Market, P.S.; Guinan, P.E.; Stephan, K.; Fox, N.I.; Svoma, B.M. Localized Climate and Surface Energy Flux Alterations across an Urban Gradient in the Central U.S. *Energies* **2014**, *7*, 1770–1791. [CrossRef]
96. Beaven, K.R. Investigating Soil Carbon, Nitrogen and Respiration Across an Intra-Urban Gradient in Mid-Missouri. Ph.D. Thesis, University of Missouri, Columbia, MO, USA, 2015.
97. Spiegel, E. Estimating Above and Below Ground Vegetation Biomass and Carbon Storage across an Intra-Urban Land-Use Gradient in Mid-Missouri. Master's Thesis, University of Missouri, Columbia, MO, USA, 2015.
98. Kellner, E.; Hubbart, J.A. A Comparison of the Spatial Distribution of Vadose Zone Water in Forested and Agricultural Floodplains a Century after Harvest. *Sci. Total Environ.* **2015**, *542*, 153–161. [CrossRef] [PubMed]

99. Zell, C.; Kellner, E.; Hubbart, J.A. Land Use Impacts on Subsurface Floodplain Storage Capacity: A Midwest Case Study of Agricultural and Remnant Hardwood Forest Land Use Types. *Environ. Earth Sci.* **2015**, *74*, 7215–7228. [CrossRef]
100. Hubbart, J.A.; Muzika, R.M.; Huang, D.; Robinson, A. Bottomland Hardwood forest influence on soil water consumption in an urban floodplain: Potential to improve flood storage capacity and reduce stormwater runoff. *Watershed Sci. Bull.* **2011**, *3*, 34–43.
101. Brown, H.L.; Bos, D.G.; Walsh, C.J.; Fletcher, T.D.; RossRakesh, S. More than money: How multiple factors influence householder participation in at-source stormwater management. *J. Environ. Plan. Manag.* **2016**, *59*, 79–97. [CrossRef]
102. Kellner, E. The Long-Term Impacts of Forest Removal on Floodplain Subsurface Hydrology. Doctoral Dissertation, University of Missouri, Columbia, MO, USA, 2015.
103. Kellner, E.; Hubbart, J.A. Agricultural and Forested Land Use Impacts on Floodplain Shallow Groundwater Temperature Regime. *Hydrol. Process.* **2015**, *30*, 625–636. [CrossRef]
104. Hooper, L. A Stream Physical Habitat Assessment in an Urbanizing Watershed of the Central USA. Master's Thesis, University of Missouri, Columbia, MO, USA, 2015.
105. Zeiger, S.J.; Hubbart, J.A. Characterizing Land Use Impacts on Channel Geomorphology and Streambed Sedimentological Characteristics. *Water* **2019**, *11*, 1088. [CrossRef]
106. Nichols, J.R. Land-Use Impacts on Aquatic Invertebrate Assemblages in a Dynamic Urbanizing Watershed of the Central U.S. Master's Thesis, University of Missouri, Columbia, MO, USA, 2012.
107. Scollan, D. A Multi-Configuration Evaluation of the Soil and Water Assessment Tool (SWAT) in a Mixed-Use Watershed in the Central USA. Master's Thesis, University of Missouri, Columbia, MO, USA, 2011.
108. Zeiger, S.J.; Hubbart, J.A. A SWAT Model Validation of Nested-Scale Contemporaneous Stream Flow, Suspended Sediment and Nutrients from a Multiple-Land-Use Watershed of the Central USA. *Sci. Total Environ.* **2016**, *572*, 232–243. [CrossRef] [PubMed]
109. Sunde, M.; He, H.S.; Hubbart, J.A.; Scroggins, C. Forecasting streamflow response to increased imperviousness in an urbanizing Midwestern watershed using a coupled modeling approach. *Appl. Geogr.* **2016**, *72*, 14–25. [CrossRef]
110. Sunde, M.G.; He, H.S.; Hubbart, J.A.; Urban, M.A. An integrated modeling approach for estimating hydrologic responses to future urbanization and climate changes in a mixed-use midwestern watershed. *J. Environ. Manag.* **2018**, *220*, 149–162. [CrossRef] [PubMed]
111. Sunde, M.G.; He, H.S.; Hubbart, J.A.; Urban, M.A. Integrating downscaled CMIP5 data with a physically based hydrologic model to estimate potential climate change impacts on streamflow processes in a mixed-use watershed. *Hydrol. Process.* **2017**, *31*, 1790–1803. [CrossRef]
112. Kellner, E.; Hubbart, J.A. Confounded by forgotten legacies: Effectively managing watersheds in the contemporary age of unknown unknowns. *Hydrol. Process.* **2017**, *31*, 2802–2808. [CrossRef]
113. Ewing, S. Landcare and community-led watershed management in Victoria, Australia. *Jawra J. Am. Water Resour. Assoc.* **1999**, *35*, 663–673. [CrossRef]
114. Tan, P.L.; Bowmer, K.H.; Mackenzie, J. Deliberative tools for meeting the challenges of water planning in Australia. *J. Hydrol.* **2012**, *474*, 2–10. [CrossRef]
115. Prato, T. Multiple-attribute evaluation of ecosystem management for the Missouri River system. *Ecol. Econ.* **2003**, *45*, 297–309. [CrossRef]
116. Borisova, T.; Racevskis, L.; Kipp, J. Stakeholder Analysis of a Collaborative Watershed Management Process: A Florida Case Study. *JAWRA* **2012**, *48*, 277–296. [CrossRef]
117. Kennen, J.G. Relation of macroinvertebrate community impairment to catchment characteristics in New Jersey Streams. *JAWRA* **1999**, *35*, 939–955. [CrossRef]
118. Sabatier, P.A.; Focht, W.; Lubell, M.; Trachtenberg, Z.; Vedlitz, A.; Matlock, M. (Eds.) *Swimming Upstream: Collaborative Approaches to Watershed Management*; MIT Press: Cambridge, MA, USA, 2005.
119. van de Meene, S.J.; Brown, R.R. Delving into the "Institutional Black Box": Revealing the Attributes of Sustainable Urban Water Management Regimes. *JAWRA* **2009**, *45*, 1448–1464. [CrossRef]

Article

Characterizing Land Use Impacts on Channel Geomorphology and Streambed Sedimentological Characteristics

Sean J. Zeiger [1,*] and Jason A. Hubbart [2,3]

1 School of Natural Resources, University of Missouri, 203-T ABNR Building, Columbia, MO 65211, USA
2 Institute of Water Security and Science, West Virginia University, 3109 Agricultural Sciences Building, Morgantown, WV 26506, USA; Jason.Hubbart@mail.wvu.edu
3 Davis College of Agriculture, Natural Resources and Design, Schools of Agriculture and Food, and Natural Resources, West Virginia University, 3109 Agricultural Sciences Building, Morgantown, WV 26506, USA
* Correspondence: ZeigerS@Missouri.edu; Tel.: +1-417-664-5321

Received: 3 April 2019; Accepted: 12 May 2019; Published: 24 May 2019

Abstract: Land use can radically degrade stream physical habitat via alterations to channel geomorphology and sedimentological characteristics. However, independent *and* combined influences such as those of agricultural and urban land use practices on channel geomorphology and substrate composition remain poorly understood. To further understanding of mixed land use influence on stream physical habitat, an intensive, 56 km hydrogeomorphological assessment was undertaken in a representative mixed land use watershed located in Midwestern USA. Sub-objectives included quantitative characterization of (1) channel geomorphology, (2) substrate frequency and embeddedness, and (3) relationships between land use, channel geomorphology, and substrate frequency and embeddedness. Channel geomorphology, and stream substrate data were directly measured at survey transects (n = 561) every 100 m of the entire 56 km distance of the reference stream. Observed data were averaged within five sub-basins (Sites #1 to #5) nested across an agricultural-urban land use gradient. Multiple regression results showed agricultural and urban land use explained nearly all of the variance in average width to depth ratios ($R^2 = 0.960$; $p = 0.020$; $n = 5$), and maximum bank angle ($R^2 = 0.896$; $p = 0.052$; $n = 5$). Streambed substrate samples of pools indicated significantly ($p < 0.001$) increased substrate embeddedness at agricultural Site #1 (80%) located in the headwaters and urban Site #5 (79%) located in the lower reaches compared to rural-urban Sites #2 to #4 (39 to 57%) located in the mid-reaches of the study stream. Streambed substrate embeddedness samples of riffles that ranged from 51 to 72% at Sites #1 and #5, and 27 to 46% at Sites #2 to #4 were significantly different between sites ($p = 0.013$). Percent embeddedness increased with downstream distance by 5% km^{-1} with the lower urban reaches indicating symptoms of urban stream syndrome linked to degraded riffle habitat. Collectively, observed alterations to channel morphology and substrate composition point to land use alterations to channel geomorphology metrics correlated with increased substrate embeddedness outside of mid-reaches where bedrock channel constraints accounted for less than 3% of substrate frequency. Results from this study show how a hydrogeomorphological assessment can help elucidate casual factors, target critical source areas, and thus, guide regional stream restoration efforts of mixed-land-use watersheds.

Keywords: physical habitat; aquatic ecology; stream health; environmental flows; land use; hydrology; hydroecology; ecohydrology

---

1. Introduction

The importance of channel geomorphology and stream substrate composition with regards to stream habitat has been well-documented [1–5]. Previous studies indicated that channel geomorphology and stream substrate are important aspects of stream physical habitat [6–8]. For example, substrate is a medium for settlement by propagules [9]. Successful propagules that mature into aquatic plants provide substrate stability, food, dissolved oxygen, physical habitat, and refugia beneficial for healthy stream ecosystems [7]. There is therefore also a direct link between suitable stream substrate composition and the abundance and distribution of macroinvertebrates [8,10], mussels [6], and fishes [11]. Given the influence on stream health, it is important to understand the dominant physical processes that control channel geomorphology and substrate composition.

Process-based understanding indicates channel geomorphology and stream substrate composition is dependent on sediment supply, and transport [12]. Surface runoff and streamflow are largely the erosive forces and transport mechanisms for overland soil erosion and stream bed and bank erosion, respectively [12]. However, a myriad of independent and interacting natural and anthropogenic factors influence surface runoff, streamflow, sediment supply and transport (e.g., meteorological conditions, hydrology, topography, soils and underlying geology, land cover, and human activities) [12]. Thus, regions associated with physiographic characteristics known to increase sediment supply and transport are areas of concern susceptible to alterations of channel geomorphology and stream substrate composition, and therefore physical habitat.

Streams of the Midwestern USA are commonly associated with increased sediment supply and degraded stream health [13]. For example, Gellis et al. [13] found channel sources of sediment accounted for the majority (>50%) of bed sediment in 79% of 99 Midwestern watersheds sampled. Claypan soils overlain by loess are especially subject to increased surface runoff and sediment supply to nearby streams [14,15]. A study by Willett et al. [15] indicated bank sediment accounted for 79% to 96% of the total in-stream sediment in two claypan watersheds of the central USA. A study by Huang [16] in mid-Missouri, showed bank erosion contributed approximately 67% of in-stream suspended sediment load in Hinkson Creek Watershed located in central claypan region of the USA. Given previous findings, it is not surprising that Midwestern streams are commonly characterized by increased frequency of fine bed sediment textures and increased substrate embeddedness related to degraded algal, macroinvertebrate, and fish communities [17–19]. Increased substrate embeddedness can influence stream community composition by reducing available riffle-pool habitat when increased deposition of fine sediments and sand fill interstitial spaces of course gravel substrate [20,21]. In Midwestern streams and other regions associated with increased sediment supply, human activity (i.e., land use) can exacerbate problems and further reduce suitable stream physical habitat and aquatic refugia.

Previous studies showed agricultural land use activities can increase sediment supply potentially leading to problems with sedimentation and substrate embeddedness [13,14,22,23]. For example, conventional tillage operations and livestock presence have been shown to increase erosion rates and sediment supply to nearby stream networks [12]. Given stream access to riparian areas, livestock can trample biomass and erode stream banks thereby increasing sediment supply [22]. Near channel removal of streambank stabilizing riparian vegetation has also been shown to increase bank erosion and alter channel geomorphology in agricultural watersheds [23].

Urban land use has also been shown to increase stream sediment supply. A literature review by Walsh et al. [24] presented consistent confirmation of urban land use influence on watershed hydrology, water chemistry, channel geomorphology, organic matter, algae, macroinvertebrates, fishes, and ecosystem processes. The term 'urban stream syndrome' is widely accepted to indicate consistent urban land use influence on stream health [24,25]. More specifically, previous studies indicated urban land use can increase sediment supply and pollutant transport when impervious surfaces and engineered water infrastructure increase the volume and velocity of surface runoff, stream power, and transport capacity [24]. Increased stream power associated with large storms can alter channel geomorphology via channel widening, channel incision, bank erosion and mass wasting [24].

For example, Gellis et al. [13] showed Tropical Storm Lee (a recent 100-year storm) caused stream bank erosion that accounted for 70% of suspended sediment concentration in the suburban and urban Upper Difficult Run, Virginia, USA. Results from Gellis et al. [13] emphasized the importance of streambank sources of sediment supply in suburban and urban watersheds. Thus, results from previous studies indicate streambank stabilization efforts are often necessary in urbanizing watersheds [13,14,24].

The combined influence of agricultural and urban land use practices on channel geomorphology and substrate composition is not well-understood. Investigations that quantitatively characterize land use influence on channel geomorphology and stream substrate composition in representative watersheds could help guide regional stream restoration efforts [4,26]. Therefore, the overall objective of this study was to quantify general trends and relationships between land use, stream hydromorphology, and substrate composition using observed data collected during a physical habitat assessment (PHA) in a representative experimental watershed located in the Midwestern USA. Sub-objectives were to (1) present observed spatial variability of channel geomorphology metrics, substrate frequency, and substrate embeddedness, and (2) quantify relationships between land use, channel geomorphology, and substrate frequency and embeddedness. Additionally, results are discussed with implications for watershed management in similar watersheds where agricultural and urban land uses have exacerbated problems with channel geomorphology alteration and streambed sedimentation.

2. Methods

2.1. Study Site

Hinkson Creek Watershed (HCW), is a rapidly urbanizing mixed-land use (forest, agriculture, urban) watershed located in the Lower Missouri Mississippi River Basin (LMMRB) (Table 1, Figure 1). Drainage area of HCW is approximately 230 km^2. Elevation ranges from 274 m above mean sea level (AMSL) in the headwaters to 177 m AMSL near the watershed outlet. At the time of this study, agricultural land use accounted for 36.7% HCW (28.4% grazing pasture and 8.3% row crop). Urban and suburban areas accounted for 29.0% of the total land use. Approximately 28 km^2 of impervious surfaces associated with urban land use were located within the municipal boundary of The City of Columbia (population 121,717; USCB [27]).

Soils in the upper elevations of HCW are generally loamy loess with a well-developed underling claypan in the argillic horizon of smectitic mineralization corresponding to the Mexico-Leonard association [28]. Claypan soils have increased surface runoff potential [29]. Soils that consist of a cherty clay solution residuum corresponding to the Weller-Bardley-Clinkenbeard (CBC) association are found in the lower reaches (upland of the alluvium). Soils in HCW are underlain by Burlington formation Mississippian series limestone in the lower reaches [28–30].

Climate in HCW is dominated by continental polar air masses in the winter and maritime and continental tropical air masses in the summer [31]. A 20-year climate record (2000–2019) obtained from the Sanborn Field climate station located on University of Missouri campus showed mean annual total precipitation was 962 mm and mean annual air temperature was 13.5 °C. A wet season occurs primarily during March through June [31,32]. The HCW main drainage, Hinkson Creek, is primarily stormflow dominated with a base flow index (ratio of base flow to total streamflow) of approximately 0.25 calculated using daily discharge data collected at a U.S. Geological Survey (USGS) gaging station (USGS 06910230, Site #4 in Figure 1) that has been intermediately monitoring stage since 1967.

During winter 2008, HCW was instrumented with a nested-scale experimental study design to investigate land use alterations to water quantity and quality [31]. Gauging sites (n = 5) partitioned the catchment into five sub-basins, each with different dominant land uses [31]. Site #5 was located near the watershed outlet, and Sites #1 to #4 were nested within (Figure 1). Site #1 (57.0% agricultural, 35.9% forested, 4.7% urban) was located in the agricultural land use dominated headwaters. Moving downstream from agricultural Site #1, agricultural land use decreased as urban land use increased. Site #3 (49.5% agricultural; 35.4% forested; 13.0% urban) was located at the rural-urban interface of HCW.

Site #5 (39.4% agricultural; 33.1% forested; 26.5% urban) is located in the lower most urbanized reaches. Results from previous studies showed land use alterations to the flow regime [28,33–35], environmental flows [36], stream temperature [37,38], suspended sediment [14,39,40], nutrients (i.e., total inorganic nitrogen species, total phosphorus) [41], and chloride [42] in HCW.

Table 1. Cumulative land use and land cover (LULC), drainage area, and stream length corresponding to each gauging site located in Hinkson Creek Watershed (HCW), Missouri, USA. Percent cumulative LULC is shown parenthetically.

Variable	Site #1	Site #2	Site #3 [km^2 (%)]	Site #4	Site #5	HCW
Agricultural	45.0 (57.0)	56.4 (54.9)	57.6 (49.5)	78.5 (44.1)	79.7 (39.4)	85.4 (36.7)
Forested	28.4 (35.9)	37.5 (36.4)	41.1 (35.4)	62.8 (34.5)	68.6 (33.1)	74.9 (32.2)
Urban	3.7 (4.7)	6.6 (6.4)	15 (13.0)	37.1 (20.4)	54.9 (26.5)	67.6 (29.0)
Wetland	1.9 (2.4)	2.4 (2.3)	2.5 (2.1)	3.6 (2.0)	4.4 (2.0)	4.9 (2.1)
Total area	79.0	102.9	116.2	182.0	207.5	232.8
Stream length §	22.8	29.8	35.4	43.6	53.0	56.1

§ Stream length is shown in km.

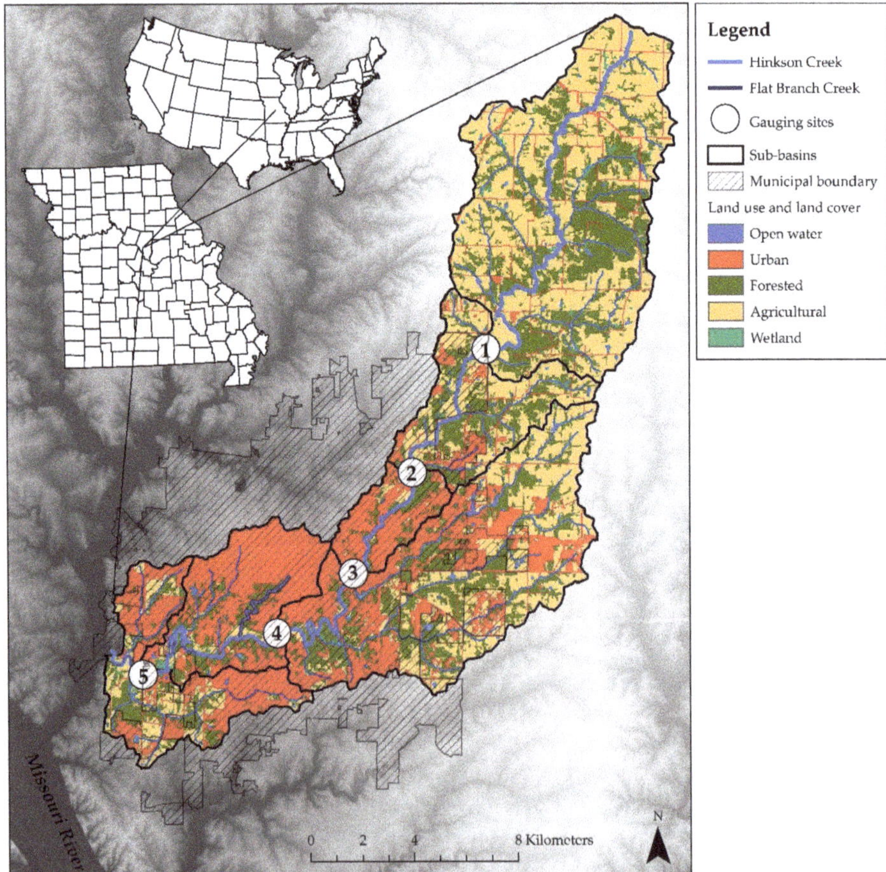

Figure 1. Land use and land cover of Hinkson Creek Watershed, Missouri, USA. Five nested gauging sites (numbered 1 to 5) and corresponding sub-basins are shown. Site #5 was located near the watershed outlet, Sites #1 to #4 nested within.

2.2. PHA Data Collection and Analysis

A physical habitat assessment (PHA) was performed in HCW during the study period (2013–2014) [43–45]. During a single survey, channel geomorphology and stream substrate data were collected at survey points (n = 561) every 100 m along the entire 56 km length of Hinkson Creek. At each survey point, data were collected (using a clinometer, extension pole, laser level, a laser range finder and/or meter stick) at principal transects that spanned from stream bank to bank perpendicular to flow. Additionally, data were collected at transects located 5 m upstream and downstream of the principal transect. Upstream and downstream transects were parallel to the principal transect. At any confluence of the main channel, three transects (upstream, downstream, and upstream in the tributary) were located equidistant from the center of the confluence.

2.2.1. Observed Channel Geomorphology

Channel width, wetted width of the stream, bankfull width, bank angle, bank height, and channel depth were measured using a laser level and/or laser range finder at each principal transect (n = 561). Following methods suggested by Harrelson et al. [46], physical indicators of bankfull level included the top of pointbars, changes in vegetation from aquatic to terrestrial, changes in slope, changes in bank material (e.g., from coarse gravel to sand), bank undercuts, or stain lines on bedrock or boulders. Bankfull width was measured parallel of the stream surface and perpendicular to stream flow from the lowest bank (i.e., bankfull bank) to the opposite bank.

Observed channel geomorphology data (collected every 100 m) dependent on drainage area were normalized as per methods used by Yanites and Tucker [47] using the following equation:

$$Ch^* = \frac{Ch_m}{kD_A^b} \tag{1}$$

where Ch^* is area-normalized channel geomorphology, Ch_m is the measured channel morphology metric, D_A is drainage area, and k and b are fitted power regression coefficients.

Channel geomorphology data were reduced by averaging within five nested HCW sub-basins (Figure 1) to quantify the change in channel geomorphology with downstream distance across a rural-urban land use gradient. Sub-basins were delineated between nested gauging sites, and thus, were not cumulative watersheds. For example, channel width at Site #2 reflected the average of all the channel width measurements collected every 100 m upstream of Site #2 and downstream of Site #1. One-way Analysis of Variance (ANOVA) and Tukey Kramer post-hoc multiple comparison tests were used to test for significant differences (CI = 95%, $p < 0.05$) in average channel geomorphology metrics (i.e., channel width, bank height, bank angle, etc.) between five HCW sub-basins [48]. Tukey Kramer post-hoc multiple comparison test was selected to elucidate site differences with a narrow confidence interval [49].

2.2.2. Observed Streambed Substrate Frequency and Embeddedness

Substrate particle size-class (Table 2) was estimated at each principal, upstream, downstream transect following methods suggested by Peck et al. [50], and Wolman [51]. At each transect, particles were sampled in five locations from bank to bank (left bank, left center, center, right center, and right bank) perpendicular to flow for a total of 8415 stream substrate samples. Substrate frequency was quantified as a percent of each individual substrate type relative to the total number of substrate samples collected within the area of interest. Additionally, substrate embeddedness of each particle was estimated as the percent vertical entrainment of a streambed substrate sample. Channel unit type (i.e., trench pool, plunge pool, impoundment pool, pool, split channel, riffle, glide, dry channel, etc.) was also recorded at each principal transect as per methods used by Peck et al. [50]. While the influence of water depth on channel unit identification was not quantified in the current study, adverse effects were assumed negligible considering the water depth recorded in the thalweg ranged from 0 to

280 cm with a median of 40 cm at the time of channel unit identification. Channel units were grouped into general categories of pool, riffle, and glide for further analysis. Observed substrate particle size frequency and percent embeddedness data were reduced by averaging in channel units (i.e., pools, riffles, glides, and dry channels), from bank to bank, and within each HCW sub-basin to quantify the change in stream substrate composition associated with riffle-pool habitat with downstream distance across a rural-urban land use gradient. Sub-basins were delineated between nested gauging sites and were thus not cumulative.

Table 2. Definition of streambed substrate variables used in this study.

Code	Description
LWD PILE	Large woody debris
VG	Vegetation
SH	Shell
CU	Culvert
BR	Bridge
RP	Rip-rap
WD	Wood
RS	Smooth Bedrock (>4000 mm)
RR	Rough Bedrock (>4000 mm)
RC	Concrete/asphalt (>4000 mm)
XB	Large boulder (1000 to 4000 mm)
SB	Small boulder (256 to 1000 mm)
CB	Cobble (64 to 256 mm)
GC	Coarse gravel (16 to 64 mm)
GF	Fine gravel (2 to 16 mm)
SA	Sand (0.06 to 2 mm)
FN	Silt/clay/muck (<0.06 mm)

3. Results

3.1. Observed Geomorphology

Results from power regression analysis showed drainage area explained 41.2% of the variance in channel width, 58.6% of the variance in bankfull width, 29.9% of the variance in bank height, and 42.1% of the variance in bankfull depth ($p < 0.001$; $n = 561$) (Figure 2). After channel width and height and bankfull width and depth metrics were area-normalized, results showed significant differences in channel morphology metrics averaged within five sub-basins of HCW (Table 3). For example, area-normalized average channel width ranged from 0.12 to 0.19 m km^{-2}. Area-normalized average bankfull width ranged from 0.21 to 0.27 m km^{-2}. Area-normalized average bank height ranged from 0.38 to 0.56 m km^{-2}. Area-normalized average bankfull depth ranged from 0.34 to 0.49 m km^{-2}. Average minimum and maximum bank angle ranged from 28.0 to 47.8 degrees. Average width to depth ratios and bed slope ranged from 6.3 to 10.3, and 0.14 to 0.22%, respectively. These results quantitatively characterized channel geomorphology metrics important for stream restoration efforts in HCW (Table 3).

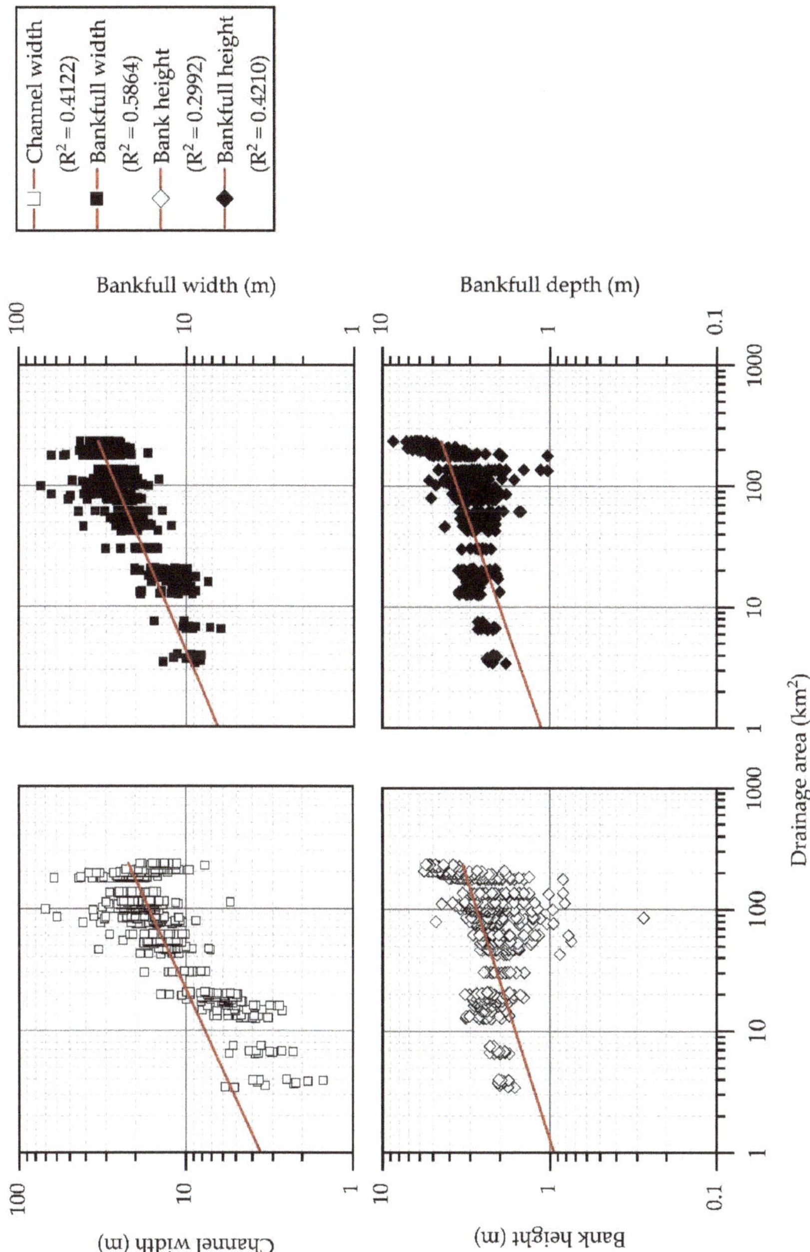

Figure 2. Power relationships between drainage area and observed channel morphological metrics in Hinkson Creek Watershed, Missouri, USA.

Table 3. Channel morphology metrics measured at survey sites (n = 561) and averaged within five nested sub-basins located in Hinkson Creek Watershed, Missouri, USA. Sub-basins were delineated between nested gauging sites and were thus not cumulative.

Variable [§]	Site #1	Site #2	Site #3	Site #4	Site #5
Channel width ($m\ km^{-2}$) [†]	0.14 [abc]	0.19 [ade]	0.15 [df]	0.16 [bg]	0.12 [cefg]
Bankfull width ($m\ km^{-2}$) [†]	0.23 [a]	0.27 [abcd]	0.23 [b]	0.23 [c]	0.21 [d]
Bank height ($m\ km^{-2}$) [†]	0.47 [abc]	0.39 [ad]	0.44 [e]	0.38 [bf]	0.56 [cdef]
Bankfull depth ($m\ km^{-2}$) [†]	0.42 [abcd]	0.36 [ae]	0.37 [bf]	0.34 [cg]	0.49 [defg]
Min bank angle (°)	28.0 [ab]	22.5 [ac]	22.0 [bd]	25.4	28.0 [cd]
Max bank angle (°)	47.8 [abc]	38.7 [a]	40.3 [b]	39.9 [c]	44.4
Width: depth (–)	6.3 [abc]	10.2 [ade]	8.5 [bdfg]	10.3 [cfh]	6.6 [egh]
Bed slope (%)	0.14 [abcd]	0.22 [aef]	0.24 [agh]	0.18 [aeg]	0.16 [dfh]

[§] Averages with corresponding letters ([a–h]) indicate significant differences (α = 0.05). [†] Channel geomorphic variables normalized by drainage area.

No significant relationships were observed between any one land use and land cover (LULC) index shown in Table 1 and average channel morphology metrics at five sub-basins in HCW ($R^2 \leq 0.593$; $p \geq 0.080$; n = 5). However, multiple linear regression (MLR) results showed agricultural and urban land use combined explained nearly all of the variance in average width to depth ratios ($R^2 = 0.960$; $p = 0.020$; n = 5), and maximum bank angle ($R^2 = 0.896$; $p = 0.052$; n = 5). Thus, results implied the expected relationships between drainage area and channel morphology were altered, at least in part, by the combined influence of agricultural and urban land use in HCW.

3.2. Observed Streambed Substrate Frequency and Embeddedness

Results from stream substrate frequency analysis quantitatively characterized bank to bank variability in streambed substrate composition important for stream physical habitat and aquatic refugia in HCW. Results showed percent fine substrate (FN) summarized for each bank to bank position along the entire drainage ranged from 11.6% at the center channel to 62.0% on the right descending bank. Sand substrate (SA) ranged from 17.3% on the left bank to 33.9% center channel. Percent fine gravel (GF), course gravel (GC), cobble (CB), and small boulder (SB) followed a similar trend as sand substrate from bank to bank. For example, GC ranged from 3.3% at the right bank to 21.3% at the center channel position. There was no obvious bank to bank trend in substrate frequency large boulders (XB), rough bedrock (RR), smooth bedrock (RS), or the rest of the substrate types presented in Table 4.

Examination of substrate frequency at five sub-basins showed the frequency of substrate types smaller than GC decreased with downstream distance from agricultural Site #1 in the headwaters to urban Site #3 at the rural-urban interface of the watershed. Continuing downstream, substrate frequency of substrate types smaller than GC increased from Site #3 to urban Site #5 located near the watershed outlet. For example, FN, SA, and GF decreased from Site #1 to Site #3 by 17%, 19%, and 3%, respectively. Continuing downstream, FN, SA, and GF increased from Site #3 to Site #5 by 19%, 12%, and 3%, respectively. Conversely, some of the larger substrate types (i.e., GC, CB, SB, and XB) increased with downstream distance from Site #1 to Site #3, and then, decreased from Site #3 to Site #5. Observed substrate composition data showed increased bed rock channel constraints at Sites #2 to #4 (9 to 11% RR and RS) located in the mid-reaches relative to Site #1 (3% RR) located in the agricultural headwaters and Site #5 (1% RR and RS) in the lower urban reaches (Table 5). These results quantifiably characterized the change of stream substrate frequency across an agricultural-urban land use gradient, and in synthesis, indicated increased substrate embeddedness in the agricultural headwaters and lower urban reaches located outside of the observed bedrock channel constraints in the mid-reaches of the study catchment.

Table 4. Bank to bank variability of streambed substrate frequency in Hinkson Creek Watershed, Missouri, USA. Each datum is a percent of a substrate type observed at a bank position along the entire 56km stream length of the reference stream. Substrate definitions are presented in Table 2.

Substrate Type	Left Bank	Left Center	Center (%)	Right Center	Right Bank
Silt/clay/muck (FN)	57.9	17.5	11.6	17.6	62.0
Sand (SA)	17.3	28.8	33.9	34.3	17.5
Fine gravel (GF)	0.2	5.1	6.1	4.1	0.4
Coarse gravel (GC)	3.5	20.1	21.3	16.8	3.3
Cobble (CB)	5.6	13.0	13.8	15.9	4.1
Small boulder (SB)	3.7	4.3	4.3	4.3	3.0
Large boulder (XB)	1.4	0.6	1.2	0.9	1.2
Rough Bedrock (RR)	4.2	5.9	3.7	2.5	1.2
Smooth Bedrock (RS)	1.5	2.0	1.4	1.1	1.1
Wood (WD)	2.9	0.4	0.6	0.4	2.5
Rip-rap (RP)	0.8	0.3	0.1	0.1	1.7
Bridge (BR)	–	–	0.1	–	–
Culvert (CU)	–	–	–	–	–
Concrete/asphalt (RC)	0.1	–	–	–	0.3
Shell (SH)	–	–	0.1	–	–
Vegetation (VG)	0.9	2.0	1.6	2.1	1.8
Large woody debris (LWD PILE)	0.1	0.1	0.1	0.1	–

Table 5. Streambed substrate frequency observed within five nested sub-basins in Hinkson Creek Watershed, Missouri, USA. Each datum is a percent of a substrate type. Sub-basins were delineated between nested gauging sites, and thus, were not cumulative watersheds excepting Site #1 located in the headwaters. Substrate definitions are presented in Table 2.

Substrate Type	Site #1	Site #2	Site #3 (%)	Site #4	Site #5
Silt/clay/muck (FN)	40	26	23	9	42
Sand (SA)	32	18	13	29	25
Fine gravel (GF)	4	1	1	3	4
Coarse gravel (GC)	9	19	16	18	14
Cobble (CB)	7	16	21	17	5
Small boulder (SB)	2	5	9	10	2
Large boulder (XB)	0	1	3	3	1
Rough Bedrock (RR)	3	10	9	1	1
Smooth Bedrock (RS)	0	0	2	8	0
Wood (WD)	2	2	0	0	1
Rip-rap (RP)	0	1	1	0	1
Bridge (BR)	–	–	0	–	–
Culvert (CU)	–	–	–	–	–
Concrete/asphalt (RC)	0	–	–	–	0
Shell (SH)	–	–	0	–	–
Vegetation (VG)	2	2	1	0	4
Large woody debris (LWD PILE)	0	0	0	0	–

Substrate embeddedness also varied between different stream habitats (i.e., riffle, glide, and pool). For example, results showed that pools (74.9% embeddedness) were associated with significantly greater substrate embeddedness than riffles (58.5% embeddedness) and glides (60.1% embeddedness) ($p = 0.0014$; CI = 95%). However, no significant differences were observed between riffles and glides ($p = 0.9367$; CI = 95%). Significant differences in substrate embeddedness were especially apparent at Sites #1 and #5 compared to the mid-reaches (i.e., Sites #2 to #4) ($p = 0.013$; CI = 95%) (Table 6). Stream banks were embedded with FN and SA in more than 88% of pools habitat at Sites #1 and #5. Even greater were differences in substrate embeddedness at center of channel samples (i.e., left center, center, and right center) of pools at Sites #1 and #5 (70 to 79%) compared to Sites #2 to #4 (39 to 57%). Center of channel samples of riffles ranged from 51 to 72% at Sites #1 and #5, and 27 to 46% at Sites #2 to #4.

Due to the observed widespread substrate embeddedness, results generally showed reduced spatial complexity of substrate composition in riffles, glides, and pools of HCW.

Table 6. Bank to bank variability of streambed substrate embeddedness in riffles, glides, and pools measured at survey sites (n = 561) and averaged within five nested sub-basins located in Hinkson Creek Watershed, Missouri, USA. Each datum shows substrate embeddedness as a percent. Sub-basins were delineated between nested gauging sites, and thus, were not cumulative watersheds excepting Site #1 located in the headwaters.

Bank Position	Site (#)	Riffles	Glides (%) *	Pools
Left bank	Site #1	83	100	88
	Site #2	76	67	71
	Site #3	76	81	67
	Site #4	72	71	72
	Site #5	85	65	88
Left center	Site #1	56	91	75
	Site #2	27	23	42
	Site #3	39	56	44
	Site #4	46	59	54
	Site #5	59	32	70
Center	Site #1	51	80	76
	Site #2	31	18	46
	Site #3	40	39	39
	Site #4	43	51	52
	Site #5	61	70	75
Right center	Site #1	52	93	79
	Site #2	30	27	57
	Site #3	34	43	47
	Site #4	41	59	52
	Site #5	72	42	76
Right bank	Site #1	70	100	90
	Site #2	77	78	80
	Site #3	70	69	86
	Site #4	65	73	77
	Site #5	88	78	90

* Percent embeddedness of glides at Site #1 indicate dry channel units.

Substrate embeddedness was significantly correlated at the 0.05 level with 4 of 8 channel morphology metrics considered in this study. For example, minimum bank angle ($R^2 = 0.8511$; $p = 0.0164$; $n = 5$), maximum bank angle ($R^2 = 0.8400$; $p = 0.0183$; $n = 5$), width to depth ratio ($R^2 = 0.7225$; $p = 0.0424$; $n = 5$), and bed slope ($R^2 = 0.7424$; $p = 0.0384$; $n = 5$) explained a substantial amount of variance in substrate embeddedness between sites in HCW. Minimum and maximum bank angle were positive correlates. Width to depth ratios and bed slope were negative correlates. Relationships between drainage area and land use were not significant at a 0.05 level. However, there was a general trend for increased substrate embeddedness in the agricultural headwaters at Site #1 and in the lower urban reaches at Site #5 (80% embedded) relative to the mid-reaches at Sites #2 to #4 (56 to 60% embedded) of HCW.

There was also a general trend for increased frequency of pool habitat coupled to decreased frequency of riffle and glide habitats in the agricultural headwaters and in the lower urban reaches of HCW (Table 7). The frequency of pools ranged from 56.1% at the urban Site #4 of HCW to 85.7% at urban Site #5. Conversely, the frequency of riffle habitat increased from 12.2% at Site #5 in the lower urban reaches to 28.0% at urban Site #4. The frequencies of pool, riffle, and glide habitats were not significantly correlated with substrate embeddedness ($R^2 \geq 0.420$; $p \leq 0.140$; $n = 5$). However, results showed strong correlations between channel unit frequency and frequency of fine substrate type ($0.927 \leq R^2 \geq 0.854$; $p \leq 0.016$; $n = 5$) highlighting observed streambed siltation in Hinkson Creek.

Table 7. Channel unit frequency measured at survey sites (n = 561) and averaged within five nested sub-basins located in Hinkson Creek Watershed, Missouri, USA.

Site (#)	Pools	Riffles (%)	Glides	Dry
Site #1	78.6	13.1	0.0	8.3
Site #2	72.1	23.6	4.3	0.0
Site #3	64.3	25.0	10.7	0.0
Site #4	56.1	28.0	15.9	0.0
Site #5	85.7	12.2	2.1	0.0

4. Discussion

4.1. Observed Geomorphology

Results from this study show how expected longitudinal change in channel geomorphology can be altered independently or by combined bedrock channel constraints and human activity. For example, the mid-reaches of HCW were associated with increased bedrock channel constraints (Figure 3), which according to Montgomery et al. [52] indicates transport capacity in excess of sediment supply. Outside of the mid-reaches (i.e., Sites #1 and #5) where less than 3% presence of channel bedrock constraints were observed, signs of channel incision were apparent. Site #1 in the agricultural headwaters and Site #5 located the lower urban reaches of HCW (Figure 3). Agricultural and urban land use explained nearly all the variance in average width to depth ratios, and maximum bank angle ($0.896 \leq R^2 \geq 0.960$; $0.020 \leq p \geq 0.052$; n = 5). At Site #1 located in the agricultural headwaters (drainage area = 79.0 km^2), width to depth ratios increased at a rate of about 0.3 km^{-1}, and maximum bank angle decreased by about 0.5 degrees km^{-1} as agricultural land use decreased from 100 to 56.9%, forested land use increased from 0 to 35.9%, and urban land use increased from 0 to 4.7% over 22.8 km stream distance (Figure 4). With stream distance from agricultural Site #1 to sub-urban Site #3 located at the rural-urban interface of HCW, (drainage area = 116.2.0 km^2), width to depth ratios increased by a rate of about 0.01 m m^{-1} km^{-1}, and maximum bank angle decreased by about 0.1 degrees km^{-1} as agricultural land use decreased to 49.5%, forested land use decreased by 0.5%, and urban land use increased to 13.0% over 12.6 km stream distance (Figure 4). Continuing downstream from sub-urban Site #3 to urban Site #5 located in the lower urban reaches (drainage area = 207.5 km^2), width to depth ratios decreased by about 0.2 km^{-1}, and maximum bank angle increased by about 0.2 degrees km^{-1} as agricultural land use decreased to 38.4%, forested land use decreased to 33.1%, and urban land use increased to 26.5% over 17.6 km stream distance (Figure 4). Bankfull depth increased by 0.13 m km^{-1} at Site #1, 0.02 m km^{-1} between Sites #1 and #3, and 0.15 m km^{-1} between Sites #3 and #5. In combination, these results indicated increased channel incision at Site #1 in the agricultural headwaters where agricultural land use accounted for greater than 50% of total catchment area, and Site #5 the lower urban reaches where urban land use accounted for greater than 20% of total catchment area in HCW. Thus, Sites #1 and #5 were considered hot spots of channel incision due to increased rates of change in channel geomorphology metrics.

Figure 3. Examples of channel characteristics associated with the agricultural headwaters (**a**), rural-urban mid-reaches (**b**,**c**), and lower urban reaches (**d**) of Hinkson Creek Watershed, USA.

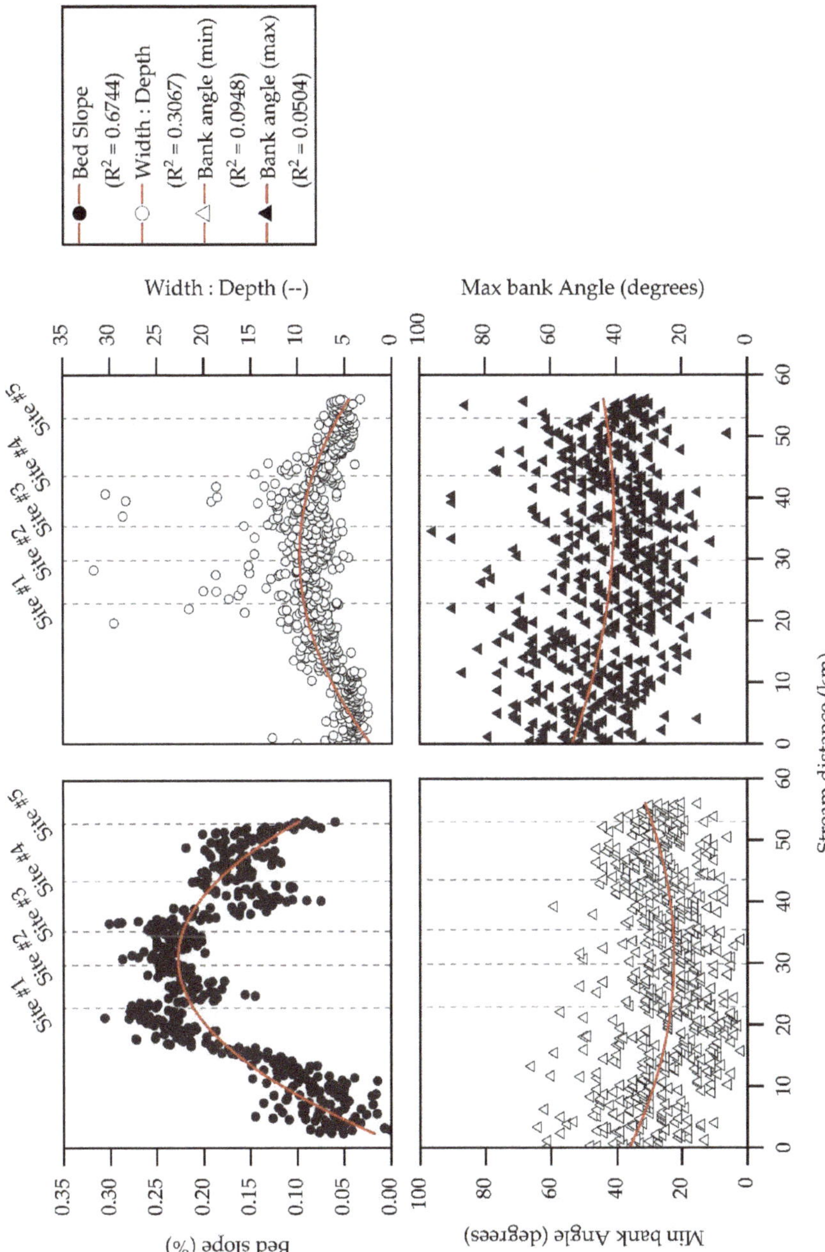

Figure 4. Trend lines associated with channel morphology variables with downstream distance (km) in Hinkson Creek Watershed, Missouri, USA. Vertical lines show location of five nested gauging sites.

Previous studies have shown agricultural land use can result in channel incision through various mechanisms including, but not limited to, channelization, deforestation, and alterations to soil and channel hydraulics [53–56]. Simon et al. [56] noted increased pore water pressure coupled to decreased sheer strength can lead to streambank erosion. Given the influence of soil hydraulic forces on streambank stability, agricultural areas are subject to increased bank failure where crop irrigation lowers water table levels in the vicinity of a stream. A study by Zaimes et al. [54] showed incised stream channels were associated with increased bank mass wasting, streambank erosion and sediment load in Beak Creek, an agricultural land use dominated Midwestern stream located in central Iowa, USA. Streambank erosion rates varied from meandering row crop fields (387 mm year^{-1}), cattle and horse pastures (295 mm year^{-1}), and meandering forest buffered reaches (142 mm year^{-1}). Results indicated that forested riparian buffers would reduce streambank erosion by 72%. Streambank stabilizing root systems associated with riparian vegetation add roughness, reduce stream power, and enhance bank accretion [56]. Results from the aforementioned studies are in agreement with results from the current work showing agricultural land use alterations to channel morphology of Midwestern streams.

Previous studies have also shown urban land use can cause channel incision via alterations to streamflow regimes [24,25,57–59]. A literature review by Walsh et al. [24] showed increased impervious surfaces associated with urban land use can cause increased volume and velocity of surface runoff and a flashy hydrologic streamflow response linked to increased bank wasting, channel incision, and scouring. For example, Jordan et al. [60] showed urban land use alterations to flow caused a 9 to 61% increase of sediment yield due to channel incision and bank erosion in Berryessa Creek, California, USA. A study by Shields et al. [59] quantified differences in channel incision, streamflow, water quality, and stream physical habitat between rural and urban catchments located in the Yazoo River basin, Mississippi, USA. Results showed urban land use was associated with decreased physical aquatic habitat, 6.4 times median rate of rise, 1.8 times channel depth, 3.5 times channel width, 2 to 3 times turbidity and suspended solids, 2 times fish species, and 4 times the amount of fish biomass per unit of effort [59]. Results from the current work are a novel addition to previous studies considering the intensive sampling regimen (n = 561) that made possible the estimation of the rate of change in channel morphology and substrate composition across an agricultural-urban land use gradient.

4.2. Observed Streambed Substrate Frequency and Embeddedness

No channel morphology variable significantly explained the variance in substrate embeddedness of pools, riffles, and glides for each bank to bank sampling position (Figure 5) highlighting the spatial complexity of streambed substrate composition in this study. Spatial complexity of substrate composition was, in part, shaped by the thalweg which meandered from bank to bank. The thalweg was generally associated with decreased substrate embeddedness due to increased stream velocity, and thus, increased sediment transport capacity. While much of the observed bank to bank variability in substrate embeddedness was attributed to thalweg position, results from the current work in combination with previous research in the region indicated longitudinal variability in substrate composition was attributed to the presence claypan soils and agricultural land use in the headwaters, increased bed slope and bedrock channel constraints in the mid-reaches, and the influence of urban land use associated with increased impervious surfaces in the lower reaches (Figure 3).

Previous studies in HCW and elsewhere have shown claypan soils and agricultural land use are associated with increased surface runoff, soil erosion, and channel sediment supply [14,15,29]. Lerch et al. [29] noted claypan soils corresponding to the Mexico-Leonard association consisting of an argillic soil horizon of smectitic mineralogy with clay content of 450–650 g kg^{-1} formed at 10 to 50 cm depth are characterized by increased surface runoff. Willett et al. [15] showed claypan soils were associated with increased bank sediment supply that accounted for 88% of total sediment supply. Streambank erosion was particularly high during winter months attributed to a combination of increased frequency and large magnitude flow events, freeze/thaw cycles, high antecedent moisture conditions, and lack of vegetation. In the current work, increased bank angle and substrate embeddedness were apparent during field sampling in the agricultural headwaters of HCW where claypan soils are present. Results from the current work are among the first to quantitatively characterize agricultural land use influence on substrate frequency and substrate embeddedness in the Central Claypan Region and point to a need to mitigate the influence claypan soils and agricultural land use on degradation of stream hydrogeomorphology in HCW. Thus, results from this study in combination with previous research have implications in other agricultural watersheds where near surface soil features (e.g., claypans, argillic horizons, or fragipans) have increased sediment supply.

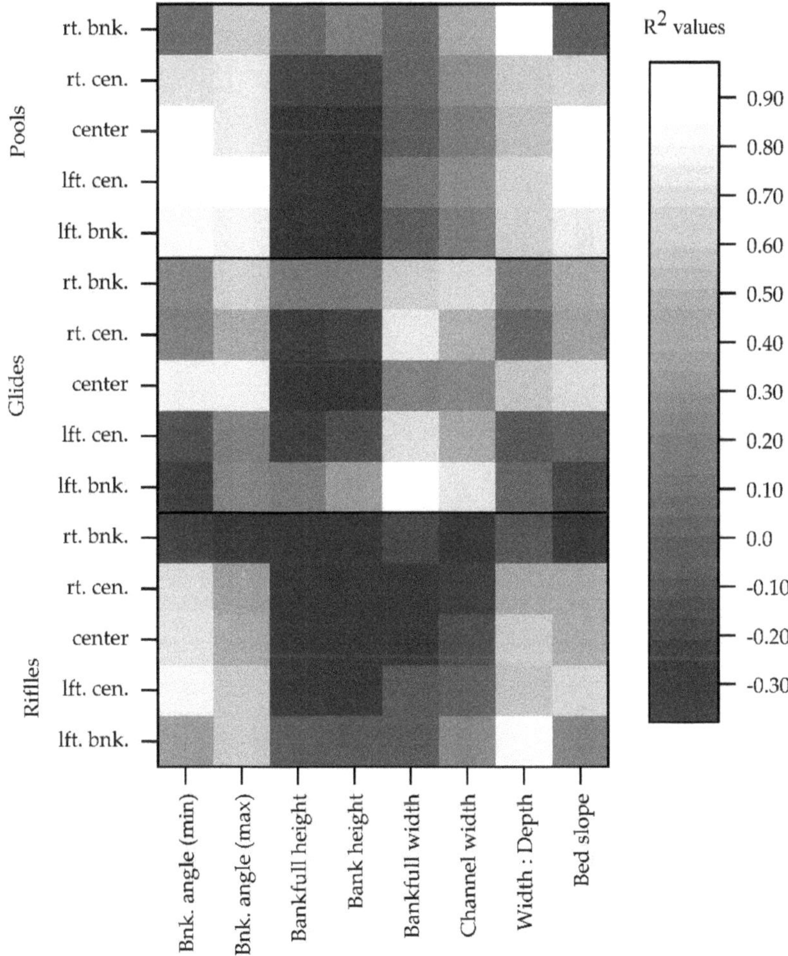

Figure 5. Explained variance between channel morphology metrics (x-axis) and substrate embeddedness from left bank (lft. bnk.) to right bank (rt. bnk.) associated with riffles, glides, and pools (y-axis) at five sub-basins of Hinkson Creek Watershed, Missouri, USA. No glide channel units were observed at Site #1, instead, percent embeddedness of dry channel units is shown.

While substrate embeddedness was greater in the agricultural headwaters, substrate embeddedness decreased as bed slope increased in the mid-reaches. Agricultural Site #1 was associated with 26 to 39% greater frequency of substrate smaller than GC, and 20 to 25% greater substrate embeddedness compared to Sites #2 to #4 in the mid-reaches of HCW (Figure 6). Bed slope, width to depth ratio, and the frequency of substrate greater than GC were inversely related to substrate embeddedness (Figure 4). Process-based understanding of the control of channel morphology on streamflow and sediment transport [52] indicated the aforementioned general trends in the observed data were physically meaningful. Given that bed slope controls velocity of streamflow, stream capacity, and stream competence, it makes sense that Sites #2 to #4 located in the mid-reaches were generally associated with less substrate embeddedness and frequency of substrate less than GC in diameter compared to Sites #1 and #5. The mid-reaches of HCW were also associated with increased bedrock substrate and channel constraints (Figure 3), which indicated transport capacity in excess of sediment supply [52]. However, outside of the mid-reaches, increased substrate embeddedness was observed at agricultural Site # 1 and urban Site #5 (Figure 3).

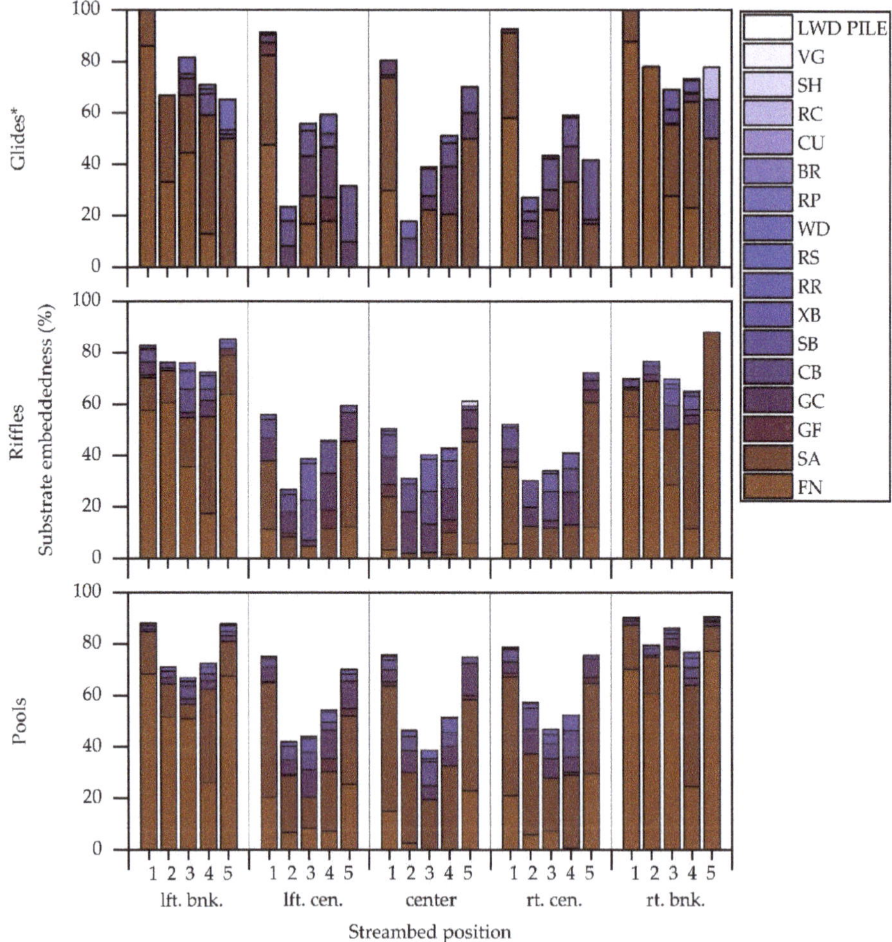

Figure 6. Substrate frequency and embeddedness from left bank (lft. bnk.) to right bank (rt. bnk.) associated with riffles, glides, and pools located within five nested sub-basins (numbered 1 to 5) of Hinkson Creek Watershed, Missouri, USA. Sub-basin #1 was located in the headwaters. Sub-basin #5 was located near the watershed outlet. No glide channel units were observed at Site #1, instead, percent embeddedness of dry channel units is shown. Substrate definitions are presented in Table 2.

The greatest rate of change in substrate embeddedness with downstream distance was observed in riffles located in the lower urban reaches where bank height exceeded 7 m (Figure 7). For example, percent embeddedness of riffle habitat decreased by at a rate of about 2% km^{-1} as agricultural land use decreased from 100 to 56.9%, forested land use increased from 0 to 35.9%, and urban land use increased from 0 to 4.7% over 22.8 km stream distance at Site #1 (Figure 7). However, percent embeddedness of riffles remained relatively constant (i.e., negligible rate of change) from agricultural Site #1 to sub-urban Site #3 located at the rural-urban interface where increased bedrock channel constraints were observed in HCW (Figure 7). Continuing downstream from sub-urban Site #3 to urban Site #4, substrate embeddedness of riffles began to increase by about 1.4% km^{-1} over approximately 8 km of stream distance. Further downstream, percent embeddedness of riffles increased rapidly by 5.3% km^{-1} between urban Sites #4 and #5 (Figure 7). Similar trends were observed in glides and pools as well (Figure 7). Thus, these results showed increased rate of change in percent embeddedness linked to degraded physical habitat (riffles, glides and pools) at agricultural Site #1 and urban Site #5, with a disproportionate rate of increase of substrate embeddedness in riffle habitat of the lower urban reaches pointing to symptoms of urban stream syndrome in HCW.

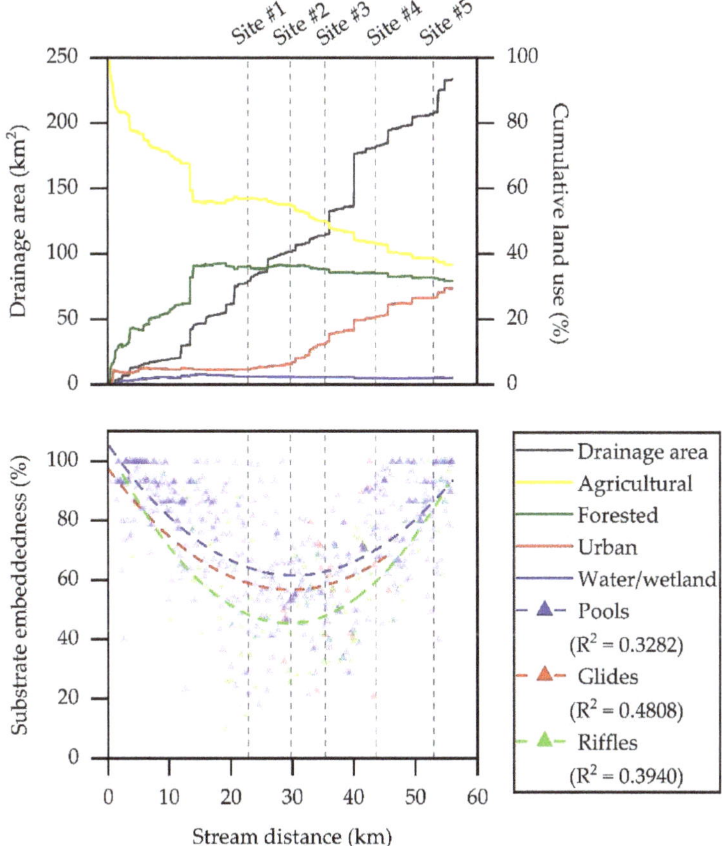

Figure 7. Watershed characteristics and trend lines associated with substrate embeddedness in pools, riffles, and glides with downstream distance (km) in Hinkson Creek Watershed, Missouri, USA. Vertical lines show location of five nested gauging sites.

The observed channel incision in the lower reaches of the current work was corroborated by other authors who also reported urban land use influence on channel morphology [24,60,61]. Blann et al. [53] discussed how increased channel incision disrupts hydrologically and ecologically important stream-floodplain-riparian connectivity. In urban catchments, channel incision is often attributed to increased impervious surfaces and engineered waterways that connect impervious surfaces directly to stream channels [24,25]. Increased impervious surfaces have been shown to increase the volume and velocity of surface runoff in HCW [36] and elsewhere [24,25,62]. Increased surface runoff commonly translates to alterations to flow regimes (i.e., flow frequency, magnitude, timing, duration and rate of change), sediment transport regimes, water quality, and aquatic ecosystem health [24,25]. Clearly, there is a need to mitigate land use alterations to channel morphology via channel stabilization efforts in HCW and similar catchments globally. However, Vietz et al. [58] noted channel stabilization methods may not be sufficient to combat channel degradation in urbanized catchments. There is also a critical need for flow regime management efforts to reduce flow magnitude, frequency, and rate of change associated with alterations to channel morphology.

It was evident in the current work that simultaneously occurring agricultural and urban land uses exacerbated problems with substrate embeddedness in HCW particularly in the headwaters and lower urban reaches were bedrock constraints were less abundant. Previous studies also showed a general trend for suspended sediment and nutrients to decrease from Site #1 in the agricultural headwaters to Site #3 at the rural-urban interface of the watershed, and then, increase from Site #3 to urban Site #5 located near the watershed outlet in HCW [14,40,41]. In fact, significant relationships were observed between substrate embeddedness values reported in the current work and concentrations of suspended sediment (R^2 = 0.798; p = 0.026), nitrite-N (R^2 = 0.975; p = 0.001), and ammonia-N (R^2 = 0.956; p = 0.002) reported in previous studies in HCW [14,41]. Notably, suspended sediment and total phosphorous yields were particularly high compared to other regions within the Mississippi River Basin. For example, observed suspended sediment yields exceeding 300 Mg km^{-2} $year^{-1}$ were 54%, 80%, and 87% greater than sediment yields from the Ohio River, the Upper Mississippi River, and the Missouri River basins, respectively [14]. Total phosphorous yields (0.979 kg ha^{-1} $year^{-1}$) were also high for the region [41]. In combination with results of LULC alterations to substrate embeddedness from the current work, results indicate increased sediment supply has degraded water quality, physical habitat, and aquatic refugia in HCW, especially in the agricultural headwaters and the lower urban reaches.

Collectively, previous studies showed streams of the Midwestern USA are commonly associated with increased sediment supply and degraded stream health [13,15,57]. For example, Gellis et al. [57] noted channel sources of sediment accounted for the majority (>50%) of bed sediment in 79% of 99 Midwestern watersheds sampled. Increased channel sediment supply and subsequent bed sedimentation can bury riffle habitat, reduce egg and fry survivorship, and lower prey densities [63,64]. Results from previous studies often focused on agricultural influence on stream bank erosion and sedimentation which was observed in the current work. However, it should be noted that urban land use may cause a greater influence on channel morphology and streambed sedimentation relative to agricultural land use. For example, substrate embeddedness was observed to decrease between Site #1 and #2 where agricultural land use accounted for 55% of total catchment area. Continuing downstream, substrate embeddedness began to increase between Sites #2 and #3 at the rural—urban interface of HCW which was associated with about 7 to 13% urban land use in this study. Thus, results from this study were in agreement with Paul and Meyer [65] who noted urban land use can account for a small percentage of total catchment area while causing a disproportionate influence on water quality and stream health relative to other land uses.

4.3. Management Implications and Future Research

The PHA and nested-scale experimental watershed study design presented in the current work is a useful model for managers that need to elucidate casual factors, target critical source areas, and thus, guide regional stream restoration efforts of mixed land use watersheds globally. The intensive field data collection associated with this assessment made it possible to estimate downstream changes in metrics important for understanding available stream physical habitat and biological refugia across an agricultural-urban land use gradient. Assuming that reduction of sediment supply below sediment transport capacity will help to decrease substrate embeddedness, management efforts might focus on reducing land use alterations to (1) erosive forces of surface runoff and streamflow, and (2) channel incision and bank losses in watersheds prone to increased sediment supply.

A complete decoupling of the simultaneously occuring natural and anthropogenic factors (legacy and ongoing) that influence channel geomorphology and streambed sedimentological characteristics was beyond the scope of the current work. Additionally, results from this study do not show temporal variablity of channel geomorphology and substrate composition. Thus, results may not reflect the total influence of past and present land use alterations that are expected to continue to alter channel geomorphology and substrate composition over long time periods. Future work focused on re-measuring channel geomorphology, substrate frequency and embeddedness may elucidate relationship between land use change and subsequent alterations to stream physical habitat.

5. Conclusions

Key findings from the current work point to (1) agricultural and urban land use alterations on channel geomorphology and stream substrate embeddedness, and (2) channel geomorphology as an indicator of stream substrate embeddedness in a mixed land use Midwestern stream. Expected relationships between drainage area and channel geomorphology were altered. Agricultural and urban land use explained nearly all of the variance in average width to depth ratios (R^2 = 0.960; p = 0.020; n = 5), and maximum bank angle (R^2 = 0.896; p = 0.052; n = 5). Also observed were reduced frequencies of riffle habitat at Site #1 in the agricultural headwaters (13.1%) and Site #5 (12.2%) in the lower urban reaches. Increased rate of change in percent embeddedness of riffle habitat exceeding 5% km^{-1} in the lower urban reaches indicated a disproportionate influence of urban land use on hydromorphology. Results showed nearly all the variability in channel unit frequency was explained by increased frequency of fine substrate type ($0.927 \leq R^2 \geq 0.854$; $p \leq 0.016$; n = 5). These results highlighted observed streambed siltation in Hinkson Creek, especially in the agricultural dominated headwaters and the lower urban reaches. Thus, results from this study point to a critical need to mitigate observed agricultural and urban land use impacts on stream hydrogeomorphology.

Given the influence of stream physical habitat on stream ecosystem health, stream physical habitat assessments are an integral component of regional stream restoration efforts. The robust observed data set collected for this study can provide critical information needed to guide regional policy development and watershed management efforts. Results of this study are particularly important for regional management efforts considering (1) rigorous hydrogeomorphic data sets are rare, and (2) the magnitude of alterations to channel geomorphology, and streambed composition in Midwestern streams of the USA. Results from this study quantitatively characterize channel geomorphology, substrate frequency, and embeddedness across 56 km stream length in the Midwestern USA where streams are commonly burdened with increased channel sediment supply, increased suspended sediment and total phosphorous, and problems associated with substrate siltation. While results from the current work are regionally applicable, key findings may also be useful to guide policy development and management decisions in physiographically similar watersheds globally.

Author Contributions: Conceptualization, J.A.H. and S.J.Z.; Formal Analysis, S.J.Z.; Investigation, J.A.H.; Data Curation, S.J.Z.; Writing-Original Draft Preparation, S.J.Z.; Writing-Review & Editing, J.A.H.; Supervision, J.A.H.; Project Administration, J.A.H.; Funding Acquisition, J.A.H.

Funding: This research was funded by the Missouri Department of Conservation and the U.S. Environmental Protection Agency Region 7 through the Missouri Department of Natural Resources (P.N: G08-NPS-17) under Section 319 of the Clean Water Act and through joint agreement of the University of Missouri, the City of Columbia, and Boone County Public works and partners of the Hinkson Creek Collaborative Adaptive Management (CAM) program. Additional funding was provided by the National Science Foundation under Award Number OIA-1458952, the USDA National Institute of Food and Agriculture, Hatch project accession number 1011536, and the West Virginia Agricultural and Forestry Experiment Station. Results presented may not reflect the views of the sponsors and no official endorsement should be inferred.

Conflicts of Interest: The authors declare no conflict of interest.

References

1. Van Looy, K.; Tonkin, J.D.; Floury, M.; Leigh, C.; Soininen, J.; Larsen, S.; Datry, T. The three Rs of river ecosystem resilience: Resources, recruitment, and refugia. *River Res. Appl.* **2019**, *35*, 107–120. [CrossRef]
2. Keppel, G.; Van Niel, K.P.; Wardell-Johnson, G.W.; Yates, C.J.; Byrne, M.; Mucina, L.; Franklin, S.E. Refugia: Identifying and understanding safe havens for biodiversity under climate change. *Glob. Ecol. Biogeogr.* **2012**, *21*, 393–404. [CrossRef]
3. Scherrer, D.; Körner, C. Topographically controlled thermal-habitat differentiation buffers alpine plant diversity against climate warming. *J. Biogeogr.* **2011**, *38*, 406–416. [CrossRef]
4. Maddock, I. The importance of physical habitat assessment for evaluating river health. *Freshw. Biol.* **1999**, *41*, 373–391. [CrossRef]
5. Poff, N.L.; Ward, J.V. Physical habitat template of lotic systems: Recovery in the context of historical pattern of spatiotemporal heterogeneity. *Environ. Manag.* **1990**, *14*, 629. [CrossRef]
6. Klos, P.Z.; Rosenberry, D.O.; Nelson, G.R. Influence of hyporheic exchange, substrate distribution, and other physically linked hydrogeomorphic characteristics on abundance of freshwater mussels. *Ecohydrology* **2015**, *8*, 1284–1291. [CrossRef]
7. Duan, X.; Wang, Z.; Zang, K. Effect of streambed sediment on benthic ecology. *Int. J. Sediment Res.* **2009**, *24*, 325–338. [CrossRef]
8. Duan, X.; Wang, Z.; Tian, S. Effect of streambed substrate on macroinvertebrate biodiversity. *Front. Environ. Sci. Eng. China* **2008**, *2*, 122–128. [CrossRef]
9. Erman, D.C.; Erman, N.A. The response of stream macroinvertebrates to substrate size and heterogeneity. *Hydrobiologia* **1984**, *108*, 75–82. [CrossRef]
10. Nichols, J.; Hubbart, J.A.; Poulton, B.C. Using macroinvertebrate assemblages and multiple stressors to infer urban stream system condition: A case study in the central US. *Urban Ecosyst.* **2016**, *19*, 679–704. [CrossRef]
11. Blann, K.L.; Anderson, J.L.; Sands, G.R.; Vondracek, B. Effects of Agricultural Drainage on Aquatic Ecosystems: A Review. *Crit. Rev. Environ. Sci. Technol.* **2009**, *39*, 909–1001. [CrossRef]
12. Neitsch, S.L.; Arnold, J.G.; Kiniry, J.R.; Williams, J.R. *Soil and Water Assessment Tool Theoretical Documentation Version 2009*; Texas Water Resources Institute: College Station, TX, USA, 2011; Available online: https://swat.tamu.edu/documentation/ (accessed on 26 August 2018).
13. Gellis, A.C.; Myers, M.K.; Noe, G.B.; Hupp, C.R.; Schenk, E.R.; Myers, L. Storms, channel changes, and a sediment budget for an urban-suburban stream, Difficult Run, Virginia, USA. *Geomorphology* **2017**, *278*, 128–148. [CrossRef]
14. Zeiger, S.J.; Hubbart, J.A. Quantifying suspended sediment flux in a mixed land use urbanizing watershed using a nested-scale study design. *Sci. Total Environ.* **2015**, *242*, 315–323. [CrossRef] [PubMed]
15. Willett, C.D.; Lerch, R.N.; Schultz, R.C.; Berges, S.A.; Peacher, R.D.; Isenhart, T.M. Streambank erosion in two watersheds of the Central Claypan Region of Missouri, United States. *J. Soil Water Conserv.* **2012**, *67*, 249–263. [CrossRef]
16. Huang, D. Quantifying Stream Bank Erosion Deposition Rates in a Central, U.S. Urban Watershed. Master's Thesis, University of Missouri, Columbia, MO, USA, 2012.
17. Waite, I.R.; Van Metre, P.C. Multistressor predictive models of invertebrate condition in the Corn Belt, USA. *Freshw. Sci.* **2017**, *36*, 901–914. [CrossRef]
18. Meador, M.R.; Frey, J.W. Relative Importance of Water-Quality Stressors in Predicting Fish Community Responses in Midwestern Streams. *JAWRA J. Am. Water Resour. Assoc.* **2018**, *54*, 708–723. [CrossRef]
19. Munn, M.D.; Waite, I.; Konrad, C.P. Assessing the influence of multiple stressors on stream diatom metrics in the upper Midwest, USA. *Ecol. Indic.* **2018**, *85*, 1239–1248. [CrossRef]
20. Poff, N.L.; Allan, J.D.; Bain, M.B.; Karr, J.R.; Prestegaard, K.L.; Richter, B.D.; Stromberg, J.C. The natural flow regime. *BioScience* **1997**, *47*, 769–784. [CrossRef]
21. Berkman, H.E.; Rabeni, C.F. Effect of siltation on stream fish communities. *Environ. Biol. Fishes* **1987**, *18*, 285–294. [CrossRef]
22. Tufekcioglu, M.; Isenhart, T.M.; Schultz, R.C.; Bear, D.A.; Kovar, J.L.; Russell, J.R. Stream bank erosion as a source of sediment and phosphorus in grazed pastures of the Rathbun Lake Watershed in southern Iowa, United States. *J. Soil Water Conserv.* **2012**, *67*, 545–555. [CrossRef]

23. Dosskey, M.G.; Vidon, P.; Gurwick, N.P.; Allan, C.J.; Duval, T.P.; Lowrance, R. The role of riparian vegetation in protecting and improving chemical water quality in streams1. *JAWRA J. Am. Water Resour. Assoc.* **2010**, *46*, 261–277. [CrossRef]
24. Walsh, C.J.; Roy, A.H.; Feminella, J.W.; Cottingham, P.D.; Groffman, P.M.; Morgan, I.I.R.P. The urban stream syndrome: Current knowledge and the search for a cure. *J. N. Am. Benthol. Soc.* **2005**, *24*, 706–723. [CrossRef]
25. Meyer, J.L.; Paul, M.J.; Taulbee, W.K. Stream ecosystem function in urbanizing landscapes. *J. N. Am. Benthol. Soc.* **2005**, *24*, 602–612. [CrossRef]
26. Poff, N.L.; Richter, B.D.; Arthington, A.H.; Bunn, S.E.; Naiman, R.J.; Kendy, E.; Henriksen, J. The ecological limits of hydrologic alteration (ELOHA): A new framework for developing regional environmental flow standards. *Freshw. Biol.* **2010**, *55*, 147–170. [CrossRef]
27. United States Census Bureau (USCB). U.S. Census Bureau: State and County Quick Facts. 2018. Available online: http://quickfacts.census.gov/qfd/states/29000.html (accessed on 13 May 2019).
28. Hubbart, J.A.; Zell, C. Considering streamflow trend analyses uncertainty in urbanizing watersheds: A base flow case study in the central United States. *Earth Interact.* **2013**, *17*, 1–28. [CrossRef]
29. Lerch, R.N.; Kitchen, N.R.; Baffaut, C.; Vories, E.D. Long-term agroecosystem research in the Central Mississippi River Basin: Goodwater Creek Experimental Watershed and regional nutrient water quality data. *J. Environ. Qual.* **2015**, *44*, 37–43. [CrossRef] [PubMed]
30. Miller, D.; Vandike, J. *Groundwater Resources of Missouri*; Missouri Department of Natural Resources Division of Geology and Land Survey Water Resources: Jefferson City, MO, USA, 1997; Volume 46, p. 136.
31. Hubbart, J.A.; Holmes, J.; Bowman, G. Integrating science based decision making and TMDL allocations in urbanizing watersheds. *Watershed Sci. Bull.* **2010**, *1*, 19–24.
32. Hubbart, J.A.; Kellner, E.; Hooper, L.; Lupo, A.R.; Market, P.S.; Guinan, P.E.; Stephan, K.; Fox, N.I.; Svoma, B.M. Localized Climate and Surface Energy Flux Alterations across an Urban Gradient in the Central U.S. *Energies* **2014**, *7*, 1770–1791. [CrossRef]
33. Zeiger, S.J.; Hubbart, J.A. Rainfall-Stream Flow Responses in a Mixed land use Municipal Watershed of the Central, U.S.A. *Environ. Earth Sci.* **2018**, *77*, 438. [CrossRef]
34. Zeiger, S.J.; Hubbart, J.A. Quantifying land use influences on event-based flow frequency, timing, magnitude, and rate of change in an urbanizing watershed of the central USA. *Environ. Earth Sci.* **2018**, *77*, 107. [CrossRef]
35. Wei, L.; Hubbart, J.A.; Zhou, H. Variable streamflow contributions in nested subwatersheds of a US midwestern urban watershed. *Water Resour. Manag.* **2018**, *32*, 213–228. [CrossRef]
36. Zeiger, S.J.; Hubbart, J.A. Assessing environmental flow targets using pre-settlement land cover: A SWAT modeling application. *Water* **2018**, *10*, 791. [CrossRef]
37. Zeiger, S.J.; Hubbart, J.A. Quantifying urban stormwater temperature surges: A central US watershed study. *Hydrology* **2015**, *2*, 193–209. [CrossRef]
38. Zeiger, S.J.; Hubbart, J.A.; Anderson, S.H.; Stambaugh, M.C. Quantifying and modelling urban stream temperature: A central US watershed study. *Hydrol. Process.* **2015**, *30*, 503–514. [CrossRef]
39. Kellner, E.; Hubbart, J.A. Improving understanding of mixed land use watershed suspended sediment regimes: Mechanistic progress through high-frequency sampling. *Sci. Total Environ.* **2017**, *598*, 228–238. [CrossRef] [PubMed]
40. Kellner, E.; Hubbart, J.A. Quantifying Urban Land-Use Impacts on Suspended Sediment Particle Size Class Distribution: A Method and Case Study. *Stormwater J.* **2014**, *15*, 40–50.
41. Zeiger, S.J.; Hubbart, J.A. Nested-Scale Nutrient Yields from a Mixed land use Urbanizing Watershed. *Hydrol. Process.* **2016**, *30*, 1475–1490. [CrossRef]
42. Hubbart, J.A.; Kellner, E.; Hooper, L.W.; Zeiger, S.J. Quantifying loading, toxic concentrations, and systemic persistence of chloride in a contemporary mixed land use watershed using an experimental watershed approach. *Sci. Total Environ.* **2017**, *581*, 822–832. [CrossRef]
43. Hubbart, J.A.; Kellner, E.; Kinder, P.; Stephan, K. Challenges in aquatic physical habitat assessment: Improving conservation and restoration decisions for contemporary watersheds. *Challenges* **2017**, *8*, 31. [CrossRef]
44. Hooper, L.; Hubbart, J.A. A rapid physical habitat assessment of wadeable streams for mixed land use watersheds. *Hydrology* **2016**, *3*, 37. [CrossRef]
45. Hooper, L. A Stream Physical Habitat Assessment of an Urbanizing Stream of the Central U.S.A. Master's Thesis, University of Missouri, Columbia, SC, USA, 2015.

46. Harrelson, C.C.; Rawlins, C.L.; Potyondy, J.P. *Stream Channel Reference Sites: An Illustrated Guide to Field Technique*; Gen. Tech. Rep. RM-245; US Department of Agriculture, Forest Service, Rocky Mountain Forest and Range Experiment Station: Fort Collins, CO, USA, 1994.
47. Yanites, B.J.; Tucker, G.E. Controls and limits on bedrock channel geometry. *J. Geophys. Res.* **2010**, *115*, F04019. [CrossRef]
48. Monk, W.A.; Wood, P.J.; Hannah, D.M.; Wilson, D.A. Macroinvertebrate community response to inter-annual and regional river flow regime dynamics. *River Res. Appl.* **2008**, *24*, 988–1001. [CrossRef]
49. Stoline, M.R. The status of multiple comparisons: Simultaneous estimation of all pairwise comparisons in one-way ANOVA designs. *Am. Stat.* **1981**, *35*, 134–141.
50. Peck, D.V.; Herlihy, A.T.; Hill, B.H.; Hughes, R.M.; Kaufmann, P.R.; Klemm, D.J.; Lazorchak, J.M.; Cappaert, M. *Environmental Monitoring and Assessment Program-Surface Waters Western Pilot Study: Field Operations Manual for Wadeable Streams*; EPA/620/R-06/003; U.S. Environmental Protection Agency, Office of Research and Development: Washington, DC, USA, 2006; pp. 128–130.
51. Wolman, M.G. A method of sampling coarse river-bed material. *Trans. Am. Geophys. Union* **1954**, *35*, 951–956. [CrossRef]
52. Montgomery, D.R.; Buffington, J.M.; Peterson, N.P.; Schuett-Hames, D.; Quinn, T.P. Stream-bed scour, egg burial depths, and the influence of salmonid spawning on bed surface mobility and embryo survival. *Can. J. Fish. Aquat. Sci.* **1996**, *53*, 1061–1070. [CrossRef]
53. Diana, M.; Allan, J.D.; Infante, D. The influence of physical habitat and land use on stream fish assemblages in southeastern Michigan. *Am. Fish. Soc. Symp.* **2006**, *48*, 359–374.
54. Zaimes, G.N.; Schultz, R.C.; Isenhart, T.M. Stream bank erosion adjacent to riparian forest buffers, row-crop fields, and continuously-grazed pastures along Bear Creek in central Iowa. *J. Soil Water Conserv.* **2004**, *59*, 19–27.
55. Urban, M.L.; Rhoads, B.L. Catastrophic Human-Induced Change in Stream-Channel Planform and Geometry in an Agricultural Watershed, Illinois, USA. *Ann. Assoc. Am. Geogr.* **2003**, *93*, 783–796. [CrossRef]
56. Simon, A.; Curini, A.; Darby, S.E.; Langendoen, E.J. Bank and near-bank processes in an incised channel. *Geomorphology* **2000**, *35*, 193–217. [CrossRef]
57. Gellis, A.C.; Fuller, C.C.; Van Metre, P.C. Sources and ages of fine-grained sediment to streams using fallout radionuclides in the Midwestern United States. *J. Environ. Manag.* **2017**, *194*, 73–85. [CrossRef]
58. Vietz, G.J.; Walsh, C.J.; Fletcher, T.D. Urban hydrogeomorphology and the urban stream syndrome: Treating the symptoms and causes of geomorphic change. *Prog. Phys. Geogr.* **2016**, *40*, 480–492. [CrossRef]
59. Shields, F.D.; Lizotte, R.E.; Knight, S.S.; Cooper, C.M.; Wilcox, D. The stream channel incision syndrome and water quality. *Ecol. Eng.* **2010**, *36*, 78–90. [CrossRef]
60. Jordan, B.A.; Annable, W.K.; Watson, C.C.; Sen, D. Contrasting stream stability characteristics in adjacent urban watersheds: Santa Clara Valley, California. *River Res. Appl.* **2010**, *26*, 1281–1297. [CrossRef]
61. Fletcher, T.D.; Vietz, G.; Walsh, C.J. Protection of stream ecosystems from urban stormwater runoff: The multiple benefits of an ecohydrological approach. *Prog. Phys. Geogr.* **2014**, *38*, 543–555. [CrossRef]
62. Trimble, S.W. Fluvial processes, morphology and sediment budgets in the Coon Creek Basin, WI, USA, 1975–1993. *Geomorphology* **2009**, *108*, 8–23. [CrossRef]
63. Walters, D.M.; Leigh, D.S.; Bearden, A.B. Urbanization, sedimentation, and the homogenization of fish assemblages in the Etowah River Basin, USA. In *The Interactions between Sediments and Water*; Springer: Dordrecht, the Netherlands, 2003; pp. 5–10.
64. Waters, T.F. *Sediment in Streams: Sources, Biological Effects, and Control*; American Fisheries Society: Bethesda, MA, USA, 1995.
65. Paul, M.J.; Meyer, J.L. Streams in the urban landscape. *Annu. Rev. Ecol. Syst.* **2001**, *32*, 333–365. [CrossRef]

Article

Efficiency Analysis of the Input for Water-Saving Agriculture in China

Yangdong Cao [1], Wang Zhang [2] and Jinzheng Ren [1,*]

[1] College of Economics and Management, China Agricultural University, Beijing 100083, China; b20163110632@cau.edu.cn

[2] Development Research Center of the Ministry of Water Resources of P.R. China, Beijing 100053, China; zhangwang@waterinfo.com.cn

* Correspondence: rjzheng@cau.edu.cn; Tel.: +86-010-6273-8506

Received: 4 December 2019; Accepted: 10 January 2020; Published: 11 January 2020

Abstract: To optimize the installation distribution of water-saving techniques and improve the efficiency of water-saving agricultural inputs, we used a three-stage data envelopment analysis (DEA) model and Chinese provincial panel data from 2014 to 2016 to analyze the input efficiency of the water-saving irrigation. This study explores the efficiency derived from the efforts of water-saving initiatives in the agricultural sector in China. We present the impacts of factors such as technology, scale, diminishing marginal revenue, and crop water requirements on the research results. We found overall efficiency of water-saving irrigation is increasing nationally. The efficiency of water-saving irrigation input will significantly increase if management and organization of the input improve. Increasing the investment in areas with increasing marginal revenue would improve the local agricultural water-saving input efficiency in areas such as Hainan, Chongqing, Guizhou, Tibet, and Qinghai; although in areas with large water requirement for major crops, such as Inner Mongolia and Xinjiang, the efficiency of water-saving irrigation is generally high. Shanxi requires a large amount of water as the efficiency of agricultural water-saving input is 0.07, which is relatively lower than the average efficiency of all regions (0.39). The cultivated area index and the GDP per capita had no significant effect on the irrigation input efficiency.

Keywords: water-saving agriculture; Chinese provincial input efficiency; three-stage DEA model; environmental variables

1. Introduction

Water resource shortages have become a serious problem in China. Chinese water resources are unevenly distributed in time and space [1]. It is estimated that the per capita water resources will be reduced to one-quarter of the world average (1760 m^3) by 2030, which is close to the lowest level of the countries with recognized water shortages. More than 400 of the 669 cities in China have insufficient water supply, 108 cities are seriously deprived of water, and the water shortage exceeds 6 billion m^3 in China annually [2]. This has affected the normal production and lives of more than 160 million people [3,4]. The amount of blue water resources in eight provinces is still unable to cover the needs of the domestic ecosystem, including in Qinghai province, which is known as the "source of rivers" [5].

According to the statistics of the Ministry of Water Resources, the agricultural sector accounts for 63.5% of the country's total water use, of which about 90% is used for farmland irrigation. A large amount of irrigation water is wasted [6,7]. Compared with the effective use coefficient of farmland irrigation in developed countries of 0.7–0.8, this value was only 0.53 in China in 2017 [8,9], and a gap still exists between the water-saving goals proposed by the National Agricultural Water Conservation Program (2012–2020). As such, further implementing water-saving measures for agriculture in China is crucial [10].

Although water-saving behavior will have huge ecological and social benefits, the effect of improving agricultural water-saving irrigation technology on individual farmer economic benefits is limited. The use of water-saving irrigation technology increases the opportunity cost of farmers, but the economic base of farmers in China is relatively weak. If no external stimulus exists, most farmers will not use water-saving technology. Therefore, the Chinese government has adopted the common worldwide practice of developing water-saving agriculture, providing policy support and guiding farmers toward water-saving behavior [11,12].

In recent years, the scale of China's farmland water conservancy investment has increased annually. Under the guidance of agricultural water-saving policies, the water-saving effect has also improved. According to the China Statistical Yearbook 2010–2019, the national fiscal expenditure for agriculture, forestry, and water affairs increased from 672.04 billion yuan in 2009 to 2.11 trillion yuan in 2018, which showed a growth rate of more than 213.75%. It is higher than the total expenditure growth rate of the state finance in the past 10 years (189.52%). Nevertheless, the growth rate of farmland investment effectiveness is not high. Water-saving irrigated areas and effective irrigated areas are presented in Figure 1.

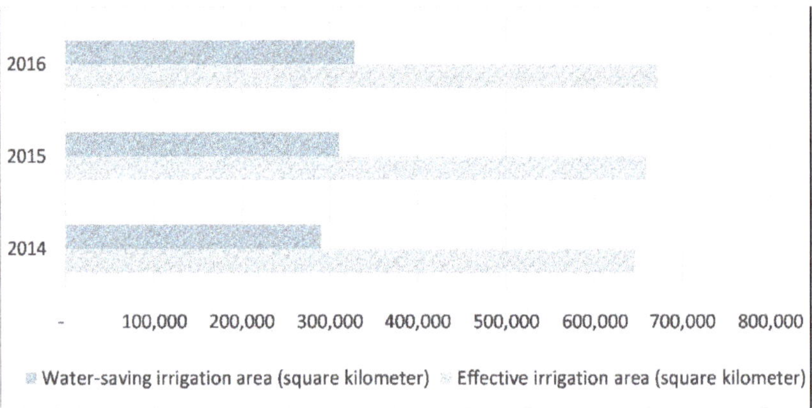

Figure 1. Water-saving irrigated and effective irrigated areas in mainland China.

Scholars have mainly studied the agricultural water-saving input from three perspectives: A macro study on the strategic regulation of water-saving agriculture and the formulation of investment policies from the perspective of the government [13–15], analyzing and evaluating the optimization and decision-making of investment plans for water-saving agriculture projects from the perspective of social capital [16,17], and examining the behaviors of farmers who use water-saving irrigation techniques and the factors affecting their behaviors [18–20].

Scholars' research on agricultural water-saving efficiency is primarily focused on two aspects, namely influencing factors and efficiency measurement. They have studied the influencing factors of agricultural water-saving efficiency from the aspects of resources, technology, and management. Additionally, they have found that low resource prices cause low agricultural water-saving efficiency [21] and participation of farmers in irrigation management improves agricultural water-saving efficiency [22]. The clear agricultural water right and water right transfer policy not only stimulate the enthusiasm of agricultural water-saving but also promote the development of water-saving technology [23]. The improvement of irrigation technology and agricultural technology is conducive to the development of agricultural water-saving efficiency [24]. Most scholars choose to use non-parametric statistical methods such as data envelope models to conduct input–output analysis on agricultural water-saving efficiency. The stochastic frontier production function is used to construct agricultural water efficiency models from the input–output perspective [25–27] and input-oriented Data Envelopment Analysis (DEA) to study regional differences and convergence of agricultural water use efficiency in China [28,29].

Precedents exist for the domestic input efficiency evaluation of agricultural water-saving. Some scholars have analyzed and evaluated the water-saving irrigation investment in different base years and determined the influence of investment efficiency based on principal component analysis [30–32]. Some scholars compared and analyzed the advantages of water-saving irrigation and developed an investment management system and investment management performance evaluation method for water-saving irrigation from a river [33,34]. Some scholars established certain of water-saving investment income models to study water-saving incentives in river basins and analyzed the mechanism of investment in the water rights market on the investment income of water-saving irrigation [35,36]. Most scholars analyzed the efficiency of agricultural water-saving investment from a country perspective, and research from the provincial unit perspective rarely explains the differences in environmental factors between regions [37].

In the existing research, the efficiency of agricultural water-saving input between provincial regions has not been measured well, making it impossible to analyze the influencing factors of agricultural water-saving input efficiency. It is necessary to determine the efficiency of China's agricultural water-saving input, ranking the areas where water technology improvement is needed the most. Determining the factors that affect agricultural water-saving input efficiency and trying to formulate proposals to improve the efficiency of agricultural water-saving inputs are of great significance to promote the rational use of water resources in China. The current forms of agricultural water-saving input are capital, labor, and equipment. Agricultural water-saving effects are primarily reflected in the size of water-saving irrigation areas. When the input and output indicators are clear, the DEA is a more accurate and commonly used tool for researching input efficiency analysis. The theoretical framework of this research is shown in the Figure 2.

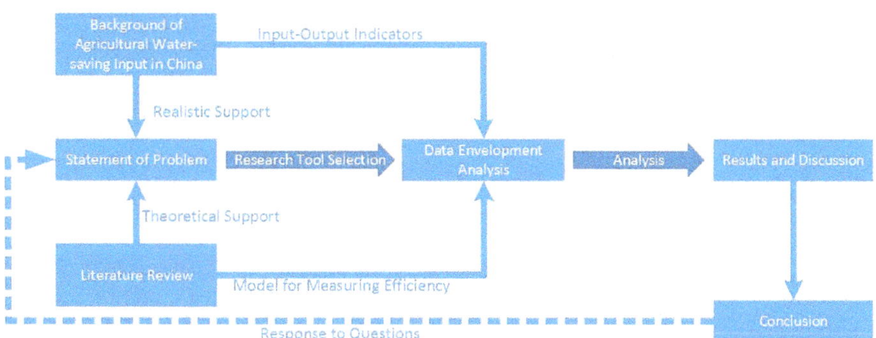

Figure 2. Theoretical framework.

To solve the problem of objective factors and statistical noise, which affect the efficiency evaluation results of decision-making units, Fried [38] published Incorporating the Operating Environment into a Nonparametric Measure of Technical Efficiency and proposed a three-stage DEA model, considering the influence of environmental factors on the evaluation results in the traditional DEA. Fried [39] published Accounting for Environmental Effects and Statistical Noise in Data Envelopment Analysis, a further revision of the traditional DEA model considering environmental factors and statistical errors.

The DEA model has 4 forms, each with its own advantages and disadvantages. Form 1 is the basic sample separation method, which decomposes the sample into subsamples according to environmental factors, and is easy to understand and apply; however, it can only be used for a categorical variable and its precision is lower than that of other DEA model forms [40]. Form 2 is the one-stage DEA model, which includes environmental variables in the traditional DEA model together with input and output factors. The one-stage DEA model is easy to interpret and apply; however, it requires prior understanding of the influence direction of the environmental variables [41]. Form 3 is the two-stage model that performs a regression evaluation on the efficiency of environmental variables, so it can accommodate continuous categorical variables without increasing decision units. This kind of DEA model does not require prior understanding of the influence direction of the environmental variables; however, if ordinary least squares (OLS) is used in the second stage, the corrected efficiency scores might be larger than 1 [39]. Form 4 is the three-stage model, which uses stochastic frontier analysis (SFA) to estimate the impact of environmental variables and the statistical noise and uses the adjusted input values in the traditional DEA model. In the three-stage DEA model, "input slack" represents the difference between the input before and after the elimination of environmental and statistical noise. Although it requires significant calculation time, this kind of model is able to capture the information contained in the input slack, which was helpful in the following analysis [42].

To address these research gaps, we used a three-stage DEA model to study the investment efficiency of national water-saving irrigation that considers the influence of objective factors and random statistical errors between different provinces on the efficiency of production units. We analyzed the scale effect of input from different regions and the impact of crop irrigation requirement on the efficiency of the water-saving irrigation input. Unlike the previous quantitative study of a single region, the calculation results of this research rank the agricultural water-saving irrigation input efficiency in 31 provinces in mainland China while depicting the changing trend of China's agricultural water-saving irrigation input efficiency. In addition, it analyzes the factors that affect input efficiency and attempts to discover ways to improve areas with low input efficiency.

2. Materials and Methods

2.1. Indicator Selection and Data Source

The purpose of this article is to explore the relationship between the efficiency of water-saving input between different provinces. All 31 provincial regions in mainland China were selected as the research objects. Due to the limitation of the area for obtaining statistical data, this study does not include areas such as Taiwan, Hong Kong, Macau, and the South China Sea Islands.

In response to the irrigation water crisis, the Chinese government implemented water-saving technological transformation through agricultural irrigation infrastructure of methods such as the old ditch pumping station throughout the country, thereby transitioning irrigation behavior from traditional irrigation to water-saving irrigation. At present, the input in agricultural water-saving not only includes funds but also the input of human resources and materials. Therefore, we chose the input indicators considering three aspects: labor, material, and funds. According to the statistics of the indicators, we used three indicators: farmland water conservancy total input, farmland water conservancy input workday, and mechanical class for farmland water conservancy to represent agricultural water-saving inputs. In terms of the agricultural water-saving output, separating the portion of the grain output that has been increased due to water-saving irrigation is difficult. Additionally, the grain output is considerably affected by natural factors. The arable land irrigated area and water-saving irrigated area of more than 6.67 square kilometers were used as the output indexes of water-saving irrigation. The index system used to determine the efficiency of Chinese provincial water-saving agriculture is shown in Figure 3.

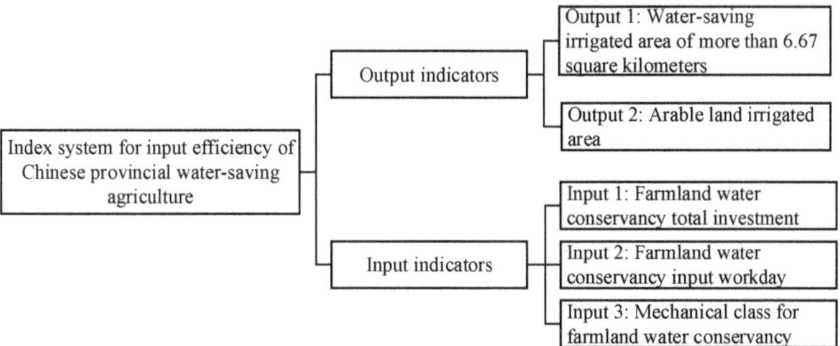

Figure 3. Index system used to determine the input efficiency of Chinese provincial water-saving agriculture.

As our research scope was 31 provinces in the Chinese mainland, their environmental factors such as climatic conditions, planting area, and economic development in different regions may affect the analysis results of the input efficiency. The assumption that the factors affecting the efficiency of agricultural water-saving input are the gross domestic product (GDP) per capita and cultivated land area is reasonable.

The State Council of China promulgated a policy to reconstruct the supporting facilities in irrigation districts and develop agricultural water-saving irrigation in 2013. Considering the time lag of the provincial data released by the Ministry of Water Resources, we used 2014–2016 as the study interval. To ensure the reliability of the analytical data, the data sources were the 2015–2017 China Statistical Yearbook, China Water Conservancy Statistical Yearbook and the National Water Development Statistics Bulletin.

2.2. Three-Stage DEA Model

2.2.1. Stage 1: Traditional DEA Model Analysis of the Original Input and Output Values

The first stage uses the initial input–output data of the decision-making units (DMUs) for traditional DEA analysis. In the literature related to the three-stage DEA model, the input-oriented model with variable scale return is mostly used as the first-stage calculation model, which is consistent with the model hypothesis. Therefore, we used the Banker, Charnes, and Cooper model (BCC) model. The dual form of the input-oriented linear programming model of one of the decision-making units can be expressed as [42]:

$$min\theta$$
$$\text{s.t.}\begin{cases} X_{ij0} \geq \sum_{j=1}^{n} X_{ij}\lambda_j \\ Y_{rj0} \geq \sum_{j=1}^{n} Y_{ij}\lambda_j \\ \sum_{j=1}^{n} \lambda_j = 1 \\ \lambda_j \geq 0 \\ j = 1,2,3\ldots,n \\ i = 1,2,3,\ldots,m \\ r = 1,2,\ldots,s \end{cases} \tag{1}$$

where θ demonstrates the comprehensive input efficiency value of each DMU; X_{ij} and Y_{ij} are the ith and rth output of the jth DMU, respectively; m, s, and n represent the number of input variables, output variables, and DMUs, respectively; and λ_j represents the j dimensional weight vector of DMU j.

The DEA-BCC model calculates the overall technical efficiency (TE), which is affected by the scale efficiency (SE) and pure technical efficiency (PTE). The relationship between these three values are expressed as [42]:

$$TE = SE \times PTE. \tag{2}$$

The efficiency evaluation results of the management unit are affected by management inefficiency, environmental factors, and statistical noise, so the last two factors need to be further separated from the results.

2.2.2. Stage 2: Statistical Noise and Exotic Environment Factors Separated from Results

The overall input efficiencies of each DMU can be calculated in the first stage. However, the input slacks of all DMUs are influenced by the management inefficiency, environmental factors, and statistical noise, so their effects on the results have to be eliminated in the second stage.

Fried [29] used SFA is to effectively separate environmental factors and statistical noise. This method is superior to simply using Tobit regression to separate environmental factors in the presence of statistical noise. The SFA regression function is constructed as [42]:

$$S_{ni} = f(Z_i; \beta_n) + v_{ni} + \mu_{ni} \tag{3}$$

where $i = 1, 2, 3, \ldots, I$ and $n = 1, 2, 3, \ldots, N$; S_{ni} is the slack variable of the nth input of the ith DMU; Z_i is the environmental variable; β_n is the coefficient of the environmental variable; $v_{ni} + \mu_{ni}$ are the mixed errors; v_{ni} is the random noise; $\mathrm{v} \sim N\left(0, \sigma_{\mathrm{v}}^2\right)$ is the influence of random interference factors on input slack variables; μ_{ni} is the management inefficiency; and $\mu \sim N^{+}\left(0, \sigma_{\mu}^2\right)$ is the influence of management factors on input slack variables.

To eliminate the influences of environmental factors and statistical noise, the input of decision-making units with better environmental conditions and statistical random variables is increased in the formula, as shown in Equation [42] (4):

$$X_{ni}^{A} = X_{ni} + \left[max\left(f\left(Z_i; \beta_n\right)\right) - f\left(Z_i; \beta_n\right)\right] + \left[max(v_{ni}) - v_{ni}\right] \tag{4}$$

where $i = 1, 2, 3, \ldots, I$ and $n = 1, 2, 3, \ldots, N$; X_{ni} is the original input; X_{ni}^{A} is the adjusted input; $max\left(f\left(Z_i; \beta_n\right)\right) - f\left(Z_i; \beta_n\right)$ is the adjustments to environment variables; and $max(v_{ni}) - v_{ni}$ is the elimination of the random errors in statistical noise.

2.2.3. Stage 3: Adjustment of the Efficiency Value

The original input of the first stage is substituted with the adjusted input data in stage 2, and then the DEA-BCC model is used again. The new efficiency value is the real efficiency that excludes exterior environmental factors and statistical noise.

2.3. Calculation of Irrigation Requirement Index

Irrigation requirement index (IR/ET_c) is the ratio of irrigation requirement (IR) to crop water requirement (ET_c), and it reflects the degree of dependence of crop growth on irrigation in different regions [43].

IR equals the difference between ET_c and effective rainfall during the growing period (P_e) [44], as shown in Equation (5):

$$IR = ET_c - P_e \tag{5}$$

Presently, the formulas for calculating effective rainfall of various crops must determine parameters suitable for local soil quality, crops, and other conditions. Studies have shown that the calculation method of effective crop rainfall is related to the selection of the calculation period length, and the accuracy can meet the research needs [45].

$$P_e = \begin{cases} P & P \leq ET_c \\ ET_c & P > ET_c \end{cases}. \quad (6)$$

ET_c is calculated as shown in Equation [46] (7):

$$ET_c = ET_0 \times K_c \quad (7)$$

Reference evapotranspiration (ET_0) is potential transpiration rate of standard reference crop, which is calculated by using the Penman–Monteith method recommended by the Food and Agriculture Organization (FAO) of the United Nations [46].

Coefficient (K_c) is the ratio of the potential evapotranspiration of a certain crop to ET_0. It reflects the difference between various crops and reference crops. The crop coefficient of several crops under standard conditions can be found in FAO-56.

3. Results and Discussion

The result is reflected in the comprehensive technical efficiency, pure technical efficiency, and scale efficiency. The slack variable of the input variable of the decision-making unit of the first stage was introduced into the SFA analysis model to analyze whether the factors significantly influenced efficiency judgment. The regression results are shown in Table 1.

Table 1. The regression results of the stochastic frontier analysis (SFA) model.

Year	Slacks Explanatory Variable	Input 1	Input 2	Input 3
2014	Constant term	(1035.5579)	4081.4798	0.007481262
	GDP per capita	0.00976548	(0.27090702)	2.5154086
	Cultivated area	0.00451017	69.011447	(0.41631877)
	σ_i^2	9,325,325.6	389,895,890	5,171,073.9
	γ_i	0.99999968	0.99999999	0.99999999
	likelihood ratio test	26.560963	14.248312	26.560963
2015	Constant term	(10.844754)	4865.9359	(1177.4461)
	GDP per capita	0.00036329	(0.28123225)	0.012144941
	Cultivated area	0.00038983	97.924711	(0.000455359)
	σ_i^2	18,050.657	413,829,350	9,919,867.8
	γ_i	0.99999999	0.99999999	0.99999999
	likelihood ratio test	8.894564	15.640943	24.943002
2016	Constant term	(6.7862996)	4597.7498	(1203.8865)
	GDP per capita	7.0433×10^{-5}	(0.27747648)	0.012064749
	Cultivated area	0.00021407	68.236315	(0.002089515)
	σ_i^2	16093.141	417683040	9998899.5
	γ_i	0.99999999	0.99999999	0.99999999
	likelihood ratio test	14.16323	14.551437	26.015291

According to the regression results of the SFA model, the likelihood ratio test values of the unilateral error of the regressions for the three input slacks with two environment variables are all under the threshold value of the mixed χ^2 distribution examination and above the 10% confidence level, implying that the regression model was not robust enough. The hypothesis that no inefficiency item exists is supported.

3.1. Comprehensive Technical Efficiency of the Agricultural Water-Saving Inputs

The comprehensive technical efficiency represents the ability of the DMU to convert inputs into outputs. The comprehensive technical efficiency of provincial input for water-saving agriculture is shown in Table 2.

Table 2. The comprehensive technical efficiency of provincial input for water-saving agriculture.

Region	2014	2015	2016	Region	2014	2015	2016
Beijing	0.234	0.098	0.232	Hubei	0.277	0.302	0.301
Tianjin	1.000	0.785	0.902	Hunan	0.084	0.093	0.107
Hebei	0.961	0.340	0.535	Guangdong	1.000	1.000	1.000
Shanxi	0.070	0.069	0.068	Guangxi	0.089	0.090	0.071
Inner Mongolia	1.000	1.000	1.000	Hainan	0.068	0.057	0.180
Liaoning	0.105	0.131	0.126	Chongqing	0.020	0.021	0.017
Jilin	1.000	0.578	0.506	Sichuan	0.404	0.399	0.562
Heilongjiang	1.000	1.000	0.686	Guizhou	0.044	0.031	0.028
Shanghai	1.000	1.000	1.000	Yunnan	0.115	0.146	0.207
Jiangsu	0.190	0.182	0.186	Tibet	0.213	0.177	0.245
Zhejiang	1.000	1.000	1.000	Shaanxi	0.207	0.309	0.365
Anhui	0.442	0.539	0.431	Gansu	0.247	0.203	0.198
Fujian	0.139	0.119	0.098	Qinghai	0.114	0.140	0.140
Jiangxi	0.061	0.054	0.049	Ningxia	0.419	0.418	0.389
Shandong	0.097	0.105	0.102	Xinjiang	1.000	1.000	1.000
Henan	0.177	0.219	0.223	China Mainland	0.412	0.374	0.386

The comprehensive technical efficiency of the national average agricultural water-saving input was 0.412 in 2014, followed by a slight decline in 2015, and the value rebounded to 0.386 in 2016. The use efficiency of China's water festival irrigation investment rebounded slightly, showing that the current input efficiency of agricultural water-saving is increases, but room for improvement remains in the management efficiency of agricultural water use reduction in China.

From a regional perspective, the Inner Mongolia Autonomous Region and the provinces of Shanghai, Zhejiang, Guangdong, and Xinjiang were at the forefront of technical efficiency. Heilongjiang province had a high overall efficiency in 2014 and 2015, but exhibited a small decline in 2016. Tianjin was more efficient in 2014 and showed a slight rebound after a slight decline in 2015. Jilin displayed a downward trend after high efficiency in 2014. The six regions of Guizhou, Chongqing, Jiangxi, Shanxi, Hainan, and Guangxi showed lower comprehensive technology efficiency than other regions. Among them, the efficiency of Hainan in 2016 significantly improved compared with the previous two years, whereas the input efficiency of the other five regions did not change much in three years, so they can share the good ideas and practices of high-efficiency provinces based on the actual situation in the region.

3.2. Pure Technical Efficiency of Agricultural Water-Saving Inputs

Pure technical efficiency is a measure of the impact of non-scale factors, such as management and technology, on the output of water-saving irrigation inputs of each DMU. The pure technical efficiency of agricultural water-saving input is shown in Table 3.

Different from the scale efficiency, pure technical efficiency measures the investment of decision-making units from a technical perspective. Under the condition of constant scale, DMU with higher pure technical efficiency has higher comprehensive efficiency. In the study of input efficiency for water-saving agriculture, technical efficiency refers to the level of management and organizational [47].

Table 3. The pure technical efficiency of the provincial input for water-saving agriculture.

Region	2014	2015	2016	Region	2014	2015	2016
Beijing	0.295	0.254	0.344	Hubei	0.279	0.303	0.303
Tianjin	1.000	0.805	0.917	Hunan	0.099	0.109	0.120
Hebei	1.000	0.388	0.568	Guangdong	1.000	1.000	1.000
Shanxi	0.112	0.117	0.106	Guangxi	0.117	0.126	0.102
Inner Mongolia	1.000	1.000	1.000	Hainan	0.252	0.270	0.301
Liaoning	0.126	0.195	0.183	Chongqing	0.103	0.104	0.082
Jilin	1.000	0.623	0.565	Sichuan	0.439	0.406	0.647
Heilongjiang	1.000	1.000	0.793	Guizhou	0.120	0.087	0.075
Shanghai	1.000	1.000	1.000	Yunnan	0.127	0.149	0.208
Jiangsu	0.192	0.191	0.189	Tibet	1.000	1.000	1.000
Zhejiang	1.000	1.000	1.000	Shaanxi	0.215	0.330	0.366
Anhui	0.459	0.541	0.431	Gansu	0.270	0.239	0.235
Fujian	0.288	0.274	0.218	Qinghai	0.446	0.528	0.514
Jiangxi	0.080	0.081	0.074	Ningxia	0.510	0.526	0.523
Shandong	0.103	0.112	0.109	Xinjiang	1.000	1.000	1.000
Henan	0.178	0.221	0.226	China Mainland	0.478	0.451	0.458

In 2014, the pure technical efficiency of the national average agricultural water-saving input was 0.478, which was the first to decline in 2015 and 2016, and then it rebounded. This shows that the input efficiency of agricultural water-saving in China was growing currently, and the efficiency of use of capital, human, and material resources needs to be further improved.

In the efficiency evaluation, the pure technical efficiency of agricultural water-saving input was found to be one, indicating that the input management of the DMU is efficient. Nationally, Inner Mongolia, Shanghai, Zhejiang, Guangdong, Tibet, and Xinjiang were at the forefront of efficiency in the assessment year. The pure technical efficiency of Heilongjiang was relatively high in 2014 and 2015 and decreased in 2016. Hebei and Tianjin showed a small rebound in 2016 after an efficiency decline in 2015. Shanxi, Jiangxi, Shandong, Hunan, Guangxi, Chongqing, Guizhou, and Yunnan showed low technical efficiency. Except for Yunnan's pure technical efficiency showing an upward trend, the remaining three low-efficiency provinces remained at a relatively low level of pure technical efficiency over the three years.

3.3. Scale Efficiency of the Agricultural Water-Saving Inputs

With a certain level of management and technology, the input efficiency is affected by the scale of input. Scale efficiency reflects the ratio of the actual input scale to the optimal input scale. The scale efficiency of the average agricultural water-saving input in mainland China is generally higher than the pure technical efficiency, at about 0.78 for the assessment period, as shown in Table 4. This means that the agricultural water-saving input is relatively high; however, room for further improvement remains. Subsidies can be adopted for different places according to actual needs.

From a national perspective, Inner Mongolia, Shanghai, Zhejiang, Guangdong, and Xinjiang were at the forefront of scale efficiency. Most provinces had a high level of input, and there is relatively low scale efficiency in Hainan, Chongqing, Guizhou, Tibet, and Qinghai. Among them, Hainan's scale efficiency considerably improved in 2016. Considering the scale of remuneration, the scale returns of Hebei, Jilin, Heilongjiang, and Sichuan decreased in 2016, while other provinces had an increasing return to scale or a constant return to scale. The redeployment of agricultural water-saving inputs between provinces may have contributed to an increase in the overall efficiency.

Table 4. The scale efficiency of provincial input for water-saving agriculture.

Region	2014	2015	2016	Region	2014	2015	2016
Beijing	0.793	0.385	0.674	Hubei	0.993	0.995	0.994
Tianjin	1.000	0.974	0.983	Hunan	0.849	0.856	0.892
Hebei	0.961	0.875	0.941	Guangdong	1.000	1.000	1.000
Shanxi	0.622	0.589	0.641	Guangxi	0.761	0.712	0.692
Inner Mongolia	1.000	1.000	1.000	Hainan	0.269	0.209	0.598
Liaoning	0.829	0.669	0.689	Chongqing	0.192	0.198	0.206
Jilin	1.000	0.928	0.897	Sichuan	0.921	0.982	0.869
Heilongjiang	1.000	1.000	0.864	Guizhou	0.370	0.355	0.371
Shanghai	1.000	1.000	1.000	Yunnan	0.906	0.984	0.998
Jiangsu	0.990	0.956	0.981	Tibet	0.213	0.177	0.245
Zhejiang	1.000	1.000	1.000	Shaanxi	0.963	0.937	0.997
Anhui	0.962	0.996	1.000	Gansu	0.915	0.851	0.845
Fujian	0.484	0.435	0.451	Qinghai	0.257	0.266	0.273
Jiangxi	0.759	0.668	0.668	Ningxia	0.822	0.796	0.743
Shandong	0.942	0.942	0.940	Xinjiang	1.000	1.000	1.000
Henan	0.990	0.989	0.989	Mainland China	0.799	0.765	0.788

3.4. *Marginal Revenue of the Agricultural Water-Saving Inputs*

We propose that the marginal revenue of input efficiency in agricultural water-saving inputs is diminishing, which means that after a regional water-saving irrigation input produces certain effects, the efficiency of the subsequent input is less than that of the previous input efficiency. We measured scale of the agricultural water-saving inputs by the ratio of local water-saving irrigation area to cultivated land area. The marginal revenue analysis was conducted by combining the input scale saturation with the comprehensive benefits of the agricultural water-saving input.

In this study, five areas with a high water-saving input efficiency and five areas with a low water-saving efficiency were selected as the research objects to analyze the diminishing marginal benefit. The data on water-saving irrigated area and cultivated area in 2016 in 10 regions are shown in Table 5.

Table 5. Agricultural water-saving situation and ranking in some areas.

Region	Water-Saving Area of the Total Cultivated Area		Agricultural Water-Saving Input Efficiency	
	Proportion	Rank of Mainland China	Value	Rank of Mainland China
Inner Mongolia	28.50%	10	1	1
Shanghai	76.10%	2	1	1
Zhejiang	54.89%	4	1	1
Guangdong	11.56%	25	1	1
Xinjiang	74.59%	3	1	1
Guangxi	23.45%	13	0.071	27
Jiangxi	17.03%	20	0.068	28
Shanxi	22.41%	16	0.049	29
Guizhou	7.14%	30	0.028	30
Chongqing	9.18%	27	0.017	31

The five regions with the largest proportions of water-saving cultivated land were Beijing, Shanghai, Xinjiang, Zhejiang, and Jiangsu, and the five regions with the smallest proportion were Tibet, Guizhou, Hubei, Hunan, and Chongqing. Among the five regions with high comprehensive efficiency of agricultural water-saving, the area with water-saving irrigation in Guangdong accounts for a small proportion of cultivated land, and the water-saving irrigation areas in the other four regions are relatively large. Of the areas where the water-saving irrigated area is relatively small, the comprehensive efficiency of agricultural water-saving is relatively low in Chongqing and Guizhou.

The results indicate that the water-saving irrigation input in 31 regions of China has not shown a significant downward trend in marginal benefits, which means that agricultural water-saving input has not reached its maximum utility in most regions. Where the proportion of the water-saving area was high, the agricultural water-saving input was highly efficient, and where the proportion of the water-saving area was low, the agricultural water-saving input was inefficient.

3.5. Irrigation Water Requirement of Crops

The 31 regions in mainland China mainland are widely distributed and have different climatic conditions, so the irrigation water requirements vary amongst the different types of major local crops in different locations. The crop requirements for agricultural irrigation in different regions affect the enthusiasm toward input in water-saving irrigation and influence the agricultural water-saving input efficiency. The irrigation requirement index indicates the degree of dependence of crops on agricultural irrigation, which is related to the water requirement characteristics of crop growth and the precipitation in the local crop growth period [44,48,49], as shown in Table 6.

Table 6. Average irrigation requirement index for the main crops in different regions.

Region	Main Crop Species	Irrigation Requirement Index	Region	Main Crop Species	Irrigation Requirement Index
Heilongjiang; Jilin; Liaoning	Middle-season rice	0.4	Shaanxi; Gansu; Shanxi	Middle-season rice	0.7
	Spring maize	0.25		Cotton	0.65
	Spring wheat	0.34		Spring maize	0.65
Beijing; Tianjin; Hebei; Shandong; Henan	Middle-season rice	0.6		Summer maize	0.4
	Cotton	0.425		Spring wheat	0.675
	Spring maize	0.425	Sichuan; Chongqing	Late-season rice	0.3
	Summer maize	0.325		Middle-season rice	0.275
	Winter wheat	0.625		Early-season rice	0.35
Jiangsu; Zhejiang; Shanghai; Hunan; Hubei; Jiangxi; Anhui	Late-season rice	0.425		Cotton	0.15
	Middle-season rice	0.375		Spring maize	0.125
	Early-season rice	0.325		Summer maize	0.125
	Cotton	0.3		Winter wheat	0.575
	Spring maize	0.2	Yunnan; Guizhou	Late-season rice	0.3
	Summer maize	0.35		Middle-season rice	0.375
Fujian; Guangdong; Guangxi; Hainan	Late-season rice	0.4		Early-season rice	0.35
	Middle-season rice	0.35		Spring maize	0.275
	Early-season rice	0.3		Summer maize	0.15
	Cotton	0.2	Qinghai; Tibet	Winter wheat	0.625
	Spring maize	0.175		Spring wheat	0.625
	Summer maize	0.2	Xinjiang	Middle-season rice	0.9
Inner Mongolia; Ningxia	Middle-season rice	0.775		Cotton	0.9
	Spring maize	0.7		Spring maize	0.85
	Spring wheat	0.6		Summer maize	0.9
				Winter wheat	0.8
				Spring wheat	0.875

According to the regional main crop irrigation requirement index, the crop requiring the most irrigation water is rice, followed by wheat and cotton. Although cotton requires more water than wheat, due to the higher amount of precipitation in the cotton growing area, the irrigation water requirement of wheat is higher than that of cotton during the growing process. Summer maize requires little irrigation, and crops grown in the dry fields in the south and northeast require no irrigation.

The comprehensive efficiency of water-saving irrigation input in Zhejiang, Shanghai, and Guangdong was higher than other regions', but their average irrigation requirement index were 0.33, 0.33, and 0.27, respectively, lower than the average irrigation requirement index (0.43). The irrigation requirement indexes in Inner Mongolia and Xinjiang were 0.69 and 0.87, respectively, significantly higher than the average. Therefore, the efficiency of agricultural water-saving input in Inner Mongolia and Xinjiang was higher due to the large water requirement of crops; without the high agricultural water-saving input efficiency, the growth needs of local crops cannot be met. For the crop water requirement in Zhejiang, Shanghai, and Guangdong, the developed social economy plays a role in increasing the efficiency of the agricultural water-saving input.

Overall, the efficiency of agricultural water-saving inputs in Guangxi, Jiangxi, Guizhou, and Chongqing was lower than other regions. The irrigation requirement index in these areas were lower than the regional average, indicating that crop irrigation in these areas requires less irrigation water than other regions. Notably, the irrigation requirement index of Shanxi was 0.61, which is higher than the regional average (0.48), but the agricultural water-saving input efficiency was low, indicating room for improvement.

According to DEA calculation results, there are six provinces with a comprehensive efficiency of less than 0.1, including Chongqing (0.02), Guizhou (0.03), Jiangxi (0.05), Shanxi (0.07), Guangxi (0.08), and Hunan (0.09). As the results of Section 3.4 reveal there is no significant downward trend in marginal benefits in China, the problem of insufficient levels of agricultural water-saving investment in these areas is widespread. Li [49] proposed that the low availability of local financial funds in agriculture is because of inadequate balancing of agricultural investment. It is of great significance to increase the proportion of local fiscal agricultural investment in public products related to agricultural production and management (such as small farmland water conservancy, research and development, and promotion of agricultural water-saving technologies), which is of great importance to improve the efficiency of agricultural water-saving investment. Therefore, it is necessary for local governments to closely consider the use and management of agricultural water-saving irrigation inputs and establish incentive and restraint mechanisms to strengthen the efficiency of local financial investment. This approach can ensure the expansion of investment scale, thereby improving agricultural water-saving investment efficiency. In addition to insufficient investment scale, the pure technical efficiency in these regions is low. Shanxi Province exhibited the highest irrigation requirement index among the six provinces, and there is an urgent need to upgrade water-saving technologies to meet the water needs of crops. Shanxi Province can learn from Gansu Province, which is also a dry farming area, to improve the efficiency of water-saving irrigation by enhancing the quality of cultivated land and increasing well water irrigation. By rationally determining the scale of planting, scientifically arranging well irrigation, and improving irrigation technology efficiency, developments to agricultural water-saving irrigation input efficiency from the perspective of improving pure technical efficiency are possible [50]. The comprehensive technical efficiency of agricultural water-saving investment in Chongqing is the lowest in the country. To improve this situation, Chongqing can learn from Sichuan Province with its similar geographical location and climatic conditions. On one hand, it actively explores agricultural credit services, establishes agricultural water-saving development funds, and broadens the sources of investment. And on the other, it is necessary to actively promote the concept of water conservation among farmers, encouraging them to actively adopt water-saving measures to improve agricultural water efficiency [51]. Guizhou, Jiangxi, Guangxi, and Hunan belong to paddy fields in southern China. Farmers' participation in agricultural water-saving irrigation is relatively poor [52]. To further enhance the efficiency of agricultural water-saving irrigation, it is essential to appropriately increase the cost of agricultural water, promote water-saving irrigation technology, and improve drainage channels [53–55]. Additionally, the reclaimed water which has been assessed quality can supplement the irrigation water, thereby the investment efficiency of agricultural water-saving irrigation will be improved [56].

4. Conclusions

In this study, the three-stage DEA model was used to analyze the input efficiency and level of water-saving agriculture of the 31 provinces in mainland China. During the second stage of the model, we found that the GDP per capita and the cultivated area do not play significant roles in the efficiency determination, which means that the result of the one-stage DEA model is the real efficiency. We analyzed the efficiency of the agricultural water-saving input in 31 regions from the perspectives of pure technical efficiency, scale efficiency, scale efficiency decreasing effect, and crop water requirement. The primary conclusions are as follows:

(1) The efficiency of agricultural water-saving input in China generally is in the stage of increasing marginal revenue, and the efficiency of agricultural water-saving input increases with increasing total input. The annual average water-saving irrigation coefficient in China was 0.39, of which the pure technical efficiency was 0.46 and the scale efficiency was 0.78. Room for improvement exists in the use rate and the scale of input. However, in different regions, agricultural water-saving investment is polarized. The comprehensive technology efficiency of eight of the regions was above 0.7, and 12 regions were below 0.15.

(2) Strengthening resource and organization management in agricultural water-saving would play a significant effect on the improvement of input efficiency, while pure technical efficiency plays a major role in the improvement of input efficiency. The pure technical efficiency of agricultural water-saving was found to be 0.46 across the whole country for three years, indicating a certain gap compared with the scale efficiency of 0.78. The main factor leading to the low efficiency of integrated technology is the low level of technical efficiency. In terms of the water-saving input in agriculture, areas with low pure technical efficiency should focus on improving resource management and the technology.

(3) Further optimizing the distribution of resources and investing in subsidies in areas with increasing scale efficiency can lead to an increase in overall efficiency, but there is a diminishing effect of scale in some regions. There is a reduced scale of input in Hebei, Jilin, Heilongjiang, and Sichuan, indicating that the agricultural water-saving inputs in these regions exceed the local resource distribution capacity. It is necessary to reduce the inputs appropriately and improve the local resource distribution capacity to further improve the efficiency of agricultural water-saving investment. The distribution of resources nationwide should be optimized, and resource subsidies should be provided to areas with high technical efficiency and lack of input, so as to maximize the overall investment efficiency of all provinces.

(4) This study analyzes irrigation input efficiency from the perspective of crop irrigation requirement, comprehensively considers the region's own water requirement and the precipitation conditions of the crop location, examines the impact of agricultural water-saving irrigation requirement, further explores the necessity of water-saving irrigation, and enhances the scientific nature of the conclusion. It is discovered that the low input efficiency of water-saving irrigation in Guangxi, Jiangxi, Guizhou, and Chongqing may be related to the low local crop irrigation requirement.

(5) In the case of input and output indicators, the impact of the regional per capita GDP on the agricultural water-saving input efficiency was not obvious. The impact of the area of cultivated land on the efficiency of agricultural water-saving input was not significant.

Due to the limitation of the DEA model, this study can only rank the regional agricultural water-saving irrigation input efficiency. If the appropriate model is used to obtain the absolute value of the input efficiency, it will help expand the research content. Limited by the available data collected from China Water Conservancy Statistical Yearbook and the National Water Development Statistics Bulletin, in the future research, we will investigate and survey the input efficiency for water-saving agriculture to collect data of some sophisticated indicators such as popularity of efficient irrigation technology, average education level of farmer households. These indicators will be used to evaluate the efficiency of water-saving agricultural inputs from the characteristics of farmers in different regions.

Author Contributions: Conceptualization, Y.C. and W.Z.; methodology, Y.C.; validation, J.R. and W.Z.; formal analysis, Y.C.; investigation, Y.C. and W.Z.; resources, Y.C. and W.Z.; data curation, Y.C. and J.R.; writing—original draft preparation, Y.C.; writing—review and editing, Y.C. and J.R.; visualization, Y.C.; supervision, W.Z. and J.R.; project administration, W.Z. All authors have read and agreed to the published version of the manuscript.

Funding: This work was supported by the Major Bidding Program of National Social Science Foundation of China. (Grant No.18ZDA074).

Conflicts of Interest: The authors declare no conflict of interest.

References

1. Blanke, A.; Rozelle, S.; Lohmar, B.; Wang, J.; Huang, J. Water saving technology and saving water in China. *Agric. Water Manag.* **2007**, *87*, 139–150. [CrossRef]
2. Li, G. Game analysis of the effect that urban water supply pricing has on the behavior of the resident resources saving. *J. Cap. Univ. Econ. Bus.* **2007**, *6*, 103–107. [CrossRef]
3. Cai, X.; Ringler, C. Balancing agricultural and environmental water needs in China: Alternative scenarios and policy options. *Water Policy* **2007**, *9*, 95–108. [CrossRef]
4. Facon, T.; Renault, D.; Rao, P.S.; Wahaj, R. High-yielding capacity building in irrigation system management: Targeting managers and operators. *Irrig. Drain.* **2008**, *57*, 288–299. [CrossRef]
5. Feng, W.; Zhong, M.; Lemoine, J.-M.; Biancale, R.; Hsu, H.-T.; Xia, J. Evaluation of groundwater depletion in North China using the Gravity Recovery and Climate Experiment (GRACE) data and ground-based measurements. *Water Resour. Res.* **2013**, *49*, 2110–2118. [CrossRef]
6. Gao, H.; Wei, T.; Lou, I.; Yang, Z.; Shen, Z.; Li, Y. Water saving effect on integrated water resource management. *Resour. Conserv. Recy.* **2014**, *93*, 50–58. [CrossRef]
7. Jiang, Y. China's water scarcity. *J. Environ. Manag.* **2009**, *90*, 3185–3196. [CrossRef]
8. Lu, Y.; Zhang, X.Y.; Chen, S.Y.; Shao, L.W.; Sun, H.Y. Changes in water use efficiency and water footprint in grain production over the past 35 years: A case study in the North China Plain. *J. Clean. Prod.* **2016**, *116*, 71–79. [CrossRef]
9. Ma, Y.; Feng, S.Y.; Song, X.F. Evaluation of optimal irrigation scheduling and groundwater recharge at representative sites in the North China Plain with SWAP model and field experiments. *Comput. Electron. Agric.* **2015**, *116*, 125–136. [CrossRef]
10. Kang, S.; Eltahir, E.A.B. North China Plain threatened by deadly heatwaves due to climate change and irrigation. *Nat. Commun.* **2018**, *9*, 2894. [CrossRef]
11. Kang, S.; Hao, X.; Du, T.; Tong, L.; Su, X.; Lu, H.; Li, X.; Huo, Z.; Li, S.; Ding, R. Improving agricultural water productivity to ensure food security in China under changing environment: From research to practice. *Agric. Water Manag.* **2017**, *179*, 5–17. [CrossRef]
12. Liu, J.; Zang, C.; Tian, S.; Liu, J.; Yang, H.; Jia, S.; You, L.; Liu, B.; Zhang, M. Water conservancy projects in China: Achievements, challenges and way forward. *Glob. Environ. Chang. Hum. Policy Dimens.* **2013**, *23*, 633–643. [CrossRef]
13. Mushtaq, S. Exploring Synergies Between Hardware and Software Interventions on Water Savings in China: Farmers Response to Water Usage and Crop Production. *Water Resour. Manag.* **2012**, *26*, 3285–3300. [CrossRef]
14. Ding, X.H.; Zhang, Z.X.; Wu, F.P.; Xu, X.Y. Study on the Evolution of Water Resource Utilization Efficiency in Tibet Autonomous Region and Four Provinces in Tibetan Areas under Double Control Action. *Sustainability* **2019**, *11*. [CrossRef]
15. Ding, X.H.; Tang, N.; He, J.H. The Threshold Effect of Environmental Regulation, FDI Agglomeration, and Water Utilization Efficiency under Double Control Actions-An Empirical Test Based on Yangtze River Economic Belt. *Water* **2019**, *11*. [CrossRef]
16. Stec, A.; Zelenakova, M. An Analysis of the Effectiveness of Two Rainwater Harvesting Systems Located in Central Eastern Europe. *Water* **2019**, *11*. [CrossRef]
17. Zhao, M.Z.; Chen, Z.H.; Zhang, H.L.; Xue, J.B. Impact Assessment of Growth Drag and Its Contribution Factors: Evidence from China's Agricultural Economy. *Sustainability* **2018**, *10*. [CrossRef]
18. Shiferaw, B.; Smale, M.; Braun, H.-J.; Duveiller, E.; Reynolds, M.; Muricho, G. Crops that feed the world 10. Past successes and future challenges to the role played by wheat in global food security. *Food Secur.* **2013**, *5*, 291–317. [CrossRef]
19. Grimaldi, M.; Pellecchia, V.; Fasolino, I. Urban Plan and Water Infrastructures Planning: A Methodology Based on Spatial ANP. *Sustainability* **2017**, *9*. [CrossRef]
20. Jing, X.E.; Zhang, S.H.; Zhang, J.J.; Wang, Y.J.; Wang, Y.Q. Assessing efficiency and economic viability of rainwater harvesting systems for meeting non-potable water demands in four climatic zones of China. *Resour. Conserv. Recy.* **2017**, *126*, 74–85. [CrossRef]

21. Zhang, X.D.; Zhu, S. The Flexible Demand Analysis of Agricultural Irrigation Water Use Based on Technical Efficiency and Shadow Price: Taking Heilongjiang Province for an Example. *Sci. Geogr. Sin.* **2018**, *38*, 1165–1173. [CrossRef]
22. Zhang, B.; Wang, X.Q. The Study of the Farmer Water Users Participating in Irrigation Management and Water-saving Irrigation: The demonstration analysis of the mode of the independent management in irrigation district in Zaohe area of Jiangsu province. *Probl. Agric. Econ.* **2004**, *3*, 48–80. [CrossRef]
23. Tian, G.L.; Gu, S.W.; Wei, D.; Shuai, M.D. A Study on the Influence of Comprehensive Reform of Agricultural Water Price on the Price of Water Right in the Process of Transaction. *Price Theory Pract.* **2017**, *2*, 66–69.
24. Luo, W.Z.; Jiang, Y.L.; Wang, X.F.; Liu, H.X.; Chen, Y.S. Farmer cognition on water-saving irrigation technology analysis in groundwater overmining area of North China: A case study of Guyuan county, Zhangjiakou city, Hebei province. *J. Nat. Resour.* **2019**, *34*, 2469–2480. [CrossRef]
25. Lei, G.R.; Hu, Z.Y.; Han, G. Technical efficiency and water saving potential in agriculture water based on SFA. *J. Econ. Water Resour.* **2010**, *1*, 55–78. [CrossRef]
26. Tian, C.H.; Zhao, T.; Xu, Z.; Shen, H. Analysis of Technical Efficiency of Agricultural Efficient Water-saving Technology Based on DEA Method in Xinjiang. *Water Sav. Irrig.* **2017**, *1*, 90–93.
27. Feng, X.; Li, J.; Guo, M.R. Studies on Efficiency of Facility Vegetable Production Considering the Water-saving in Beijing and Its Development Countermeasures. *China Veg.* **2017**, *1*, 55–60.
28. Wang, X.; Lu, Q. The Empirical Analysis of Regional Differences in Chinese Agricultural Water Use Efficiency and Convergence Test. *Soft Sci.* **2014**, *28*, 133–137.
29. Wu, X.; Zhu, M.L.; He, C. Research on the Efficiency of Agricultural High-efficiency Water-saving Technology Based on Super Efficiency DEA Method. *Rural Econ. Sci. Technol.* **2015**, *26*, 63–65.
30. Benfica, R.; Cunguara, B.; Thurlow, J. Linking agricultural investments to growth and poverty: An economywide approach applied to Mozambique. *Agric. Syst.* **2019**, *172*, 91–100. [CrossRef]
31. Berbel, J.; Gutierrez-Martin, C.; Rodriguez-Diaz, J.A.; Camacho, E.; Montesinos, P. Literature Review on Rebound Effect of Water Saving Measures and Analysis of a Spanish Case Study. *Water Resour. Manag.* **2015**, *29*, 663–678. [CrossRef]
32. Cai, X.M.; Rosegrant, M.W.; Ringler, C. Physical and economic efficiency of water use in the river basin: Implications for efficient water management. *Water Resour. Res.* **2003**, *39*. [CrossRef]
33. Gleick, P.H.; Christian-Smith, J.; Cooley, H. Water-use efficiency and productivity: Rethinking the basin approach. *Water Int.* **2011**, *36*, 784–798. [CrossRef]
34. Huffaker, R.; Whittlesey, N. The allocative efficiency and conservation potential of water laws encouraging investments in on-farm irrigation technology. *Agric. Econ.* **2000**, *24*, 47–60. [CrossRef]
35. Orfanou, A.; Pavlou, D.; Porter, W.M. Maize Yield and Irrigation Applied in Conservation and Conventional Tillage at Various Plant Densities. *Water* **2019**, *11*. [CrossRef]
36. Playan, E.; Slatni, A.; Castillo, R.; Faci, J.M. A case study for irrigation modernisation: II Scenario analysis. *Agric. Water Manag.* **2000**, *42*, 335–354. [CrossRef]
37. Qureshi, M.E.; Grafton, R.Q.; Kirby, M.; Hanjra, M.A. Understanding irrigation water use efficiency at different scales for better policy reform: A case study of the Murray-Darling Basin, Australia. *Water Policy* **2011**, *13*, 1–17. [CrossRef]
38. Fried, H.O.; Schmidt, S.S.; Yaisawarng, S. Incorporating the operating environment into a nonparametric measure of technical efficiency. *J. Product. Anal.* **1999**, *12*, 249–267. [CrossRef]
39. Fried, H.O.; Lovell, C.A.K.; Schmidt, S.S.; Yaisawarng, S. Accounting for environmental effects and statistical noise in data envelopment analysis. *J. Product. Anal.* **2002**, *17*, 157–174. [CrossRef]
40. Banker, R.D.; Kauffman, R.J.; Morey, R.C. Measuring gains in operational efficiency from information technology: A study of the Positran deployment at Hardee's Inc. *J. Manag. Inf. Syst.* **1990**, *7*, 29–54. [CrossRef]
41. Banker, R.D.; Morey, R.C. Efficiency Analysis for Exogenously Fixed Inputs and Outputs. *Oper. Res.* **1986**, *34*, 513–521. [CrossRef]
42. Yang, H.; Pollitt, M. Incorporating both undesirable outputs and uncontrollable variables into DEA: The performance of Chinese coal-fired power plants. *Eur. J. Oper. Res.* **2009**, *197*, 1095–1105. [CrossRef]
43. Liu, Y.; Wang, L. Spatial distribution characteristics of irrigation water requirement for main crops in China. *Trans. Chin. Soc. Agric. Eng.* **2009**, *25*, 6–12. [CrossRef]

44. Liu, Y.; Teixeira, J.L.; Zhang, H.J.; Pereira, L.S. Model validation and crop coefficients for irrigation scheduling in the North China plain. *Agric. Water Manag.* **1998**, *36*, 233–246. [CrossRef]
45. Liu, Y.; Cai, J.B.; Cai, L.G.; Pereira, L.S. Analysis of irrigation scheduling and water balance for an Irrigation district at lower reaches of the Yellow River. *J. Hydraul. Eng.* **2005**, *36*, 701–708. [CrossRef]
46. Allen, R.G.; Smith, M.; Perrier, A.; Pereira, L.S. An update for the definition of reference evapotranspiration. *ICID Bull.* **1994**, *43*, 1–34.
47. Sun, Y.; Jiang, N.; Cui, Y. Research on Economic, Social and Environmental Efficiency of Environmental Protection Investment: Based on Three-stage DEA Model. *Sci. Technol. Manag. Res.* **2019**, *39*, 219–226. [CrossRef]
48. Shen, X.B.; Lin, B.Q. The shadow prices and demand elasticities of agricultural water in China: A StoNED-based analysis. *Resour. Conserv. Recy.* **2017**, *127*, 21–28. [CrossRef]
49. Pan, J.F.; Liu, Y.Z.; Zhong, X.H.; Lampayan, R.M.; Singleton, G.R.; Huang, N.R.; Liang, K.M.; Peng, B.L.; Tian, K. Grain yield, water productivity and nitrogen use efficiency of rice under different water management and fertilizer-N inputs in South China. *Agric. Water Manag.* **2017**, *184*, 191–200. [CrossRef]
50. Li, P.L. Evaluation of Efficiency of Local Financial Investing in Agriculture. *Econ. Manag.* **2011**, *25*, 70–76.
51. Li, G.F.; Zhou, D.Y.; Shi, M.J. Technical efficiency of crop irrigation and its determinants in the arid areas of Northwest China. *J. Nat. Resour.* **2019**, *34*, 853–866. [CrossRef]
52. Li, W.; Yu, F.W. Analysis of factors influencing agricultural water performance in western China. *Res. Dev.* **2008**, *6*, 60–63. [CrossRef]
53. Wang, Y.H.; Tao, Y.; Kang, J.N. Influencing factors of irrigation governance in China's rural areas. *Resour. Sci.* **2019**, *41*, 1769–1779. [CrossRef]
54. Wang, X.J.; Li, Z. Analysis of irrigation water efficiency and influencing factors. *Chin. Rural. Econ.* **2005**, *7*, 11–18.
55. Song, C.X.; Ma, H.Y.; Huang, J.K.; Wang, J.X. Impact of climate change and farmer adaptation on wheat irrigation efficiency. *J. Agrotech. Econ.* **2014**, *2*, 4–16. [CrossRef]
56. Rak, J.R.; Pietrucha-Urbanik, K. An Approach to Determine Risk Indices for Drinking Water–Study Investigation. *Sustainability* **2019**, *11*, 3189. [CrossRef]

Article

Climatic Trends of West Virginia: A Representative Appalachian Microcosm

Evan Kutta [1,*] and Jason Hubbart [1,2,*]

[1] Institute of Water Security and Science, West Virginia University, Agricultural Sciences Building, Morgantown, WV 26506, USA

[2] Davis College, Schools of Agriculture and Food, and Natural Resources, West Virginia University, 3109 Agricultural Sciences Building, Morgantown, WV 26506, USA

* Correspondence: Evan.Kutta@mail.wvu.edu (E.K.) and Jason.Hubbart@mail.wvu.edu (J.H.); Tel.: +1-304-293-8851 (E.K.)

Received: 18 March 2019; Accepted: 24 May 2019; Published: 28 May 2019

Abstract: During the late 19th and very early 20th centuries widespread deforestation occurred across the Appalachian region, USA. However, since the early 20th century, land cover rapidly changed from predominantly agricultural land use (72%; 1909) to forest. West Virginia (WV) is now the USA's third most forested state by area (79%; 1989–present). It is well understood that land cover alterations feedback on climate with important implications for ecology, water resources, and watershed management. However, the spatiotemporal distribution of climatic changes during reforestation in WV remains unclear. To fill this knowledge gap, daily maximum temperature, minimum temperature, and precipitation data were acquired for eighteen observation sites with long periods of record (POR; ≥77 years). Results indicate an increasingly wet and temperate WV climate characterized by warming summertime minimum temperatures, cooling maximum temperatures year-round, and increased annual precipitation that accelerated during the second half (1959–2016) of the POR. Trends are elevation dependent and may be accelerating due to local to regional ecohydrological feedbacks including increasing forest age and density, changing forest species composition, and increasing globally averaged atmospheric moisture. Furthermore, results imply that excessive wetness may become the primary ecosystem stressor associated with climate change in the USA's rugged and flood prone Appalachian region. The Appalachian region's physiographic complexity and history of widespread land use changes makes climatic changes particularly dynamic. Therefore, mechanistic understanding of micro- to mesoscale climate changes is imperative to better inform decision makers and ensure preservation of the region's rich natural resources.

Keywords: climate change; Appalachia; reforestation; land use-land cover; land-atmosphere coupling

1. Introduction

During the late 19th to early 20th century West Virginia (WV) was the frontier of American industrial capitalism where the rich natural resources of the Mountain State, especially timber and coal, were exploited for the benefit of aggressively populating eastern cities [1,2]. In 1880, it was estimated that two thirds of WV was covered by ancient growth hardwood forest [2], but by 1909 approximately 72% of WV was agriculture or pasture lands [3]. Widespread logging created a landscape characterized by vast acreages of dried slash that were easily ignited by passing locomotives [3] and burned with unusual intensity during dry periods [4]. By the 1920s, most lumbermen abandoned WV for lands further south and west [2] and extensive fire prevention efforts resulted in the redevelopment of vast hardwood forests [4]. When the first United States (US) Forest Service forest survey of WV was completed in 1949 an estimated 64% of WV lands were forested, which increased to 75% by the 1961 forest survey [5] and was relatively steady (75% to 79%) through 2012 [6]. Despite relatively steady areal coverage since the 1961 forest survey, timber volume increased through 2012 consistent with a maturing forest ecosystem [6]. Given the magnitude of reforestation throughout the broader Appalachian region, WV is an ideal location to assess how the regrowth of native forests may have influenced (or been influenced by) local to regional climatic trends.

Changes in land use and land cover (LULC) alters radiative (e.g., absorption and reflection) and non-radiative (e.g., surface roughness and evapotranspiration) climate forcing that influence spatial and temporal patterns of temperature and precipitation [7,8]. Few studies have focused on climatic influence of reforestation in the northeast US and studies utilizing observations are less frequent than modeling efforts [7,9]. Considering results of model and observation based analyses are contradictory [7,9], spatiotemporal analysis of the climatic influence of extensive reforestation over a long period of record (POR) is needed in WV and other similar regions globally. For example, spatiotemporal patterns in minimum temperature, maximum temperature, and precipitation trends are valuable to agricultural interests [10], natural resource managers (i.e., water, land, energy, wildlife; [11,12]), and other public and private stakeholders. More specifically, spatiotemporal climatic analyses are needed to better understand and prepare for terrestrial ecosystem vulnerabilities that may impact habitat suitability, human health, and food security [12]. Impacts to human health and food security are particularly relevant to West Virginia given the broader Appalachian region classification as a food desert (i.e., low income and limited access to fresh produce) by the US Department of Agriculture, Food and Nutrition Services [13]. Quantifying spatiotemporal climatic changes within WV will guide future climatic analyses, inform natural resource management decisions, and help preserve WV's diverse terrestrial ecosystem.

West Virginia's climate, water resources, and ecology are intricately linked and all three are influenced by WV's complex terrain. Elevation gradually increases from the Ohio River (~300 m) on the western border of WV to the ridges of the Allegheny Mountains (>1000 m) in the eastern third of WV. Annual average precipitation is approximately 100 cm across the far western portions of the state (Wheeling, WV; 102.6 cm), gradually increasing to near 150 cm at the highest elevations (Snowshoe Mountain, WV; 152.1 cm), and abruptly decreases to near 90 cm immediately east of the Allegheny Mountains (Franklin, WV; 91.5 cm; [14]). Complex climatic gradients resulting from WV's rugged terrain make the Monongahela National Forest, located in the Allegheny Mountains, one of the most ecologically diverse forests in the US National Forest System [15]. Stands of red spruce and populations of snowshoe hare, more typical of northern boreal forests, occur at higher elevations in WV. Localized cold air traps (i.e., deep rock crevices, talus slopes, and caves) provide refugium for organisms adapted to survival in cold and dark environments (e.g., bacteria and fungi; [16]). Drier east-facing slopes located east of the Allegheny ridges support prickly pear cactus and rare plant species including the shale barren rock cress [15]. Lowland locations, including mountain valleys, are characterized by stands of mixed hardwoods growing in relatively deep and rich soils more typical of southern Appalachia. Coupling of WV's climatic gradients, biodiversity, and physiography suggests an opportunity to manage WV's water resources using an integrated and multidisciplinary approach. However, before management practices can be advanced, investigation of spatiotemporal climatic changes during and after severe and widespread anthropogenic disturbance (e.g., deforestation and reforestation) is needed.

The overarching objective of the current work was to leverage state and federally funded WV climate observation datasets to quantify spatiotemporal characteristics of long-term (≥77 years) minimum temperature, maximum temperature, and daily precipitation trends across West Virginia, USA. Sub objectives included (a) assessing statistically significant ($p < 0.05$) trends using the Mann–Kendall trend test, (b) performing analyses during the first (1900–1958) and second (1959–2016) half of the time series, and (c) discussing the ecological and topographic dependence of climatic trends across WV.

2. Materials and Methods

Daily summaries of total precipitation, maximum temperature, and minimum temperature were acquired from the National Center for Environmental Information (NCEI), a branch of the National Oceanic and Atmospheric Administration (NOAA; [17]). Stations were selected with start dates as early as 1900, but no later than 1930, with a nearly continuous time series through the end of 2016. These temporal criteria ensured analyses included the rapid transition from primarily agricultural and pasture lands in the early 20th century to primarily forested land coverage by the middle 20th century and subsequent forest maturation [3,6]. All leap days (i.e., February 29th) were excluded from analyses and daily data were post-processed to ensure all missing dates were included resulting in a continuous time series with 365 days in each year. Annual averages of daily summaries of maximum temperature, minimum temperature, and total precipitation were estimated and counts of missing data points were tabulated annually for each of the 18 selected stations. The spatial distribution of the 18 selected stations is shown in Figure 1 and additional site-specific data are included in Table 2. It was assumed that observers and their equipment followed station siting guidelines established by the United States Weather Bureau in 1890 (Fiebrich 2009). The World Meteorological Organization's guidelines on missing data in the calculation of climatological normals recommends not calculating monthly averages when five or more daily values are missing [18], which scales to 60 or more daily values for annual averages. For the purposes of this work years with data gaps exceeding 15% of possible daily observations (>54 daily observations) were removed from analyses. Of 117 possible annual averages or totals, each observation location had between 77 and 110 annual averages for maximum and minimum temperature and between 84 and 110 annual precipitation totals.

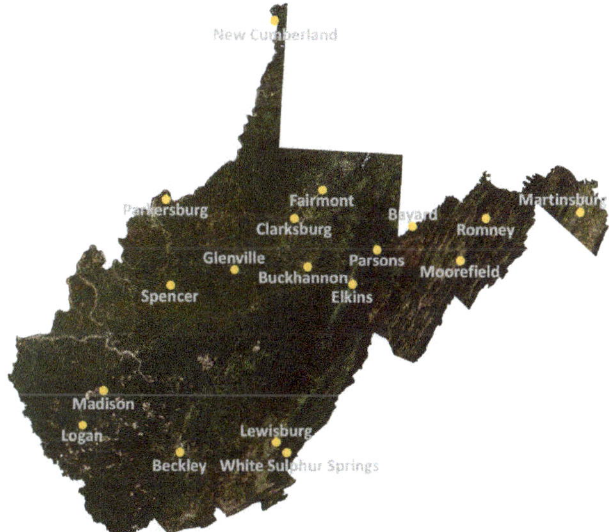

Figure 1. Observed climate record site locations in West Virginia, USA. Satellite image credit: National Aeronautics and Space Administration (NASA; [19]).

Investigating temporal trends of the central tendency of a data series is of interest in the context of identifying a changing climate [20,21]. Sen's slope estimator paired with the Mann–Kendall trend test is a robust, non-parametric method to estimate the linear trend and assess statistical significance [22] and was thus deemed appropriate for this work. Spatial characteristics of temporal trends were estimated and statistical significance ($\alpha = 0.05$) assessed over the entire time series (1900–2016) and during the first (1900–1958) and second (1959–2016) halves of each time series at each observation location. Analyses for each half of the time series were performed because the data series was sufficiently long and the first half corresponded with reforestation whereas the second half corresponded with forest maturation and globally averaged warming exceeding 0.65 °C [4–6,23]. Except for New Cumberland (Table 1, Figure 1), each observation location had a more complete time series of each variable during the second half (93.8%) rather than the first half (71.7%) of the POR suggesting early data gaps (1900–1930) could have influenced results. As a result, the data should not be considered stationary and readers should use caution when extrapolating results into the future. Changes in temporal characteristics (i.e., seasonality) were estimated for all 365 days using a continuous time series of daily data averaged across all eighteen locations for the entire POR and the second half of the POR (1959–2016). Seasonal analyses were not performed for the first half of the POR due to larger and more frequent data gaps. For plotting purposes, centered three-week moving averages of daily Sen's slope, upper and lower 95% confidence interval values were calculated to smooth day-to-day variability and better show seasonal changes of each variable. Centered moving averages (CMAs) represent a series of arithmetic means throughout a time series that is centered on the middle value (i.e., day 11 of the 21-day average). Three-week CMAs were selected because meteorological and astronomical definitions of seasonality are static in time and space and may be inadequate considering a swiftly changing climate [24]. Ultimately, the methods described in the current work are easily applied and provide valuable information about spatiotemporal changes in climate that are needed for effective decision making.

Table 1. Monthly averages of total precipitation (Precip; cm), maximum temperature (Tmax; °C), minimum temperature (Tmin; °C), and average temperature (Tavg; °C) averaged across eighteen observation locations in West Virginia, USA (Table 2) between 1900 and 2016.

	Jan	Feb	Mar	Apr	May	Jun	Jul	Aug	Sep	Oct	Nov	Dec
Precip	8.5	7.4	9.7	9.1	10.5	11.1	11.7	10.1	8.2	7.4	7.3	8.3
Tmax	5.2	6.6	12.2	18.4	23.7	27.5	29.4	28.7	25.6	19.6	12.6	6.6
Tavg	−0.3	0.7	5.5	11.0	16.3	20.6	22.7	22.0	18.5	12.3	6.2	1.3
Tmin	−5.8	−5.3	−1.1	3.6	8.8	13.6	16.0	15.2	11.4	4.9	−0.1	−4.1

3. Results

3.1. Climate during Period of Record

Monthly and daily averages of precipitation and maximum, minimum, and daily average temperatures are included in Table 1 and Figure 2, respectively. West Virginia's (WV) wettest month of the year was July when an average of 11.7 cm of precipitation was observed during the period of record (Table 1). July was also WV's warmest month of the year with a daily average temperature of 22.7 °C and average maximum and minimum temperatures of 29.4 °C and 16.0 °C, respectively. Alternatively, January was the coolest month with a daily average temperature of −0.3 °C and daily maximum and minimum temperatures of 5.2 °C and −5.8 °C, respectively. February, October, and November were similarly dry with each month averaging less than 7.5 cm of precipitation. At daily resolution, the warmest average maximum temperature was 29.9 °C on July 18th and the coolest average minimum temperature was −6.9 °C on January 28th (Figure 2). Maximum daily average precipitation occurred on July 10th with 0.49 cm day^{-1} and minimum daily average precipitation occurred on November 21st with 0.14 cm day^{-1} (not shown). Between 1900 and 2016, WV's spatially averaged climate was cold with hot summers and no dry season (i.e., Köppen-Geiger Dfa; [25]).

3.2. Spatial Climatic Changes

The observation location, elevation, and POR for each selected site as well as trends in minimum temperature, maximum temperature, and total annual precipitation are summarized in Table 2 and shown spatially in Figure 3. Unless otherwise stated, estimated trends reported parenthetically represent trends averaged across observation sites with either positive or negative trends.

Table 2. Eighteen observation stations in West Virginia, USA. Elevation (m), period of record (POR), and estimated linear slope for minimum temperature (°C/decade), maximum temperature (°C/decade), and total annual precipitation (cm/decade). Higher and lower elevation averages refer to above or below the median elevation (323 m), respectively.

City	Elev	Period of Record	Minimum 1900–2016	Minimum 1900–1958	Minimum 1959–2016	Maximum 1900–2016	Maximum 1900–1958	Maximum 1959–2016	Precipitation 1900–2016	Precipitation 1900–1958	Precipitation 1959–2016
Martinsburg	165	1926–2016	0.12	0.01	0.32	−0.06	−0.40	0.12	−0.30	−5.10	1.20
New Cumb.	205	1902–2016	0.17	0.30	0.22	0.02	0.50	−0.29	0.49	1.34	3.21
Romney	206	1900–2016	−0.02	0.08	0.05	−0.06	0.16	−0.09	−0.05	−0.89	1.30
Madison	208	1909–2016	0.04	0.03	0.31	0.04	−0.07	0.24	3.05	3.03	4.44
Glenville	229	1900–2016	−0.02	0.15	0.42	−0.21	0.00	0.15	−0.54	1.44	2.83
Moorefield	254	1901–2016	0.01	0.29	−0.23	−0.12	0.15	−0.12	0.30	−1.56	1.90
Parkerburg	254	1926–2016	−0.09	−0.03	0.05	0.01	0.17	0.19	1.52	3.19	3.71
Spencer	294	1906–2016	−0.07	0.40	−0.07	−0.10	−0.01	−0.13	0.36	−0.77	4.48
Clarksburg	302	1923–2016	0.11	−0.05	0.30	−0.10	−0.04	−0.03	1.42	−2.20	4.30
Fairmont	345	1906–2016	0.10	0.32	0.27	−0.21	−0.24	0.11	0.79	0.28	2.16
Logan	367	1902–2016	0.11	0.00	0.52	−0.09	−0.10	0.16	0.18	−3.01	3.66
Buckhannon	433	1909–2016	−0.09	−0.07	0.14	−0.30	0.15	0.02	−1.17	−1.50	0.69
Parsons	543	1900–2016	0.06	0.39	0.07	−0.15	0.23	−0.05	0.94	0.86	1.23
White SS	597	1921–2016	−0.01	0.17	−0.04	−0.16	0.29	−0.40	1.15	2.84	2.14
Elkins	604	1926–2016	−0.07	−0.31	0.31	0.04	0.14	0.21	−0.51	−8.34	1.33
Lewisburg	696	1901–2016	−0.04	0.07	−0.06	−0.06	0.20	−0.29	0.19	−3.17	2.38
Beckley	711	1907–2016	−0.03	0.03	0.02	0.01	−0.16	−0.24	−1.77	−1.37	1.62
Bayard	722	1903–2016	0.02	−0.14	0.06	−0.20	−0.17	−0.07	0.06	−1.84	2.33
Lower Elevation Average			0.03	0.13	0.15	−0.06	0.05	0.00	0.69	−0.17	3.04
Higher Elevation Average			0.01	0.05	0.14	−0.12	0.04	−0.06	−0.02	−1.69	1.95
Average			0.02	0.09	0.15	−0.09	0.04	−0.03	0.34	−0.93	2.50

White SS = White Sulphur Springs, New Cumb. = New Cumberland, Elev = elevation, blue or brown shading = significant decrease ($p < 0.05$), red or green shading = significant increase ($p < 0.05$).

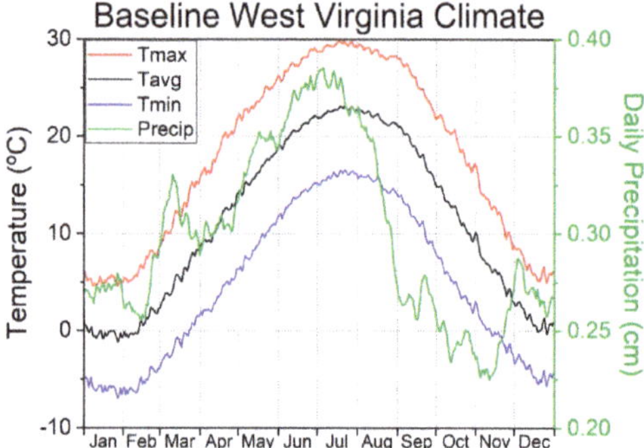

Figure 2. Daily average maximum temperature (Tmax), minimum temperature (Tmin), average temperature (Tavg), and precipitation (Precip) over a long period of record (1900–2016) averaged across eighteen observation locations in West Virginia, USA. Daily average precipitation data were smoothed with a centered three-week moving average for plotting purposes.

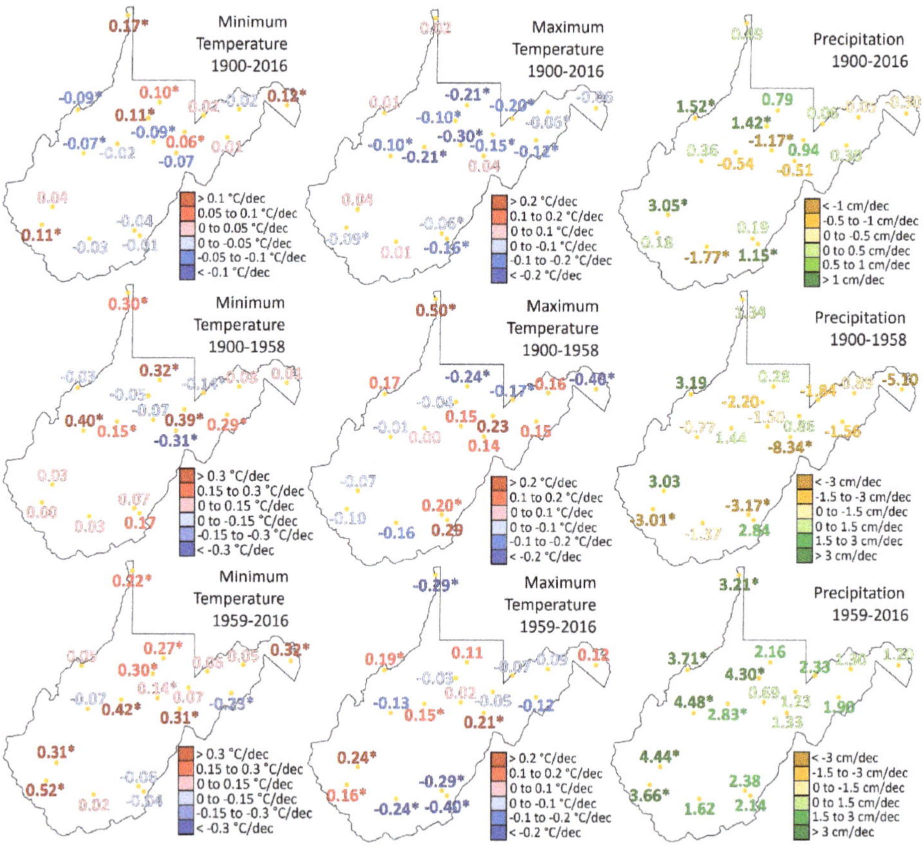

Figure 3. Sen's slope estimator for minimum temperature (left), maximum temperature (middle), and precipitation (right) across the state of West Virginia, USA between 1900–2016 (top), 1900–1958 (middle), and 1959–2016 (bottom). * Indicates statistically significant trends ($p < 0.05$).

3.2.1. Entire Time Series (1900–2016)

Half of the eighteen observation sites in WV indicated a decreasing trend in minimum temperature (–0.05 °C/decade), while the other half indicated a greater increasing trend (0.08 °C/decade). Warming minimum temperatures were generally observed across the northern and southwestern portions of WV, but the two lowest elevation observation sites had the largest rate of warming (≥0.12 °C/decade) and trends were statistically significant ($p < 0.00$). Alternatively, thirteen observations sites spatially and topographically distributed across WV recorded decreasing trends (–0.14 °C/decade) in maximum temperatures. Three of the four largest negative trends (≤–0.20 °C/decade) were observed across north-central WV above the median station elevation (323 m) and all four of the largest negative trends were statistically significant ($p < 0.00$). Total annual precipitation increased (0.87 cm/decade) at twelve observation sites and decreased (–0.72 cm/decade) at the other six sites. The two sites with the largest increase (>1.5 cm/decade) and decrease (<–1 cm/decade) in annual precipitation were below and above the median elevation, respectively and all four trends were statistically significant ($p < 0.03$). In synthesis, annually averaged minimum temperatures increased more at lower elevations, maximum temperatures decreased more at higher elevations, and precipitation increased at lower elevations between 1900 and 2016 across WV.

3.2.2. First Half (1900–1958)

Twelve observation sites were characterized by increasing minimum temperatures (0.17 °C/decade) and the remaining six sites had decreasing minimum temperatures (−0.12 °C/decade). Just two sites below the median elevation had decreasing minimum temperatures and the two largest decreases (≤0.14 °C/decade), both statistically significant ($p < 0.04$), were found at higher elevations (604 and 722 m). Six of the twelve sites with positive minimum temperature trends were statistically significant ($p < 0.02$) and four of the six sites, including the largest positive trend (0.40 °C/decade), were below the median elevation. Ten of the observation sites indicated warming maximum temperature trends (0.20 °C/decade), the remaining eight sites indicated cooling (−0.15 °C/decade). Half of sites with observed warming and cooling maximum temperatures were above or below the median elevation. Seven observation sites indicated increasing trends in total annual precipitation (1.85 cm/decade) and eleven indicated decreasing trends (−2.70 cm/decade). Three of the eleven sites with decreasing trends were significant ($p < 0.03$) and all were above the median elevation. In summary, annually averaged minimum temperatures increased particularly at lower elevations, maximum temperatures increased, and precipitation decreased in WV between 1900 and 1958.

3.2.3. Second Half (1959–2016)

Fourteen climate observation sites indicated increasing minimum temperatures (0.22 °C/decade) and the remaining four sites indicated decreasing minimum temperatures (−0.10 °C/decade). Eight observation sites indicated increasing trends in annually averaged maximum temperatures (0.15 °C/decade) and ten observation sites indicated decreasing trends (−0.17 °C/decade). Half of all sites with observed warming or cooling minimum and maximum temperatures were above or below the median elevation, but sites with warming minimum (maximum) temperatures warmed 0.04 °C/decade (0.05 °C/decade) slower at higher elevation stations. Additionally, sites with cooling minimum (maximum) temperatures cooled −0.10 °C/decade (−0.08 °C/decade) faster at lower (higher) elevations. All eighteen observation sites indicated increasing trends (2.50 cm/decade) in total annual precipitation, but lower elevation sites had trends 1.09 cm/decade larger than higher elevation sites. Additionally, statistically significant ($p < 0.05$) increases in total annual precipitation were confined to observation locations below 400 meters in elevation across the western half of WV. As a result, minimum temperatures increased particularly at lower elevations, maximum temperatures decreased particularly at higher elevations, and precipitation increased state-wide with significant ($p < 0.05$) increases at lower elevations across western WV between 1959 and 2016.

3.3. Temporal Trends (1900–2016)

3.3.1. Entire Time Series (1900–2016)

Centered three-week moving averages (CMAs) of state-averaged, daily precipitation and temperature trends are indicative of seasonal trends in WV (Figure 4). Daily precipitation trends were generally small (92% between ±0.0005 cm year^{-1}), except for increases during May, decreases during June, and increases during July when 12 of the 17 significant ($p < 0.05$) daily trends occurred (Figure 4a). All but four of the 23 significant daily trends ($p < 0.05$) in average daily temperature were negative and the significant positive trends occurred in June, August, and December (Figure 4b). Daily average temperature trends were most negative during January (−0.021 °C year^{-1}), October (−0.008 °C year^{-1}), and March (−0.008 °C year^{-1}) whereas April (0.006 °C year^{-1}) and December (0.003 °C year^{-1}) were most positive. Nineteen of the 26 significant ($p < 0.05$) daily minimum temperature trends were positive, especially during June, July, and August when sixteen significantly positive trends occurred ($p < 0.05$; Figure 4c). All seven significantly decreasing ($p < 0.05$) daily minimum temperature trends occurred between January 22nd and April 8th. All significant daily maximum temperature trends (n = 73) were negative, but CMAs were briefly positive during April, late November, and early December (Figure 4d). Between May and October, WV's growing season, 79% of the significant ($p < 0.05$) negative maximum temperature trends occurred. In summary, daily precipitation trends were largest during May, June, and July; average temperatures decreased most during January; minimum temperatures increased most during June, July, and August; and maximum temperatures decreased year-round between 1900 and 2016.

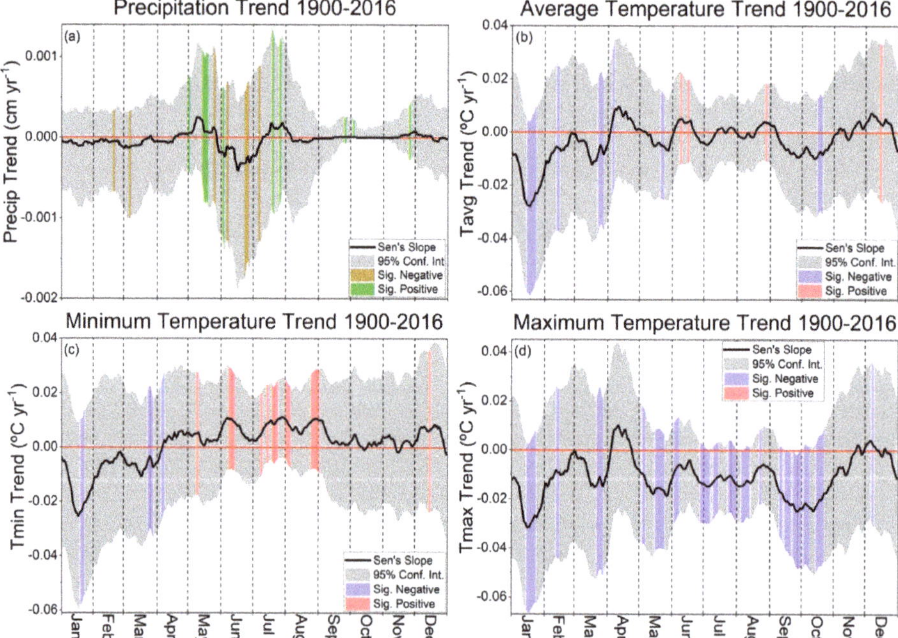

Figure 4. Centered three-week moving averages of (**a**) daily precipitation, (**b**) average temperature, (**c**) minimum temperature, and (**d**) maximum temperature trends averaged across all eighteen stations between 1900 and 2016. Grey shading represents the 95% confidence interval. Brown and green (red and blue) shading represents statistically significant ($p < 0.05$) daily precipitation (temperature) trends.

3.3.2. Second Half (1959–2016)

Between 1959 and 2016, 312 of the 365 (85%) daily precipitation trends were between ±0.001 cm year^{-1} (not shown) and twelve days (four negative; eight positive) were characterized by significant trends ($p < 0.05$; Figure 5a). Average temperatures cooled between late January and early March when five of the nine (56%) significantly negative ($p < 0.05$) daily trends occurred (Figure 5b). Average temperatures warmed at an average rate of 0.015 °C year^{-1} during June, July, and August (JJA) when eleven of the sixteen (69%) significantly positive ($p < 0.05$) daily trends occurred. Similarly, minimum temperatures warmed an average of 0.022 °C year^{-1} during JJA when thirteen of the seventeen (76%) significantly positive ($p < 0.05$) daily trends occurred (Figure 5c). Alternatively, maximum temperatures cooled an average of −0.018 °C year^{-1} between January 1st and July 1st when twenty of the twenty three (87%) significantly negative trends ($p < 0.05$) occurred. Between May and October, WV's growing season, average trends in maximum and minimum temperatures were −0.016 °C year^{-1} and 0.014 °C year^{-1} respectively. In summary, daily precipitation trends were largest between March and August; average temperatures increased most during JJA; minimum temperatures increased most during JJA; and maximum temperatures decreased between January 1st and July 1st between 1959 and 2016.

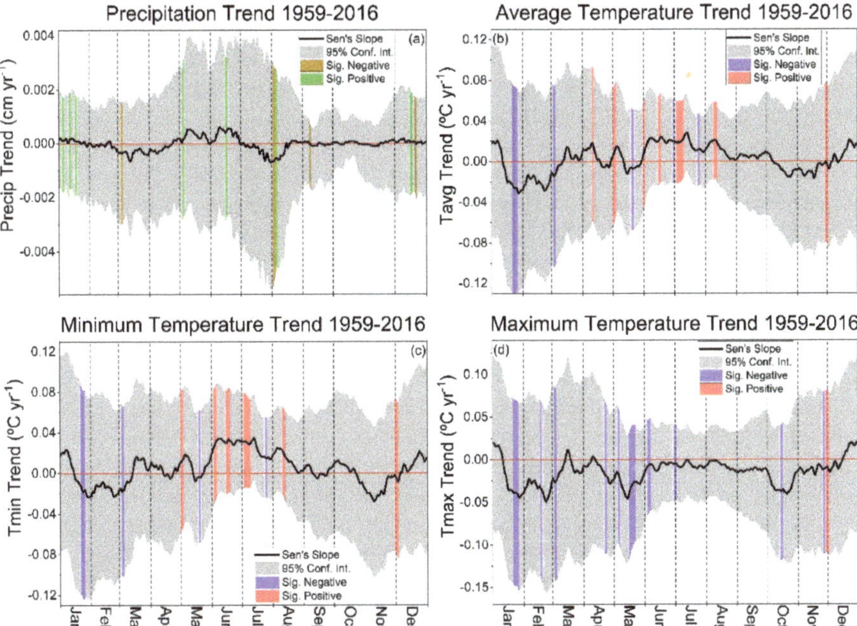

Figure 5. Centered three-week moving averages of (**a**) daily precipitation, (**b**) average temperature, (**c**) minimum temperature, and (**d**) maximum temperature trends averaged across all eighteen stations between 1959 and 2016. Grey shading represents the 95% confidence interval. Brown and green (red and blue) shading represents statistically significant ($p < 0.05$) daily precipitation (temperature) trends.

4. Discussion

Extensive LULC changes occurred across West Virginia (WV) and the broader Appalachian region between 1900 and 2016 that greatly contrast global LULC changes [26]. WV's dominant LULC transitioned from agricultural land (72%) to native forest cover (64%) between 1909 and 1949 [5]. Forest cover continued to increase in areal coverage (79%) and density (i.e., timber volume) through 2012 [6]. LULC changes influence biogeophysical regulation of climate associated with changes in albedo, turbulent fluxes of energy, mass, and momentum at the land–atmosphere interface, and the hydrologic cycle [26–29]. Simulations of coupled land–atmosphere interactions indicated that temperate forests are characterized by lower albedo and evapotranspiration values relative to irrigated croplands resulting in significantly warmer temperatures [29]. However, near surface eddy-covariance observations were compared between forest cover and an adjacent, non-irrigated grass field in nearby North Carolina, USA [30]. Results of Reference [30] showed the cooling effect of greater bulk aerodynamic conductance in forests was about three times larger than warming attributable to albedo effects. Therefore, the rapid transition from non-irrigated agricultural to forested land uses in WV may have resulted in a net cooling effect.

Net cooling in WV was characterized by decreasing state-average maximum temperatures (−1.0 °C) and increasing minimum temperatures (+0.4 °C) between 1900 and 2016 [9], but spatial and seasonal changes were not addressed. At the regional scale, changes in maximum and minimum temperatures contrasted the reported warming of 0.64 °C and 0.94 °C across the broader Northeast US region that included WV, respectively [31]. However, temperature trends in WV were more consistent with the Southeast US region during the period of study where maximum and minimum temperatures increased 0.16 °C and 0.76 °C, respectively [31]. In WV, minimum temperatures increased more at lower elevation locations (Table 1) where old-growth hardwood forests were harvested, conceivably with relatively less effort than WV's rugged mountainous terrain [2]. Between 1900 and 2016, increasing daily minimum temperature trends were consistently positive during JJA when 84% of significant ($p < 0.05$) increasing daily trends occurred. Similarly, 76% of the seventeen significantly positive ($p < 0.05$) daily average trends between 1959 and 2016 occurred during JJA. Warming summertime minimum temperatures in WV's humid subtropical climate suggests warmer dew point temperatures that may alter dewfall dynamics [32] and contribute to WV's increasingly humid climate [9]. Maximum temperatures cooled significantly ($p < 0.05$) on 72 days between 1900 and 2016 (Figure 4) and annual trends were larger at higher elevations between 1959 and 2016 (Table 2) where forest cover may be less disturbed and more mature [3]. Cooling high-elevation maximum temperatures may be explained by topographically induced circulations focusing afternoon cloud development along high-elevation ridges [33] that could be more frequent in an increasingly humid climate [9]. Further investigation into the mechanism(s) responsible for differential temperature trends at low and high elevation locations is needed, but warmer minimum and cooler maximum temperatures are consistent with attenuation of the diurnal temperature range due to forest cover [34] or cloud cover [35]. In addition to LULC changes in WV, increased forest coverage and maturity, and agricultural intensification (i.e., fertilization and irrigation; [36]) upwind of WV (i.e., Midwestern US) likely contributed to WV's increasingly humid WV climate [37]. Additional contributing mechanisms include (but are not limited to) increases in globally averaged atmospheric moisture associated with warming air and sea surface temperatures [38] or regional precipitation recycling [39]. Irrespective of the forcing mechanism(s), increased cloud cover aligns with an increasingly wet, temperate, and humid WV climate [35] and suggests regional climate forcing associated with coupled land–atmosphere interactions exceeded radiative forcing in WV (i.e., greenhouse gases; [40]).

Extrapolating annual trends over the entire POR then averaging across all eighteen locations indicated total annual precipitation increased by 3.8%, which is consistent with the national average reported in the Fourth National Climate Assessment [31]. However, precipitation increased at all locations between 1959 and 2016 when linear trends averaged across all sites indicated a 14.2 cm (13.2%) increase in precipitation relative to the epoch average. This precipitation trend was more than triple the rate of increase during the entire POR suggesting precipitation trends may be accelerating as a result of global scale increases in water vapor content [41] and super Clausius–Clapeyron scaling (SCCS) of precipitation extremes [42]. Significant ($p < 0.05$) increasing precipitation trends between 1959 and 2016 were estimated at lower elevation sites across western WV where forest harvesting, regrowth, and maturation was widespread [2] implying greater feedbacks associated with LULC change. The methods documented in the current work could be used to address the need for data-driven assessments of feedbacks between climate and LULC change in many locations, globally [43]. Understanding of LULC feedbacks on precipitation are particularly important in WV where some of the most extreme rainfall accumulations in the world were observed at time scales less than 6 h [44]. Thus, WV may be particularly vulnerable to changes in the frequency and magnitude (i.e., SCCS) of precipitation extremes and the impact of LULC changes (e.g., reforestation) on precipitation extremes and SCCS should be investigated.

The combination of an increasingly wet and temperate climate should help to secure WV's water resource quantity, but water quality problems exist in WV including (but not limited to) LULC change [45], acid mine drainage [46], and pathogenic water contamination [47]. Additionally, WV's increasingly temperate climate combined with decreasing temperature variance [9] may prolong episodes of excessive soil wetness and humid conditions that increases the vulnerability of terrestrial ecosystems to pathogens and fungal-like oomycetes such as *Phytophtora* and *Pythium* [48]. Similarly, increasing precipitation variance [9] indicates wet years are becoming more extreme, which may exacerbate vulnerabilities associated with pests and pathogens [49–53]. As a result, further investigation into changing climate variability may help to better explain observed changes in ecosystem biodiversity, carbon dynamics, and hydrology at a range of spatiotemporal scales [54]. However, increasing average precipitation (i.e., freshwater availability) signals potential for increasing agricultural productivity to ameliorate the food desert crisis in WV through conservation practices sponsored by the United States Department of Agriculture's (USDA) Environmental Quality Incentives Program (EQIP; [55]). In particular EQIP's High Tunnel System Initiative allows farmers to more efficiently deliver water and nutrients to plants, better protect plants from pests and pathogens, and minimize LULC change by increasing productivity per unit land area. Ultimately, climatic changes in WV may be distinct from the Northeast US climate region and Appalachia's complex topography may therefore necessitate refined climatic regions based on climate trends observed over a long POR and physiographic properties [56], including (but not limited to) elevation, slope and aspect, historic LULC changes, and forest age and composition.

5. Conclusions

Long term changes in West Virginia's LULC associated with widespread deforestation in the early 20th century, rapid forest regrowth through the mid-20th century, and subsequent forest maturation resulted in climatic changes distinct from the broader Northeast US climate region. Overall climatic trends averaged across the state suggest an increasingly temperate climate with increasing summertime minimum temperatures, particularly at lower elevations, and decreasing maximum temperatures year-round, particularly at higher elevations. Total annual precipitation is increasing and trends appear to be accelerating, particularly across the lower elevations of western WV where forest regrowth and maturation was particularly widespread. The mechanisms forcing observed changes in WV climate require much more investigation. This article therefore serves as an alert for greatly needed climate investigations in the Appalachian region to provide science-based guidance for future land-use/ land management activities in the region. This is critical since a wetter and more temperate climate may improve some ecosystems services (i.e., water security) while others may become increasingly stressed by increasing vulnerability to invasive pests and pathogens. Additionally, observed climatic changes support increasing agricultural productivity to address the food desert crisis through sustainable agricultural practices that minimize LULC changes (i.e., high tunnel systems). Therefore, climate observations over a long period of record should be used to identify spatiotemporal patterns of dynamic climatic changes at local to regional scales to better understand climatic changes, inform decision makers, and ensure preservation of WV's rich natural resources.

Author Contributions: Conceptualization, J.H. and E.K.; methodology, E.K. and J.H.; formal analysis, E.K.; investigation, E.K.; data curation, E.K.; writing—original draft preparation, E.K.; writing—review and editing, J.H.; visualization, E.K. and J.H.; supervision, J.H.; project administration, J.H.; funding acquisition, J.H.

Funding: This research was funded by the National Science Foundation, award number OIA-1458952, The United States Department of Agriculture's National Institute of Food and Agriculture, Hatch project 1011536, and the West Virginia Agricultural and Forestry Experiment Station.

Acknowledgments: Additional thanks are due to feedback from anonymous reviews and administrative support from Laura Tinney and Felix Greiner.

Conflicts of Interest: The authors declare no conflict of interest. The funders had no role in the design of the study; in the collection, analyses, or interpretation of data; in the writing of the manuscript, or in the decision to publish the results.

References

1. Lewis, R.L. Appalachian Restructuring in Historical Perspective: Coal, Culture and Social Change in West Virginia. *Urban. Stud.* **1993**, *30*, 299–308. [CrossRef]
2. Lewis, R.L. *Transforming the Appalachian Countryside Railroads, Deforestation, and Social Change in West Virginia*; The University of North Carolina Press: Chapel Hill, NC, USA, 1998; ISBN 9780807847060.
3. Brooks, A.B. *West Virginia Geological Survey, Vol. 5: Forestry and Wood Industries*; Acme Publishing Co.: Morgantown, WV, USA, 1911.
4. Brose, P.; Schuler, T.; Van Lear, D.; Berst, J. Bringing fire back: The changing regimes of the Appalachian mixed-oak forests. *J. For.* **2001**, *99*, 30–35. [CrossRef]
5. Bones, J.T. *The Forest Resources of West Virginia*; Resource Bulletin Me-56; U.S. Department of Agriculture, Forest Service, Northeastern Forest Experiment Station: Broomall, PA, USA, 1978. [CrossRef]
6. Morin, R.S.; Cook, G.W.; Barnett, C.J.; Butler, B.J.; Crocker, S.J.; Hatfield, M.A.; Kurtz, C.M.; Lister, T.W.; Luppold, W.G.; McWilliams, W.H.; et al. *West Virginia Forests 2013*; Resource Bulletin NRS-105; U.S. Department of Agriculture, Forest Service, Northeastern Forest Experiment Station: Newtown Square, PA, USA, 2016. [CrossRef]
7. Burakowski, E.A.; Ollinger, S.V.; Bonan, G.B.; Wake, C.P.; Dibb, J.E.; Hollinger, D.Y. Evaluating the climate effects of reforestation in new England using a weather research and forecasting (WRF) model multiphysics ensemble. *J. Clim.* **2016**, *29*, 5141–5156. [CrossRef]
8. Campbell, G.; Norman, J. An. *Introduction to Environmental Biophysics*; Springer Science & Business Media, Inc.: New York, NY, USA, 1998; ISBN 0-387-94937-2.
9. Kutta, E.; Hubbart, J.A. Changing climatic averages and variance: Implications for mesophication at the eastern edge of North America's Eastern deciduous forest. *Forests* **2018**, *9*. [CrossRef]
10. Hoffmann, M.P.; Smith, D.L. Feeding Our Great Cities: Climate Change and Opportunities for Agriculture in Eastern Canada and the Northeastern US. Available online: http://www2.ca.uky.edu/Anr/PDF/US-CanadaClimateChangeWhitePaper.pdf (accessed on 14 December 2018).
11. Darvill, R.; Lindo, Z. The inclusion of stakeholders and cultural ecosystem services in land management trade-off decisions using an ecosystem services approach. *Landsc. Ecol.* **2016**, *31*, 533–545. [CrossRef]
12. Gómez-Aíza, L.; Martínez-Ballesté, A.; Álvarez-Balderas, L.; Lombardero-Goldaracena, A.; García-Meneses, P.M.; Caso-Chávez, M.; Conde-Álvarez, C. Can wildlife management units reduce land use/land cover change and climate change vulnerability? Conditions to encourage this capacity in Mexican municipalities. *Land Use Policy* **2017**, *64*, 317–326. [CrossRef]
13. United States Department of Agriculture Food and Nutrition Service. Food Desert Locator: Release No. 0191.11. Available online: https://www.usda.gov/media/blog/2011/05/03/interactive-web-tool-maps-food-deserts-provides-key-data (accessed on 15 January 2018).
14. Arguez, A.; Durre, I.; Applequist, S.; Squires, M.; Vose, R.; Yin, X. NOAA's U.S. Climate Normals (1981–2010). Available online: https://www.ncdc.noaa.gov/cdo-web/datatools/normals (accessed on 14 August 2018).
15. United States Department of Agriculture Food and Nutrition Service. Monongahela National Forest Land and Resource Management Plan. Available online: https://www.fs.usda.gov/Internet/FSE_DOCUMENTS/stelprdb5330420.pdf (accessed on February 15 2018).
16. Edenborn, H.M.; Sams, J.I.; Kite, J.S. Thermal Regime of a Cold Air Trap in Central Pennsylvania, USA: The Trough Creek Ice Mine. *Permafr. Periglac. Process.* **2012**, *23*, 187–195. [CrossRef]
17. NOAAs NCEI Climate Data Online. Available online: https://www.ncdc.noaa.gov/ghcn-daily-description (accessed on October 6 2017).

18. *The Role of Climatological Normals in a Changing Climate*; World Meteorological Organization (WMO): Geneva, Switzerland, 2007.
19. Allen, J. NASA Image based on data from the MODIS Rapid Response Team, Goddard Space Flight Center. Available online: https://earthobservatory.nasa.gov/images/8344/mining-permits-across-west-virginia (accessed on 14 March 2018).
20. Lanzante, J.R. Resistant, robust and non-parametric techniques for the analysis of climate data: Theory and examples, including applications to historical radiosonde station data. *Int. J. Climatol.* **1996**, *16*, 1197–1226. [CrossRef]
21. Wilks, D.S. *Statistical Methods in the Atmospheric Sciences*; Academic Press: Cambridge, MA, USA, 2011; ISBN 9780444515544.
22. Gocic, M.; Trajkovic, S. Analysis of changes in meteorological variables using Mann-Kendall and Sen's slope estimator statistical tests in Serbia. *Glob. Planet. Chang.* **2013**, *100*, 172–182. [CrossRef]
23. Hayhoe, K.; Wuebbles, D.J.; Easterling, D.R.; Fahey, D.W.; Doherty, S.; Kossin, J.P.; Sweet, W.V.; Vose, R.S.; Wehner, M.F. Chapter 2: Our Changing Climate. In *Impacts, Risks, and Adaptation in the United States: The Fourth National Climate Assessment, Volume II*. Available online: https://data.globalchange.gov/report/nca4/chapter/our-changing-climate (accessed on 31 December 2018).
24. Kutta, E.; Hubbart, J.A. Reconsidering meteorological seasons in a changing climate. *Clim. Chang.* **2016**, *137*, 511–524. [CrossRef]
25. Peel, M.C.; Finlayson, B.L.; McMahon, T.A. Updated world map of the Köppen-Geiger climate classification. *Hydrol. Earth Syst. Sci.* **2007**, *11*, 1633–1644. [CrossRef]
26. Bonan, G.B. Forests and climate change: Forcings, feedbacks, and the climate benefits of forests. *Science* **2008**, *320*, 1444–1449. [CrossRef]
27. Pitman, A.J. Review the Evolution of and Revolution in Land Surface Schemes. *Int. J. Climatol.* **2003**, *510*, 479–510. [CrossRef]
28. Hubbart, J.A.; Link, T.E.; Gravelle, J.A.; Elliot, W.J. Timber harvest impacts on water yield in the continental/maritime hydroclimatic region of the United States. *For. Sci.* **2007**, *53*, 169–180.
29. Pitman, A.J.; Brovkin, V.; van den Hurk, B.J.J.M.; Strengers, B.J.; Ganzeveld, L.; Gayler, V.; Claussen, M.; Müller, C.; Seneviratne, S.I.; Bonan, G.B.; et al. Uncertainties in climate responses to past land cover change: First results from the LUCID intercomparison study. *Geophys. Res. Lett.* **2009**, *36*. [CrossRef]
30. Juang, J.-Y.; Stoy, P.; Novick, K.; Katul, G.; Siqueira, M. Separating the effects of albedo from eco-physiological changes on surface temperature along a successional chronosequence in the southeastern United States. *Geophys. Res. Lett.* **2007**, *34*. [CrossRef]
31. U.S. Global Change Research Program—Executive summary. In *Climate Science Special Report: Fourth National Climate Assessment, Volume I*. Available online: https://science2017.globalchange.gov/ (accessed on 14 December 2018).
32. Richards, K. Urban and rural dewfall, surface moisture, and associated canopy-level air temperature and humidity measurements for Vancouver, Canada. *Boundary-Layer Meteorol.* **2005**, *114*, 143–163. [CrossRef]
33. Jr Houze, R.A. Orographic effects on precipitating clouds. *Rev. Geophys.* **2012**, *50*, 1–47. [CrossRef]
34. Zheng, D.; Chen, J.; Song, B.; Xu, M.; Sneed, P.; Jensen, R. Effects of silvicultural treatments on summer forest microclimate in southeastern Missouri Ozarks. *Clim. Res.* **2000**, *15*, 45–59. [CrossRef]
35. Dai, A.; Trenberth, K.E.; Karl, T.R. Effects of clouds, soil moisture, precipitation, and water vapor on diurnal temperature range. *J. Clim.* **1999**, *12*, 2451–2473. [CrossRef]
36. Mueller, N.D.; Butler, E.E.; Mckinnon, K.A.; Rhines, A.; Tingley, M.; Holbrook, N.M.; Huybers, P. Cooling of US Midwest summer temperature extremes from cropland intensification. *Nat. Clim. Chang.* **2016**, *6*, 317–322. [CrossRef]
37. Kutta, E.; Hubbart, J.A. Observed climatic changes in West Virginia and opportunities for agriculture. *Reg. Environ. Chang.* **2019**. [CrossRef]
38. Held, I.M.; Soden, B.J. Robust responses of the hydrological cycle to global warming. *J. Clim.* **2006**, *19*, 5686–5699. [CrossRef]
39. Dirmeyer, P.A.; Brubaker, K.L. Characterization of the Global Hydrologic Cycle from a Back-Trajectory Analysis of Atmospheric Water Vapor. *J. Hydrometeorol.* **2007**, *8*, 20–37. [CrossRef]
40. Deser, C.; Phillips, A.S.; Alexander, M.A.; Smoliak, B.V. Projecting North American climate over the next 50 years: Uncertainty due to internal variability. *J. Clim.* **2014**, *27*, 2271–2296. [CrossRef]

41. Kevin, E.T.; Aiguo, D.; Roy, M.R.; David, B.P. The Changing Character of Precipitation. *Bull. Am. Meteorol. Soc.* **2003**, *84*, 1205–1218. [CrossRef]
42. Lenderink, G.; Barbero, R.; Loriaux, J.M.; Fowler, H.J. Super-Clausius-Clapeyron scaling of extreme hourly convective precipitation and its relation to large-scale atmospheric conditions. *J. Clim.* **2017**, *30*, 6037–6052. [CrossRef]
43. Alkama, R.; Cescatti, A. Climate change: Biophysical climate impacts of recent changes in global forest cover. *Science* **2016**, *351*, 600–604. [CrossRef] [PubMed]
44. Villarini, G.; Steiner, M.; Smith, J.A.; Ntelekos, A.A.; Baeck, M.L. Extreme rainfall and flooding from orographic thunderstorms in the central Appalachians. *Water Resour. Res.* **2011**, *47*, 1–24. [CrossRef]
45. Kellner, E.; Hubbart, J.A. Flow class analyses of suspended sediment concentration and particle size in a mixed-land-use watershed. *Sci. Total Environ.* **2019**, *648*, 973–983. [CrossRef]
46. Zipper, C.E.; Barton, C.B.; Davis, V.; Franklin, J.A.; Burger, J.A.; Skousen, J.G.; Angel, P.N. Restoring Forests and Associated Ecosystem Services on Appalachian Coal Surface Mines [electronic resource]. *Environ. Manage.* **2011**, *47*, 751–765. [CrossRef]
47. Petersen, F.; Hubbart, J.A.; Kellner, E.; Kutta, E. Land-use-mediated Escherichia coli concentrations in a contemporary Appalachian watershed. *Environ. Earth Sci.* **2018**, *77*. [CrossRef]
48. Hubbart, J.A.; Guyette, R.; Muzika, R.M. More than Drought: Precipitation Variance, Excessive Wetness, Pathogens and the Future of the Western Edge of the Eastern Deciduous Forest. *Sci. Total Environ.* **2016**, *566*, 463–467. [CrossRef]
49. Anagnostakis, S.L. Chestnut Blight: The Classical Problem of an Introduced Pathogen. *Mycologia* **1987**, *79*, 23. [CrossRef]
50. Cale, J.A.; Teale, S.A.; Johnston, M.T.; Boyer, G.L.; Perri, K.A.; Castello, J.D. New ecological and physiological dimensions of beech bark disease development in aftermath forests. *For. Ecol. Manage.* **2015**, *336*, 99–108. [CrossRef]
51. Kauffman, B.W.; Clatterbuck, W.K.; Liebhold, A.M.; Coyle, D.R. *Gypsy Moth in the Southeastern U.S.: Biology, Ecology, and Forest Management Strategies*; SREF-FH-008; Southern Regional Extension Forestry: Athens, GA, USA, 2017.
52. Herms, D.A.; McCullough, D.G. Emerald Ash Borer Invasion of North America: History, Biology, Ecology, Impacts, and Management. *Annu. Rev. Entomol.* **2014**, *59*, 13–30. [CrossRef]
53. Kim, J.; Hwang, T.; Schaaf, C.L.; Orwig, D.A.; Boose, E.; Munger, J.W. Increased water yield due to the hemlock woolly adelgid infestation in New England. *Geophys. Res. Lett.* **2017**, *44*, 2327–2335. [CrossRef]
54. Greenland, D.; Goodin, D.G.; Smith, R.C. *Climate Variability and Ecosystem Response at Long-Term Ecological Research Sites*; Oxford University Press, Inc.: Oxford, UK, 2003; ISBN 0195150597.
55. United States Department of Agriculture. High Tunnel System Initiative. Available online: https://www.nrcs.usda.gov/wps/portal/nrcs/detailfull/national/programs/?cid=stelprdb1046250 (accessed on March 14 2018).
56. Raitz, K.B.; Ulack, R. *Appalachia A regional geography Land, People, and Development*; Westview Press, Inc.: Boulder, CO, USA, 1984.

Article

Twenty-First Century Streamflow and Climate Change in Forest Catchments of the Central Appalachian Mountains Region, US

Brandi Gaertner [1], Rodrigo Fernandez [2] and Nicolas Zegre [2,*]

[1] Environmental Science Department, Alderson Broaddus University, 101 College Hill Drive, Philippi, WV 26416, USA; gaertnerba@ab.edu

[2] Forestry & Natural Resources, West Virginia University, 334 Percival Hall, Morgantown, WV 26506, USA; rodrigofernrey@gmail.com

* Correspondence: nicolas.zegre@mail.wvu.edu; Tel.: +1-304-293-0049

Received: 30 November 2019; Accepted: 3 February 2020; Published: 8 February 2020

Abstract: Forested catchments are critical sources of freshwater used by society, but anthropogenic climate change can alter the amount of precipitation partitioned into streamflow and evapotranspiration, threatening their role as reliable fresh water sources. One such region in the eastern US is the heavily forested central Appalachian Mountains region that provides fresh water to local and downstream metropolitan areas. Despite the hydrological importance of this region, the sensitivity of forested catchments to climate change and the implications for long-term water balance partitioning are largely unknown. We used long-term historic (1950–2004) and future (2005–2099) ensemble climate and water balance data and a simple energy–water balance model to quantify streamflow sensitivity and project future streamflow changes for 29 forested catchments under two future Relative Concentration Pathways. We found that streamflow is expected to increase under the low-emission pathway and decrease under the high-emission pathway. Furthermore, despite the greater sensitivity of streamflow to precipitation, larger increases in atmospheric demand offset increases in precipitation-induced streamflow, resulting in moderate changes in long-term water availability in the future. Catchment-scale results are summarized across basins and the region to provide water managers and decision makers with information about climate change at scales relevant to decision making.

Keywords: climate change; Appalachian Mountains; streamflow sensitivity; water security; water balance partitioning; Budyko

1. Introduction

Forested headwater catchments play a critical role in provisioning freshwater to humanity [1–3], but anthropogenic climate change can alter the amount of precipitation (P) partitioned into streamflow (Q), evapotranspiration (E), and storage [4]. Changes in Q from headwater catchments is of particular concern given the importance of mountain regions in generating fresh water [5,6], which is used by society for drinking and water-intensive production [7–9]. One such headwaters area in the eastern US is the central Appalachian Mountains region (Figure 1). This region provides water to local metropolitan areas including Pittsburgh, Pennsylvania; Charlotte, North Carolina; and Memphis, Tennessee and to downstream cities such as Washington, DC; Cincinnati, Ohio; Louisville, Kentucky; Atlanta, Georgia; and New Orleans, Louisiana [10]. Given the hydrological importance of this region, gaining an insight into the potential impacts of anthropogenic climate change on future Q is critical for developing policies and practices that enhance water security throughout the region [11,12].

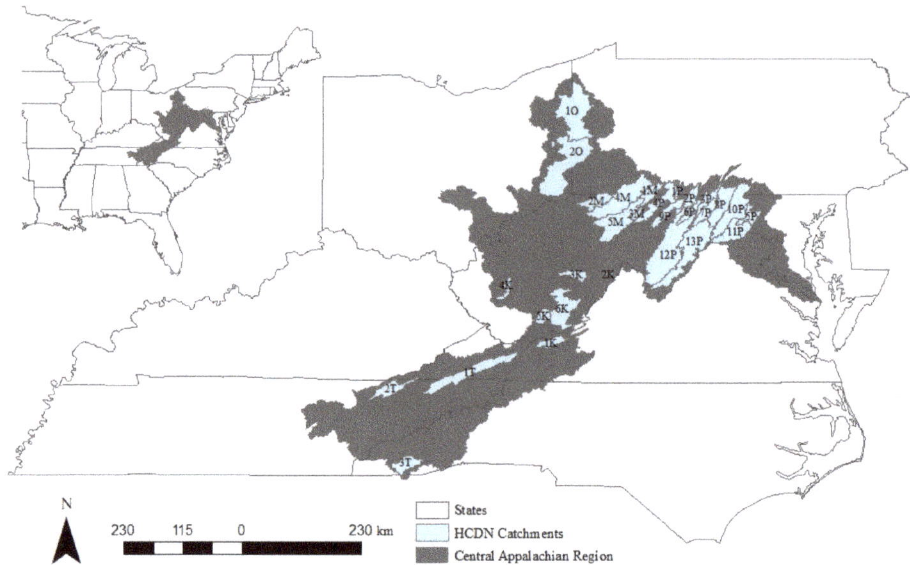

Figure 1. Location map of the U.S. Geological Survey Hydro-Climatic Data Network (HCDN) catchments studied in the central Appalachian Mountains region.

Water availability at the land surface is the result of the balance between atmospheric water demand [12–14], described here as potential evapotranspiration (PET) and P. Atmospheric water demand is driven primarily by solar radiation, which supplies the energy required to vaporize liquid water [15,16]. The ratio between PET and P, called the aridity index (AI), describes how much P is partitioned and returned back to the atmosphere [17–19]. AI is a first-order hydro-climatological control since it describes the absolute amount of water that can be lost to evaporation.

Over the last several decades, this region has become warmer and wetter [20,21], theoretically consistent with water cycle intensification [22]. Air temperatures have increased, on average, by 1 °C, or 0.09 °C/decade [23]. Precipitation has increased, on average, by ~10 mm/decade [24], as have heavy downpours (events that are exceeded 1% of the time in any given year) which have increased by 71% in the Northeast and 37%in the Midwest over the last three to five decades [25]. In West Virginia, a small state that is geographically central to the region and representative of the topographical, ecological, and climatological gradients of the area [26], minimum and maximum air temperatures have increased by 0.15 °C and 0.03 °C/decade, respectively, while P has increased by 25 mm/decade since 1959 [26]. During the last half of the 20th century, mean annual Q, low flow, and base flow have all increased throughout the larger mid-Atlantic region [27–29]. Since 1951, the monthly Q throughout the region has increased by 1 to 2 mm/year, with the largest increases during the fall and winter seasons due to increases in precipitation [29]. Coincidentally, the forest growing season length throughout the region has increased on average by 22 days since 1983 [30] as has E, which has increased between 0.4 to 4 mm/year [30–32].

Fernandez and Zegre [16] quantified future changes in P and energy balance components for 420 counties and 13 states that comprise the Appalachian region, USA. The study found that both PET and P are projected to increase through the 21st century. P has been projected to become more variable through time; under Relative Concentration Pathways (RCP) 4.5 [33], where global atmospheric CO_2 emissions are stabilized by mid-century, P is projected to increase during the first quarter of the 21st century (2006–2025), progressively decrease through mid-21st century (2026–2075), and then increase again late in the century (2076–2099). Under RCP8.5, where global atmospheric CO_2 continues to increase unabated, P is projected to decrease early in the century, progressively increase through the mid-century, and continue to increase in the northern extent and decrease in the southern extent of the Appalachians. PET, on the other hand, is projected to progressively increase throughout the 21st century by 5%–20% (RCP4.5) and 20%–35% (RCP8.5)—a rate greater than P [16]. AI is projected to increase at higher elevations, shrinking the areas of headwaters [16] that have historically been characterized as energy limited (i.e., P exceeds PET) [34,35] and are critical for maintaining downstream water availability [2,36]. Furthermore, P and PET are projected to become seasonally desynchronized, with P increasing during the dormant season and PET increasing during the growing season. Collectively, these changes have the potential to increase the occurrences of floods and droughts throughout the region [37–40], increasing the challenges of managing water resources under greater uncertainty and risk [36].

Despite the importance of this region as a freshwater source to local and downstream ecosystems and economies [10,39,41], the sensitivity of forested catchments to climate change, and the implications for long-term water balance partitioning, are largely unknown. Previous studies have focused on single catchments, mixed-use catchments, or larger regional scales (e.g., mid-Atlantic, mid-western, northeastern US; Chesapeake Bay) [37–39,42,43] but the uniqueness of place and variations in topography, geography, land use, and landcover limit inference for catchment-scale decision making.

In this study, we quantified water balance sensitivity to changes in future climate using a simple Budyko-based energy model [14] to provide insight into how fresh water provisioning from forested catchments throughout the region might change due to anthropogenic climate change. Specifically, we quantified streamflow changes at the catchment scale and summarized them across basins and the region to provide watershed and natural resource managers and decision makers with information about climate change at scales relevant to decision making.

The objectives of this study are therefore to (1) quantify streamflow sensitivity to historical changes in climate in minimally impacted catchments; (2) quantify future climate change at the catchment-scale throughout the region; (3) model future Q under two Relative Concentration Pathways to understand how future climate change might impact water resources availability throughout the region; and (4) evaluate spatial patterns of future streamflow sensitivity to climate.

2. Materials and Methods

2.1. Data

Mean daily streamflow were extracted for 29 catchments located across five dominant river basins across the region (Figure 1): the Monongahela, Upper Ohio, Kanawha, and Tennessee Rivers, which drain west to the Mississippi River and Gulf of Mexico, and the Potomac River, which drains east to Washington D.C. and the Chesapeake Bay. The 29 catchments represent 39% of the total area of the five river basins, with five catchments in the Monongahela, two in the Upper Ohio, six catchments in the Kanawha, three in the Tennessee, and 13 in the Potomac Basin (Figure 1). This largely forested region covers approximately 125,000 km^2 and spans across eight states: West Virginia (WV), eastern Ohio (OH), southwestern Pennsylvania (PA), eastern Kentucky (KY), western Virginia (VA), western North Carolina (NC), Maryland (MD), and western Tennessee (TN). Criteria for selecting the catchments were based on (1) inclusion in the U.S. Geological Survey Hydro-Climatic Data Network (HCDN) [44,45] and (2) a representative range of catchment areas (34–8101 km^2) (Table 1). The HCDN consists of streamflow gaging station data for minimally impacted catchments (<10% human influence, e.g., reservoirs, diversion, land use change, or severe ground-water pumping) identified as a subset of the national-scale USGS gauging stations and developed specifically for the purpose of studying the variation in surface-water conditions throughout the United States [45]. The HCDN data set has been used in many hydro climatic studies [28,46–51]. We extracted Q data from the national USGS Water Data dataset (https://waterdata.usgs.gov/nwis) from 1965–2004. While the World Meteorological Organization (WMO) recommends 30 years of continuous data to calculate "climate normals" [52], recent Budyko-based studies have used shorter time periods (e.g., 5–27 years) [51,53], based on the requirement of steady-state water balance conditions. [51].

A regional land cover analysis using the 2011 National Land Cover Database (NLCD) [54] was used to assess whether catchments still met the definition of being "minimally impacted" in the extended analysis period beyond the original NLCD. Q data were normalized by area (km^2), converted to millimeters per year, and averaged to annual values using the USGS water-year (1 October–30 September).

Historic and future air temperature and P were extracted from the Multivariate Adaptive Constructed Analogs version 2 (MACAv2-METDATA) dataset [55]. MACA data are downscaled and bias corrected from Global Climate Models (GCMs) to a spatial resolution of gridded 4 km using non-parametric-quantile mapping and a constructed Analogs method [16,56]. The gridded climate values were averaged within the watershed boundaries. MACA only developed RCPs for 4.5 and 8.5, excluding 2.6 and 6.0 [3,57]. RCP4.5 represents a scenario in which greenhouse gas emissions are stabilized by the mid-21st century [57]. In RCP8.5, greenhouse gas emissions continue to rise with business as usual throughout the 21st century. In this study, 17 GCM models included in MACA were used to create ensemble climate data for the historic (1965–2004) period and for four approximate quarter-century periods of the 21st century: quarter 1, 2006–2025; quarter 2, 2026–2050; quarter 3, 2051–2075; and quarter 4, 2076–2099. Historical and future potential evapotranspiration (PET) were calculated for the region based on the Penman–Montieth method combined with the regional MACA dataset [16]. E was estimated using the annual water balance, $E = P - Q + \Delta S$, based on the USGS water year, where P is the long-term average annual precipitation, Q is the long-term average streamflow, and ΔS is watershed storage, which is assumed to approach zero over a long time period.

GCMs have known uncertainties and statistical biases that limit their direct use for hydrological assessments especially at short time scales, such as daily to weekly [58–60]. To tackle the issue of statistical biases, several correction techniques have been developed to adjust modelled variables to observations [61]. The MACA dataset has been developed using the data from CMIP 5 models and using a comprehensive six step downscaling and bias correcting approach and has been validated against observations [56]. The information from GCMs can also be informative if applied for long-term assessments [62]. Since our sensitivity analysis based on the Budyko model is a long-term water balance analysis, we considered the MACA dataset a viable tool to analyze possible change scenarios for the study catchments.

2.2. *Quantifying Historical Streamflow Sensitivity to Climate*

Streamflow sensitivity [35] is an elasticity-based method [63] based on the formulation of the Budyko equation [14] by Choudhury [64]. It includes the adjustable parameter, n, which accounts for non-climatic factors that influence E, such as slope, vegetation, geology, and soils using catchment-specific E variability [65]. The n value is calculated from the intersection of the AI (PET/P) and evaporative index (ET/P) and plotted in the Budyko space based on the specific catchment parameters. The Budyko analysis incorporates intrinsic limitations due to the simplistic design structure and few required datasets. However, the approach is useful because it uses available data, and the macro-climatological processes within the Budyko analysis has been shown to meet the requirements for the long-term water balance and climatologic processes [66,67].

$$E = \frac{E \times PET}{(P^n + PET^n)^{1/n}} \quad (1)$$

where E is long-term evapotranspiration, PET is long-term potential evapotranspiration, P is long-term precipitation, and n is a dimensionless parameter describing catchment properties that modify the partitioning of P into E and Q.

Table 1. Historical (1950–2004) water balance, Budyko, and streamflow sensitivity coefficients for selected USGS gaging stations and HCDN catchments in the central Appalachian Mountains region, US. P: precipitation; Q: streamflow; E: evapotranspiration; PET: potential evapotranspiration.

Basin/Station Name											Sensitivity Coefficeints		
	ID	Station Number	Area	P	Q	E	PET	PET/P	E/P	n	ΔP/P	ΔPET/PET	Δn/n
			km^2	mm	mm	mm	mm	-	-	-	-	-	-
Monongahela													
Casselman River at Grantsville, MD	1M	3078000	163	1172	678	494	1228	1.05	0.42	0.78	1.36	−0.36	0.65
West Fork River at Enterprise, WV	2M	3061000	1965	1170	529	641	1310	1.12	0.55	1.06	1.57	−0.57	0.79
Youghiogheny River near Oakland, MD	3M	3075500	347	1275	824	451	1203	0.94	0.35	0.69	1.28	−0.28	0.55
Laurel Hill Creek at Ursina, PA	4M	3080000	313	1275	778	497	1203	0.94	0.39	0.76	1.33	−0.33	0.58
Cheat River near Parsons, WV	5M	3069500	1869	1329	870	459	1215	0.91	0.35	0.68	1.27	−0.27	0.54
Basin average			932	1244	736	508	1232	0.99	0.41	0.79	1.36	−0.36	0.62
Ohio													
Little Shenango River at Greenville, PA	1O	3102500	269	1012	490	522	1253	1.24	0.52	0.91	1.48	−0.48	0.81
Little Beaver Creek near East Liverpool, OH	2O	3109500	1284	1047	376	671	1316	1.26	0.64	1.26	1.76	−0.76	0.97
Basin average			777	1029	433	596	1285	1.25	0.58	1.08	1.62	−0.62	0.89
Kanawha													
Wolf Creek near Narrows, VA	1K	3175500	577	1002	481	520	1414	1.41	0.52	0.85	1.46	−0.46	0.87
Greenbrier River at Durbin, WV	2K	3180500	344	1246	719	527	1187	0.95	0.42	0.83	1.37	−0.37	0.61
Williams River at Dyer, WV	3K	3186500	331	1479	949	530	1224	0.83	0.36	0.75	1.30	−0.30	0.52
Big Coal River at Ashford, WV	4K	3198500	1012	1196	490	706	1376	1.15	0.59	1.17	1.66	−0.66	0.85
Bluestone River at Pipestem, WV	5K	3179000	1020	1018	423	595	1381	1.36	0.58	1.02	1.59	−0.59	0.94
Greenbrier River at Alderson, WV	6K	3183500	3531	1080	520	560	1304	1.21	0.52	0.93	1.49	−0.49	0.80
Basin average			1136	1170	597	573	1314	1.15	0.50	0.92	1.48	−0.48	0.76
Tennessee													
North Fork Holston River near Saltsville, VA	1T	3488000	572	1155	481	674	1435	1.24	0.58	1.08	1.62	−0.62	0.89
Clinch River above Tazewell, TN	2T	3528000	3816	1275	485	791	1433	1.12	0.62	1.30	1.75	−0.75	0.87
Little Tennessee River near Prentiss, NC	3T	3500000	362	1695	960	735	1443	0.85	0.43	0.92	1.41	−0.41	0.57
Basin average			1584	1375	642	733	1437	1.07	0.55	1.10	1.59	−0.59	0.78

Table 1. *Cont.*

Basin/Station Name	ID	Station Number	Area	P	Q	E	PET	PET/P	E/P	n	Sensitivity Coefficeints		
											ΔP/P	ΔPET/PET	Δn/n
Potomac													
Wills Creek near Cumberland, MD	1P	1601500	639	988	494	494	1352	1.37	0.50	0.82	1.44	−0.44	0.83
Pototmac River near at Paw Paw, WV	2P	1610000	8101	988	398	591	1352	1.37	0.60	1.05	1.62	−0.62	0.96
Cacapon River near Great Cacapon, WV	3P	1611500	1748	998	321	677	1387	1.39	0.68	1.29	1.84	−0.83	1.09
Patterson Creek near Headsville, WV	4P	1604500	572	1071	283	788	1303	1.22	0.74	1.75	2.16	−1.16	1.08
Bennett Creek at Park Mills, MD	5P	1634500	163	1040	399	641	1413	1.36	0.62	1.11	1.67	−0.67	0.98
South Branch Potoamc River near Springfield, WV	6P	1608500	3783	973	347	626	1364	1.40	0.64	1.17	1.73	−0.73	1.04
Conococheague Creek and Fairview, MD	7P	1614500	1279	1004	452	552	1404	1.40	0.55	0.92	1.52	−0.52	0.90
Marsh Run at Grimes, MD	8P	1617800	49	1014	223	791	1427	1.41	0.78	1.76	2.26	−1.26	1.31
North Branch Potomac River at Steyer, MD	9P	1595000	189	941	322	619	1342	1.43	0.66	1.20	1.76	−0.76	1.08
Catoctin Creek near Middletown, MD	10P	1637500	173	1063	419	644	1424	1.34	0.61	1.09	1.65	−0.65	0.96
Goose Creek near Leesburg, VA	11P	1644000	860	1081	366	715	1441	1.33	0.66	1.27	1.80	−0.80	1.04
North Fork Shenandoah River at Cootes Store, VA	12P	1632000	544	995	348	647	1382	1.39	0.65	1.20	1.75	−0.75	1.05
Cedar Creek near Winchester, VA	13P	1643500	264	1040	366	674	1413	1.36	0.65	1.22	1.76	−0.76	1.03
Basin average			1413	1015	364	651	1385	1.37	0.64	1.22	1.76	−0.76	1.03
Regional average			1246	1125	510	615	1342	1.22	0.56	1.06	1.61	−0.61	0.87
±std. dev.			1671	167	198	98	82	0.19	0.11	0.26	0.23	0.23	0.20

Equation (1) can be used to quantify the change in E due to changes in climate (P, PET) and n [35]:

$$\Delta E = \frac{\partial E}{\partial P}\Delta P + \frac{\partial E}{\partial PET}\Delta PET + \frac{\partial E}{\partial n}\Delta n \tag{2}$$

The respective partial differentials are given by Equations (3a)–(3c), which provide insight into how changes in climate (P, PET) and land cover (n) affect E.

$$\frac{\partial E}{\partial P} = \frac{E}{P}\left(\frac{PET^n}{P^n + PET^n}\right) \tag{3a}$$

$$\frac{\partial E}{\partial PET} = \frac{E}{PET}\left(\frac{P^n}{P^n + PET^n}\right) \tag{3b}$$

$$\frac{\partial E}{\partial n} = \frac{E}{n}\left(\frac{\ln(P^n + PET^n)}{n}\right) - \frac{(P^n \ln P + PET^n \ln PET)}{P^n + PET^n} \tag{3c}$$

It is assumed that the water balance changes over time are from one steady state to another steady state [35]; i.e., that transient changes in storage can be ignored [67]. Based on this assumption, Q was calculated by

$$\Delta Q = \Delta P - \Delta E \tag{4}$$

By combining Equations (2) and (4), ΔQ is given by

$$\Delta Q = \left(1 - \frac{\partial E}{\partial P}\right)\Delta P - \frac{\partial E}{\partial PET}\Delta PET - \frac{\partial E}{\partial n}\Delta n \tag{5}$$

Relative ΔQ is then solved by

$$\frac{\Delta Q}{Q} = \left[\frac{P}{Q}\left(1 - \frac{\partial E}{\partial P}\right)\right]\frac{\Delta P}{P} - \left[\frac{PET}{Q}\frac{\partial E}{\partial PET}\right]\frac{\Delta PET}{PET} - \left[\frac{n}{Q}\frac{\partial E}{\partial n}\right]\frac{\Delta n}{n} \tag{6}$$

The terms in square brackets are the sensitivity coefficients expressing the effect of changing P and PET on relative ΔQ (Table 1), where ΔP/P represents the theoretical sensitivity of streamflow to changes in precipitation, ΔPET/PET represents the sensitivity of streamflow to changes in energy (PET), and Δn/n represents the change in Q following a change in watershed characteristics. An increase in ΔP/P will increase Q, while an increase in ΔPET/PET and Δn/n will decrease Q.

2.3. Modeling Future Streamflow

The sensitivity of future Q to future changes in P and PET was modeled using historical sensitivity coefficients and future climates based on ensemble RCP4.5 and RCP8.5 data. Future changes in P and PET were calculated relative to the historical period (H):

$$\frac{\Delta P}{P}RCP, x = \frac{P_{RCP,x} - P_H}{P_H} \tag{7}$$

$$\frac{\Delta PET}{PET}RCP, x = \frac{PET_{RCP,x} - PET_H}{PET_H} \tag{8}$$

Substituting future values of P and PET into Equation (5) with the historical sensitivity coefficients for P and PET (Equations (3a)–(3c)), future ΔQ was quantified by

$$\Delta Q = \left(1 - \frac{\Delta P}{P}\right)\Delta P_{RCP,x} - \frac{\Delta PET}{PET}\Delta PET_{RCP,x} - \frac{\Delta n}{n}\Delta n \tag{9}$$

Because our analysis focused on HCDN catchments, we assumed that catchment properties (parameter n) do not change in the future, thereby setting the n sensitivity coefficient and dn to

0. We recognize that this assumption is likely an oversimplification of future landscape conditions, particularly in light of changes in forest structure, age, productivity, and growing season length [32,68,69] in relatively undisturbed catchments throughout the region. Future analysis should consider dn to more thoroughly account for ecosystem changes important to the partition of P into E and Q.

3. Results

3.1. Historic Climate, Water Balance Components, and Streamflow Sensitivity throughout the Central Appalachian Mountains Region

Long-term average annual climate, water balance, and Budyko components over the historical period (1965–2004) are summarized in Table 1. Annual precipitation averaged 1125 ± 167 mm across the region. P was greatest in the Tennessee basin ($\overline{x}$ = 1375 mm), followed by the Monongahela ($\overline{x}$ = 1244 mm), Kanawha ($\overline{x}$ = 1170 mm), Ohio ($\overline{x}$ = 1029 mm), and the Potomac ($\overline{x}$ = 1015 mm). Annual Q averaged 510 ± 198 mm across the region and was greatest in the Monongahela basin ($\overline{x}$ = 736 mm), followed by the Tennessee ($\overline{x}$ = 642 mm), Kanawha ($\overline{x}$ = 597 mm), Ohio ($\overline{x}$ = 433 mm) and the Potomac ($\overline{x}$ = 364 mm). Annual E and PET averaged 615 ± 98 mm and 1342 ± 83 mm, respectively, generally following a south-to-north gradient. E was greatest in the Tennessee basin ($\overline{x}$ = 733 mm), followed by the Potomac ($\overline{x}$ = 651 mm), Ohio ($\overline{x}$ = 596 mm), Kanawha ($\overline{x}$ = 573 mm), and Monongahela ($\overline{x}$ = 508 mm). PET averaged 1437 mm in the Tennessee, 1385 mm in the Potomac, 1314 mm in the Kanawha, 1285 mm in the Ohio, and 1232 mm in the Monongahela.

The aridity index, AI, averaged 1.22 ± 0.19 across the region, and was greatest in the Potomac ($\overline{x}$ = 1.37) and smallest in the Monongahela basin ($\overline{x}$ = 0.99) (Table 1). The evaporation ratio, E/P, which represents the proportion of P returned to the atmosphere through actual E, averaged 0.56 ± 0.11 across the region. Similar to AI, E/P was greatest in the Potomac ($\overline{x}$ = 0.64) and smallest in the Monongahela basin ($\overline{x}$ = 0.41) (Figure 2). The catchment-specific landscape parameter, n, averaged 1.61 ± 0.23 across the region, and was greatest in the Potomac ($\overline{x}$ = 1.76) and smallest in the Monongahela basin ($\overline{x}$ = 0.79) (Figure 3).

Streamflow sensitivity coefficients, which quantify the historical sensitivity of Q to changes in climate and catchment properties, averaged 1.61 ± 0.23 (ΔP/P), 0.61 ± 0.23 (ΔPET/PET), and 0.87 ± 0.20 Δn/n) across the region (Table 1). These imply that Q is more sensitive to ΔP than ΔPET and Δn. Based on this, the theoretical results from Equation (6) predict that, for example, a 10% increase in P would increase Q by 16%, while a 10% increase in PET would decrease Q by 6%, and a 10% increase in landcover change would decrease Q by 8.7% [35] (Figure 4).

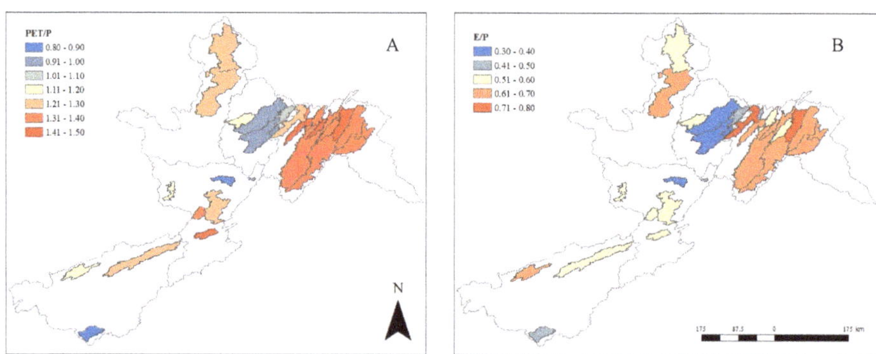

Figure 2. Historical (2965–2005) (**A**) aridity index (PET/P) and (**B**) evaporative index (E/P).

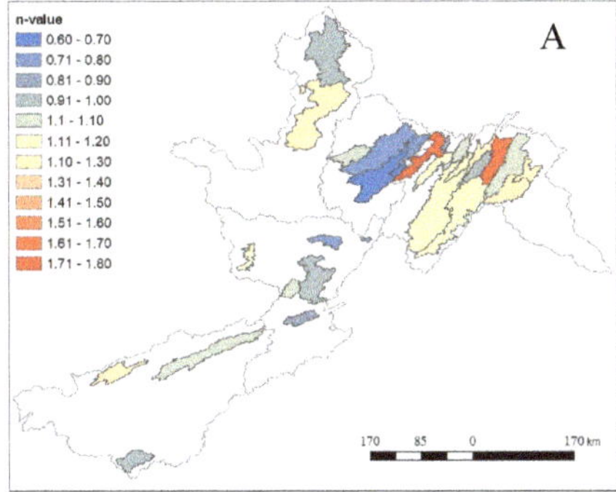

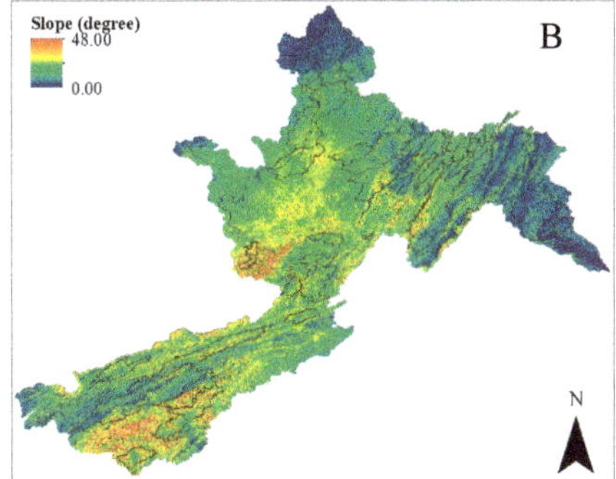

Figure 3. Landcover characteristics displayed as (**A**) the n value and (**B**) slope in degrees.

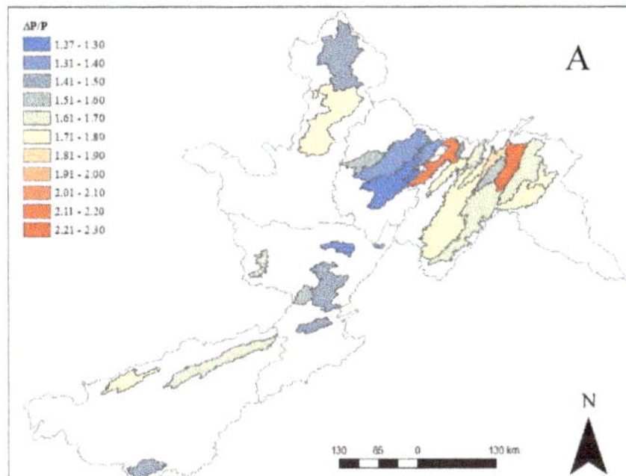

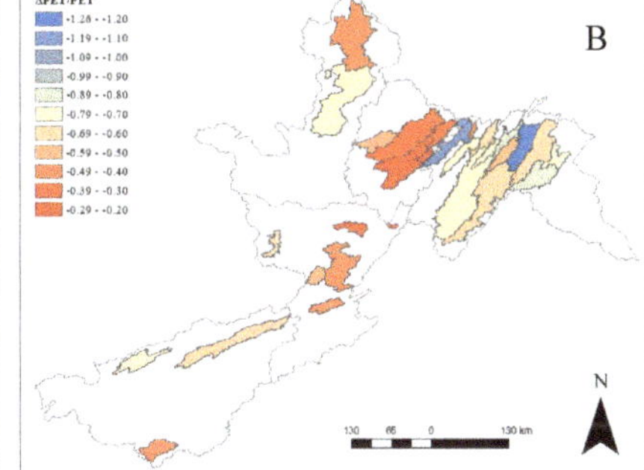

Figure 4. Sensitivity of streamflow to changes in (**A**) precipitation and (**B**) PET using data from 1965–2004.

3.2. *Twenty-First Century Climate and Streamflow*

Future climates (P, PET, AI) are summarized by RCP in Tables 2 and 3. For each pathway, P, PET, and AI at the catchment scale were projected to increase over the 21st century, albeit at different rates. Relative to the historic period, P was projected to increase in all catchments on average by 4% and 8% over the century for RCP4.5 and RCP8.5, respectively. Changes were greatest during the last quarter (2075–2099) of the 21st century, with P increasing between 1%–14% for RCP4.5 and 5%–15% for RCP8.5. PET is projected to progressively increase throughout the 21st century by 10% and 15%, on average, for RCP4.5 and RCP8.5, respectively.

Similar to P, the greatest changes in PET were in Quarter 4, increasing between 11%–17% for RCP4.5 and 5%–15% for RCP8.5. Changes in AI largely follow patterns in P and PET, increasing on average by 4% for RCP4.5 and 9% for RCP8.5 throughout the region. By the end of the century, AI was projected to be greatest over the study period, with changes ranging between 0.2%–11% (RCP4.5) and 8%–22% (RCP8.5).

Future Q are summarized in Table 4 and Figures 5 and 6. Q is projected to increase across catchments, basins, and the region under RCP4.5, with 93%–100% of catchments experiencing positive changes over the century. Changes across catchments were generally similar in magnitude when averaged across basins and ranged between 9% to 13% and. Under RCP8.5, Q was projected to be more variable in the future, decreasing by as much as 9% and increasing by as much as 9% (Table 4). Q increased early (2005–2025) in the century, with 28/29 catchments experiencing greater Q relative to the historic period. Q progressively decreased throughout the century, with nearly half of the catchments experiencing decreases in Q by mid- and late-century periods.

Table 2. Future (2005–2099) long-term annual precipitation (P), potential evapotranspiration (PET), and aridity index (AI) and changes by quarter for RCP4.5 for selected HCDN catchments.

RCP4.5	Q1 (2005–2025)						Q2 (2026–2050)						Q3 (2051–2075)						Q4 (2076–2099)					
Basin/Station Name	P	DP	PET	DPET	AI	DAI	P	DP	PET	DPET	AI	DAI	P	DP	PET	DPET	AI	DAI	P	DP	PET	DPET	AI	DAI
	mm	%	mm	%	-	%	mm	%	mm	%	-	%	mm	%	mm	%	-	%	mm	%	mm	%	-	%
Monongahela																								
Casselman River at Grantsville, MD	1217	4	1302	6	1.07	2	1230	5	1349	10	1.10	5	1245	6	1389	13	1.12	7	1266	8	1396	14	1.10	5
West Fork River at Enterprise, WV	1212	4	1385	6	1.14	2	1228	5	1430	9	1.16	4	1247	7	1467	12	1.18	5	1260	8	1475	13	1.17	5
Youghiogheny River near Oakland, MD	1324	4	1275	6	0.96	2	1340	5	1321	10	0.99	5	1359	7	1359	13	1.00	6	1374	8	1367	14	0.99	5
Laurel Hill Creek at Ursina, PA	1298	2	1318	10	1.02	8	1313	3	1364	13	1.04	10	1331	4	1402	17	1.05	12	1347	6	1410	17	1.05	11
Cheat River near Parsons, WV	1379	4	1288	6	0.93	2	1394	5	1336	10	0.96	5	1418	7	1372	13	0.97	6	1430	8	1381	14	0.97	6
Basin average	1286	3	1314	7	1.03	3	1301	5	1360	10	1.05	6	1320	6	1398	14	1.06	7	1335	7	1406	14	1.06	6
Ohio																								
Little Shenango River at Greenville, PA	1058	4	1328	6	1.26	1	1066	5	1372	10	1.29	4	1084	7	1412	13	1.30	5	1090	8	1421	13	1.30	5
Little Beaver Creek near East Liverpool, OH	1084	4	1394	6	1.29	2	1099	5	1438	9	1.31	4	1116	7	1476	12	1.32	5	1123	7	1485	13	1.32	5
Basin average	1071	4	1361	6	1.27	2	1083	5	1405	9	1.30	4	1100	7	1444	12	1.31	5	1106	7	1453	13	1.31	5
Kanawha																								
Wolf Creek near Narrows, VA	1040	4	1486	5	1.43	1	1047	4	1537	9	1.47	4	1063	6	1570	11	1.48	5	1076	7	1579	12	1.47	4
Greenbrier River at Durbin, WV	1298	4	1259	6	0.97	2	1316	6	1309	10	0.99	4	1331	7	1345	13	1.01	6	1349	8	1354	14	1.00	5
Williams River at Dyer, WV	1533	4	1293	6	0.84	2	1548	5	1341	10	0.87	5	1569	6	1376	12	0.88	6	1586	7	1385	13	0.87	6
Big Coal River at Ashford, WV	1235	3	1451	5	1.18	2	1242	4	1500	9	1.21	5	1261	5	1531	11	1.21	6	1274	7	1543	12	1.21	5
Bluestone River at Pipestem, WV	1058	4	1453	5	1.37	1	1062	4	1504	9	1.42	4	1079	6	1537	11	1.42	5	1090	7	1547	12	1.42	5
Greenbrier River at Alderson, WV	1123	4	1373	5	1.22	1	1127	4	1422	9	1.26	5	1143	6	1455	12	1.27	6	1157	7	1465	12	1.27	5
Basin average	1214	4	1386	5	1.17	2	1223	5	1436	9	1.20	5	1241	6	1469	12	1.21	5	1255	7	1479	13	1.21	5

Table 2. *Cont.*

RCP4.5	Q1 (2005–2025)						Q2 (2026–2050)						Q3 (2051–2075)						Q4 (2076–2099)					
Basin/Station Name	P	DP	PET	DPET	AI	DAI	P	DP	PET	DPET	AI	DAI	P	DP	PET	DPET	AI	DAI	P	DP	PET	DPET	AI	DAI
Tennessee																								
North Fork Holston River near Saltsville, VA	1195	3	1516	6	1.27	2	1196	4	1569	9	1.31	6	1222	6	1597	11	1.31	5	1236	7	1608	12	1.30	5
Clinch River above Tazewell, TN	1319	3	1503	5	1.14	1	1309	3	1556	9	1.19	6	1342	5	1582	10	1.18	5	1354	6	1595	11	1.18	5
Little Tennessee River near Prentiss, NC	1753	3	1512	5	0.86	1	1745	3	1567	9	0.90	6	1793	6	1599	11	0.89	5	1819	7	1612	12	0.89	4
Basin average	1422	3	1510	5	1.09	2	1417	3	1564	9	1.13	6	1452	6	1593	11	1.13	5	1469	7	1605	12	1.12	5
Potomac																								
Wills Creek near Cumberland, MD	1025	4	1429	6	1.39	2	1038	5	1475	9	1.42	4	1048	6	1515	12	1.44	6	1070	8	1522	13	1.42	4
Pototmac River near at Paw Paw, WV	994	1	1436	6	1.44	6	1005	2	1482	10	1.47	8	1017	3	1522	13	1.50	9	1037	5	1529	13	1.47	8
Cacapon River near Great Cacapon, WV	1040	4	1465	6	1.41	1	1049	5	1510	9	1.44	4	1062	6	1550	12	1.46	5	1084	9	1558	12	1.44	3
Patterson Creek near Headsville, WV	1112	4	1379	6	1.24	2	1126	5	1426	9	1.27	4	1139	6	1466	12	1.29	6	1156	8	1473	13	1.27	5
Bennett Creek at Park Mills, MD	1084	4	1489	5	1.37	1	1098	6	1538	9	1.40	3	1110	7	1576	12	1.42	5	1135	9	1584	12	1.40	3
South Branch Potoamc River near Springfield, WV	1013	4	1441	6	1.42	2	1025	5	1488	9	1.45	4	1036	7	1529	12	1.47	5	1057	9	1536	13	1.45	4
Conococheague Creek and Fairview, MD	1046	4	1481	5	1.42	1	1057	5	1527	9	1.45	3	1070	7	1567	12	1.46	5	1092	9	1575	12	1.44	3
Marsh Run at Grimes, MD	1059	4	1505	5	1.42	1	1066	5	1551	9	1.46	4	1080	6	1592	12	1.47	5	1104	9	1599	12	1.45	3
North Branch Potomac River at Steyer, MD	980	4	1419	6	1.45	2	992	5	1466	9	1.48	4	1003	7	1505	12	1.50	5	1019	8	1513	13	1.48	4
Catoctin Creek near Middletown, MD	1109	4	1501	5	1.35	1	1115	5	1548	9	1.39	4	1130	6	1589	12	1.41	5	1158	9	1597	12	1.38	3
Goose Creek near Leesburg, VA	1129	4	1518	5	1.34	1	1136	5	1566	9	1.38	3	1153	7	1606	11	1.39	5	1179	9	1614	12	1.37	3
North Fork Shenandoah River at Cootes Store, VS	1037	4	1458	6	1.41	1	1052	6	1509	9	1.43	3	1064	7	1548	12	1.45	5	1085	9	1556	13	1.44	3
Cedar Creek near Winchester, VA	1136	9	1513	7	1.33	-2	1137	9	1560	10	1.37	1	1155	11	1600	13	1.39	2	1181	14	1609	14	1.36	0.2
Basin average	1059	4	1464	6	1.38	1.40	1069	5	1511	9	1.42	4	1082	7	1551	12	1.44	5	1104	9	1559	13	1.41	4
Regional average	1169	4	1420	6	1	2	1178	5	1468	9	1	4	1196	6	1505	12	1	6	1213	8	1513	13	1	5
±std. dev	171	1.3	82	0.8	0.2	1.5	169	1.3	83	0.9	0.2	1.6	175	1.2	82	1.1	0.2	1.6	175	1.5	83	1.1	0.2	1.8

Table 3. Future (2005–2099) long-term annual precipitation (P), potential evapotranspiration (PET), and aridity index (AI) and changes by quarter for RCP8.5 for selected HCDN catchments.

RCP8.5	Q1 (2005–2025)						Q2 (2026–2050)						Q3 (2051–2075)						Q4 (2076–2099)					
Basin/Station Name	P	DP	PET	DPET	AI	DAI	P	DP	PET	DPET	AI	DAI	P	DP	PET	DPET	AI	DAI	P	DP	PET	DPET	AI	DAI
	mm	%	mm	%	-	%	mm	%	mm	%	-	%	mm	%	mm	%	-	%	mm	%	mm	%	-	%
Monongahela																								
Casselman River at Grantsville, MD	1223	4	1322	8	1.08	3	1227	5	1375	12	1.12	7	1248	6	1471	20	1.18	12	1276	9	1547	26	1.21	15
West Fork River at Enterprise, WV	1223	5	1406	7	1.15	4	1223	5	1456	11	1.19	7	1241	6	1549	18	1.25	12	1272	9	1621	24	1.27	15
Youghiogheny River near Oakland, MD	1328	4	1295	8	0.97	4	1334	5	1345	12	1.01	7	1355	6	1437	20	1.06	13	1385	9	1510	26	1.09	16
Laurel Hill Creek at Ursina, PA	1308	3	1339	11	1.02	9	1309	3	1390	16	1.06	13	1327	4	1484	23	1.12	19	1358	7	1558	30	1.15	22
Cheat River near Parsons, WV	1386	4	1310	8	0.95	4	1390	5	1361	12	0.98	8	1409	6	1455	20	1.03	13	1438	8	1529	26	1.06	17
Basin average	1294	4	1334	8	1.03	5	1297	4	1385	13	1.07	8	1316	6	1479	20	1.13	14	1346	8	1553	26	1.16	17
Ohio																								
Little Shenango River at Greenville, PA	1069	6	1346	7	1.26	2	1066	5	1399	12	1.31	6	1074	6	1495	19	1.39	13	1115	10	1571	25	1.41	14
Little Beaver Creek near East Liverpool, OH	1096	5	1413	7	1.29	2	1097	5	1465	11	1.34	6	1106	6	1561	19	1.41	12	1141	9	1635	24	1.43	14
Basin average	1083	5	1379	7	1.27	2	1082	5	1432	11	1.32	6	1090	6	1528	19	1.40	12	1128	10	1603	25	1.42	14
Kanawha																								
Wolf Creek near Narrows, VA	1034	3	1507	7	1.46	3	1049	5	1558	10	1.49	5	1066	6	1646	16	1.54	9	1080	8	1725	22	1.60	13
Greenbrier River at Durbin, WV	1294	4	1280	8	0.99	4	1308	5	1332	12	1.02	7	1334	7	1424	20	1.07	12	1353	9	1500	26	1.11	16
Williams River at Dyer, WV	1536	4	1314	7	0.86	3	1541	4	1364	11	0.89	7	1561	6	1454	19	0.93	13	1586	7	1527	25	0.96	16
Big Coal River at Ashford, WV	1232	3	1477	7	1.20	4	1235	3	1525	11	1.23	7	1249	4	1616	17	1.29	13	1272	6	1690	23	1.33	16
Bluestone River at Pipestem, WV	1053	3	1476	7	1.40	3	1067	5	1526	11	1.43	6	1076	6	1619	17	1.50	11	1092	7	1699	23	1.56	15
Greenbrier River at Alderson, WV	1117	3	1395	7	1.25	4	1130	5	1444	11	1.28	6	1144	6	1531	17	1.34	11	1160	7	1605	23	1.38	15
Basin average	1211	3	1408	7	1.19	4	1222	4	1458	11	1.22	6	1238	6	1548	18	1.28	11	1257	7	1624	24	1.32	15
Tennessee																								
North Fork Holston River near Saltsville, VA	1194	3	1539	7	1.29	4	1202	4	1588	11	1.32	6	1211	5	1676	17	1.38	11	1226	6	1753	22	1.43	15
Clinch River above Tazewell, TN	1318	3	1528	7	1.16	3	1316	3	1576	10	1.20	6	1322	4	1659	16	1.26	12	1343	5	1726	20	1.29	14
Little Tennessee River near Prentiss, NC	1766	4	1533	6	0.87	2	1765	4	1586	10	0.90	6	1780	5	1675	16	0.94	11	1780	5	1753	22	0.98	16
Basin average	1426	4	1533	7	1.11	3	1428	4	1583	10	1.14	6	1437	5	1670	16	1.19	11	1449	5	1744	21	1.23	15
Potomac																								
Wills Creek near Cumberland, MD	1032	4	1449	7	1.40	3	1037	5	1501	11	1.45	6	1056	7	1595	18	1.51	10	1080	9	1669	23	1.55	13
Pototmac River near Great Cacapon, WV	999	1	1455	8	1.46	6	1005	2	1508	12	1.50	10	1023	4	1601	18	1.56	14	1046	6	1675	24	1.60	17
Cacapon River near Great Cacapon, WV	1044	5	1483	7	1.42	2	1050	5	1537	11	1.46	5	1069	7	1629	17	1.52	10	1096	10	1703	23	1.55	12
Patterson Creek near Headsville, WV	1116	4	1399	7	1.25	3	1122	5	1452	11	1.29	6	1143	7	1547	19	1.35	11	1168	9	1621	24	1.39	14
Bennett Creek at Park Mills, MD	1078	4	1509	7	1.40	3	1095	5	1564	11	1.43	5	1123	8	1656	17	1.47	9	1143	10	1733	23	1.52	12
South Branch Potoamc River near Springfield, WV	1017	4	1460	7	1.44	2	1024	5	1515	11	1.48	6	1043	7	1609	18	1.54	10	1066	10	1685	24	1.58	13
Conococheague Creek and Fairview, MD	1049	4	1500	7	1.43	2	1058	5	1554	11	1.47	5	1078	7	1647	17	1.53	9	1105	10	1722	23	1.56	12
Marsh Run at Grimes, MD	1061	5	1524	7	1.44	2	1068	5	1579	11	1.48	5	1088	7	1673	17	1.54	9	1119	10	1748	22	1.56	11
North Branch Potomac River at Steyer, MD	981	4	1439	7	1.47	3	989	5	1492	11	1.51	6	1009	7	1586	18	1.57	10	1030	9	1661	24	1.61	13
Catoctin Creek near Middletown, MD	1109	4	1521	7	1.37	2	1119	5	1577	11	1.41	5	1140	7	1669	17	1.46	9	1172	10	1745	23	1.49	11
Goose Creek near Leesburg, VA	1126	4	1538	7	1.37	2	1138	5	1594	11	1.40	5	1164	8	1686	17	1.45	9	1191	10	1763	22	1.48	11
North Fork Shenandoah River at Cootes Store, VS	1032	4	1480	7	1.43	3	1048	5	1536	11	1.47	6	1074	8	1631	18	1.52	9	1093	10	1710	24	1.56	13
Cedar Creek near Winchester, VA	1131	9	1532	8	1.35	0	1141	10	1589	12	1.39	2	1163	12	1680	19	1.44	6	1198	15	1755	24	1.46	8
Basin average	1060	4	1484	7	1.40	3	1069	5	1538	11	1.44	6	1090	7	1632	18	1.50	10	1116	10	1707	23	1.53	12
Regional average	1174	4	1441	7	1.25	3.2	1180	5	1493	11	1.29	6	1198	6	1585	18	1.35	11	1222	8	1659	24	1.38	14
±std. dev.	167	1	83	1	0.18	1.5	165	1	84	1	0.19	2	165	2	83	2	0.19	2	162	2	83	2	0.19	3

Table 4. Future (2005–2099) changes in long-term annual streamflow (ΔQ) due to changes in precipitation (ΔQ_P) and potential evapotranspiration (ΔQ_{PET}) relative to the historical period (1950–2004) based on RCP4.5 and RCP8.5.

Basin /Station Name	Q1 (2005–2025)			Q2 (2026–2050)			Q3 (2051–2075)			Q4 (2076–2099)			Q1 (2005–2025)			Q2 (2026–2050)			Q3 (2051–2075)			Q4 (2076–2099)		
	DQ_P %	D Q_{PET} %	DQ %	DQ_P %	D Q_{PET} %	DQ %	DQ_P %	D Q_{PET} %	DQ %	DQ_P %	D Q_{PET} %	DQ %	DQ_P %	D Q_{PET} %	DQ %	DQ_P %	D Q_{PET} %	DQ %	DQ_P %	D Q_{PET} %	DQ %	DQ_P %	D Q_{PET} %	DQ %
Monongahela																								
Casselman River at Grantsville, MD	5	−2	3	7	−4	3	8	−5	4	11	−5	6	6	−3	3	6	−4	2	9	−7	2	12	−9	3
West Fork River at Enterprise, WV	6	−3	2	8	−5	3	10	−7	4	12	−7	5	7	−4	3	7	−6	1	10	−10	−1	14	−14	0
Youghiogheny River near Oakland, MD	5	−2	3	7	−3	4	8	−4	5	10	−4	6	5	−2	3	6	−3	3	8	−5	3	11	−7	4
Laurel Hill Creek at Ursina, PA	2	−3	−1	4	−4	0	6	−5	0	7	−6	2	3	−4	0	4	−5	−2	5	−8	−2	9	−10	−1
Cheat River near Parsons, WV	5	−2	3	6	−3	4	9	−4	5	10	−4	6	5	−2	3	6	−3	3	8	−5	2	10	−7	3
Basin average	5	−2	2	6	−4	3	8	−5	3	10	−5	5	5	−3	2	6	−5	1	8	−7	1	11	−9	2
Ohio																								
Little Shenango River at Greenville, PA	7	−3	4	8	−5	3	11	−6	4	11	−6	5	8	−4	5	8	−6	2	9	−9	0	15	−12	3
Little Beaver Creek near East Liverpool, OH	6	−5	2	9	−7	2	12	−9	2	13	−10	3	8	−6	3	9	−9	0	10	−14	−4	16	−19	−3
Basin average	6	−4	3	8	−6	3	11	−8	3	12	−8	4	8	−5	4	8	−7	1	9	−12	−2	16	−15	0
Kanawha																								
Wolf Creek near Narrows, VA	6	−2	3	7	−4	3	9	−5	4	11	−5	5	5	−3	2	7	−5	2	9	−8	2	11	−10	1
Greenbrier River at Durbin, WV	6	−2	3	8	−4	4	9	−5	4	11	−5	6	5	−3	2	7	−5	2	10	−7	2	12	−10	2
Williams River at Dyer, WV	5	−2	3	6	−3	3	8	−4	4	9	−4	5	5	−2	3	5	−3	2	7	−6	2	9	−7	2
Big Coal River at Ashford, WV	5	−4	2	6	−6	0	9	−7	2	11	−8	3	5	−5	0	5	−7	−2	7	−12	−4	11	−15	−5
Bluestone River at Pipestem, WV	6	−3	3	7	−5	2	9	−7	3	11	−7	4	5	−4	1	8	−6	1	9	−10	−1	12	−14	−2
Greenbrier River at Alderson, WV	6	−3	3	6	−4	2	9	−6	3	11	−6	4	5	−3	2	7	−5	2	9	−9	0	11	−11	0
Basin average	6	−3	3	7	−4	2	9	−6	3	11	−6	5	5	−3	2	6	−5	1	9	−9	0	11	−11	0

Table 4. *Cont.*

Basin /Station Name	Q1 (2005–2025) DQ$_P$ %	D Q$_{PET}$ %	DQ %	Q2 (2026–2050) DQ$_P$ %	D Q$_{PET}$ %	DQ %	Q3 (2051–2075) DQ$_P$ %	D Q$_{PET}$ %	DQ %	Q4 (2076–2099) DQ$_P$ %	D Q$_{PET}$ %	DQ %	Q1 (2005–2025) DQ$_P$ %	D Q$_{PET}$ %	DQ %	Q2 (2026–2050) DQ$_P$ %	D Q$_{PET}$ %	DQ %	Q3 (2051–2075) DQ$_P$ %	D Q$_{PET}$ %	DQ %	Q4 (2076–2099) DQ$_P$ %	D Q$_{PET}$ %	DQ %
Tennessee																								
North Fork Holston River near Saltsville, VA	6	−3	2	6	−6	0	9	−7	2	11	−7	4	6	−4	1	7	−7	0	8	−10	−3	10	−14	−4
Clinch River above Tazewell, TN	6	−4	2	5	−6	-2	9	−8	1	11	−9	2	6	−5	1	6	−7	−2	6	−12	−5	9	−15	−6
Little Tennessee River near Prentiss, NC	5	−2	3	4	−4	1	8	−4	4	10	−5	5	6	−3	3	6	−4	2	7	−7	0	7	−9	−2
Basin average	6	−3	2	5	−5	0	9	−6	2	11	−7	4	6	−4	2	6	−6	0	7	−10	−2	9	−13	−4
Potomac																								
Wills Creek near Cumberland, MD	5	−2	3	7	−4	3	9	−5	3	12	−5	6	6	−3	3	7	−5	2	10	−8	2	13	−10	3
Pototmac River near at Paw Paw, WV	1	−4	-3	3	−6	-3	5	−8	−3	8	−8	0	2	−5	−3	3	−7	−4	6	−11	−6	10	−15	−5
Cacapon River near Great Cacapon, WV	8	−5	3	9	−7	2	12	−10	2	16	−10	6	9	−6	3	10	−9	1	13	−15	−1	18	−19	−1
Patterson Creek near Headsville, WV	8	−7	2	11	−11	0	14	−14	−1	17	−15	2	9	−8	1	10	−13	−3	15	−22	−7	20	−28	−9
Bennett Creek at Park Mills, MD	7	−4	3	9	−6	3	11	−8	4	15	−8	7	6	−5	2	9	−7	2	13	−12	2	16	−15	1
South Branch Potoamc River near Springfield, WV	7	−4	3	9	−7	3	11	−9	3	15	−9	6	8	−5	3	9	−8	1	12	−13	−1	17	−17	−1
Conococheague Creek and Fairview, MD	6	−3	3	8	−5	3	10	−6	4	13	−6	7	7	−4	3	8	−6	2	11	−9	2	15	−12	3
Marsh Run at Grimes, MD	10	−7	3	11	−11	0	15	−15	0	20	−15	5	10	−9	2	12	−13	−1	16	−22	−5	23	−28	−5
North Branch Potomac River at Steyer, MD	7	−4	3	9	−7	2	12	−9	2	15	−10	5	7	−5	2	9	−8	0	13	−14	−1	17	−18	−1
Catoctin Creek near Middletown, MD	7	−4	4	8	−6	2	10	−8	3	15	−8	7	7	−4	3	9	−7	2	12	−11	1	17	−15	2
Goose Creek near Leesburg, VA	8	−4	4	9	−7	2	12	−9	3	16	−10	7	8	−5	2	9	−9	1	14	−14	0	18	−18	0
North Fork Shenandoah River at Cootes Store, VS	7	−4	3	10	−7	3	12	−9	3	16	−9	6	6	−5	1	9	−8	1	14	−13	0	17	−18	−1
Cedar Creek near Winchester, VA	16	−5	11	16	−8	9	19	−10	9	24	−11	13	15	−6	9	17	−9	8	21	−14	6	27	−18	8
Basin average	8	−4	3	9	−7	2	12	−9	2	15	−10	6	8	−5	2	9	−8	1	13	−14	−1	17	−18	0
Regional average	6	−3	3	8	−6	2	10	−7	3	13	−8	5	7	−4	2	8	−7	1	10	−11	−1	14	−14	0
±std. dev	3	1	4	3	2	5	3	3	5	4	3	6	2	2	4	3	3	5	3	4	7	4	5	10

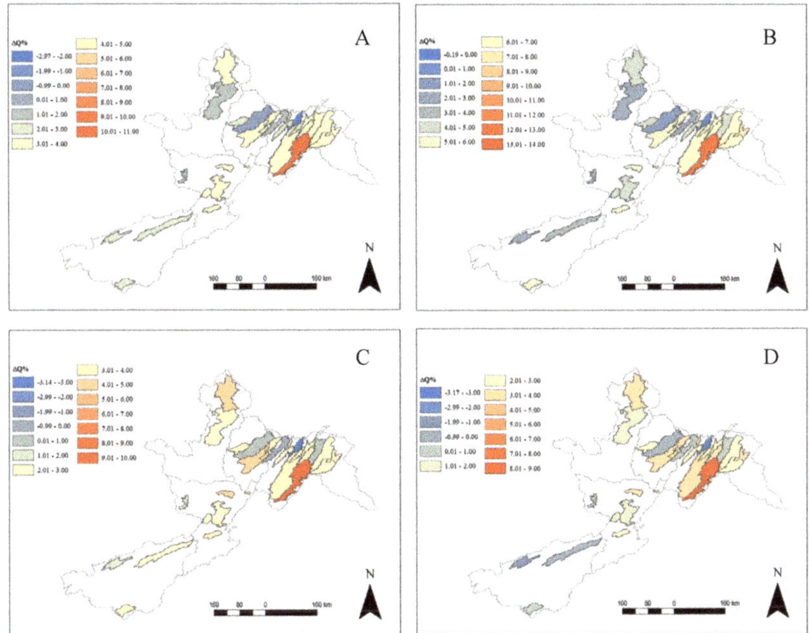

Figure 5. Changes in future streamflow for four future time periods based on RCP4.5 relative to the historic period (1950–2004). (**A**) 2005–2025; (**B**) 2026–2050; (**C**) 2051–2075; (**D**) 2076–2099.

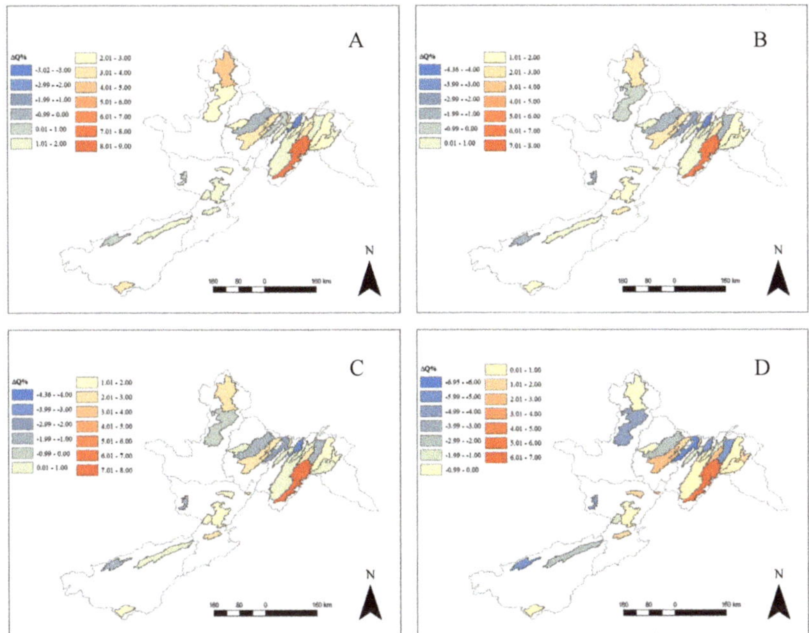

Figure 6. Changes in future streamflow for four future time periods based on RCP8.5 relative to the historic period (1950–2004). (**A**) 2005–2025; (**B**) 2026–2050; (**C**) 2051–2075; (**D**) 2076–2099.

Streamflow changes attributed to changes in P (ΔQ_P) were greater than changes attributed to PET(ΔQ_{PET}) under RCP4.5 (Figure 5), but ΔQ_{PET} increased to comparable magnitudes of ΔQ_P late in the century under RCP8.5 (Table 4 and Figure 6). ΔQ_P increased by 1%–24% under RCP4.5 and 2%–27% under RCP8.5, while ΔQ_{PET} increased by 2%–15% and 2%–28% for RCPs 4.5 and 8.5, respectively (Figure 7).

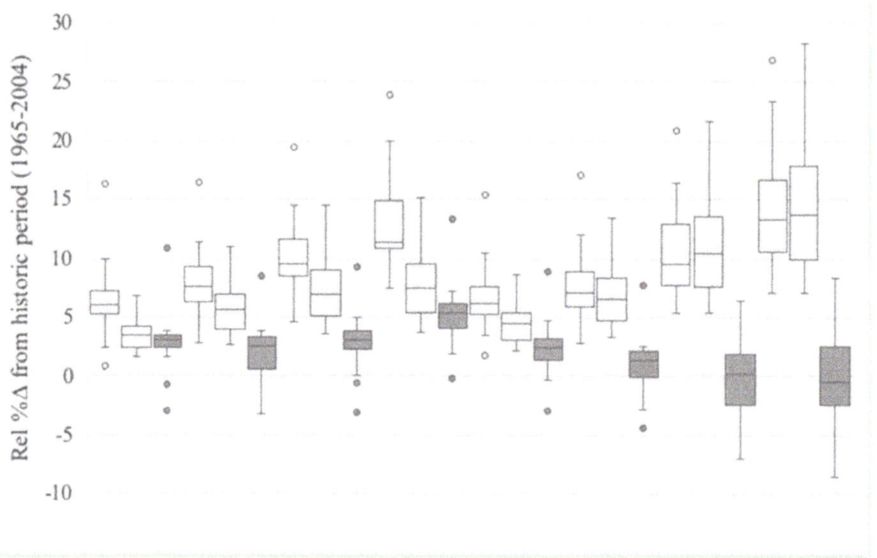

Figure 7. Box plots of future (2005–2099) changes in long-term annual streamflow (ΔQ) attributed to changes in precipitation (ΔQ_P) and potential evapotranspiration (ΔQ_{PET}) for RCP4.5 and RCP8.5. Streamflow changes are relative to the historical period (1950–2004).

4. Discussion

4.1. Historic Climate, Water Balance Components, and Streamflow Sensitivity

The catchment-scale Q sensitivity to P was greater than 1 and generally increased with increasing E/P, while the sensitivity of Q to PET was negative and the absolute value generally increased with increasing E/P (Figures 2 and 4). This implies that sensitivity to climate conditions was greater in arid catchments (those in the Potomac basin) than humid catchments (those in the Monongahela basin), which is consistent with other energy-limited regions [70,71]. The sensitivity to both P and PET increases with increasing n value since the higher value of n generally corresponds to greater E for a given P and PET (Figure 2). The n value is a function of precipitation, topography, and slope [47,65,72,73], indicating lower water availability in regions with higher n value.

The catchment sensitivities to both P and PET generally follow the continental divide of the Appalachian Mountains, which occurs between the Monongahela and Potomac where runoff moves to the Chesapeake Bay, due to rainfall partitioning properties occurring in the headwater region (Figures 3 and 4). Catchments situated closer to the divide (Monongahela, Kanawha, eastern Tennessee) tend to have lower sensitivity, except for the northern catchment in Ohio (1O), likely due to its proximity to the Great Lakes and therefore its lower PET. On the other hand, catchments east (leeward) of the divide tend to have greater sensitivity due to orographic lift, rainfall partitioning, and higher PET. Furthermore, high-elevation headwater catchments in the Monongahela, eastern Kanawha, and eastern Tennessee have lower sensitivity to P due to the low PET attributed to decreased temperatures and therefore lower aridity at high altitudes, which is consistent with other research conducted in the

Appalachian Mountains [16]. Sensitivity to both P and PET increases with decreasing latitude, as PET rises due to increased dependence on sunshine hours and therefore radiation and temperature.

While P, soil storage, and slope are important factors contributing to rainfall partitioning and Q sensitivity [65], landcover has important implications for rainfall partitioning in the region. Forests of the eastern US are highly dynamic [68,74] due to natural (e.g., chestnut blight) and anthropogenic (e.g., forest succession, harvesting, industrialization) drivers. Furthermore, forests of the region are changing due to the rapid increase in unconventional gas development and legacy of surface coal mining throughout the region [75–77], which will likely increase regional streamflow (Δn/n in Table 1).

Forest disturbances due to harvesting [78,79], coal mining [76,80], urbanization [81], unconventional gas development, and afforestation [74] alter the amount of P partitioned into Q. Deforestation generally increases Q over the short-term by decreasing canopy interception and E [82], although the response of Q to forest removal also depends on the amount of water stored in a catchment [83]. As forests regrow, Q can return to similar pre-disturbance levels [84], but because tree water use and canopy interception differ among species (e.g., [85,86]), post-disturbance forest composition can alter Q over the long-term [53,87]. In the case of the forested catchments examined in this study, increases in deforestation will decrease n and E and increase Q (Table 1), while afforestation would show inverse relationships. Furthermore, forest disturbances over the 20th century have shifted the forest composition of the region from dominance by shade-intolerant and fire-adapted xerophytic species, which have adapted to live in areas with little water and low direct solar radiation (e.g., oaks (*Quercus*), chestnut (*Castanea*)), to shade-tolerant and fire-intolerant mesophytic species, which have adapted to variations in water content and high direct solar radiation (e.g., maple (*Acer*), cherry (*Prunus*)) [88]. Tree water use is generally higher for these species [85], which increases E and decreases Q [89]. Furthermore, the tree water use and carbon dynamics of these species are also more sensitive to variations in climate [86,90], and the sensitivity of Q to climate change can increase with the increasing density of mesophytic species [87]. Changes in forest age, productivity, and succession [91], as well as changes in growing season length [26,30,32], are important factors for water balance partitioning. Therefore, increases in mesophication, forest productivity, and growing season length will increase n and E and subsequently decrease Q.

4.2. *Twenty-First Century Climate and Streamflow*

Future climate changes projected for catchments in the central Appalachian Mountains region are consistent with other studies that show increases in both P and atmospheric water demand (e.g., [42,92,93]). Changes were greater in magnitude and variability with the more severe RCP compared to the more conservative pathway with lower atmospheric CO_2 equivalents [21,23,24]. Early in the 21st century, relative changes in P and PET were similar in magnitude, but by the century's end under RCP8.5, increases in PET were projected to outpace P increases in all catchments. The larger increases in PET move the heavily forested catchments of the region towards greater aridity [16], potentially jeopardizing their role as reliable sources of freshwater to downstream communities [2,89]. The projection of P and PET represents some level of uncertainty due to the physics and resolution of the downscaled model. PET exhibits greater uncertainty than P for the region, which is likely due to its greater dependency on temperature and solar radiation [15]. Uncertainty increases in lower portions of the regions and in Q3 for RCP4.5 and Q4 for RCP8.5 due to the degree of change that occurs during these time periods [16].

Under RCP4.5, ΔQ_P values were consistently greater than ΔQ_{PET}, with the net effect of increasing Q in 27 of 29 catchments for each future quarter period (Figure 5). Under RCP8.5, however, Q was more variable, decreasing by as much as 9% in quarter 4 for Patterson Creek and increasing by as much as 9% in the first quarter for Cedar Creek (both are located in the Potomac basin) (Figure 6). Across the region, Q was projected to increase in 28 of 29 catchments early in the century, but by the century's end, Q was projected to decrease in more than half of the catchments, moving closer to the historic long-term regional average. In this case, the disproportionately large increases in PET offset potentially larger increases in Q due to ΔP [43] (Figures 5 and 6). An important caveat of our work was that we assumed no changes in catchment properties in the future (Δn = 0, Equation (9)); i.e., a transition from one steady state to another [35]. While quantifying future land use change and its role on Q sensitivity was beyond the scope of this paper, it is important for managers and decision makers to consider disturbance and climate change together.

While our analysis suggests relatively moderate Q changes throughout the 21st century at the long-term annual scale, other studies show greater variability across the hydrologic regime. The largest increases in P and Q are projected to occur during winter and early spring [39,41,43], when atmospheric demand and forest water use are low [16], increasing the frequency of flooding [23]. P, on the other hand, is projected to decrease during summer when PET and human and ecosystem water use are high, which decreases summer flow [39]. Throughout the greater mid-Atlantic region that includes our study catchments, peak flow events with a 1% exceedance probability are projected to increase by between 10%–20%, while low flows are projected to increase by as much as 14% [37,38,40].

5. Conclusions

Three clear patterns arise from our analysis on streamflow sensitivity to the changing climate in the central Appalachian region of the United States.

1. Catchments are more sensitive to P than PET throughout the region and increased with increasing E/P, implying that arid catchments were more sensitive to change. Sensitivity also increased with increasing distance from the continental divide, with catchments on the leeward (eastern) side of the divide more sensitive to change. Furthermore, sensitivity increased with decreasing elevation and decreasing latitude due to greater dependence on temperature and radiation in those catchments.

2. Future P and PET were greater in magnitude and variability with the more severe RCP (8.5). Early in the 21st century (2005–2025), changes in P and PET will remain consistent with each other; however, by 2099 under RCP 8.5, PET increases will outpace changes in P, which implies that business as usual CO_2 emissions could trigger increased aridity in the region and threaten water resource sustainability.

3. Under RCP4.5, future Q will increase in 27 of the 29 catchments for each future quarter period. However, under RCP 8.5, Q was more variable, especially in the catchments with high sensitivity in the Potomac basin. In the first half of the century (2005–2050), Q is projected to increase in 28/29 catchments, but by 2099, Q is projected to decrease in more than half of the catchments.

Our study contributes to the growing body of recent research that shows that anthropogenic climate change is altering freshwater provisioning throughout the region, posing considerable challenges for water resources management and water security. Floods, droughts, and low flow directly impact society through their immediate effects (e.g., pollution, inundation) and indirectly through ecosystem degradation, human health, work productivity, and the disruption of supply chains and the economy. Flooding poses significant risks to water quality and infrastructure. Increases in climate-driven droughts and low flows, due to the lack of precipitation or increase in atmospheric demand, degrades water quality and aquatic habitats by concentrating pollutants, increasing the costs of water treatment. As with much of the US, the central Appalachian region's roads, bridges, culverts, dams, and water treatment facilities are outdated, poorly maintained, and at a high risk of failure. Future changes in climate and streamflow will only increase the costs and damages to critical infrastructure. The changes in climate and hydrology found in this study suggest considerable challenges for managing the region's water quality, quantity, and security over the 21st century.

Author Contributions: Conceptualization, B.G., R.F., and N.Z.; Methodology, B.G. and R.F.; Formal Analysis, B.G. and N.Z.; Investigation, B.G. and N.Z.; Resources, N.Z.; Data Curation, R.F.; Writing—Original Draft Preparation, B.G. and N.Z.; Writing—Review & Editing, N.Z.; Visualization, B.G. and N.Z.; Supervision, N.Z.; Project Administration, N.Z.; Funding Acquisition, N.Z. All authors have read and agreed to the published version of the manuscript.

Funding: This research was funded by the National Science Foundation, grant number OIA-148952 and the USDA National Institute of Food and Agriculture Hatch project, grant number 1004360, both to Zegre. The dataset METDATA was produced by Northwestern University with funding from the NSF Idaho EPScoR Program, the National Science Foundation award number EPS-0814387, and the National Institute for Food and Agriculture competitive grant award 2011-68002-30191.

Conflicts of Interest: The authors declare no conflict of interest.

References

1. Creed, I.F.; Jones, J.A.; Archer, E.; Claassen, M.; Ellison, D.; McNulty, S.G.; van Noordwijk, M.; Vira, B.; Wei, X.; Bishop, K.; et al. Managing Forests for Both Downstream and Downwind Water. *Front. For. Glob. Chang.* **2019**, *2*. [CrossRef]
2. Duan, K.; Sun, G.; Caldwell, P.V.; McNulty, S.G.; Zhang, Y. Implications of Upstream Flow Availability for Watershed Surface Water Supply Across the Conterminous United States. *JAWRA J. Am. Water Resour. Assoc.* **2018**, *54*, 694–707. [CrossRef]
3. Dudley, N.; Stolton, S. *Running Pure: The Importance of Forest Protected Areas to Drinking Water*; World Bank/WWF Alliance for Forest Conservation and Sustainable Use: Gland, Switzerland; Washington, DC, USA, 2003.
4. Rodriguez-Iturbe, I.; Porporato, A. *Ecohydrology of Water Controlled Ecosystems: Soil Moisture and Plant Dynamics*; Cambridge University Press: Cambridge, UK, 2004.
5. Viviroli, D.; Dürr, H.H.; Messerli, B.; Meybeck, M.; Weingartner, R. Mountains of the world, water towers for humanity: Typology, mapping, and global significance. *Water Resour. Res.* **2007**, *43*. [CrossRef]
6. Viviroli, D.; Weingartner, R. The hydrological significance of mountains: from regional to global scale. *Hydrol. Earth Syst. Sci.* **2004**, *8*, 1017–1030. [CrossRef]
7. Marston, L.; Ao, Y.; Konar, M.; Mekonnen Mesfin, M.; Hoekstra Arjen, Y. High-Resolution Water Footprints of Production of the United States. *Water Resour. Res.* **2018**, *54*, 2288–2316. [CrossRef]
8. Hoekstra, A.Y.; Mekonnen, M.M. The water footprint of humanity. *PNAS* **2012**, *109*, 3232–3237. [CrossRef]
9. Rushforth, R.R.; Ruddell, B.L. A spatially detailed blue water footprint of the United States economy. *Hydrol. Earth Syst. Sci.* **2018**, *22*, 3007–3032. [CrossRef]
10. Brooks, R.P.; Limpisathian, P.W.; Gould, T.; Mazurczyk, T.; Sava, E.; Mitsch, W.J. Does the Ohio River Flow All the Way to New Orleans? *JAWRA J. Am. Water Resour. Assoc.* **2018**, *54*, 752–756. [CrossRef]
11. Bates, B.; Kundzewicz, Z.W.; Wu, S.; Palutikof, J. *Climate Change and Water: Technical Paper vi*; Intergovernmental Panel on Climate Change (IPCC): Geneva, Switzerland, 2008.
12. Vorosmarty, C.J.; Green, P.; Salisbury, J.; Lammers, R.B. Global Water Resources: Vulnerability from Climate Change and Population Growth. *Science* **2000**, *289*, 284–288. [CrossRef]
13. Döll, P.; Fiedler, K.; Zhang, J. Global-scale analysis of river flow alterations due to water withdrawals and reservoirs. *Hydrol. Earth Syst. Sci.* **2009**, *13*, 2413–2432. [CrossRef]
14. Budyko, M. *Climate and Life*; Academic: San Diego, CA, USA, 1974.
15. Guo, D.; Westra, S.; Maier, H.R. Sensitivity of potential evapotranspiration to changes in climate variables for different Australian climatic zones. *Hydrol. Earth Syst. Sci.* **2017**, *21*, 2107–2126. [CrossRef]
16. Fernandez, R.; Zegre, N. Seasonal Changes in Water and Energy Balances over the Appalachian Region and Beyond throughout the Twenty-First Century. *J. Appl. Meteorol. Clim.* **2019**, *58*, 1079–1102. [CrossRef]
17. Milly, P.C.D.; Dunne, K.A. Potential evapotranspiration and continental drying. *Nat. Clim. Chang.* **2016**, *6*, 946–949. [CrossRef]
18. Gudmundsson, L.; Greve, P.; Seneviratne, S.I. The sensitivity of water availability to changes in the aridity index and other factors—A probabilistic analysis in the Budyko space. *Geophys. Res. Lett.* **2016**, *43*, 6985–6994. [CrossRef]
19. Arora, V.K. The use of the aridity index to assess climate change effect on annual runoff. *J. Hydrol.* **2002**, *265*, 164–177. [CrossRef]
20. Huntington, T.G.; Richardson, A.D.; McGuire, K.J.; Hayhoe, K. Climate and hydrological changes in the northeastern United States: recent trends and implications for forested and aquatic ecosystemsThis article is one of a selection of papers from NE Forests 2100: A Synthesis of Climate Change Impacts on Forests of the Northeastern US and Eastern Canada. *Can. J. For. Res.* **2009**, *39*, 199–212.

21. Hayhoe, K.; Wake, C.; Anderson, B.; Liang, X.-Z.; Maurer, E.; Zhu, J.; Bradbury, J.; DeGaetano, A.; Stoner, A.; Wuebbles, D. Regional climate change projections for the Northeast USA. *Mitig. Adapt. Strateg. Glob. Chang.* **2008**, *13*, 425–436. [CrossRef]
22. Huntington, T.G. Evidence for intensification of the global water cycle: Review and synthesis. *J. Hydrol.* **2006**, *319*, 83–95. [CrossRef]
23. Melillo, J.M.; Richmond, T.T.C.; Yohe, G. *Chapter 16: Northeast*; U.S. Global Change Research Program: Washington, DC, USA, 2014.
24. Pachauri, R.K.; Allen, M.R.; Minx, J.C. *Climate Change 2014: Synthesis Report*; Cambridge University Press: Cambridge, UK, 2014.
25. Walsh, K.J.E.; McBride, J.L.; Klotzbach, P.J.; Balachandran, S.; Camargo, S.J.; Holland, G.; Knutson, T.R.; Kossin, J.P.; Lee, T.-C.; Sobel, A.; et al. Tropical cyclones and climate change. *Wiley Interdiscip. Rev. Clim. Chang.* **2016**, *7*, 65–89. [CrossRef]
26. Kutta, E.; Hubbart, J. Climatic Trends of West Virginia: A Representative Appalachian Microcosm. *Water* **2019**, *11*, 1117. [CrossRef]
27. Douglas, E.M.; Vogel, R.M.; Kroll, C.N. Trends in floods and low flows in the United States: impact of spatial correlation. *J. Hydrol.* **2000**, *240*, 90–105. [CrossRef]
28. McCabe, G.J.; Wolock, D.M. A step increase in streamflow in the conterminous United States. *Geophys. Res. Lett.* **2002**, *29*, 2185. [CrossRef]
29. McCabe, G.J.; Wolock, D.M. Independent effects of temperature and precipitation on modeled runoff in the conterminous United States. *Water Resour. Res.* **2011**, *47*. [CrossRef]
30. Gaertner, B.A.; Zegre, N.; Warner, T.; Fernandez, R.; He, Y.; Merriam, E.R. Climate, forest growing season, and evapotranspiration changes in the central Appalachian Mountains, USA. *Sci. Total. Environ.* **2019**, *650*, 1371–1381. [CrossRef] [PubMed]
31. Hwang, T.; Band, L.E.; Miniat, C.F.; Song, C.; Bolstad, P.V.; Vose, J.M.; Love, J.P. Divergent phenological response to hydroclimate variability in forested mountain watersheds. *Glob. Chang. Boil.* **2014**, *20*, 2580–2595. [CrossRef] [PubMed]
32. Hwang, T.; Martin, K.L.; Vose, J.M.; Wear, D.; Miles, B.; Kim, Y.; Band, L.E. Nonstationary Hydrologic Behavior in Forested Watersheds Is Mediated by Climate-Induced Changes in Growing Season Length and Subsequent Vegetation Growth. *Water Resour. Res.* **2018**, *54*, 5359–5375. [CrossRef]
33. van Vuuren, D.P.; Edmonds, J.; Kainuma, M.; Riahi, K.; Thomson, A.; Hibbard, K.; Hurtt, G.C.; Kram, T.; Krey, V.; Lamarque, J.-F.; et al. The representative concentration pathways: an overview. *Clim. Chang.* **2011**, *109*, 5. [CrossRef]
34. Donohue, R.J.; Roderick, M.L.; McVicar, T.R. Roots, storms and soil pores: Incorporating key ecohydrological processes into Budyko's hydrological model. *J. Hydrol.* **2012**, *436–437*, 35–50. [CrossRef]
35. Roderick, M.L.; Farquhar, G.D. A simple framework for relating variations in runoff to variations in climatic conditions and catchment properties. *Water Resour. Res.* **2011**, *47*, W00G07. [CrossRef]
36. Devineni, N.; Lall, U.; Etienne, E.; Shi, D.; Xi, C. America's water risk: Current demand and climate variability. *Geophys. Res. Lett.* **2015**, *42*, 2285–2293. [CrossRef]
37. Baran, A.A.; Moglen, G.E.; Godrej, A.N. Quantifying Hydrological Impacts of Climate Change Uncertainties on a Watershed in Northern Virginia. *J. Hydrol. Eng.* **2019**, *24*, 05019030. [CrossRef]
38. Demaria, E.M.C.; Palmer, R.N.; Roundy, J.K. Regional climate change projections of streamflow characteristics in the Northeast and Midwest U.S. *J. Hydrol. Reg. Stud.* **2016**, *5*, 309–323. [CrossRef]
39. Stagge, J.H.; Moglen, G.E. Water Resources Adaptation to Climate and Demand Change in the Potomac River. *J. Hydrol. Eng.* **2017**, *22*, 04017050. [CrossRef]
40. Wu, S.-Y. Potential impact of climate change on flooding in the Upper Great Miami River Watershed, Ohio, USA: a simulation-based approach. *Hydrol. Sci. J.* **2010**, *55*, 1251–1263. [CrossRef]
41. Najjar, R.G.; Pyke, C.R.; Adams, M.B.; Breitburg, D.; Hershner, C.; Kemp, M.; Howarth, R.; Mulholland, M.R.; Paolisso, M.; Secor, D.; et al. Potential climate-change impacts on the Chesapeake Bay. *Estuarine Coast. Shelf Sci.* **2010**, *86*, 1–20. [CrossRef]
42. Hayhoe, K.; Wake, C.; Huntington, T.; Luo, L.; Schwartz, M.; Sheffield, J.; Wood, E.; Anderson, B.; Bradbury, J.; DeGaetano, A.; et al. Past and future changes in climate and hydrological indicators in the US Northeast. *Clim. Dyn.* **2007**, *28*, 381–407. [CrossRef]

43. Hawkins, T.W.; Austin, B.J. Simulating streamflow and the effects of projected climate change on the Savage River, Maryland, USA. *J. Water Clim. Chang.* **2012**, *3*, 28–43. [CrossRef]
44. Lins, H.F. *USGS Hydro-Climatic Data Network 2009 (HCDN–2009)*; U.S. Geologic Survey: Reston, VA, USA, 2012; 4p.
45. Slack, J.R.; Landwehr, J.M. *Hydro-climatic Data Network (HCDN): A U.S. Geological Survey Streamflow Data Set for the United States for the Study of Climate Variations, 1874–1988*; U.S. Geological Survey: Reston, VA, USA, 1998.
46. Wang, D.; Hejazi, M. Quantifying the relative contribution of the climate and direct human impacts on mean annual streamflow in the contiguous United States. *Water Resour. Res.* **2011**, *47*. [CrossRef]
47. Sankarasubramanian, A.; Vogel, R.M.; Limbrunner, J.F. Climate elasticity of streamflow in the United States. *Water Resour. Res.* **2001**, *37*, 1771–1781. [CrossRef]
48. Small, D.; Islam, S.; Vogel, R.M. Trends in precipitation and streamflow in the eastern US: Paradox or perception? *Geophys. Res. Lett.* **2006**, *33*. [CrossRef]
49. Krakauer, N.; Fung, I. Mapping and attribution of change in streamflow in the coterminous United States. *Hydrol. Earth Syst. Sci.* **2008**, *12*, 1111–1120. [CrossRef]
50. Luce, C.; Staab, B.; Kramer, M.; Wenger, S.; Isaak, D.; McConnell, C. Sensitivity of summer stream temperatures to climate variability in the Pacific Northwest. *Water Resour. Res.* **2014**, *50*, 3428–3443. [CrossRef]
51. Patterson, L.A.; Lutz, B.; Doyle, M.W. Climate and direct human contributions to changes in mean annual streamflow in the South Atlantic, USA. *Water Resour. Res.* **2013**, *49*, 7278–7291. [CrossRef]
52. Trewin, B.C. *The Role of Climatological Normals in a Changing Climate*; World Meteorological Organization: Geneva, Switzerland, 2007.
53. Jones, J.A.; Creed, I.F.; Hatcher, K.L.; Warren, R.J.; Adams, M.B.; Benson, M.H.; Boose, E.; Brown, W.A.; Campbell, J.L.; Covich, A.; et al. Ecosystem processes and human influences regulate streamflow response to climate change at long-term ecological research sites. *BioScience* **2012**, *62*, 390–404. [CrossRef]
54. Homer, C.; Dewitz, J.; Fry, J.; Coan, M.; Hossain, N.; Larson, C.; Herold, N.; McKerrow, A.; VanDriel, J.N.; Wickham, J. Completion of the 2001 national land cover database for the counterminous United States. *Photogram. Eng. Remote Sens.* **2007**, *73*, 337.
55. Abatzoglou, J.T. Development of gridded surface meteorological data for ecological applications and modelling. *Int. J. Clim.* **2013**, *33*, 121–131. [CrossRef]
56. Abatzoglou, J.T.; Brown, T.J. A comparison of statistical downscaling methods suited for wildfire applications. *Int. J. Clim.* **2012**, *32*, 772–780. [CrossRef]
57. Meinshausen, M.; Smith, S.J.; Calvin, K.; Daniel, J.S.; Kainuma, M.; Lamarque, J.-F.; Matsumoto, K.; Montzka, S.; Raper, S.; Riahi, K. The RCP greenhouse gas concentrations and their extensions from 1765 to 2300. *Clim. Chang.* **2011**, *109*, 213. [CrossRef]
58. Rocheta, E.; Sugiyanto, M.; Johnson, F.; Evans, J.; Sharma, A. How well do general circulation models represent low-frequency rainfall variability? *Water Resour. Res.* **2014**, *50*, 2108–2123. [CrossRef]
59. Tamaddun, K.A.; Kalra, A.; Kumar, S.; Ahmad, S. CMIP5 Models' Ability to Capture Observed Trends under the Influence of Shifts and Persistence: An In-Depth Study on the Colorado River Basin. *J. Appl. Meteorol. Clim.* **2019**, *58*, 1677–1688. [CrossRef]
60. Tamaddun, K.A.; Kalra, A.; Ahmad, S. Spatiotemporal Variation in the Continental US Streamflow in Association with Large-Scale Climate Signals Across Multiple Spectral Bands. *Water Resour. Manag.* **2019**, *33*, 1947–1968. [CrossRef]
61. Papadimitriou, L.V.; Koutroulis, A.G.; Grillakis, M.G.; Tsanis, I.K. The effect of GCM biases on global runoff simulations of a land surface model. *Hydrol. Earth Syst. Sci.* **2017**, *21*, 4379–4401. [CrossRef]
62. Mote, P.; Brekke, L.; Duffy, P.B.; Maurer, E. Guidelines for constructing climate scenarios. *Eos Trans. Am. Geophys. Union* **2011**, *92*, 257–258. [CrossRef]
63. Wang, X. Advances in separating effects of climate variability and human activity on stream discharge: An overview. *Adv. Water Resour.* **2014**, *71*, 209–218. [CrossRef]
64. Choudhury, B. Evaluation of an empirical equation for annual evaporation using field observations and results from a biophysical model. *J. Hydrol.* **1999**, *216*, 99–110. [CrossRef]
65. Padrón, R.S.; Gudmundsson, L.; Greve, P.; Seneviratne, S.I. Large-Scale Controls of the Surface Water Balance Over Land: Insights From a Systematic Review and Meta-Analysis. *Water Resour. Res.* **2017**, *53*, 9659–9678. [CrossRef]

66. Budyko, M. *Climate and Life, 508 pp*; Academic Press: Cambridge, MA, USA, 1974; Volume 508, pp. 72–191.
67. Li, L.J.; Zhang, L.; Wang, H.; Wang, J.; Yang, J.W.; Jiang, D.J.; Li, J.Y.; Qin, D.Y. Assessing the impact of climate variability and human activities on streamflow from the Wuding River basin in China. *Hydrol. Process.* **2007**, *21*, 3485–3491. [CrossRef]
68. Roman, D.; Novick, K.; Brzostek, E.; Dragoni, D.; Rahman, F.; Phillips, R. The role of isohydric and anisohydric species in determining ecosystem-scale response to severe drought. *Oecologia* **2015**, *179*, 641–654. [CrossRef]
69. Caldwell, P.V.; Miniat, C.F.; Elliott, K.J.; Swank, W.T.; Brantley, S.T.; Laseter, S.H. Declining water yield from forested mountain watersheds in response to climate change and forest mesophication. *Glob. Chang. Boil.* **2016**, *22*, 2997–3012. [CrossRef]
70. Sankarasubramanian, A.; Vogel, R.M. Hydroclimatology of the continental United States. *Geophys. Res. Lett.* **2003**, *30*. [CrossRef]
71. Gao, G.; Fu, B.; Wang, S.; Liang, W.; Jiang, X. Determining the hydrological responses to climate variability and land use/cover change in the Loess Plateau with the Budyko framework. *Sci. Total. Environ.* **2016**, *557*, 331–342. [CrossRef]
72. Milly, P.C.D. Climate, soil water storage, and the average annual water balance. *Water Resour. Res.* **1994**, *30*, 2143–2156. [CrossRef]
73. Cooper, M.G.; Schaperow, J.R.; Cooley, S.W.; Alam, S.; Smith, L.C.; Lettenmaier, D.P. Climate Elasticity of Low Flows in the Maritime Western U.S. Mountains. *Water Resour. Res.* **2018**, *54*, 5602–5619. [CrossRef]
74. Kutta, E.; Hubbart, J.A. Changing Climatic Averages and Variance: Implications for Mesophication at the Eastern Edge of North America's Eastern Deciduous Forest. *Forests* **2018**, *9*, 605. [CrossRef]
75. Drohan, P.J.; Brittingham, M.; Bishop, J.; Yoder, K. Early Trends in Landcover Change and Forest Fragmentation Due to Shale-Gas Development in Pennsylvania: A Potential Outcome for the Northcentral Appalachians. *Environ. Manag.* **2012**, *49*, 1061–1075. [CrossRef]
76. Miller, A.; Zégre, N. Landscape-Scale Disturbance: Insights into the Complexity of Catchment Hydrology in the Mountaintop Removal Mining Region of the Eastern United States. *Land* **2016**, *5*, 22. [CrossRef]
77. Sayler, K.L. Land Cover Trends: Central Appalachians. Available online: https://pubs.usgs.gov/ds/844/pdf/ds844.pdf (accessed on 30 November 2019).
78. Kochenderfer, J.N.; Adams, M.B.; Miller, G.W.; Helvey, D.J. *Factors affecting large peakflows on Appalachian watersheds: lessons from the Fernow Experimental Forest*; U.S. Department of Agriculture, Forest Service, Northern Research Station: Newtown Square, PA, USA, 2007.
79. Kochenderfer, J.N.; Edwards, P.; Wood, F. Hydrologic Impacts of Logging an Appalachian Watershed Using West Virginia's Best Management Practices. *North. J. Appl. For.* **1997**, *14*, 207–218.
80. Messinger, T. *Comparison of Storm Response of Streams in Small, Unmined and Valley-Filled Watersheds, 1999–2001, Ballard Fork, West Virginia*; USGS: Reston, VA, USA, 2003.
81. Roy, A.H.; Dybas, A.L.; Fritz, K.M.; Lubbers, H.R. Urbanization affects the extent and hydrologic permanence of headwater streams in a midwestern US metropolitan area. *J. North Am. Benthol. Soc.* **2009**, *28*, 911–928. [CrossRef]
82. Bosch, J.M.; Hewlett, J.D. A review of catchment experiments to determine the effect of vegetation changes on water yield and evapotranspiration. *J. Hydrol.* **1982**, *55*, 3–23. [CrossRef]
83. Evaristo, J.; McDonnell, J.J. Global analysis of streamflow response to forest management. *Nature* **2019**, *570*, 455–461. [CrossRef]
84. Hornbeck, J.W.; Adams, M.B.; Corbett, E.S.; Verry, E.S.; Lynch, J.A. Long-term impacts of forest treatments on water yield: a summary for northeastern USA. *J. Hydrol.* **1993**, *150*, 323–344. [CrossRef]
85. Wullschleger, S.D.; Hanson, P.J.; Todd, D.E. Transpiration from a multi-species deciduous forest as estimated by xylem sap flow techniques. *For. Ecol. Manag.* **2001**, *143*, 205–213. [CrossRef]
86. Ford, C.R.; Laseter, S.H.; Swank, W.T.; Vose, J.M. Can forest management be used to sustain water-based ecosystem services in the face of climate change? *Ecol. Appl.* **2011**, *21*, 2049–2067. [CrossRef]
87. Young, D.; Zégre, N.; Edwards, P.; Fernandez, R. Assessing streamflow sensitivity of forested headwater catchments to disturbance and climate change in the central Appalachian Mountains region, USA. *Sci. Total. Environ.* **2019**. [CrossRef]
88. Nowacki, G.J.; Abrams, M.D. The Demise of Fire and "Mesophication" of Forests in the Eastern United States. *BioScience* **2008**, *58*, 123–138. [CrossRef]

89. Caldwell, P.; Muldoon, C.; Ford-Miniat, C.; Cohen, E.; Krieger, S.; Sun, G.; McNulty, S.; Bolstad, P.V. *Quantifying the Role of National Forest System Lands in Providing Surface Drinking Water Supply for the Southern United States*; U.S. Department of Agriculture Forest Service, Southern Research Station: Asheville, NC, USA, 2014.
90. Brzostek, E.R.; Dragoni, D.; Schmid, H.P.; Rahman, A.F.; Sims, D.; Wayson, C.A.; Johnson, D.J.; Phillips, R.P. Chronic water stress reduces tree growth and the carbon sink of deciduous hardwood forests. *Glob. Chang. Boil.* **2014**, *20*, 2531–2539. [CrossRef]
91. Schuler, T.M.; McGill, D.W. *Long-Term Assessment of Financial Maturity, Diameter-Limit Selection in the Central Appalachians*; USDA Forest Service: Washington, DC, USA, 2007; 16p.
92. Kang, H.; Sridhar, V. Assessment of Future Drought Conditions in the Chesapeake Bay Watershed. *JAWRA J. Am. Water Resour. Assoc.* **2018**, *54*, 160–183. [CrossRef]
93. Fernandez, R.; Sayama, T. Comparison of future runoff projections using Budyko framework and global hydrologic model: conceptual simplicity vs process complexity. *Hydrol. Res. Lett.* **2015**, *9*, 75–83. [CrossRef]

Article

Net Ecosystem Production of a River Relying on Hydrology, Hydrodynamics and Water Quality Monitoring Stations

Fernando Rojano [1,*], David H Huber [1,2], Ifeoma R Ugwuanyi [2], Vadesse Lhilhi Noundou [2], Andrielle Larissa Kemajou-Tchamba [2] and Jesus E Chavarria-Palma [1,2]

1 Gus R. Douglass Institute, West Virginia State University, Institute, WV 25112, USA; huberdh@wvstateu.edu (D.H.H.); jchavarriap@wvstateu.edu (J.E.C.-P.)

2 Department of Biology, West Virginia State University, Institute, WV 25112, USA; iru1@scarletmail.rutgers.edu (I.R.U.); vadln@udel.edu (V.L.N.); akemajoutchamba@wvstateu.edu (A.L.K.-T.)

* Correspondence: fernando.rojano@wvstateu.edu

Received: 27 January 2020; Accepted: 8 March 2020; Published: 12 March 2020

Abstract: Flow and water quality of rivers are highly dynamic. Water quantity and quality are subjected to simultaneous physical, chemical and biological processes making it difficult to accurately assess lotic ecosystems. Our study investigated net ecosystem production (NEP) relying on high-frequency data of hydrology, hydrodynamics and water quality. The Kanawha River, West Virginia was investigated along 52.8 km to estimate NEP. Water quality data were collected along the river using three distributed multiprobe sondes that measured water temperature, dissolved oxygen, dissolved oxygen saturation, specific conductance, turbidity and ORP hourly for 71 days. Flows along the river were predicted by means of the hydrologic and hydrodynamic models in Hydrologic Simulation Program in Fortran (HSPF). It was found that urban local inflows were correlated with NEP. However, under hypoxic conditions, local inflows were correlated with specific conductance. Thus, our approach represents an effort for the systematic integration of data derived from models and field measurements with the aim of providing an improved assessment of lotic ecosystems.

Keywords: basin; hydrologic model; reaeration rates; stream metabolism; watershed

1. Introduction

Physicochemical and hydro-morphological properties of a river combined with biological communities [1,2] cause simultaneous physical, chemical and biological processes. Consequently, spatiotemporal variation is under an endless search for equilibrium by means of interactions between biotic and abiotic factors [3,4]. Those interactions can be assessed using the net ecosystem production (NEP) [5], which has been successfully applied to rivers for several decades [6]. NEP is the combined representation of gross primary production and ecosystem respiration. However, NEP must balance reaeration rates with photosynthetic production, respiratory consumption and all processes that can cause changes to dissolved oxygen [7–9]. Among the complex interaction of nutrients, biomass and trophic structure, NEP can be used, for example, as a means to explain carbon fluxes [10,11], given the potential of rivers to store, mineralize and transport carbon to coastal areas [9]. To accurately estimate NEP, data associated with flow and water quality must be reliable [12]. Nonetheless, different deployed instruments are needed. To succeed in an analysis over time and space, our study alternatively proposes the use of models to estimate flow as an adequate approach to reduce instrumentation. It also proposes an integration of hydrologic and hydrodynamic models with water quality data which can also be implemented as an automated procedure to estimate NEP.

Development, calibration and validation of a hydrologic model to represent a drainage area along a river within the spatiotemporal domain must consider phenomena such as infiltration, evaporation and streamflow [13–15]. The hydrologic model took account of information about land use and topography specifications to handle hydro-climatic conditions in order to determine streamflow [16]. Consequently, streamflow can be coupled to a hydrodynamic model to predict flow along the river. In our study, the hydrologic and hydrodynamic models were implemented in the Hydrologic Simulation Program in Fortran (HSPF). Given that there was a set of unknown parameters governing the HSPF model, our study integrated a process for parameters calibration using the non-sorted genetic algorithm II (NSGA-II) [17,18]. This approach increased reliability under uncertainty of new hydrologic scenarios [19–21] and supported the use of an optimal solution [20,22,23] as the best set of parameters for the HSPF model. To guarantee satisfactory predictions, HSPEXP+ 2.0 was used to assess the calibrated HSPF model, as in the works of Xie et al. [20] and Lampert and Wu [24]. In this way, the streams module of the HSPF model can provide reliable predictions of the flow and velocity variables at specific locations along the river. Flow and velocity variables were input data to estimate reaeration rates. Nonetheless, an enhanced calculation of reaeration rates can be achieved by means of a set of equations [25] using a standardized Schmidt number [26,27] which was empirically estimated as a function of water temperature [28,29].

To provide an improved assessment of NEP, high-frequency data of water quality is now feasible [30], which seems promising and advantageous with respect to the periodic collection of water samples on a daily or lower frequency. Periodic observations of water quality at high-frequency could incorporate all processes such as cycles of nutrients (e.g., nitrogen, phosphorous), dissolved oxygen balance, sorption/desorption, volatilization, ionization, oxidation, biodegradation, hydrolysis and photolysis [31]. As result of all these processes, dissolved oxygen has been given special attention because it is the key variable for estimating NEP [32]. Dissolved oxygen is also closely related to water mixing, gas exchanges at the air–water interface, water temperature, flow, velocity and irradiance [27]. In addition, dissolved oxygen is subjected to spatial variability according to specifications of land use and local inflows [33–37]. Therefore, our study uses high-frequency data at various locations along the river as a way to contribute to the analysis of spatiotemporal impact of physicochemical properties of water on NEP.

This study was conducted within the Appalachian Region which is subjected to various water related stressors such as mining, urban settlements and industry. The Kanawha River, West Virginia, was chosen because it merges multiple inflows along the river at different rates and locations, disturbing water quantity and quality. Those inflows include tributaries, creeks, combined sewer overflows (CSO) and national pollutant discharge elimination system (NPDES). Therefore, NEP can be used as a proxy to assess those stressors and also to depict their variability over time and space.

In our study, hydrologic and hydrodynamic models and water quality data were integrated using a series of steps to estimate NEP under a high-frequency approach. Those steps were defined in this research as follows: (1) implementing a hydrologic model for drainage area along the river; (2) linking a hydrologic model with a hydrodynamic model; (3) collecting data about water temperature, dissolved oxygen and dissolved oxygen saturation, specific conductance, turbidity and ORP using monitoring stations installed along the river; (4) analyzing NEP under a spatiotemporal approach; and (5) assessing the impact of water quality on NEP and local inflows.

2. Materials and Methods

2.1. Watershed Description

The study area within the Appalachian Region was defined by 2995 km^2 draining water along the Kanawha River located in West Virginia, USA. Elk, Coal and Pocatalico Rivers (Figure 1) are tributaries of the Kanawha River; these rivers delimited the drainage area at the location of the flow gages F2, F4 and F5, respectively. The Kanawha River had F1 and F3 flow gages located at the upstream and center of the study area. Locations for the five flow gages are provided in Table 1. Along the Kanawha River, the start and end limits were defined at the highest and lowest elevation of 190 m (38.1381° N, 81.2144° W) and 172 m (38.4828° N, 81.8258° W) above sea level, respectively.

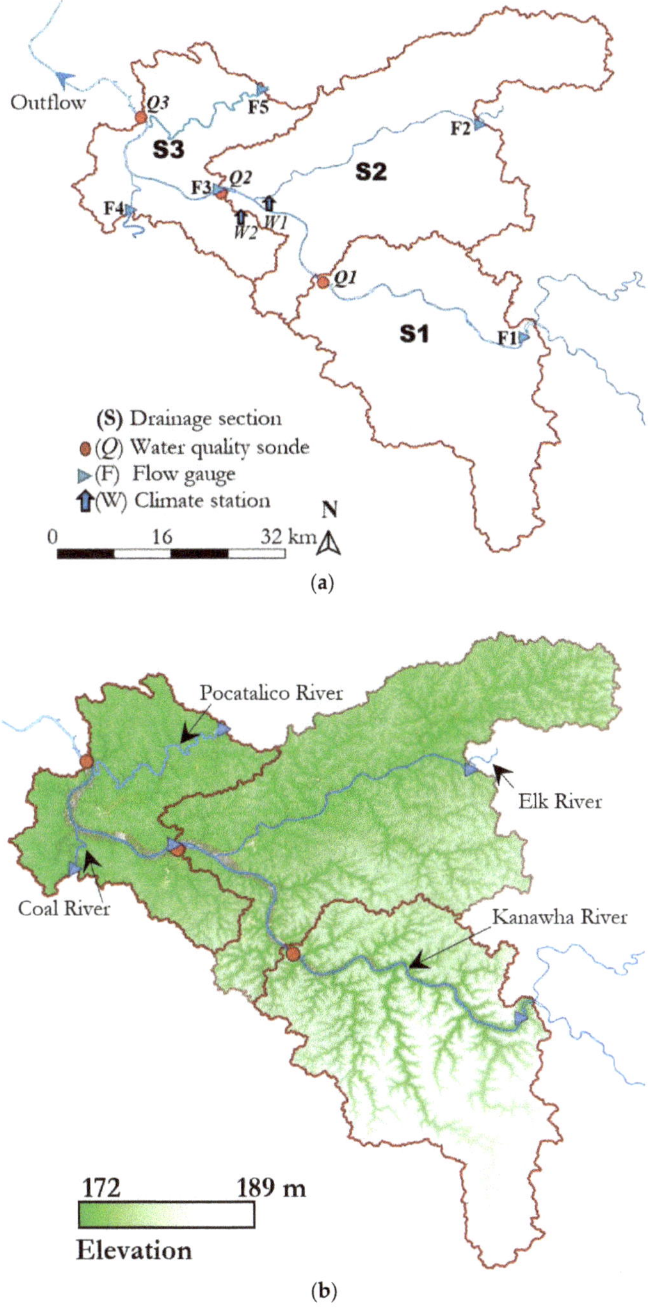

Figure 1. Description of (**a**) the study area and (**b**) drainage area of the Kanawha River, West Virginia, which were delimited by flow gages and water quality monitoring stations specified in Table 1.

Table 1. Location of flow gages, weather stations and water quality sondes indicated in Figure 1.

		Latitude (N)	Longitude (W)	Elevation (m)
Flow gage	F1	38.1381°	81.2144°	190
	F2	38.4714°	81.2839°	186
	F3	38.3714°	81.7022°	173
	F4	38.3389°	81.8412°	177
	F5	38.5261°	81.6310°	183
Weather stations	W1	38.3794°	81.5900°	279
	W2	38.3131°	81.7192°	277
Water quality sondes	Q1	38.2244°	81.5356°	181
	Q2a	38.3638°	81.6630°	173
	Q2b	38.3625°	81.6642°	173
	Q2c	38.3625°	81.6642°	170.5
	Q3	38.4828°	81.8258°	172

Land topography is predominantly hilly dissected terrain. It contains forest, urban, barren and agricultural land representing 86%, 5.7%, 4.8% and 3%, respectively, where barren land was mainly characterized by mining activities, and the rest of the study area was dedicated to wetland. A detailed description of land use is shown in Table 2. Hourly discharge at start of the Kanawha River (F1) had minimum, average and maximum flows of 35.1, 387.2 and 3645 m^3/s, respectively, during a period of observation of 913 days. The study area along the Kanawha River started with drainage areas at F1 and F2 of 21,680.8 and 2965.5 km^2, respectively. Inflows of tributaries at F4 and F5 comprised drainage areas of 2232.6 and 616.4 km^2, respectively. The study area (S1, S2 and S3) was focused on 2995 km^2, which was bounded by the locations along the Kanawha River of the first flow gage (F1) and three sondes (Q1, Q2 and Q3). This study area provided conditions to conduct an analysis of the flow dynamics and impact of local inflows on water quality.

Table 2. General description of land use for study area indicated in Figure 1.

		S1 (km^2)	S2 (km^2)	S3 (km^2)
Land Use	Urban	33.33	88.32	49.1
	Agriculture	14.68	39.6	36.34
	Forest	899.7	1307	359.83
	Wetland	9.5	7.69	5.18
	Barren	131.75	11.11	1.71

The study area was characterized by annual average minimum and maximum dry bulb temperatures of 6.7 and 18.9 °C, respectively, an annual average precipitation of 1107 mm, and an annual average snowfall of 838 mm. Snow counted as precipitation, which was melted following the heat balance approach relying on precipitation, air temperature, solar radiation, wind velocity and dew point [13]. Climate data about precipitation, dry and dew point temperatures and wind speed were provided by two weather stations (Figure 1). Solar radiation was not available in the two weather stations, so it was retrieved as an average for the study area from the national solar radiation database [38].

2.2. HSPF Model Description

The HSPF model was used for predicting flows of a watershed that joins modeling of watersheds and streams [39–41]. In addition, the HSPF model can be used to incorporate the transport of pollutants and nutrients. Information about land use, topography and climate assisted in the estimation of flow in streams (Table 3) that follows a hydrodynamic pattern according to the terrain slope. Before water arrives at the streams, there are regular paths of water flow within the watershed which can be summarized in Figure 2. For instance, after rainfall there is an immediate interception by the canopy (CEPSC). The remaining water enters soil, which has a capacity to infiltrate and store within the upper zone (UZS) and lower zone (LZS). The excess of water could continue to reach active groundwater (AGWS) and base flow (BASETP) or enter deeper aquifers.

Table 3. Mainstream and tributaries specifications.

Drainage Section	Stream	Length, km	Elevation Change, m	Average Slope, m	Average Width, m
1	Kanawha River	38.5	9 *	0.00005	183
2	Kanawha River	23.5	7.8 *	0.00004	183
	Elk River	41.5	13	0.00031	76
3	Kanawha River	29.3	0.61	0.00003	183
	Coal River	18.7	4.57	0.00024	52
	Pocatalico River	41	10.4	0.00025	31

* Includes a dam with an average level change of 7.1 m.

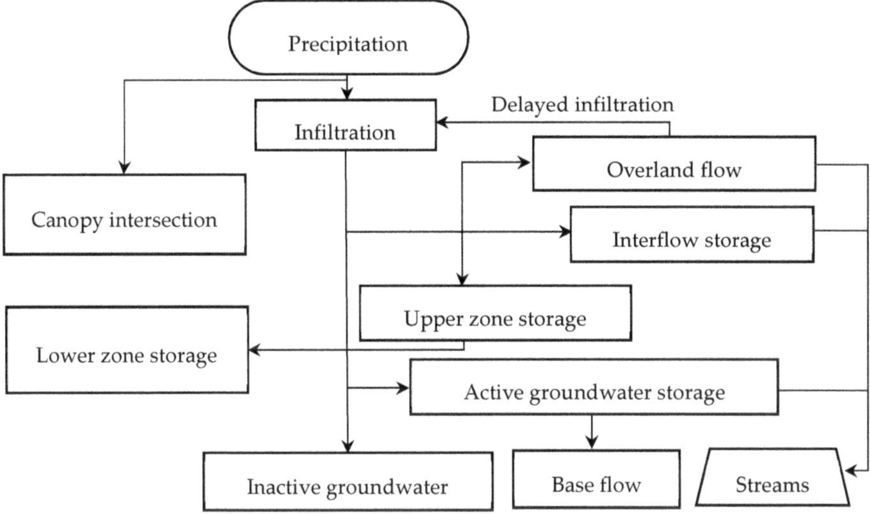

Figure 2. Water paths and storages within a watershed occurring in the pervious land used in the HSPF model.

Concurrently, all of the watershed is subjected to evapotranspiration, where rates depend mostly on solar radiation, air temperature and humidity, whereas underground water is subjected only to evaporation. Each stage follows specific equations to determine flow rates which are defined by a set of parameters according to specifications of the watershed such as average slope (SLSUR) and mean elevation (MLS). Other parameters must be found during a process of adjustment such as groundwater recession flow (KVARY) and infiltration (INFILT). Descriptions of all parameters in the HSPF model are shown in Table 4, where seven values were deduced from the input data related to land use and topography and thirteen parameters must be calibrated in order to reduce error in flow predictions.

Modeling of streams relied on input data from inflows, river network configuration and a one dimensional approach of the river under a fully advective flow. The one dimensional approach required a homogeneous river transect, a representative Manning's coefficient (equal to 0.1) and a slope excluding dam elevation. Then, drainage sections with their corresponding streams were used to generate a HPSF model that was able to predict average flow and water velocity of streams. The governing equations used within the frame of a HSPF model can be found in Duda et al. [13].

Table 4. Parameters required to build up a Hydrologic Simulation Program in Fortran (HSPF) model.

Parameter	Units	Meaning	Range or Averaged Data
LZSN	(mm)	Lower zone nominal storage	51–381
INFILT	(mm/h)	Index to soil infiltration capacity	0.02–13
LSUR	(m)	Average length of the assumed overland flow plane	69.2 *
SLSUR	(-)	Average slope of the overland flow plane	0.107 *
KVARY	(1/mm)	Variable groundwater recession flow	0–0.2
AGWRC	(1/day)	Basic groundwater recession rate	0.85–0.99
PETMAX	(°C)	Air temperature below which evapotranspiration (ET) will arbitrarily be reduced below the value obtained from the input time series	4.4 *
PETMIN	(°C)	Temperature below which E-T will be zero regardless of the value in the input time series	1.7 *
INFEXP	(-)	Exponent in the infiltration equation	2 **
DEEPFR	(-)	Fraction of groundwater inflow which will enter deep (inactive) groundwater	0–0.5
BASETP	(-)	Base flow evapotranspiration	0–0.2
AGWETP	(-)	Fraction of remaining potential E-T which can be satisfied from active groundwater storage	0–0.2
CEPSC	(mm)	Interception storage capacity	0–10
UZSN	(mm)	Upper zone nominal storage	1–51
NSUR	(-)	Manning's n for overland flow	0.1–0.5
INTFW	(-)	Interflow inflow parameter	1–10
IRC	(1/day)	Interflow recession constant	0.3–0.85
LZETP	(-)	Lower zone E-T parameter. It is an index to the density of deep-rooted vegetation.	0–0.9
LAT	(°)	Latitude of the pervious land segment (PLS)	38 *
MELEV	(m)	Mean elevation of the PLS	181 *

* Average value was deduced from the watershed data in BASINS 4.1. ** Value recommended by [13].

2.3. Multiobjective Calibration of the HSPF Parameters

Among the various tools available in the search for adequate parameters defining the water dynamics of the HSPF model, we chose NSGA-II to look at the solution of two optimized objectives. The procedure to implement NSGA-II in MATLAB consisted of an iterative evaluation of different scenarios of the hydrologic model. Different scenarios were obtained using initial random values within the range stated in Table 3 regarding the parameters LZSN, INFILT, KVARY, AGWRC, DEEPFR, BASETP, AGWETP, CEPSC, UZSN, INTFW, IRC, LZETP and NSUR. The iterative evaluation was accomplished for 400 sets of parameters that were evaluated in the HSPF model (Figure 3). The next generation was deduced by creating 400 new sets of parameters that had a crossover and mutation probability of 0.9 and 0.1, respectively. The NSGA-II considered 1000 generations to define final calibrated values of the 13 parameters. To identify the best set of solutions, NSGA-II implemented two objectives: Nash–Sutcliffe model efficiency (NSE) and the percent bias coefficient (PBIAS) as the criteria to evaluate the error between flow measurements and HSPF model flow predictions. Identification of the optimal solution was accomplished by means of the Pareto front, which has the best solution when the magnitude is minimum for NSE [42] and PBIAS [43].

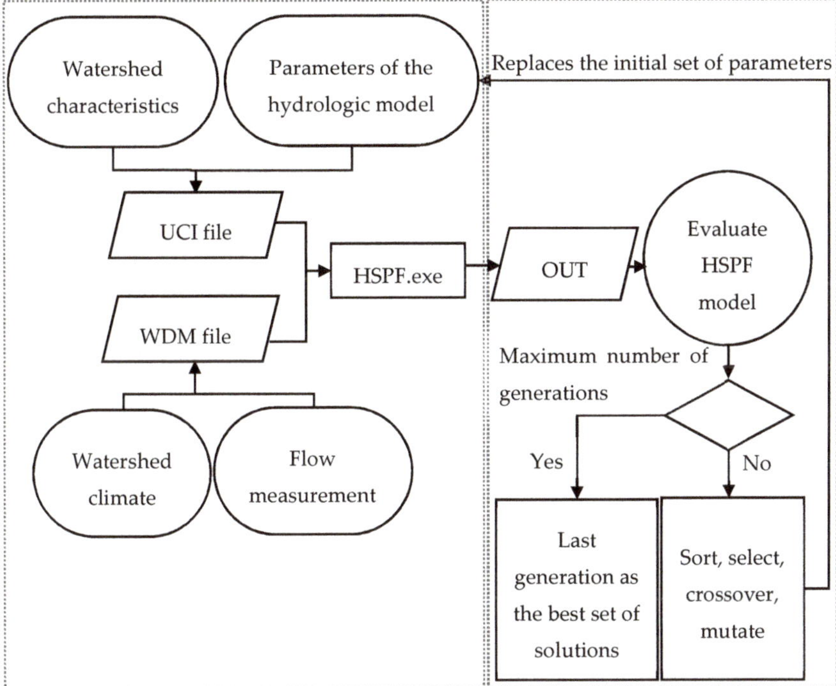

Figure 3. Relation between the HSPF model and non-sorted genetic algorithm II (NSGA-II) to find the best optimal solution.

2.4. Field Measurements of Water Quality

The high-frequency monitoring system consisted of three Eureka Manta 2 multiprobe sondes. Each sonde measured water temperature (±0.1 °C), dissolved oxygen (±0.2 mg/L), dissolved oxygen saturation (±1%), specific conductance (±1 of reading), turbidity (±3% of reading) and oxidative-reductive potential (±20 mV) with a time step of 1 h. Three locations along the river indicated in Figure 1, defined by Q1, Q2 and Q3, were monitored. That configuration facilitated our estimate of water quality changes as water moved downstream. Differences of sensor measurements served to further assess impact of local inflows comprising point and nonpoint sources of water pollution due to drained water along the river.

2.5. Net Ecosystem Production

NEP provides an assessment of rivers that encompasses physical and chemical characteristics. Physical characteristics include slope, width, depth and flow together with chemical characteristics such as nutrients, organic matter and water chemistry. In addition, other factors can be intrinsically intervening in NEP dynamics such as the effects of dams, riparian vegetation and pollution. NEP can also be seen as the balance of autotrophic and heterotrophic elements of the river [7]. Specifically, NEP can be evaluated through Equation (1) of Odum, 1956 [5].

$$\frac{dO}{dt} = \text{NEP} + k(C_s - C) + P \tag{1}$$

where NEP is the gross primary production minus ecosystem respiration. k is oxygen reaeration coefficient. C_s is dissolved oxygen saturation and C is dissolved oxygen observed. P is the drainage accrual and accounts for all processes happening in the river together with dissolved oxygen of local inflows. Some of those processes include horizontal and vertical advection, photochemical oxidation of organic matter and nonaerobic consumption of oxygen during the time step of observation [7].

Estimation of the k value was determined by means of k_{600} (Equation (2)) which can be obtained by using one of three candidate Equations (4)–(6) and the Schmidt number. According to Raymond et al. [25], those three equations had the best fit with respect to field measurements. Those three equations also relied on the Schmidt number (Equation (3)) to estimate the mass transfer rates under momentum. The Schmidt number is the ratio of kinematic viscosity to the diffusion coefficient, which in turn can be determined as a function of temperature.

$$k_{600} = \left(\frac{600}{Sc}\right)^{-0.5} \times k \quad (2)$$

$$Sc = A + BT + CT^2 + DT^3 \quad (3)$$

$$k_{600} = 5037 \times (VS)^{0.89} \times H^{0.54} \quad (4)$$

$$k_{600} = 5937(1 - 2.54Fr^2) \times (VS)^{0.54} \times H^{0.58} \quad (5)$$

$$Fr = V/\sqrt{gH}$$

$$k_{600} = 4725 \times (VS)^{0.86} \times Q^{-0.14} \times H^{0.66} \quad (6)$$

where V is water velocity, S is slope and H is depth of the river. g is the gravity force. Sc is the Schmidt number and T is temperature. Fr is the Froude number. Constant values are A = 1568, B = −86.04, C = 2.142 and D = −0.0216 [25].

3. Results and Discussion

3.1. Input Data and Calibration

Data for all flow gages and climate stations were retrieved for the period from 1 October 2015 to 31 March 2018. To match all data, a common time step of one hour was adopted for all variables. Data from two climate stations were averaged instead of segmenting the watershed according to the area of influence, since both stations were in proximity. An example of average precipitation is shown in Figure 4. It should be noted that peaks related to precipitation might not coincide with peaks on measurements of flow gages due to local inflows of tributaries to the Kanawha River. The estimation of evapotranspiration rates was deduced by following the Turc method [44] and adding this data to the HSPF model. Climate data used in this model were compared by means of coefficient of variation (CV) with NASA data sources (i.e., NLDAS and AIRS) in daily time step for precipitation, temperature, dew-point temperature and evapotranspiration; results are presented in Table 5. Data of flow gages originally obtained with a time step of 15 minutes were converted to a 1 h time step using a moving average filter.

Table 5. Average daily data comparison between NASA data sources and climate station.

	Precipitation, mm		**Temperature, °C**		**Dew Point Temperature [1], °C**		**Evapotranspiration (EVT), mm**		
Source	Station	NLDAS	Station	NLDAS	Station	NLDAS and AIRS	Station	NLDAS	Potential EVT
Mean	2.54	2.91	12.78	12.88	7.72	10.11	0.965	1.677	4.318
CV	2.292	2.085	0.299	0.302	0.380	0.368	0.964	0.706	0.533

[1] Dew point was estimated from average daily temperature (NLDAS) and relative humidity (AIRS) datasets.

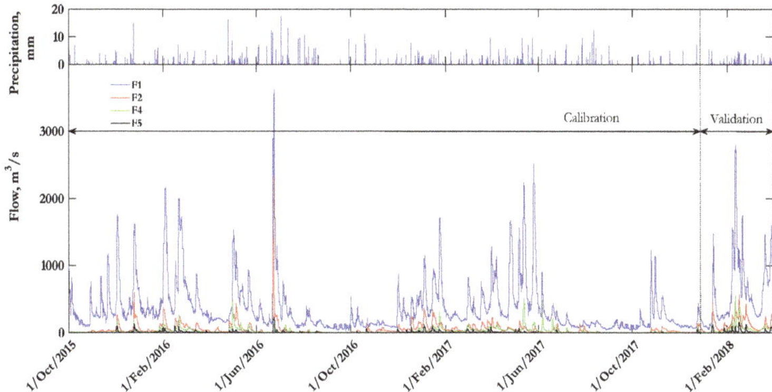

Figure 4. Observed precipitation and flows in the Kanawha River and its tributaries during 1 October 2015 to 31 March 2018.

The calibration of the HSPF model was conducted for 2413 km^2, corresponding to S1 and S2 drainage sections. The HSPF model was subjected to an iterative evaluation using the NSGA-II. The best optimal solution was deduced with the minimum Euclidean distance from the origin to the NSE and PBIAS scores. It should be pointed out that NSE and PBIAS were applied to the outflow F3 comprising inflows F1 and F2 together with predicted drained water at S1 and S2. Nonetheless, inflows F1 and F2 greatly contributed to the outflow predictions, given that the drainage area from F1 and F2 to F3 increased by 10%. This means that drainage changed from 24,646.3 to 27,060 km^2 at the location of gage F3. Such conditions enhanced NSE and PBIAS scores, which were 0.96 and 1.97%, respectively, when comparing predictions and observations from 1 October 2015 to 11 January 2018 in flow gage F3 (Figure 5). NSE and PBIAS scores can be categorized as acceptable [43]; however, those results should be weighed based on the aggregated water between inflow and outflow of the drainage area along the river.

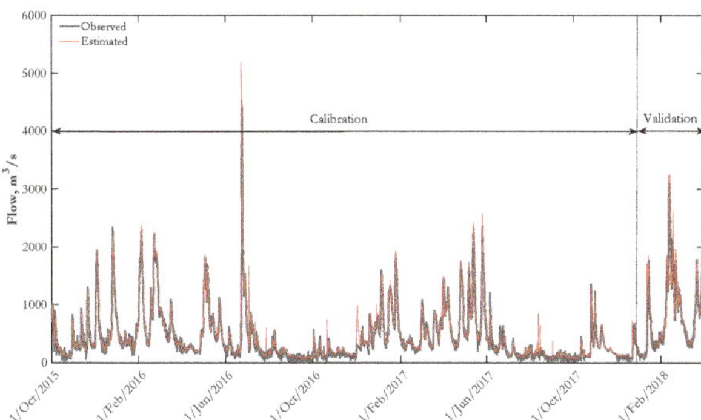

Figure 5. Observed and HPSF predictions of the flows at F3 during the stage of calibration and validation.

To confirm the adequacy of these parameters (Table 6) as the best optimal solution identified by NSGA-II, the calibrated HSPF model was analyzed by means of the HSPEXP+ 2.0 program in order to fulfill overall criteria regarding water balance (Table 7). It was found that error between predictions and measurements increased as the flow decreased; even so, the criteria were satisfied. Subsequently, the HSPF model predicted a water budget that was distributed as follows: 4.3% to surface flow, 17.9% to interflow, 32.5% to base flow and deep aquifers and 45.3% to evapotranspiration. The rate of evapotranspiration dominated water balance and was driven by BASETP and AGWETP and

LZETP parameters. Water also accumulated in soil at rates determined by USZN, LZSN and INFILT parameters; however, a significant volume of water moved down to the base flow and deep aquifers.

Table 6. Calibrated parameters of the HSPF model identified by the NSGA-II.

Parameter	Value	Parameter	Value
LZSN	51	CEPSC	9.7
INFILT	6.1	UZSN	4.4
KVARY	0.098	NSUR	0.109
AGWRC	0.98	INTFW	2.647
DEEPFR	0.494	IRC	0.302
BASETP	0.198	LZETP	0.709
AGWETP	0.105		

Table 7. HSPF model performance through HSPEXP+ 2.0.

Measure		HSPEXP Limit
Error in total volume, %	1.71	10
Error of highest 10% flows, %	−0.21	15
Error of highest 25% flows, %	−0.49	10
Error of highest 50% flows, %	−0.03	10
Error of lowest 50% flows, %	9.59	10
Error of lowest 25% flows, %	13.29	15
Error of lowest 10% flows, %	14.72	20

Flow estimations using the calibrated HSPF model at the S1 and S2 drainage sections can be considered reliable as they were validated by flow measurements at gage F3. However, flow estimations at the outlet (Q3) of the S3 drainage area entirely relied on the accuracy of the calibrated HSPF model. The HSPF model validation found a significant contribution of the tributaries to the Kanawha River. For instance, from the total amount of water added within the S2 section, 78% of the water was contributed by the Elk River based on flow measurements at F2. In the same way, from the total amount of water added within the S3 section, 91% of the water was contributed by Coal and Pocatalico Rivers, according to measurements at F4 and F5. Flow dynamics at Q1, Q2 and Q3, during the period from 11 January 2018 to 31 March 2018 (Figure 6), were based on the combined effects of inflows and drainage areas along the river. S1, S2 and S3 involved local inflows such as rainfalls, CSO and NPDES. The CV for flow data was computed having 0.77, 0.73 and 0.73 for locations Q1, Q2 and Q3, respectively. These CV values verified that flow dynamics were similar only in the Q2 and Q3 locations.

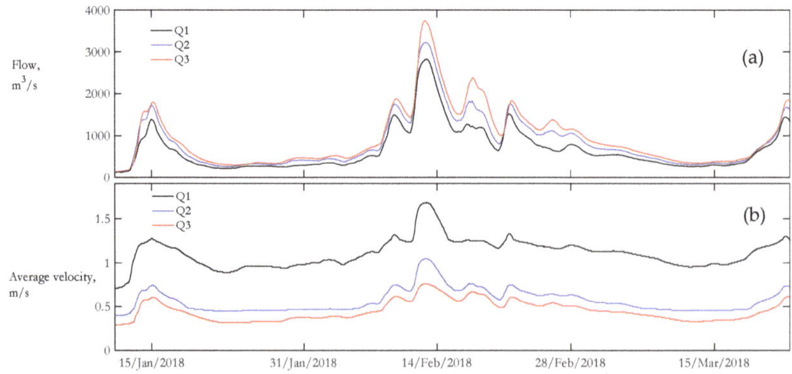

Figure 6. HSPF model predictions about (**a**) flow and (**b**) average water velocity at the locations Q1, Q2 and Q3 along the Kanawha River.

3.2. Water Quality

Water quality data were collected for 71 days with a time step of 1 h covering the period from 11 January 2018 to 22 March 2018 (Supplementary Data). Erroneous readings were discarded. Measurements for the three locations are shown in Figures 7 and 8 and available in Huber et al. [45]. The CV among all sensor readings (Table 8) showed that minimum and maximum scores were for dissolved oxygen saturation and turbidity, respectively. The same type of sensor readings among the three locations also had minimum and maximum differences of the CV for temperature and dissolved oxygen saturation. In summary, we found that dissolved oxygen saturation had minimum dispersion among all the sensors at the same location and maximum dispersion among the three locations.

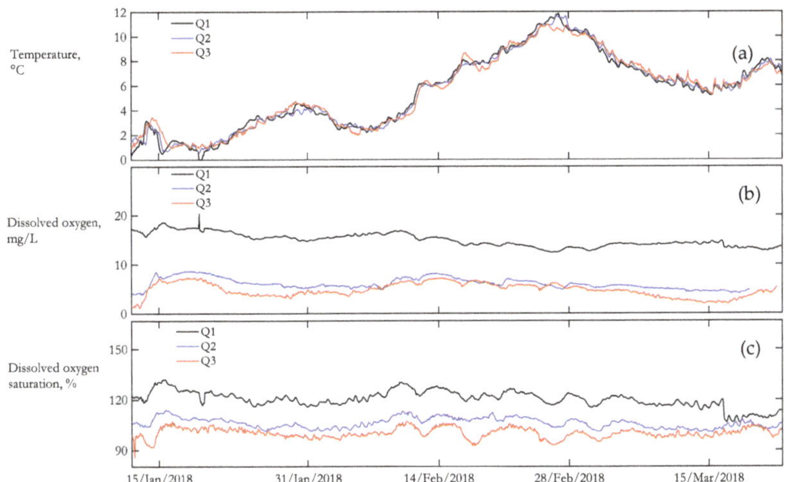

Figure 7. Sensor measurements of (**a**) temperature, (**b**) dissolved oxygen and (**c**) dissolved oxygen saturation at three locations (Q1, Q2 and Q3) of the Kanawha River.

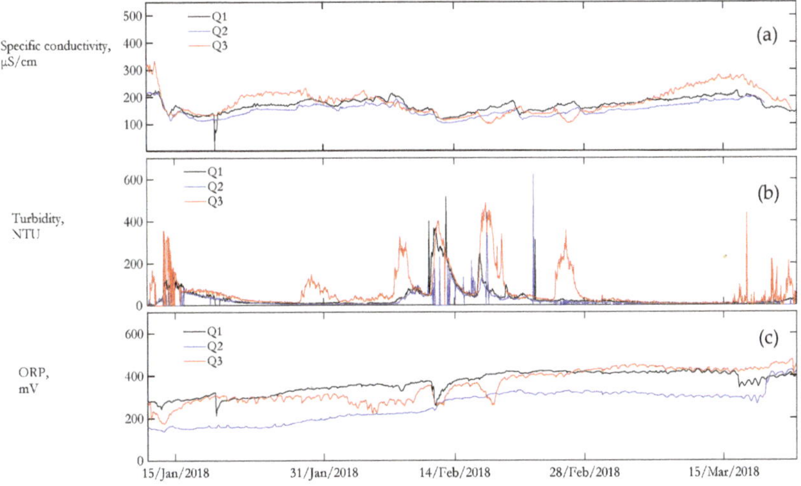

Figure 8. Sensor measurements of (**a**) specific conductance, (**b**) turbidity and (**c**) oxidation–reduction potential (ORP) at three locations (Q1, Q2 and Q3) of the Kanawha River.

Table 8. Coefficient of variation for water quality measurements from 11 January 2018 to 22 March 2018.

Location	Temperature	Dissolved Oxygen	Dissolved Oxygen Saturation	Specific Conductance	Turbidity	ORP
Q1	0.551	0.097	0.042	0.142	1.549	0.134
Q2	0.543	0.192	0.029	0.280	1.628	0.281
Q3	0.529	0.304	0.030	0.266	1.311	0.208

3.3. Net Ecosystem Production

The reaeration rates (k) were calculated using Equations (4)–(6) that consequently helped to estimate NEP through Equation (1). The time series of hydrodynamics (Figure 6) and water quality data (Figure 7) were used at Q1, Q2 and Q3 locations to estimate NEP (Figure 9). Hydrodynamics of the Kanawha River showed higher flows and lower water velocities as water moved downstream, from Q1 to Q2 and to Q3. From water quality monitoring stations (Figure 7), a significant decay of dissolved oxygen and its saturation were observed in Q2 and Q3. For instance, a length of 23.5 km along the river, the distance between Q1 and Q2, had an average decay from 14.9 to 6.1 mg/L. In the following 29.3 km, the distance between Q2 and Q3, had an additional decay from 6.1 to 4.8 mg/L. Those dissolved oxygen decays reduced the NEP estimations from Q1 to Q2 by 93% and from Q2 to Q3 by 95%.

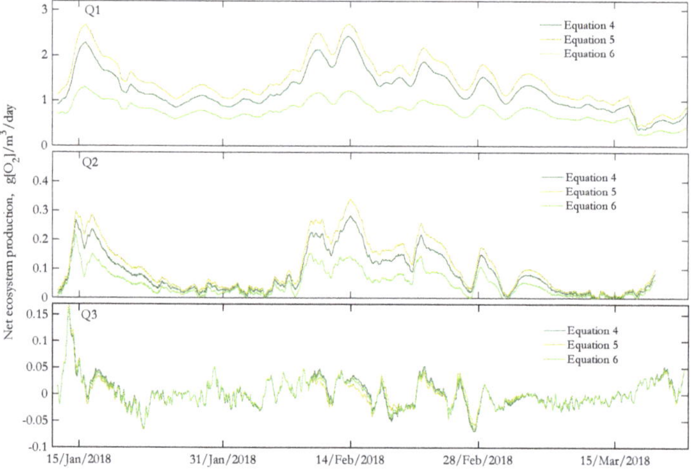

Figure 9. Comparison of the net ecosystem production using Equations (4)–(6) to estimate reaeration rates k at Q1, Q2 and Q3 locations along the Kanawha River, from 11 January 2018 to 22 March 2018.

3.4. Spatial and Temporal Variability of NEP

Repeatability and reproducibility are issues around NEP estimations when limited datasets are available in either time or space. These issues frequently happen over sub-daily patterns of dissolved oxygen [46]. In practice, NEP estimations are subjected to periods of observations, the choice of location along the river and the choice of the equation to estimate reaeration rates. Our study provided insight into NEP resolution over space and time based on field surveys and estimating reaeration rates through Equation (4).

For large rivers, either temporal or permanent local gradients may be observed as a result of flow regimes generating specific hydrodynamics [47,48] and consequently variability in NEP. Those NEP estimations will be the consequence of the level of turbulence in local mixing and the exchange rate of gases in the air–water boundary layer [49]. Nonetheless, NEP variability is also a consequence of physicochemical properties, such as organic matter [50], nutrient regimes [35], water temperature [51] and flow [10]. Our study conducted a river transect examination of NEP using horizontal and vertical profiles that occur near Q2. The horizontal profile was based on the Q2a and Q2b locations which are separated by 179 m, whereas the vertical profile was based on the Q2b and Q2c locations which are separated by 1.5 m. Three repetitions were conducted at each location, and water velocity was deduced from the HSPF streams module (Table 9). The horizontal profile of NEP was not homogeneous since different conditions were observed in the field; however, there was a prevailing lower NEP in Q2b, which mainly occurred due to water temperature and dissolved oxygen measurements. In contrast, we found that the vertical profile generated lower NEP amounts in Q2c with respect to Q2b, which can be inferred as less prevailing irradiance, as it was 2.5 m deep. Still, such a discrepancy in NEP at the vertical profile was lower than the horizontal profile as a consequence of the distance between locations. Thus, these findings illustrated that spatial heterogeneity of NEP is driven by transport-reaction phenomena [52] due to local gradients created by hydrodynamics and their corresponding water quality.

Table 9. River transect examination using average net ecosystem production (NEP) with N number of observations.

Location	Repetition	N	Depth, m	Velocity, m/s	Water Temperature, °C	Dissolved Oxygen, mg/L	NEP, $g[O_2]/m^3/day$
Q2a	1	36	1	0.12	28.3 (±0.9)	8.6 (±0.3)	4.3
	2	54	1	0.08	26.7 (±1.2)	9.2 (±0.5)	3.3
	3	163	1	0.05	26 (±0.5)	8.3 (±0.1)	1.7
Q2b	1	63	1	0.11	27.8 (±0.8)	8.2 (±0.2)	1.8
	2	49	1	0.08	26 (±0.5)	8.2 (±0.1)	0.9
	3	73	1	0.05	27.3 (±1.7)	8.7 (±0.3)	1.2
Q2c	1	62	2.5	0.06	26.7 (±1.5)	8.4 (±0.3)	0.8
	2	60	2.5	0.04	26.3 (±1.0)	8.4 (±0.1)	0.7
	3	70	2.5	0.05	26.6 (±1.3)	8.4 (±0.2)	0.9

NEP in rivers is the result of a dynamic interaction between biotic [53] and abiotic factors [54,55]. Among those biotic factors, autotrophs and heterotrophs are continuously balanced to determine NEP dynamics through the year [56]. Autotrophy impacting NEP along the river is a consequence of nutrient loads from local inflows to the mainstream such as wastewater treatment plants, CSO and NPDES [57]. For instance, it has been found that spatial heterogeneity of NEP can be caused by watersheds comprising urban areas [58]. Our study had an urban area between Q2 and Q3, causing differentiated NEP estimations along the river (Figure 9). A driving variable causing a decay of NEP (Table 10) was dissolved oxygen, which mainly declined due to the various local inflows mixed with the mainstream. It can be interpreted that water residence times of 0.25 ± 0.03 days and 0.62 ± 0.12 days for the Q1–Q2 and Q2–Q3 river sections, respectively, along with local inflows with different water quality did not help to keep the same NEP estimations as observed in Q1. From our study, we can also claim that the balance between autotrophs and heterotrophs was significantly impaired as water moved downstream. We estimated a decay of 1.18 ± 0.38 $g[O_2]/m^3/day$ from NEP_{Q1} to NEP_{Q2} and an additional decay of 0.08 ± 0.12 $g[O_2]/m^3/day$ from NEP_{Q2} to NEP_{Q3}. The latter one was a consequence of prevailing hypoxic conditions observed in Q3.

Table 10. Net ecosystem production (NEP) at three locations of the Kanawha River using N number of observations.

Location	N	Velocity, m/s	Water Temperature, °C	Dissolved Oxygen, mg/L	NEP, $g[O_2]/m^3/day$
Q1	1682	1.11 (±0.16)	5.5 (±3)	14.9 (±1.4)	1.24 (±0.4)
Q2	1600	0.6 (±0.13)	5.4 (±3)	6.1 (±1.2)	0.09 (±0.07)
Q3	1674	0.46 (±0.11)	5.4 (±3)	4.8 (±1.5)	0.004 (±0.03)

3.5. Impact of Water Quality on NEP and Local Inflows

Additional water quality data were retrieved that can potentially affect NEP dynamics. In particular, the effects of specific conductance [59] and turbidity [60] on NEP were evaluated using the Spearman coefficient. We found through the Spearman coefficient that NEP and turbidity were positive at the three locations along the river (Table 11). We also found that NEP and specific conductance had negative values at the three locations along the river. Nevertheless, it can be deduced that specific conductance and turbidity could play a significant role in determining NEP if the dissolved oxygen measurements do not reach hypoxic conditions.

Table 11. Spearman correlation of net ecosystem production (NEP) with water quality data at three locations of the Kanawha River.

	Specific Conductance	**Turbidity**
NEP_{Q1}	−0.61	0.81
NEP_{Q2}	−0.57	0.41
NEP_{Q3}	−0.01	0.06

Because water quality and consequently NEP changed due to local inflows along the river, we computed volumes of aggregated water to the Kanawha River between paired locations $W_{Q1\text{–}Q2}$ and $W_{Q2\text{–}Q3}$ as well as their corresponding changes on NEP. We also conducted the same calculations for changes in specific conductance, turbidity and ORP measurements. For the aforementioned calculations, we considered average travel times obtained from the HSPF streams module. Then, Spearman correlations were calculated (Table 12). The segment of the river between Q1 and Q2 showed that $W_{Q1\text{–}Q2}$ was mainly correlated with NEP and turbidity. For the segment of the river between Q2 and Q3, we found that $W_{Q2\text{–}Q3}$ was mainly correlated with specific conductance and NEP. From these correlations, we can state that NEP can be used as an indicator to assess water quality of local inflows, as it merges various properties of the river related to hydrodynamics and water quality data along the river. However, a more reliable assessment could be achieved if hypoxic conditions are avoided.

Table 12. Spearman correlations between local inflows ($W_{Q1\text{–}Q2}$ and $W_{Q2\text{–}Q3}$) along the Kanawha River and their corresponding changes on the net ecosystem production (NEP), specific conductance, turbidity and ORP.

Local Inflows	**NEP**	**Specific Conductance**	**Turbidity**	**ORP**
$W_{Q1\text{–}Q2}$	−0.71	−0.28	−0.55	0.42
$W_{Q2\text{–}Q3}$	−0.42	−0.61	0.15	−0.29

4. Conclusions

In order to estimate flows along the Kanawha River, our study had to consider a drainage area of 2995 km^2. A HSPF model was developed and then calibrated by means of NSGA-II in order to identify the best optimal solution. The streams module of the HSPF model served for hydrodynamic modeling, which provided data about flow and average water velocity at Q1, Q2 and Q3 locations along the Kanawha River. In addition, water quality data were collected for 71 days by placing sondes in the same three locations to hourly log dissolved oxygen concentration, dissolved oxygen saturation, water temperature, specific conductance, turbidity and ORP. Flow and average velocity data were used to estimate reaeration rates (k). Then, k values were used together with water quality data to estimate NEP. It was found that NEP greatly depends on the specific location within the river, as it was observed during a river transect examination. Our study also identified a decreasing NEP as water moved downstream, starting from NEP_{Q1} equal to 1.24 (±0.4) g[O_2]/m^3/day to NEP_{Q2} equal to 0.09 (±0.07) g[O_2]/m^3/day and to NEP_{Q3} equal to 0.004 (±0.03) g[O_2]/m^3/day. Such decay was attributed to local inflows ($W_{Q1\text{–}Q2}$ and $W_{Q2\text{–}Q3}$), which were computed and correlated with their corresponding changes in water quality and NEP. The best Spearman coefficient (ρ = −0.71) was between W_{Q1-Q2} and NEP. However, under hypoxic conditions, the best Spearman coefficient (ρ = −0.61) was between W_{Q2-Q3} and specific conductance. These findings showed that spatial and temporal analyses of NEP were adequately addressed through datasets of hydrology, hydrodynamics and high-frequency data from water quality monitoring stations. Our study can also be useful for further research where assessment of local inflows to the mainstream should be accomplished by means of NEP. These advances encourage us to count more on field surveys, given that the scope of NEP dynamics in rivers depends on multiple scenarios related to flow and water quality conditions.

Supplementary Materials: The Supplementary Data are available online at http://www.mdpi.com/2073-4441/12/3/783/s1, Data: water quality data.

Author Contributions: Conceptualization, Software, F.R.; Investigation and Data Curation, D.H.H., I.R.U., V.L.N., A.L.K.-T. and J.E.C.-P.; Writing Original Draft Preparation, Writing-Review and Editing, D.H.H. and F.R. All authors have read and agreed to the published version of the manuscript.

Funding: This research is supported by the National Science Foundation under Award No. OIA-1458952. Any opinions, findings and conclusions, or recommendations expressed in this material are those of the author(s) and do not necessarily reflect the views of the National Science Foundation.

Conflicts of Interest: The authors declare no conflict of interest.

References

1. Izagirre, O.; Bermejo, M.; Pozo, J.; Elosegi, A. RIVERMET©: An Excel-based tool to calculate river metabolism from diel oxygen-concentration curves. *Environ. Model. Softw.* **2007**, *22*, 24–32. [CrossRef]
2. Palmer, M.; Ruhi, A. Linkages between flow regime, biota, and ecosystem processes: Implications for river restoration. *Science* **2019**, *365*, eaaw2087. [CrossRef] [PubMed]
3. Uehlinger, U. Annual cycle and inter-annual variability of gross primary production and ecosystem respiration in a floodprone river during a 15-year period. *Freshw. Biol.* **2006**, *51*, 938–950. [CrossRef]
4. Savoy, P.; Appling, A.P.; Heffernan, J.B.; Stets, E.G.; Read, J.S.; Harvey, J.W.; Bernhardt, E.S. Metabolic rhythms in flowing waters: An approach for classifying river productivity regimes. *Limnol. Oceanogr.* **2019**, *64*, 1835–1851. [CrossRef]
5. Odum, H.T. Primary production in flowing waters. *Limnol. Oceanogr.* **1956**, *1*, 102–117. [CrossRef]
6. Staehr, P.A.; Testa, J.M.; Kemp, W.M.; Cole, J.J.; Sand-Jensen, K.; Smith, S.V. The metabolism of aquatic ecosystems: History, applications, and future challenges. *Aquat. Sci.* **2012**, *74*, 15–29. [CrossRef]
7. Staehr, P.A.; Bade, D.; Van de Bogert, M.C.; Koch, G.R.; Williamson, C.; Hanson, P.; Cole, J.J.; Kratz, T. Lake metabolism and the diel oxygen technique: State of the science. *Limnol. Oceonogr. Methods* **2010**, *8*, 628–644. [CrossRef]
8. Lovett, G.M.; Cole, J.J.; Pace, M.L. Is net ecosystem production equal to ecosystem carbon accumulation? *Ecosystems* **2006**, *9*, 152–155. [CrossRef]
9. Hall, R.O.; Tank, J.L.; Baker, M.A.; Rosi-Marshall, E.J.; Hotchkiss, E.R. Metabolism, Gas Exchange, and Carbon Spiraling in Rivers. *Ecosystems* **2016**, *19*, 73–86. [CrossRef]
10. Mejia, F.H.; Fremier, A.K.; Benjamin, J.R.; Bellmore, J.R.; Grimm, A.Z.; Watson, G.A.; Newsom, M. Stream metabolism increases with drainage area and peaks asynchronously across a stream network. *Aquat. Sci.* **2019**, *81*, 1–17. [CrossRef]
11. Beaulieu, J.J.; Arango, C.P.; Balz, D.A.; Shuster, W.D. Continuous monitoring reveals multiple controls on ecosystem metabolism in a suburban stream. *Freshw. Biol.* **2013**, *58*, 918–937. [CrossRef]
12. Blersch, S.S.; Blersch, D.M.; Atkinson, J.F. Metabolic Variance: A Metric to Detect Shifts in Stream Ecosystem Function as a Result of Stream Restoration. *J. Am. Water Resour. Assoc.* **2019**, *55*, 608–621. [CrossRef]
13. Duda, P.B.; Hummel, P.R.; Donigian, A.S., Jr.; Imhoff, J.C. BASINS/HSPF: Model Use, Calibration, and Validation. *Trans. ASABE* **2012**, *55*, 1523–1547. [CrossRef]
14. Chen, W.; He, B.; Nover, D.; Duan, W.; Luo, C.; Zhao, K.; Chen, W. Spatiotemporal patterns and source attribution of nitrogen pollution in a typical headwater agricultural watershed in Southeastern China. *Environ. Sci. Pollut. Res.* **2018**, *25*, 2756–2773. [CrossRef] [PubMed]

15. Fu, B.; Merritt, W.S.; Croke, B.F.W.; Weber, T.R.; Jakeman, A.J. A review of catchment-scale water quality and erosion models and a synthesis of future prospects. *Environ. Model. Softw.* **2019**, *114*, 75–97. [CrossRef]
16. Wagena, M.B.; Collick, A.S.; Ross, A.C.; Najjar, R.G.; Rau, B.; Sommerlot, A.R.; Fuka, D.R.; Kleinman, P.J.A.; Easton, Z.M. Impact of climate change and climate anomalies on hydrologic and biogeochemical processes in an agricultural catchment of the Chesapeake Bay watershed, USA. *Sci. Total. Environ.* **2018**, *637*, 1443–1454. [CrossRef]
17. Deb, K.; Pratap, A.; Agarwal, S.; Meyarivan, T. A fast and elitist multiobjective genetic algorithm: NSGA-II. *IEEE Trans. Evol. Comput.* **2002**, *6*, 182–197. [CrossRef]
18. Panagopoulos, Y.; Makropoulos, C.; Kossida, M.; Mimikou, M. Optimal implementation of irrigation practices: Cost-effective desertification action plan for the Pinios basin. *J. Water Resour. Plan. Manag.* **2014**, *140*, 05014005. [CrossRef]
19. Hassanzadeh, Y.; Afshar, A.A.; Pourreza-Bilondi, M.; Memarian, H.; Besalatpour, A.A. Toward a combined Bayesian frameworks to quantify parameter uncertainty in a large mountainous catchment with high spatial variability. *Environ. Monit. Assess.* **2019**, *191*, 1–22. [CrossRef]
20. Xie, H.; Shen, Z.; Chen, L.; Lai, X.; Qiu, J.; Wei, G.; Dong, J.; Peng, Y.; Chen, X. Parameter estimation and uncertainty analysis: A comparison between continuous and event-based modeling of streamflow based on the Hydrological Simulation Program-Fortran (HSPF) model. *Water* **2019**, *11*, 171. [CrossRef]
21. Zhang, R.; Moreira, M.; Corte-Real, J. Multi-objective calibration of the physically based, spatially distributed SHETRAN hydrological model. *J. Hydroinform.* **2015**, *18*, 428–445. [CrossRef]
22. Dumedah, G.; Berg, A.A.; Wineberg, M.; Collier, R. Selecting Model Parameter Sets from a Trade-off Surface Generated from the Non-Dominated Sorting Genetic Algorithm-II. *Water Resour. Manag.* **2010**, *24*, 4469–4489. [CrossRef]
23. Mostafaie, A.; Forootan, E.; Safari, A.; Schumacher, M. Comparing multi-objective optimization techniques to calibrate a conceptual hydrological model using in situ runoff and daily GRACE data. *Comput. Geosci.* **2018**, *22*, 789–814. [CrossRef]
24. Lampert, D.J.; Wu, M. Development of an open-source software package for watershed modeling with the Hydrological Simulation Program in Fortran. *Environ. Model. Softw.* **2015**, *68*, 166–174. [CrossRef]
25. Raymond, P.A.; Zappa, C.J.; Butman, D.; Bott, T.L.; Potter, J.; Mulholland, P.; Laursen, A.E.; McDowell, W.H.; Newbold, D. Scaling the gas transfer velocity and hydraulic geometry in streams and small rivers. *Limnol. Oceanogr. Fluids Environ.* **2012**, *2*, 41–53. [CrossRef]
26. Hall, R.O.; Ulseth, A.J. Gas Exchange in Streams and Rivers. *WIREs Water* **2019**, 1–18. [CrossRef]
27. Zappa, C.J.; McGillis, W.R.; Raymond, P.A.; Edson, J.B.; Hintsa, E.J.; Zemmelink, H.J.; Dacey, J.W.H.; Ho, D.T. Environmental turbulent mixing controls on air-water gas exchange in marine and aquatic systems. *Geophys. Res. Lett.* **2007**, *34*, 1–6. [CrossRef]
28. Wanninkhof, R. Relationship between wind speed and gas exchange over the ocean revisited. *Limnol. Oceanogr. Methods* **2014**, *12*, 351–362. [CrossRef]
29. Hill, N.B.; Riha, S.J.; Walter, M.T. Temperature dependence of daily respiration and reaeration rates during baseflow conditions in a northeastern U.S. stream. *J. Hydrol. Reg. Stud.* **2018**, *19*, 250–264. [CrossRef]
30. Engel, F.; Attermeyer, K.; Ayala, A.I.; Fischer, H.; Kirchesch, V.; Pierson, D.C.; Weyhenmeyer, G.A. Phytoplankton gross primary production increases along cascading impoundments in a temperate, low-discharge river: Insights from high frequency water quality monitoring. *Sci. Rep.* **2019**, *9*, 1–13. [CrossRef]
31. Bernhardt, E.S.; Heffernan, J.B.; Grimm, N.B.; Stanley, E.H.; Harvey, J.W.; Arroita, M.; Appling, A.P.; Cohen, M.J.; McDowell, W.H.; Hall, R.O.; et al. The metabolic regimes of flowing waters. *Limnol. Oceanogr.* **2018**, *63*, S99–S118. [CrossRef]
32. Demars, B.O.L.; Thompson, J.; Manson, J.R. Stream metabolism and the open diel oxygen method: Principles, practice, and perspectives. *Limnol. Oceanogr. Methods* **2015**, *13*, 356–374. [CrossRef]
33. Mansoor, S.Z.; Louie, S.; Lima, A.T.; Van Cappellen, P.; MacVicar, B. The spatial and temporal distribution of metals in an urban stream: A case study of the Don River in Toronto, Canada. *J. Great Lakes Res.* **2018**, *44*, 1314–1326. [CrossRef]
34. Abdul-Aziz, O.I.; Ahmed, S. Relative linkages of stream water quality and environmental health with the land use and hydrologic drivers in the coastal-urban watersheds of southeast Florida. *GeoHealth* **2017**, *1*, 180–195. [CrossRef]

35. Fuß, T.; Behounek, B.; Ulseth, A.J.; Singer, G.A. Land use controls stream ecosystem metabolism by shifting dissolved organic matter and nutrient regimes. *Freshw. Biol.* **2017**, *62*, 582–599. [CrossRef]
36. O'Brien, J.M.; Warburton, H.J.; Elizabeth Graham, S.; Franklin, H.M.; Febria, C.M.; Hogsden, K.L.; Harding, J.S.; McIntosh, A.R. Leaf litter additions enhance stream metabolism, denitrification, and restoration prospects for agricultural catchments. *Ecosphere* **2017**, *8*, 1–17. [CrossRef]
37. Griffiths, N.A.; Tank, J.L.; Royer, T.V.; Roley, S.S.; Rosi-Marshall, E.J.; Whiles, M.R.; Beaulieu, J.J.; Johnson, L.T. Agricultural land use alters the seasonality and magnitude of stream metabolism. *Limnol. Oceanogr.* **2013**, *58*, 1513–1529. [CrossRef]
38. National Solar Radiation Database. Available online: https://nsrdb.nrel.gov (accessed on 27 May 2019).
39. Berndt, M.E.; Rutelonis, W.; Regan, C.P. A comparison of results from a hydrologic transport model (HSPF) with distributions of sulfate and mercury in a mine-impacted watershed in northeastern Minnesota. *J. Environ. Manag.* **2016**, *181*, 74–79. [CrossRef]
40. Bello, A.A.D.; Haniffah, M.R.M.; Hanapi, M.N.; Usman, A.B. Identification of critical source areas under present and projected land use for effective management of diffuse pollutants in an urbanized watershed. *Int. J. River Basin Manag.* **2019**, *17*, 171–184. [CrossRef]
41. Borah, D.K.; Bera, M. Watershed-Scale Hydrologic and Nonpoint-Source Pollution Models: Review of Applications. *Trans. ASAE* **2013**, *47*, 789–803. [CrossRef]
42. Nash, J.E.; Sutcliffe, J.V. River flow forecasting through conceptual models part I—A discussion of principles. *J. Hydrol.* **1970**, *10*, 282–290. [CrossRef]
43. Moriasi, D.N.; Arnold, J.G.; Van Liew, M.W.; Bingner, R.L.; Harmel, R.D.; Veith, T.L. Model evaluation guidelines for systematic quantification of accuracy in watershed simulations. *Am. Soc. Agric. Biol. Eng.* **2007**, *50*, 885–900.
44. Pandey, P.K.; Dabral, P.P.; Pandey, V. Evaluation of reference evapotranspiration methods for the northeastern region of India. *Int. Soil Water Conserv. Res.* **2016**, *4*, 52–63. [CrossRef]
45. Huber, D.H.; Ugwuanyi, I.R.; Lhilhi, N.V.; Kemajou, T.; Andrielle, L.; Chavarria-Palma, J.E. Water Quality for Kanawha River WV. Available online: https://doi.org/10.6084/m9.figshare.9786212.v1 (accessed on 22 September 2019).
46. Appling, A.P.; Hall, R.O.; Yackulic, C.B.; Arroita, M. Overcoming Equifinality: Leveraging Long Time Series for Stream Metabolism Estimation. *J. Geophys. Res. Biogeosci.* **2018**, *123*, 624–645. [CrossRef]
47. Tang, G.; Zhu, Y.; Wu, G.; Li, J.; Li, Z.L.; Sun, J. Modelling and analysis of hydrodynamics and water quality for rivers in the northern cold Region of China. *Int. J. Environ. Res. Public Health* **2016**, *13*, 408. [CrossRef]
48. Falkowski, T.; Ostrowski, P.; Siwicki, P.; Brach, M. Channel morphology changes and their relationship to valley bottom geology and human interventions; a case study from the Vistula Valley in Warsaw, Poland. *Geomorphology* **2017**, *297*, 100–111. [CrossRef]
49. Read, J.S.; Hamilton, D.P.; Desai, A.R.; Rose, K.C.; MacIntyre, S.; Lenters, J.D.; Smyth, R.L.; Hanson, P.C.; Cole, J.J.; Staehr, P.A.; et al. Lake-size dependency of wind shear and convection as controls on gas exchange. *Geophys. Res. Lett.* **2012**, *39*, 1–5. [CrossRef]
50. Halbedel, S.; Büttner, O.; Weitere, M. Linkage between the temporal and spatial variability of dissolved organic matter and whole-stream metabolism. *Biogeosciences* **2013**, *10*, 5555–5569. [CrossRef]
51. Demars, B.O.L.; Russell, M.J.; Ólafsson, J.S.; Gíslason, G.M.; Gudmundsdóttir, R.; Woodward, G.; Reiss, J.; Pichler, D.E.; Rasmussen, J.J.; Friberg, N. Temperature and the metabolic balance of streams. *Freshw. Biol.* **2011**, *56*, 1106–1121. [CrossRef]
52. Reichert, P.; Uehlinger, U.; Acuña, V. Estimating stream metabolism from oxygen concentrations: Effect of spatial heterogeneity. *J. Geophys. Res. Biogeosci.* **2009**, *114*, 1–15. [CrossRef]
53. Huber, D.H.; Ugwuanyi, I.R.; Malkaram, S.A.; Montenegro-Garcia, N.A.; Lhilhi, N.V.; Chavarria-Palma, J.E. Metagenome Sequences of Sediment from a Recovering Industrialized Appalachian River in West Virginia. *Genome Announc.* **2018**, *6*, e00350-18. [CrossRef] [PubMed]
54. Bernot, M.J.; Sobota, D.J.; Hall, R.O.; Mulholland, P.J.; Dodds, W.K.; Webster, J.R.; Tank, J.L.; Ashkenas, L.R.; Cooper, L.W.; Dahm, C.N.; et al. Inter-regional comparison of land-use effects on stream metabolism. *Freshw. Biol.* **2010**, *55*, 1874–1890. [CrossRef]
55. Uehlinger, U.; Kawecka, B.; Robinson, C.T. Effects of experimental floods on periphyton and stream metabolism below a high dam in the Swiss Alps (River Spöl). *Aquat. Sci.* **2003**, *65*, 199–209. [CrossRef]

56. Izagirre, O.; Agirre, U.; Bermejo, M.; Pozo, J.; Elosegi, A. Environmental controls of whole-stream metabolism identified from continuous monitoring of Basque streams. *J. N. Am. Benthol. Soc.* **2008**, *27*, 252–268. [CrossRef]
57. Sánchez-Pérez, J.M.; Gerino, M.; Sauvage, S.; Dumas, P.; Maneux, É.; Julien, F.; Winterton, P.; Vervier, P. Effects of wastewater Treatment Plant pollution on in-stream ecosystems functions in an agricultural watershed. *Ann. Limnol.* **2009**, *45*, 79–92. [CrossRef]
58. Kaushal, S.S.; Delaney-Newcomb, K.; Findlay, S.E.G.; Newcomer, T.A.; Duan, S.; Pennino, M.J.; Sivirichi, G.M.; Sides-Raley, A.M.; Walbridge, M.R.; Belt, K.T. Longitudinal patterns in carbon and nitrogen fluxes and stream metabolism along an urban watershed continuum. *Biogeochemistry* **2014**, *121*, 23–44. [CrossRef]
59. Grace, M.R.; Giling, D.P.; Hladyz, S.; Caron, V.; Thompson, R.M.; Mac Nally, R. Fast processing of diel oxygen curves: Estimating stream metabolism with base (BAyesian single-station estimation). *Limnol. Oceanogr. Methods* **2015**, *13*, 103–114. [CrossRef]
60. Tang, S.; Sun, T.; Shen, X.M.; Qi, M.; Feng, M.L. Modeling net ecosystem metabolism influenced by artificial hydrological regulation: AN application to the Yellow River Estuary, China. *Ecol. Eng.* **2015**, *76*, 84–94. [CrossRef]

Article

Assessing Regional Scale Water Balances through Remote Sensing Techniques: A Case Study of Boufakrane River Watershed, Meknes Region, Morocco

Mohammed El Hafyani [1,*], Ali Essahlaoui [1], Anton Van Rompaey [2], Meriame Mohajane [1,3], Abdellah El Hmaidi [1], Abdelhadi El Ouali [1], Fouad Moudden [1] and Nour-Eddine Serrhini [4]

1 Department of Geology, Laboratory of Geoengineering and Environment, Research Group "Water Sciences and Environment Engineering", Faculty of Sciences, Moulay Ismail University, Zitoune, Meknès BP 11201, Morocco; a.essahlaoui@fs-umi.ac.ma (A.E.); mohajane.meriame13@gmail.com (M.M.); a.elhmaidi@fs-umi.ac.ma (A.E.H.); a.elouali@fs-umi.ac.ma (A.E.O.); moudden2000@yahoo.fr (F.M.)

2 Department Earth and Environmental Science, Geography and Tourism Research Group, KU Leuven, Celestijnenlaan 200E, 3001 Heverlee, Belgium; anton.vanrompaey@kuleuven.be

3 Department of Biology, Research Group "Soil and Environment Microbiology Unit", Faculty of Sciences, Moulay Ismail University, Zitoune, Meknès BP 11201, Morocco

4 Sebou Hydraulic Basin Agency, Fès BP 2101, Morocco; serininour@yahoo.fr

* Correspondence: m.elhafyani@edu.umi.ac.ma; Tel.: +212-7063-5119

Received: 15 December 2019; Accepted: 16 January 2020; Published: 21 January 2020

Abstract: This paper aims to develop a method to assess regional water balances using remote sensing techniques. The Boufakrane river watershed in Meknes Region (Morocco), which is characterized by both a strong urbanization and a rural land use change, is taken as a study case. Firstly, changes in land cover were mapped by classifying remote sensing images (Thematic Mapper, Enhanced Thematic Mapper Plus and Operational Land Imager) at a medium scale resolution for the years 1990, 2003 and 2018. By means of supervised classification procedures the following land cover categories could be mapped: forests, bare soil, arboriculture, arable land and urban area. For each of these categories a water balance was developed for the different time periods, taking into account changing management and consumption patterns. Finally, the land cover maps were combined with the land cover specific water balances resulting in a total water balance for the selected catchment. The procedure was validated by comparing the assessments with data from water supply stations and the number of licensed ground water extraction pumps. In terms of land use/land cover changes (LULCC), the results showed that urban areas, natural vegetation, arboriculture and cereals increased by 183.74%, 12.55%, 34.99 and 48.77% respectively while forests and bare soils decreased by 78.65% and 16.78% respectively. On the other hand, water consumption has been increased significantly due to the Meknes city growth, the arboriculture expansion and the new crops' introduction in the arable areas. The increased water consumption by human activities is largely due to reduced water losses through evapotranspiration because of deforestation. Since the major part of the forest in the catchment has disappeared, a further increase of the water consumption by human activities can no longer be offset by deforestation.

Keywords: Boufakrane river watershed; remote sensing; LULCC; water balances

1. Introduction

Worldwide, freshwater consumption is increasing due to population growth and livelihoods and land use change. The total volume of water is estimated at about 1.4 billion km^3. Only 2.5% of this volume (35 million km^3) can be considered as fresh water, with an unequal distribution [1]. Moreover, about 66% of the world population suffers from water scarcity [1]. Therefore, water is considered as a major challenge both as a resource and as a risk.

The Mediterranean is one of the most affected regions by water scarcity because of several factors including, extreme natural inter-annual variability, seasonality of water resources and decreasing stream flows forecast in coming decades [2]. Identified as a climate change "hot spot" [3], the Mediterranean area experiences severe episodic droughts, and has a high anthropogenic demand for surface and ground water [4,5]. Unfortunately, in most Mediterranean countries, where water demand is expected to increase significantly due to population growth, agricultural development and scarcity of water resources, could lead to poverty for millions of people [1], and influencing socio-economic development [6–8]. Therefore, water availability has been one of the most important environmental scientific research [9–11] to assess and develop adequate adaptation strategies.

Land use land cover changes refer to the modification of the Earth's surface by human activities [12]. Recently with the development of space technology, remote sensing integrating with GIS has become an effective tool for mapping and spatial–temporal monitoring of land use change [13–19], it provides a detailed understanding of the different functioning of ecosystems [20–25]. Therefore these techniques show a strong interest in deriving information about the Earth's surface, both in space and time [26,27]. Around the world, several studies have been carried out with the aim of studying land use changes through satellite image processing tools in different climatic contexts [28–32], and especially in semi-arid areas [14,16,33]. Understanding the relationship between the biosphere aspects of the hydrological cycle (BAHC) and the land use and land cover changes has been taken as the core plan of the International Geosphere-Biosphere Program (IGBP) and the International Human Dimension Program (IHDP) [34]. Additionally, understanding the impact of land use/land cover (LULC) changes on hydrological process is considered as one core problem in the LUCC research areas established by IGBP and IHDP. In this sense, many authors made several efforts to understand and evaluate the hydrological responses to different LULC patterns' changes. For example, Woldesenbet, Tekalegn Ayele et al., [35], based on an integrated approach of hydrological modeling and partial least squares regression (PLSR), quantified the contributions of changes in individual LULC classes to changes in hydrological components with application in two watersheds, namely Lake Tana and Beles in the Upper Blue Nile Basin in Ethiopia.

Van Ty, Tran et al., [36] applied an interdisciplinary scenario analysis approach to assess the potential impacts of climate, land use/cover and population changes on future water availability and demand in the Srepok River basin, a trans-boundary basin. They found that surface runoff will be increased with increased future rainfall and concluded that LULC change is found to have the largest impact on increased water demand, and thus reduced future water availability. Li, Zhihui et al., [37] studied the impacts of land use changes on surface runoff and water yield with scenario-based land use change in the upper and middle reaches of the Heihe River Basin. Their results showed an expansion of the forestland and grassland with the increase in water utilization ratio.

In Morocco, the challenges of water scarcity, harsh climatic conditions, population growth, intense urbanization and socio-economic development issues, considerable efforts are made to a wiser water resources use. In addition, the uneven spatiotemporal distribution of rainfall and water resources requires the implementation of an optimal management system of these scarce resources, and to relieve the pressure exerted on it.

Given this context, the specific objectives were: (i) to study the land use/land cover changes in the Boufakrane river watershed; (ii) to quantify the water availability and demand and (iii) to establish the relationship between land use/land cover, and water availability.

To the best of our knowledge, this paper presents the first detailed study using a combination of remote sensing and reference data to explore the relationship between landscape dynamics, water availability, and consumption in this study area. Analyzing water demand using satellites data as prevention, warning, monitoring and modeling tools, is of crucial importance for an effective control of water consumption over the Boufakrane river watershed.

2. Materials and Methods

2.1. Study Area

The study area is the Boufakrane river watershed, which is located in Meknes region in Morocco, between longitudes 6°0′00″ and 6°6°15′00″ S and latitudes 37°00′00″ and 37°45′00″ N (Figure 1). The study site total area is of nearly 39,948 ha. It crosses the western part of the plain of Saïss, bounded to the South by the Middle Atlas mountain range, to the North by the Pre-Rif wrinkles, and to the West by the Beht river and its Paleozoic outcrops that is made from the Western Meseta. The topographical rating decreases from South to North and ranges from 1383 to 277 m. Geologically, the basin includes various formations stretching from the Paleozoic to the Quaternary, the majority of which are lacustrine limestone and fauves of the Plio-Quaternary sand [38].

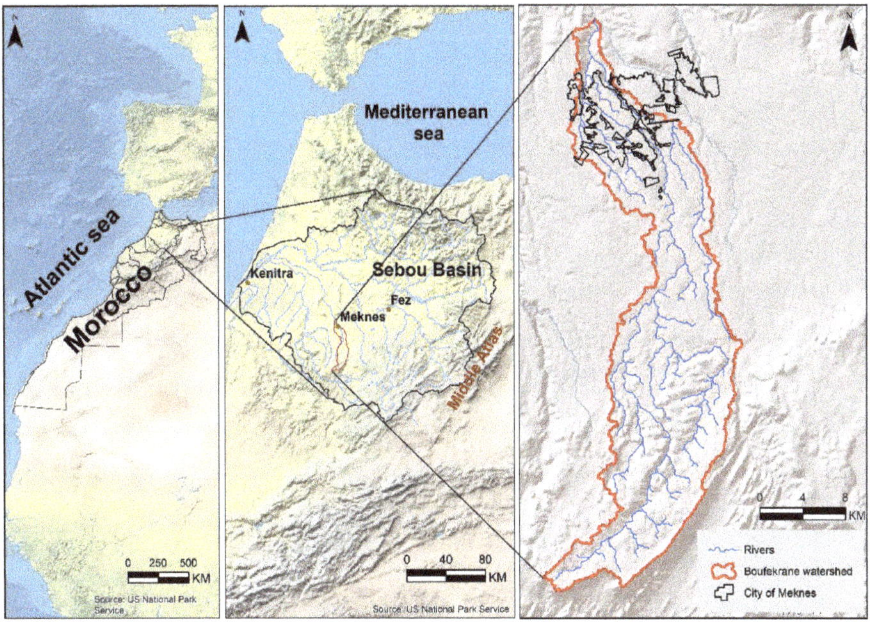

Figure 1. Study Area.

From the climatic point of view, the selected study area is characterized by a semi-arid climate, with an average annual rainfall of 500 mm recorded in the Meknes station, and a dry season extending from June to October. The area is experiencing very important economic activity, consisting of agriculture (olive groves, arboriculture, cereal farming and legumes). It is characterized by very important agricultural activities with an acceptable soil quality and a very large amount of water resources. Recently, the management of these water resources is no longer sustainable due to the overexploitation of water resources. The latter, is mainly induced by the population growth, the development of industry and agricultural activities, and the urban extension at the expense of the agricultural areas. At the downstream of the Boufakrane river watershed, the city of Meknes—one of the largest cities of Morocco with a very high population density—is located. According to the national census of 2014, its population is estimated at 630,079 inhabitants [39]. This city is known for its historical riches, its mosaic culture and its considerable economic potential in the agricultural, industrial, touristic and commercial fields.

2.2. Methodology

The data used in this work is composed of three Landsat images including Thematic Mapper (TM), Enhanced Thematic Mapper Plus (ETM+) and Operational Land Imager (OLI) of the years 1990 (TM), 2003 (ETM+) and 2018 (OLI). The images were radiometrically corrected (Figure 2). Images were used to identify six main land cover types—built-up areas, natural vegetation, cereals and agricultural bare soil, arboriculture, forest and bare soil—using a maximum likelihood classification. More detailed explanations about this classification algorithm can be found in the literature [40–42]. All images were captured in July in order to avoid the misclassification between agriculture area and soil bare area. Sixty-three points were taken during the sampling campaign conducted in many parts of the study area during May 2018 to validate the classification results (Figure 3). The evaluation of the classification accuracy is achieved by comparing the obtained classes using the classification algorithm to known classes at sampled reference locations. These collected data are further characterized in a confusion matrix [43,44]. The kappa index was used to measure the concordance between the field truth points and the different classification outputs. The values of this index are ranked from 0 to 1, where 0 indicates a no concordance, and 1 is a significance of a perfect concordance [45,46]. For the next steps of the methodology applied in this work, the information about the water consumption, water demand and the population of the city of Meknes were obtained from several documents and reports including, census data and FAO reports [39,47,48], as well as crops' water consumption data [12,49].

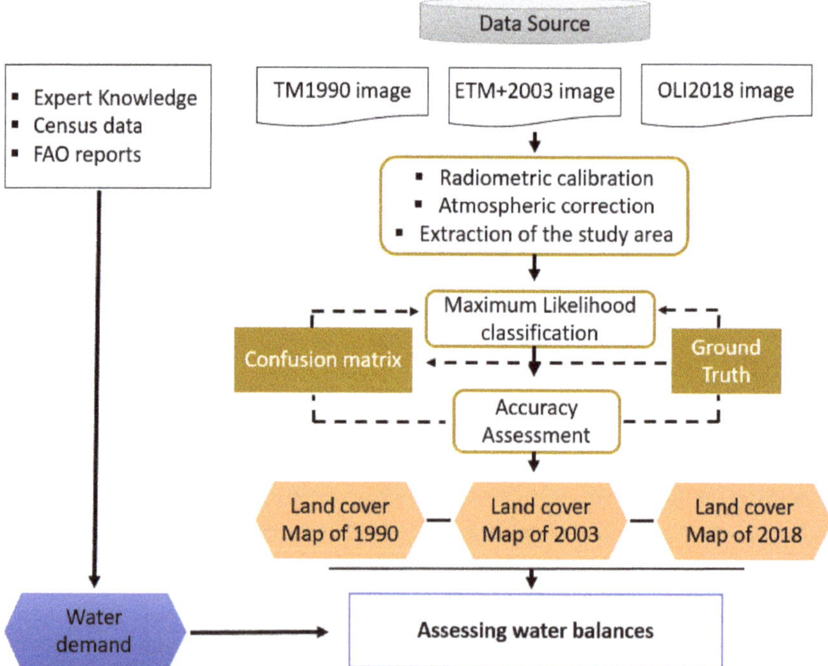

Figure 2. Flowchart of the implemented methodology.

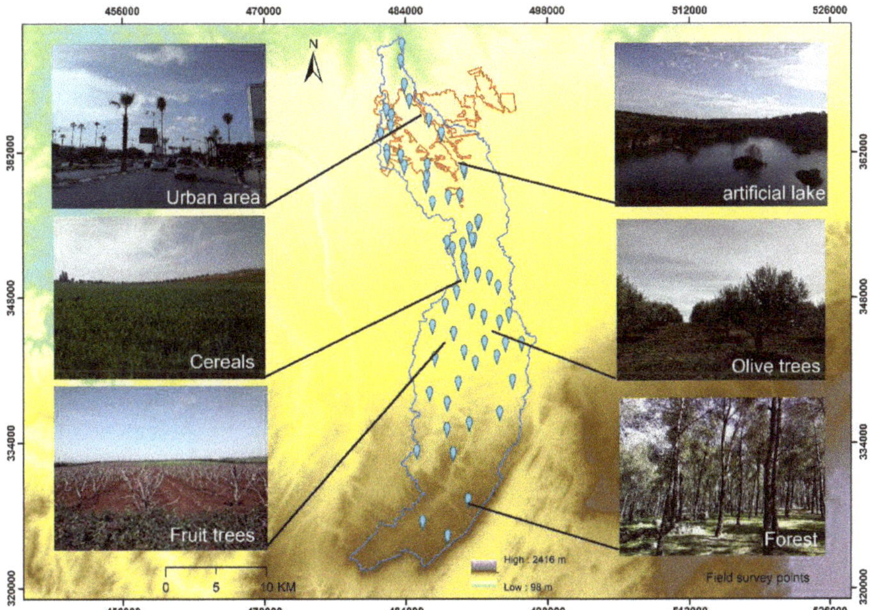

Figure 3. Field survey points.

3. Results and Discussion

3.1. Land Use Land Cover Changes (LULCC)

Land use/land cover changes (LULCC) were estimated during the period 1990–2018 using the supervised classification and based on multi-date images (Table 1, Figure 4). Cereal and agricultural bare soil, arboriculture and urban areas were increased respectively by 48%, 34% and 183%, whereas the forest has decreased by 78% over the same period (Table 1). The increase in the agricultural areas (cereals and arboriculture areas) is illustrated in the agricultural development that the region has known, as well as the integration of new technologies in this area, with the new subsidies that the State has provided to farmers within the framework of the Moroccan Green Plan [50].

These results confirm the evolution of agriculture in Morocco, according to the Moroccan Agriculture Ministry statistics, the areas of olive groves in Morocco increased from 251,061 ha to 397,178 ha during the period 2001–2018 for irrigated areas, and from 374,146 ha to 676,314 ha for rain-fed areas, during the same period with an overall yield of 1,912,237 Tons in 2018 for all the country. For cereals class, the areas decreased from 4,826,100 ha in 2001 to 4,486,778 ha in 2018 for both rain-fed and irrigated ones. While production increased from 4,471,530 tons to 10,261,554 tons during the same period [51].

In terms of urbanization, the city of Meknes showed an increase by three times in the period 1990–2018, with a rate of 183% (Table 1). The results showed a strong spread of construction in this period, while the surface of the city has tripled from 25 to 73 km^2.

In Meknes, the majority of urban growth occurred in the city's surroundings and it is generally, presented in the economic housing category that recently has met the needs of the population. According to the classification results, the evolution was generally noted in the South and the North-East where the landscape is characterized by a very low relief, unlike the wrinkly Northern part of the city, which has remained almost stable throughout this period.

Table 1. Area change in different land use/land cover (LULC) categories from 1990 to 2018.

	Area 1990 (ha)	Area 2003 (ha)	Area 2018 (ha)	Chang 1990–2018	
				(ha)	(%)
Built-up area	2577	3758	7314	+4736	+183
Natural vegetation	941	1022	1059	+118	+12
Cereals and agricultural bare soil	9720	11,420	14,262	+4541	+48
Arboriculture	5733	6707	7740	+2006	+34
Forest	7026	3026	1500	−5526	−78
Bare soil	16,217	14,535	13,499	−2718	−16

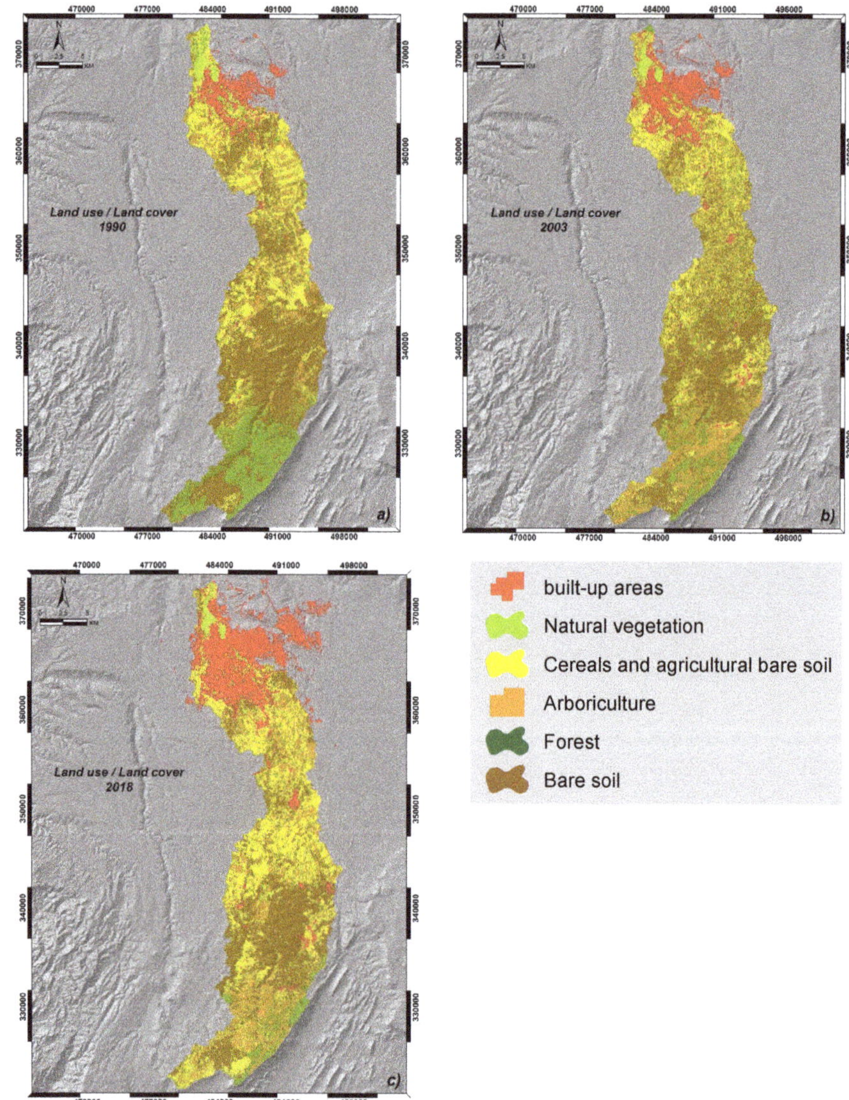

Figure 4. LULC map, (**a**) LULC map for 1990, (**b**) LULC map for 2003 and (**c**) LULC map for 2018.

3.2. Overall Accuracy

The evaluation of accuracy of LULC classification was made by ground-truth points collected from reference data including field survey data, and high-resolution images from Google Earth, and by using the confusion matrix. The results show an overall accuracy of 67.55%, 70.37% and 87.59%, respectively in 1990, 2003 and 2018, whereas the kappa index is 0.56, 0.56 and 0.83, respectively (Table 2). The classification of 2018 showed a perfect agreement, while those of 1990 and 2003 showed moderate agreement.

Regarding the confusion matrix, the validation of the classification by field data shows confusion between a few classes, such as: arboriculture, cereals and agriculture bare soil and bare soil. This is essentially due to the similar spectral response between cereals and arboricultural areas (Figure 5), as well as the leaf area of a tree, per unit of soil area or pixel unit.

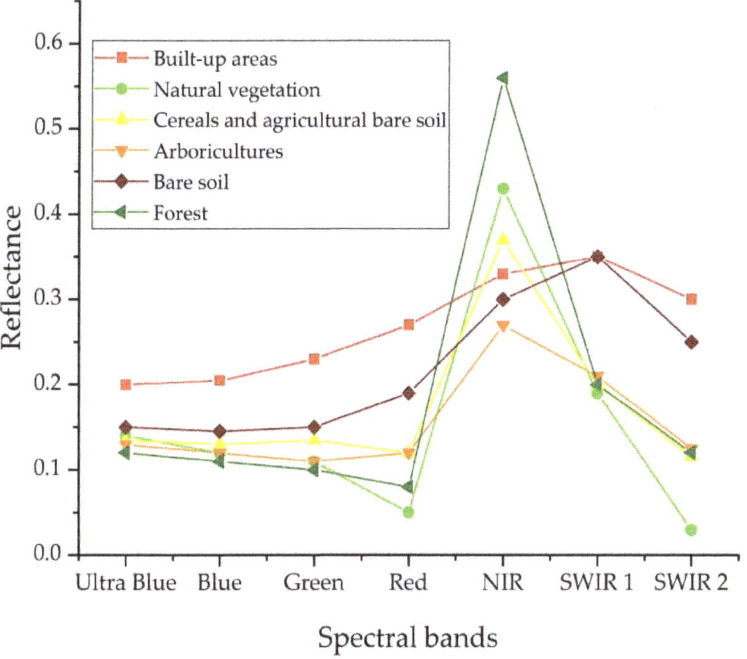

Figure 5. Spectral profiles of different classes.

In Table 2, a classification confusion matrix for the image captured in 1990 was presented. The overall accuracy of this classification was 67.55%, and its kappa index was 0.56. It showed that there was confusion during the classification process. For example, 28.16% of the points sampled in cereals and agricultural bare soil were classified as arboriculture areas, and 11.87% as bare soil areas. The same is shown in the other two time periods (2003 and 2018), but with a decrease observed in term of confusion. Conversely, the classification showed a strong performance in mapping urban area by clearly distinguishing it from the other LULC classes. Results show that the classification accuracy of sampled urban areas was 99.00%, 96.01% and 96.18 for 2018, 2003 and 1990 respectively.

Table 2. Confusion matrix for the three times (1990, 2003 and 2018).

	Ground Truth (%)																		Total (%)		
Class	**Cereals and Agricultural Bare Soil**			**Arboriculture**			**Forest**			**Urban Area**			**Bare Soil**			**Natural Vegetation**					
Year	**1990**	**2003**	**2018**	**1990**	**2003**	**2018**	**1990**	**2003**	**2018**	**1990**	**2003**	**2018**	**1990**	**2003**	**2018**	**1990**	**2003**	**2018**	**1990**	**2003**	**2018**
Cereals and agricultural bare soil	57.83	76.45	82.14	19.57	19.57	2.46	1.95	0	0.19	0.52	0.78	0.56	17	21.63	7.77	0.26	0	0	30.67	40.06	30.67
Arboriculture	28.16	4.74	7.7	55	26.24	86.26	9.53	11.09	27.63	3.3	2.86	0.15	1.61	1.94	0.33	0.77	1.02	8.18	19.42	6.83	19.42
Forest	2.13	1.31	0.59	4.51	18.85	0.03	81.13	88.91	72.18	0.67	0	0.11	0.98	0.48	0.36	2.3	5.12	0	3.74	5.61	3.74
Urban area	0	0.09	0.59	0.42	0.09	0.03	0	0	0	96.18	96.01	99	0.05	0.01	0	0	0	0	5.04	5.02	5.04
Bare soil	11.87	16.76	8.63	16.17	21.16	1.2	7.39	0	0	0	0	0.19	80.36	75.94	91.55	0	0	0	38.79	39.38	38.79
Natural vegetation	0.01	0.61	0.64	4.33	8.21	10.01	0	0	0	0	0.35	0	0	0	0	96.88	93.86	91.82	2.34	3.1	2.34

3.3. Urban Growth of Meknes City 1990–2018

This part was developed generally to have an idea about urbanization trends in the city of Meknes. The results show a very important urban expansion, generally in the areas surrounding the city (Figure 6). The "Cities without slums" program that was launched in Morocco in 2004 has made significant progress in reducing slums initially targeted and improving housing conditions for low-income households [48]. This ambitious governmental strategy resulted in in-depth revision of public housing policies, opening of new construction sites, reforms and the definition of innovative and proactive programs able to respond, a more appropriate way of social housing promotion, eliminating and preventing of insalubrious housing and anticipation of urban development. About 26 projects with 44,400 dwellings on a surface of 822 ha, were authorized between 2008 and 2012, and subsequent constructions had even exceeded the population needs [48].

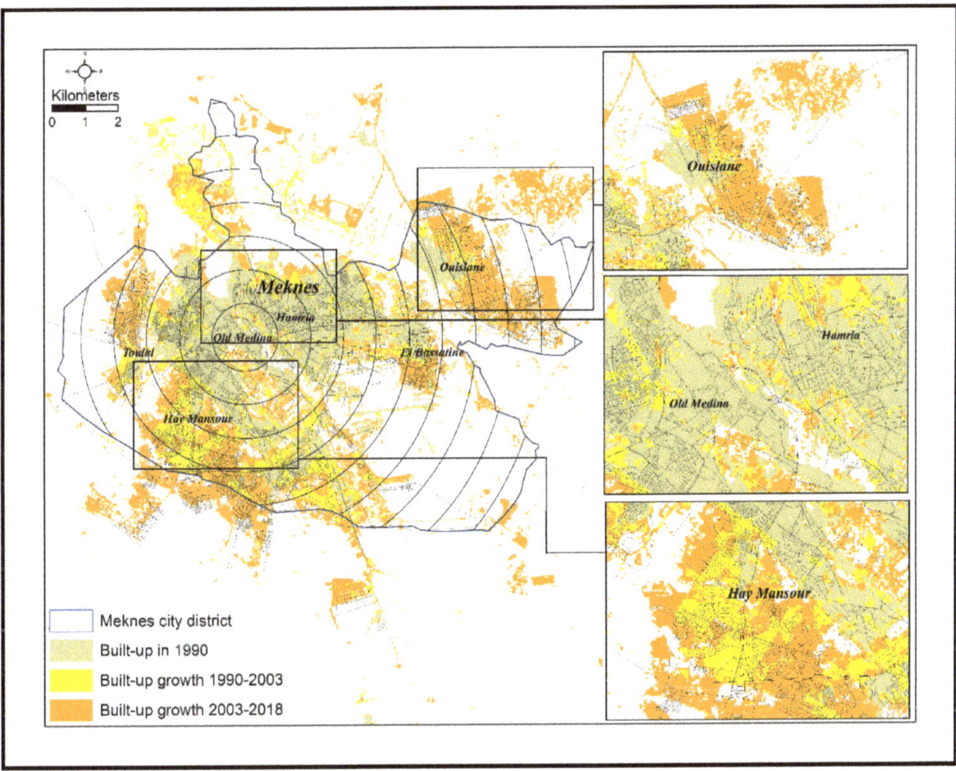

Figure 6. Observed urban expansion between 1990 and 2018 in Meknes city.

Historically speaking, about 1590 ha of agricultural areas in Boufakrane river watershed were transformed into urban areas during the last 30 years. This can be explained by the socio-economic boom of the city of Meknes combined to the improvement of infrastructure, the population and urbanization growth, the medical care networks, the communication, transport and mobility facilities. Furthermore, the industrialization and the installation of various economic activities in its edge designed the immigration flow from the surrounding areas.

Meknes city agglomeration is comprised of clearly individualized urban units, which are namely: the old Medina, the imperial city, the new city and the recent urban expansions. Each of them is related to a specific historical period.

Historically, Meknes city experienced the apparition of new established neighborhoods in its surroundings especially in the South-Western and the North-Eastern parts such as, Hay Mansour (Figure 6). Unfortunately, these areas correspond to very fertile agricultural soils (Saïss plain) are affected and will continue to be so in the upcoming years by various urban development projects.

3.4. Water Availability and Water Demand

Based on the latest United Nations statistics, Morocco is characterized by a very high human potential with a total population of 36,465,862 dwellers with a median age of 28.3 years. The percentage of the urban population is estimated to 60.3% (22,093,561 people in 2019), in a total area of 446,300 Km^2. According to the last Moroccan census conducted in 2014, the population of the city of Meknes; including its municipalities, Al Machouar, Ouislane and Toulal, increased by 25.4%, from 471,908 inhabitants to 632,079 inhabitants during the period 1990–2014 [39]. Therefore as the population grew, the water demand has increased in Meknes, from 19.30 to 29.8 million m^3 during the same period (Figure 7). So far, the changes that occurred in the land use of the Boufakrane river watershed and the increase in agricultural areas (cereals and arboriculture) resulted in an increase in the water demand during the same period from 91.92 million to 96.12 million m^3. Projected data shows that the water demand for the population, cereals and tree crops will increase, respectively by 8.54%, 16% and 10.68%, with a global increase of 8.82% (Figure 7), and consequently water demand will be of 104.27 Mm^3 in 2030. Thus, with this alarming situation and unsustainable management of this resource, many problems will be encountered in terms of water availability.

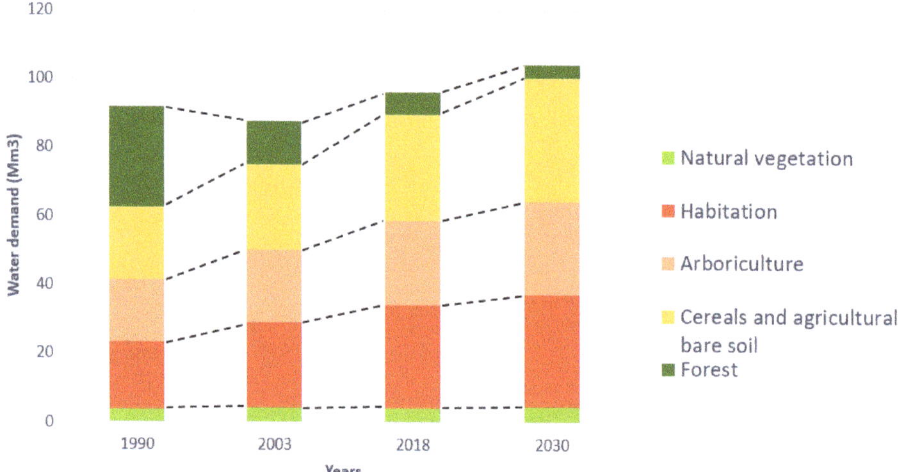

Figure 7. Water demand for different classes.

At the same time, groundwater resources are also affected by this change. Even with the large quantity of water stored in the aquifer system of this watershed, the piezometric history of the aquifer shows a general decline with about 1 m/year in average (Figure 8). As a result, during the last 30 years, this level has declined by almost 20 m. This has led to the disappearance of this groundwater in some places, and consequently to the transformation of the agricultural system from an irrigated system to a rain-fed system.

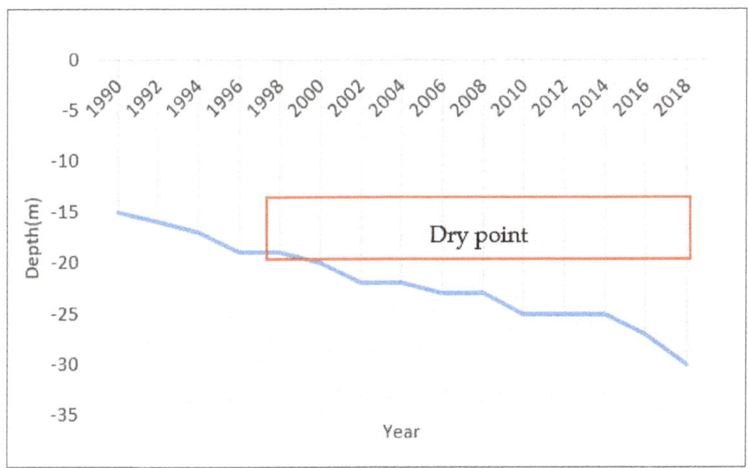

Figure 8. Depth of the water table piezometric surface.

In this situation, it is mandatory to set up a sustainable management system, including the intervention of the various stakeholders in the region. To find solutions and re-orient the usage of this resource in the right path by implementing an irrigation system adapted to the available amount of water, and to adopt more smart cropping systems fitting to the local context.

In terms of urban demography, this sector has known an enormous evolution, especially the city of Meknes in the downstream part of the watershed. This is generally due to several reasons, including; the industrialization and the installation of diverse economic activities in the peripheries of the city, the development and the improvement of transport network and mobility, as well as the enhancement of the life quality in the city with the availability of medical and communication services, and infrastructure development. This development had an effect on water consumption in the city. Yet, according to the local water and electricity distribution agency [52], for 2018, the consumed amount of water in the city is about 30.28 million m^3 with an increase of 58% during the same period. With a total number of 180,609, the consumers are largely private consumers while only 715 of them are big consumers and 969 are local governments, organizations and communities.

3.5. Results' Interpretation

The combination of satellite image processing techniques, surveys and field data to study the relationship of land use change and its relation to water availability, demand and consumption has been very effective in assessing water balances at the regional scale. In particular, the classification used in this work showed a high degree of accuracy, and clearly described the different changes in the study area during the last 28 years. Furthermore, the combination of satellite and field data allows us to assess the water balances in the Boufakrane river watershed.

Throughout the literature analysis, for several works about the same research topic, remote sensing has shown a strong performance in the study of land use/cover changes [14,16], the development and generation of land-use change scenarios for hydrological modeling [53] and urban dynamics [29,54–56]. For example, Ashraf M. Dewan et al. [29] studied land-use changes in Greater Dhaka, Bangladesh for a period of 28 years, based on remote sensing images to promote sustainable urbanization. They found that the Dhaka region had experienced rapid changes in land use, particularly in the urban area.

In a related context, A. El Garouani et al. [57] carried out a study to understand the relationship between urbanization and land use/land cover changes and their impacts on the urban landscape during the period 1984–2013 in the Fez region, Morocco. The results showed a strong performance of the classification with an overall accuracy ranging from 78% to 87%. In terms of land use, the urban area increased with 121%, whereas agriculture and forests decreased with 11% and 3% respectively.

Additionally, previous studies have been carried out, aimed at using geospatial tools for spatio-temporal monitoring of land use/land cover changes in areas with similar landscape. For example C. Höpfner and D. Scherer [58], has used this technique to analyze the spatio-temporal dynamics of land use and vegetation in north-west Morocco. The approach showed that with this technique, it is possible to derive conclusions about land-use dynamics in a largely unknown region with ground truth knowledge. While the validation was very powerful, it showed 80.24% overall accuracy, and a Kappa coefficient of 0.74.

In fact, the method used in this work is a temptation to link satellite and field data for the evaluation of the water balance. After the calculation of the different classes' surfaces extracted by the supervised classification. The water demand for each class is estimated based on field surveys with farmers, reports from the Regional Directorate of Agriculture, and the reports from the Food and Agriculture Organization (FAO). Concerning water demand for the population, the demand was estimated based on reports from the Meknes Water Distribution Agency.

4. Conclusions

In this study, a methodology of land use/land cover changes, availability and water demand based on the Landsat satellites images and field data in the Boufakrane river watershed is presented. It aims to link the image processing approach and water demand analysis. The classification used in this work has been very effective in improving land use changes that the region has experienced, and it showed high accuracy. The results showed that urban areas, natural vegetation, arboriculture and cereals increased by 183.74%, 12.55%, 34.99 and 48.77% respectively while forests and bare soils decreased by 78.65% and 16.78% respectively. However, the development of agriculture had an effect on water demand, since it increased by 4.56%, from 91.92 million to 96.12 million m^3 in the same period. Projected data showed that water demand will increase by 8.82% with 104.27 Mm^3 in 2030, which will create in the near future many problems in term of water availability, which requires intervention from all water stakeholders in the region as well as policy makers to find real solutions for sustainable management. The fully adopted methodology used in this research topic was based on free data and open source software, which can be applied to other areas to generate the same output based on the same approach. Moreover, this work represents new approaches to be urgently adopted in the region, and presents a strong base for that. This was possible by depicting the land use/land cover changes experienced by the region in the last 30 years then by providing accurate information on the management of water resources system in the region. To sum up, this research can be of valuable contribution in decision-making regarding the planning and the implementation of sustainable agricultural policies, and the elaboration of strategies for efficient and sustainable integrated water resources management. It can be also useful for assessing water balances taking into account the land use/land cover changes and its relationship with water demand. Ultimately, the developed approach can be of precious usefulness in other areas with similar background.

Author Contributions: Conceptualization, M.E.H., A.E. and A.V.R.; Data curation, M.E.H.; Formal analysis, M.E.H., A.E. and A.V.R.; Methodology, M.E.H., A.E. and A.V.R.; Project administration, A.E. and A.V.R.; Resources, A.E.H., A.E.O., N.-E.S. and F.M.; Software, M.E.H.; Supervision, A.E. and A.V.R.; Validation, M.E.H., A.E. and A.V.R.; Visualization, M.E.H.; Writing–original draft, M.E.H.; Writing—review & editing, M.E.H., A.E., A.V.R., M.M. and F.M. All authors have read and agreed to the published version of the manuscript.

Funding: This research received no external funding.

Acknowledgments: The authors acknowledge the financial support of VLIR-UOS for the help of the equipment and missions at the KU Leuven, Belgium. Thanks are also due to the anonymous reviewers for their valuable comments on this article, which allowed us to improve the scientific quality of this research.

Conflicts of Interest: The authors declare that they have no conflict of interest.

References

1. Cuenca, J.C.; de Murcia, R.; del Campo García, A.; España, D.R.M. Euro-Mediterranean Information System on Know How in the Water Sector. Available online: https://www.riob.org/fr/node/3071 (accessed on 18 January 2020).
2. Martinez, M.G.; Poole, N. The development of private fresh produce safety standards: Implications for developing Mediterranean exporting countries. *Food Policy* **2004**, *29*, 229–255. [CrossRef]
3. Cramer, W.; Guiot, J.; Fader, M.; Garrabou, J.; Gattuso, J.P.; Iglesias, A.; Lange, M.A.; Lionello, P.; Llasat, M.C.; Paz, S.; et al. Climate change and interconnected risks to sustainable development in the Mediterranean. *Nat. Clim. Chang.* **2018**, *8*, 972–980. [CrossRef]
4. Anderson, R.G.; Jin, Y.; Goulden, M.L. Assessing regional evapotranspiration and water balance across a Mediterranean montane climate gradient. *Agric. For. Meteorol.* **2012**, *166*, 10–22. [CrossRef]
5. Bolle, H.-J. Climate, climate variability, and impacts in the Mediterranean area: An overview. In *Mediterranean Climate*; Springer: Berlin/Heidelberg, Germany, 2003; pp. 5–86.
6. Ouatiki, H.; Boudhar, A.; Tramblay, Y.; Jarlan, L.; Benabdelouhab, T.; Hanich, L.; El Meslouhi, M.; Chehbouni, A. Evaluation of TRMM 3B42 v7 rainfall product over the Oum Er Rbia watershed in Morocco. *Climate* **2017**, *5*, 1. [CrossRef]
7. Schilling, J.; Freier, K.P.; Hertig, E.; Scheffran, J. Climate change, vulnerability and adaptation in North Africa with focus on Morocco. *Agric. Ecosyst. Environ.* **2012**, *156*, 12–26. [CrossRef]
8. Jellali, M.M. Développement des ressources en eau au Maroc. In *Aspects Économiques de la Gestion de l'Eau dans le Bassin Méditerranéen*; CIHEAM: Bari, Italy, 1997; pp. 51–68.
9. Giorgi, F.; Lionello, P. Climate change projections for the Mediterranean region. *Glob. Planet. Chang.* **2008**, *63*, 90–104. [CrossRef]
10. Marcè, R.; RODRÍGUEZ-ARIAS, M.À.; García, J.C.; Armengol, J.O.A.N. El Niño Southern Oscillation and climate trends impact reservoir water quality. *Glob. Chang. Biol.* **2010**, *16*, 2857–2865. [CrossRef]
11. Rivas-Tabares, D.; Tarquis, A.M.; Willaarts, B.; De Miguel, Á. An accurate evaluation of water availability in sub-arid Mediterranean watersheds through SWAT: Cega-Eresma-Adaja. *Agric. Water Manag.* **2019**, *212*, 211–225. [CrossRef]
12. FAO (Ed.) *Forests and Agriculture: Land-Use Challenges and Opportunities*; FAO: Rome, Italy, 2016.
13. Bakr, N.; Afifi, A.A. Quantifying land use/land cover change and its potential impact on rice production in the Northern Nile Delta, Egypt. *Remote Sens. Appl. Soc. Environ.* **2019**, *13*, 348–360. [CrossRef]
14. Barakat, A.; Ouargaf, Z.; Khellouk, R.; El Jazouli, A.; Touhami, F. Land use/land cover change and environmental impact assessment in béni-mellal district (Morocco) using remote sensing and gis. *Earth Syst. Environ.* **2019**, *3*, 113–125. [CrossRef]
15. Barakat, A.; Khellouk, R.; El Jazouli, A.; Touhami, F.; Nadem, S. Monitoring of forest cover dynamics in eastern area of Béni-Mellal Province using ASTER and Sentinel-2A multispectral data. *Geol. Ecol. Landsc.* **2018**, *2*, 203–215. [CrossRef]
16. El Jazouli, A.; Barakat, A.; Khellouk, R.; Rais, J.; El Baghdadi, M. Remote sensing and GIS techniques for prediction of land use land cover change effects on soil erosion in the high basin of the Oum Er Rbia River (Morocco). *Remote Sens. Appl. Soc. Environ.* **2019**, *13*, 361–374. [CrossRef]
17. Mohajane, M.; Essahlaoui, A.; Oudija, F.; El Hafyani, M.; Hmaidi, A.E.; El Ouali, A.; Randazzo, G.; Teodoro, A.C. Land Use/Land Cover (LULC) Using Landsat Data Series (MSS, TM, ETM+ and OLI) in Azrou Forest, in the Central Middle Atlas of Morocco. *Environments* **2018**, *5*, 131. [CrossRef]
18. Rogan, J.; Chen, D. Remote sensing technology for mapping and monitoring land-cover and land-use change. *Prog. Plan.* **2004**, *61*, 301–325. [CrossRef]
19. Wu, Q.; Li, H.Q.; Wang, R.S.; Paulussen, J.; He, Y.; Wang, M.; Wang, B.H.; Wang, Z. Monitoring and Predicting Land Use Change in Beijing Using Remote Sensing. *Landsc. Urban Planing* **2006**, *78*, 322–333. [CrossRef]
20. Balthazar, V.; Vanacker, V.; Molina, A.; Lambin, E.F. Impacts of forest cover change on ecosystem services in high Andean mountains. *Ecol. Indic.* **2015**, *48*, 63–75. [CrossRef]
21. Gerard, F.; Petit, S.; Smith, G.; Thomson, A.; Brown, N.; Manchester, S.; Wadsworth, R.; Bugar, G.; Halada, L.; Bezak, P.; et al. Land cover change in Europe between 1950 and 2000 determined employing aerial photography. *Prog. Phys. Geogr.* **2010**, *34*, 183–205. [CrossRef]

22. Muttitanon, W.; Tripathi, N.K. Land use/land cover changes in the coastal zone of Ban Don Bay, Thailand using Landsat 5 TM data. *Int. J. Remote Sens.* **2005**, *26*, 2311–2323. [CrossRef]
23. Pervez, M.S.; Henebry, G.M. Assessing the impacts of climate and land use and land cover change on the freshwater availability in the Brahmaputra River basin. *J. Hydrol. Reg. Stud.* **2015**, *3*, 285–311. [CrossRef]
24. Essahlaoui, A.; Teodoro, A.C.; Mohajane, M. Modeling and mapping of soil salinity in Tafilalet plain (Morocco). *Arab. J. Geosci.* **2019**, *12*, 35.
25. Mohajane, M.; Essahlaoui, A.; Oudija, F.; El Hafyani, M.; Cláudia Teodoro, A. Mapping forest species in the central middle atlas of Morocco (Azrou Forest) through remote sensing techniques. *ISPRS Int. J. Geo-Inf.* **2017**, *6*, 275. [CrossRef]
26. Liang, D.; Zuo, Y.; Huang, L.; Zhao, J.; Teng, L.; Yang, F. Evaluation of the consistency of MODIS Land Cover Product (MCD12Q1) based on Chinese 30 m GlobeLand30 datasets: A case study in Anhui Province, China. *ISPRS Int. J. Geo-Inf.* **2015**, *4*, 2519–2541. [CrossRef]
27. Satyanarayana, B.; Thierry, B.; Seen, D.L.; Raman, A.V.; Muthusankar, G. Remote sensing in mangrove research–relationship between vegetation indices and dendrometric parameters: A Case for Coringa, East Coast of India. In Proceedings of the 22nd Asian Conference on Remote Sensing, Singapore, 5–9 November 2001; pp. 5–9.
28. Reis, S. Analyzing land use/land cover changes using remote sensing and GIS in Rize, North-East Turkey. *Sensors* **2008**, *8*, 6188–6202. [CrossRef] [PubMed]
29. Dewany, A.M. Yamaguchi, Land use and land cover change in Greater Dhaka, Bangladesh: Using remote sensing to promote sustainable urbanization. *Appl. Geogr.* **2009**, *29*, 390–401. [CrossRef]
30. Baldi, G.; Paruelo, J.M. Land-use and land cover dynamics in South American temperate grasslands. *Ecol. Soc.* **2008**, *13*, 6. [CrossRef]
31. Rawat, J.S.; Manish, K. Monitoring land use/cover change using remote sensing and GIS techniques: A case study of Hawalbagh block, district Almora, Uttarakhand, India. *Egypt. J. Remote Sens. Space Sci.* **2015**, *18*, 77–84. [CrossRef]
32. Shalaby, A.; Tateishi, R. Remote sensing and GIS for mapping and monitoring land cover and land-use changes in the Northwestern coastal zone of Egypt. *Appl. Geogr.* **2007**, *27*, 28–41. [CrossRef]
33. Diouf, A.; Lambin, E.F. Monitoring land-cover changes in semi-arid regions: Remote sensing data and field observations in the Ferlo, Senegal. *J. Arid Environ.* **2001**, *48*, 129–148. [CrossRef]
34. Seitzinger, S.P.; Gaffney, O.; Brasseur, G.; Broadgate, W.; Ciais, P.; Claussen, M.; Erisman, J.W.; Kiefer, T.; Lancelot, C.; Monks, P.S.; et al. International Geosphere–Biosphere Programme and Earth system science: Three decades of co-evolution. *Anthropocene* **2015**, *12*, 3–16. [CrossRef]
35. Woldesenbet, T.A.; Elagib, N.A.; Ribbe, L.; Heinrich, J. Hydrological responses to land use/cover changes in the source region of the Upper Blue Nile Basin, Ethiopia. *Sci. Total Environ.* **2017**, *575*, 724–741. [CrossRef]
36. Van Ty, T.; Sunada, K.; Ichikawa, Y.; Oishi, S. Scenario-based impact assessment of land use/cover and climate changes on water resources and demand: A case study in the Srepok River Basin, Vietnam—Cambodia. *Water Resour. Manag.* **2012**, *26*, 1387–1407.
37. Li, Z.; Deng, X.; Wu, F.; Hasan, S. Scenario analysis for water resources in response to land use change in the middle and upper reaches of the Heihe River Basin. *Sustainability* **2015**, *7*, 3086–3108. [CrossRef]
38. Essahlaoui, A.; Sahbi, H.; Bahi, L.; El-Yamine, N. Reconnaissance de la structure géologique du bassin de Saiss occidental, Maroc, par sondages électriques. *J. Afr. Earth Sci.* **2001**, *32*, 777–789. [CrossRef]
39. Population Légale d'après les Résultats du RGPH 2014 sur le Bulletin Officiel N°. Available online: https://rgph2014.hcp.ma/downloads/Resultats-RGPH-2014_t18649.html (accessed on 18 January 2020).
40. Erbek, F.S.; Özkan, C.; Taberner, M. Comparison of maximum likelihood classification method with supervised artificial neural network algorithms for land use activities. *Int. J. Remote Sens.* **2004**, *25*, 1733–1748. [CrossRef]
41. Otukei, J.R.; Blaschke, T. Land cover change assessment using decision trees, support vector machines and maximum likelihood classification algorithms. *Int. J. Appl. Earth Obs. Geoinf.* **2010**, *12*, S27–S31. [CrossRef]
42. Paola, J.D.; Schowengerdt, R.A. A review and analysis of backpropagation neural networks for classification of remotely-sensed multi-spectral imagery. *Int. J. Remote Sens.* **1995**, *16*, 3033–3058. [CrossRef]
43. Lewis, H.G.; Brown, M. A generalized confusion matrix for assessing area estimates from remotely sensed data. *Int. J. Remote Sens.* **2001**, *22*, 3223–3235. [CrossRef]
44. Card, D.H. Using known map category marginal frequencies to improve estimates of thematic map accuracy. *Photogramm. Eng. Remote Sens.* **1982**, *48*, 431–439.

45. Titus, K.; Mosher, J.A.; Williams, B.K. Chance-corrected classification for use in discriminant analysis: Ecological applications. *Am. Midl. Nat.* **1984**, *111*, 1–7. [CrossRef]
46. Bwangoy, J.R.B.; Hansen, M.C.; Roy, D.P.; De Grandi, G.; Justice, C.O. Wetland mapping in the Congo Basin using optical and radar remotely sensed data and derived topographical indices. *Remote Sens. Environ.* **2010**, *114*, 73–86. [CrossRef]
47. Agence Urbaine. Projet de Plan D'action de Développement Local Meknès; Meknès, Maroc, 2009. Available online: https://www.hcp.ma/region-meknes/attachment/428604/ (accessed on 18 January 2020).
48. Ministère Délègue Charge de l'Habitat et de l'Habitat et de l'Urbanisme, 2004, «Villes sans Bidonvilles». Available online: https://mirror.unhabitat.org/downloads/docs/11592_4_594598.pdf (accessed on 18 January 2020).
49. FAO. *Food and Agricultural Organization of the United Nations FAO AQUASTAT Database*; FAO: Rome, Italy, 2014; Available online: https://www.fao.orgnraquastat.pdf (accessed on 11 November 2018).
50. MAPM. *Ministère de l'Agriculture et de la Pèche Maritime, Rapport Plan Maroc Vert*; MAPM: Rabat, Morocco, May 2011.
51. Plan Bleu. Environnement et Développement en Méditerranée. Les Notes du Plan Bleu, No 11. 2009. Available online: https://planbleu.org/sites/default/files/publications/soed2009-fr.pdf (accessed on 18 January 2020).
52. Régie Autonome de Distribution d'Eau et d'Électricité de Meknès, Rapport de Gestion 2016 & Comptes Annuels. Available online: https://www.radeema.ma/documents/56925/56948/RG2016+web.pdf/2bc8116b-3429-459f-997a-783226946cbc (accessed on 18 January 2020).
53. Oñate-Valdivieso, F.; Sendra, J.B. Application of GIS and remote sensing techniques in generation of land use scenarios for hydrological modeling. *J. Hydrol.* **2010**, *395*, 256–263. [CrossRef]
54. Haque, M.M.; Egodawatta, P.; Rahman, A.; Goonetilleke, A. Assessing the significance of climate and community factors on urban water demand. *Int. J. Sustain. Built Environ.* **2015**, *4*, 222–230. [CrossRef]
55. Xiao, J.; Shen, Y.; Ge, J.; Tateishi, R.; Tang, C.; Liang, Y.; Huang, Z. Evaluating urban expansion and land use change in Shijiazhuang, China, by using GIS and remote sensing. *Landsc. Urban Plan.* **2006**, *75*, 69–80. [CrossRef]
56. Fung-Loy, K.; Van Rompaey, A.; Hemerijckx, L.M. Detection and Simulation of Urban Expansion and Socioeconomic Segregation in the Greater Paramaribo Region, Suriname. *Tijdschr. Voor Econ. En Soc. Geogr.* **2019**, *110*, 339–358. [CrossRef]
57. El Garouani, A.; Mulla, D.J.; El Garouani, S.; Knight, J. Analysis of urban growth and sprawl from remote sensing data: Case of Fez, Morocco. *Int. J. Sustain. Built Environ.* **2017**, *6*, 160–169. [CrossRef]
58. Höpfner, C.; Scherer, D. Analysis of vegetation and land cover dynamics in north-western Morocco during the last decade using MODIS NDVI time series data. *Biogeosciences* **2011**, *8*, 3359–3373. [CrossRef]

Article

A Comparison and Validation of Saturated Hydraulic Conductivity Models

Kaylyn S. Gootman [1], Elliott Kellner [1,2] and Jason A. Hubbart [1,2,3,*]

1 Institute of Water Security and Science, West Virginia University, Agricultural Sciences Building, Morgantown, WV 26506, USA; kaylyn.gootman@mail.wvu.edu (K.S.G.); elliott.kellner@mail.wvu.edu (E.K.)

2 Division of Plant and Soil Sciences, Davis College of Agriculture, Natural Resources and Design, West Virginia University, Agricultural Sciences Building, Morgantown, WV 26506, USA

3 Division of Forestry and Natural Resources, Davis College of Agriculture, Natural Resources and Design, West Virginia University, Agricultural Sciences Building, Morgantown, WV 26506, USA

* Correspondence: jason.hubbart@mail.wvu.edu; Tel.: +1-304-293-2472

Received: 26 June 2020; Accepted: 16 July 2020; Published: 18 July 2020

Abstract: Saturated hydraulic conductivity (K_{sat}) is fundamental to shallow groundwater processes. There is an ongoing need for observed and model validated K_{sat} values. A study was initiated in a representative catchment of the Chesapeake Bay Watershed in the Northeast USA, to collect observed K_{sat} and validate five K_{sat} pedotransfer functions. Soil physical characteristics were quantified for dry bulk density (*bdry*), porosity, and soil texture, while K_{sat} was quantified using piezometric slug tests. Average *bdry* and porosity ranged from 1.03 to 1.30 g/cm^3 and 0.51 to 0.61, respectively. Surface soil (0–5 cm) *bdry* and porosity were significantly ($p < 0.05$) lower and higher, respectively, than deeper soils (i.e., 25–30 cm; 45–50 cm). *bdry* and porosity were significantly different with location ($p < 0.05$). Average soil composition was 92% sand. Average K_{sat} ranged from 0.29 to 4.76 m/day and significantly differed ($p < 0.05$) by location. Four models showed that spatial variability in farm-scale K_{sat} estimates was small (CV < 0.5) and one model performed better when K_{sat} was 1.5 to 2.5 m/day. The two-parameter model that relied on silt/clay fractions performed best (ME = 0.78 m/day; SSE = 20.68 m^2/day^2; RMSE = 1.36 m/day). Results validate the use of simple, soil-property-based models to predict K_{sat}, thereby increasing model applicability and transferability.

Keywords: saturated hydraulic conductivity; pedotransfer function; model validation; Chesapeake Bay Watershed; experimental watershed study

1. Introduction

Saturated hydraulic conductivity (K_{sat}) is an important hydraulic parameter [1–4], as K_{sat} represents the ability of soils to transmit water throughout the saturated zone, which is essential for relating water transport rates to hydraulic gradients [5–7]. Accurate K_{sat} estimates are needed to characterize and predict how soil–water dynamics influence local water balances [8–10]. Thus, K_{sat} estimates can inform resource management decisions related to water conservation, irrigation systems, fertilizer application, drainage, solute mitigation, and plant growth [11–13]. Generally, K_{sat} is measured through field and laboratory techniques (e.g., pumping, permeameter, and slug tests) [14–17] that are relatively simple to complete [18–20]. However, performing a sufficient number of field-based tests may be too expensive, in terms of duration and cost [21,22]. Additionally, field-based K_{sat} estimates can be limited by incomplete aquifer geometry information, while laboratory methods can present problems with obtaining representative sample numbers. These challenges suggest the need for methods to estimate K_{sat} that are accurate and efficient [23–25].

An alternative to direct K_{sat} measurements is to predict K_{sat} with pedotransfer functions [26] that utilize soil physical characteristics such as bulk density [27], porosity [28], particle size fractions [11,29], along with empirical methods that utilize multiple physical properties [30–35]. Gathering the required data for physical models is simpler and routinely done, as characterizing soil physical properties is generally less complex than pumping or permeameter tests [36–38]. Numerous studies have developed pedotransfer functions to estimate K_{sat} based on soil properties, as physical characteristics are easily measured and not hydraulic-boundary-dependent [13,32,34]. Many of these simple models utilize one or more soil particle size fractions to directly estimate K_{sat} [39–41] or characterize relationships between additional soil properties and K_{sat} [11,27]. For example, Saxton et al. [11] showed that the relationships between soil water content and textures resulted in K_{sat} predictions that compared well with independent K_{sat} measurements. More complex models, such as Jabro [27], rely on a combination of soil characteristics and particle size fractions to estimate K_{sat}. More recently, Vienken and Dietrich [14] showed that grain-size data could be adequately used to predict K_{sat} for initial site assessments, further supporting the use of soil-property-based K_{sat} models.

Although these studies predicted K_{sat} using different, widely accepted grain-size metrics and soil characteristics, the results show that soil-property-based K_{sat} models warrant validation to demonstrate their applicability and transferability. Five previously published models [11,27,39–41] present an opportunity for a model validation study, as these models predict K_{sat} from either one or two particle size fractions and bulk density. More complex models, such as those that utilize grain size distributions [14,30], may result in more accurate K_{sat} predictions but require additional analyses for adequate model parameterization. These additional steps may not be achievable/affordable for practitioners interested in efficient K_{sat} predictions [8,42]. However, particle size fractions and bulk density are typically included in soil characterizations and are relatively easy to obtain [13,32], thus supporting the evaluation and comparison of soil-property-based K_{sat} models. An applied model comparison can determine the accuracy and efficiency of K_{sat} predictions generated from relatively simple pedotransfer functions.

The importance of accurate K_{sat} predictions extends beyond validating soil property based K_{sat} models. K_{sat} is commonly used to parameterize process based hydrologic models since K_{sat} governs permeability and contributes to shallow groundwater-surface water exchange processes [17,29]. Many near surface process models (e.g., HEC-HMS) use K_{sat} as a calibration parameter to understand how water flows change within a specific drainage area [43–45]. Additionally, groundwater (e.g., MODFLOW) and solute transport (e.g., ReacTran modeling depends on measures of soil K_{sat} at specific spatial scales of interest [46–49]. Thus, accurate K_{sat} measurements are a prerequisite for simulating flow and transport processes at local levels and up-scaling to regional watersheds [14,50,51].

The primary objective of the current work was to collect observed K_{sat} values from a representative watershed in the northeastern USA. The sub-objectives were to use observed data to validate the predictive accuracy of the Puckett et al. [39], Jabro [27], Campbell [40], Smettem and Bristow [41], and Saxton et al. [11] K_{sat} models for coarse-grained, alluvial floodplain soils with a history of agricultural production. Validated model accuracy will improve end-user confidence in model estimated K_{sat} for use in hydrologic and constituent transport predictions, thus expanding characterizations of local soil–water dynamics. Additionally, these results can increase the applicability of K_{sat} models for practitioners and improve understanding of local soil–water dynamics.

2. Methods

2.1. Site Description

This study was conducted at eight co-located stream stage and piezometer monitoring sites using an experimental watershed study design [52–57] along Moore's Run, located within Reymann Memorial Farm (RMF; 39°6′12.7″ N, 78°35′8.19″ W) near Wardensville, West Virginia, USA (Figure 1). Moore's Run flows southeast approximately 42 km to the Cacapon River and is in the upper Chesapeake Bay Watershed (CBW) (Figure 1) [58–60]. The Cacapon River drainage area is approximately 65.2 km^2 (Table 1). Land use and land cover (LULC) throughout the upper Cacapon River drainage area is 77.8%

forest and 14.9% agriculture (Table 1) [61,62]. Land cover across the contributing drainage area for each study site (n = 8) is similarly dominated by forest and agriculture LULC (Table 1).

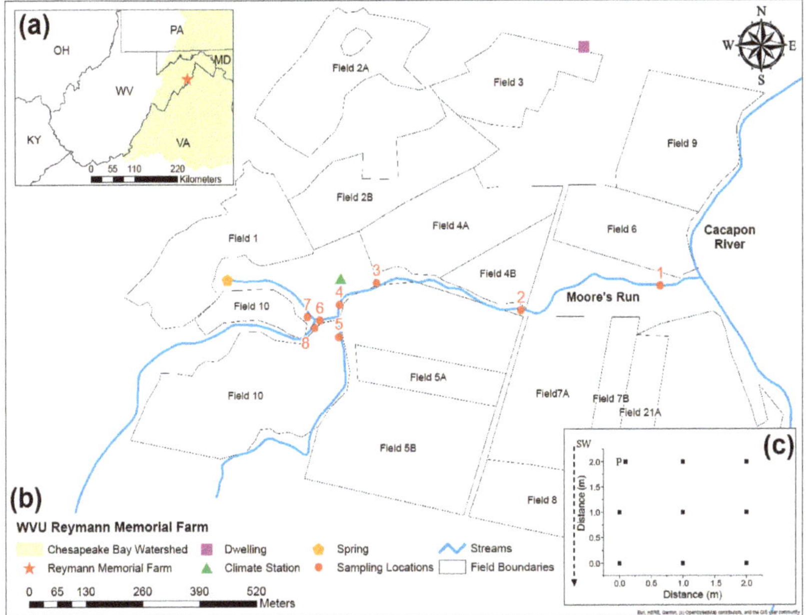

Figure 1. (**a**) The location of Reymann Memorial Farm within the upper Chesapeake Bay Watershed and (**b**) the locations of eight co-located, nested piezometers throughout the Moore's Run drainage area, Wardensville, West Virginia, USA [61,62]. (**c**) Soil core grid design of the current study in Wardensville, West Virginia, USA. SW represents an example stilling well location and P represents an example piezometer location in relation to Moore's Run. The dashed arrow indicates the flow direction towards the Cacapon River.

Table 1. Cumulative land use and land cover (LULC) [61] and drainage area (km^2) corresponding with each study site (n = 8) sub-basin from the studied portion of Moore's Run and a sub-watershed of the upper Cacapon River, located near Wardensville, West Virginia, USA [1]. Percent cumulative land use type is displayed parenthetically.

LULC [km^2 (%)]	RMF1	RMF2	RMF3	RMF4	RMF5	RMF6	RMF7	RMF8	Cacapon
Agriculture	3.8 (10.5)	3.1 (8.8)	3.1 (8.8)	3.1 (8.8)	2.9 (8.5)	0.2 (20.3)	<0.1 (42.8)	0.1 (16.3)	9.7 (14.9)
Forested	31.1 (86.5)	30.8 (88.3)	30.8 (88.3)	30.8 (88.3)	30.2 (88.5)	0.6 (79.2)	<0.1 (57.2)	0.4 (83.4)	50.8 (77.8)
Mixed Development	0.7 (2.0)	0.7 (1.9)	0.7 (1.9%)	0.7 (1.9)	0.7 (1.9)	<0.1 (0.2)	0 (0)	<0.1 (0.3)	3.1 (4.7)
Open Water	0.4 (1.0)	0.4 (1.1)	0.4 (1.1)	0.4 (1.1)	0.4 (1.1)	<0.1 (0.3)	0 (0)	0 (0)	1.7 (2.5)
Total area	35.9 (100)	34.9 (100)	34.9 (100)	34.9 (100)	34.1 (100)	0.8 (100)	<0.1 (100)	0.5 (100)	65.2 (100)

[1] Site specific drainage areas were delineated using a 1–3 m statewide mosaic for West Virginia [62].

The dominant soils in the study area are Basher fine sandy loam in the floodplains and Monongahela silt loam (3–8% slopes) along the stream terraces, with dry bulk density and K_{sat} ranging from 1.34 to 1.53 (g/cm^3) and 0.19–1.99 (m/day), respectively [63–66]. Natural Resources Conservation Service (NRCS) methods used to derive K_{sat} typically include double ring infiltrometers, constant head well permeameters, and empirical relationships between dry bulk density and particle size distributions [27,63]. Both soil types are moderately well drained [67]. Particle size fractions for Basher fine sandy loam range from 50 to 90% sand, 6 to 48% silt, and 5 to 12% clay, while average particle size fractions for Monongahela silt loam range from 20 to 27% sand, 51 to 54% silt, and 19 to 26% clay [63].

Regional geology consists of Silurian limestone, Devonian sandstone, Devonian shale, and Quaternary alluvium [68,69]. Soil textures include sand, sandy loam, silty clay loam, silt loam, and loam [63]. A hydroclimate monitoring station was installed at RMF in 2019 as a part of an ongoing nested-scale experimental watershed study (Figure 1) [52–56,70–72]. Average daily air temperature (August 2019 to April 2020) was 9.43 °C, while daily precipitation totals ranged from 0 mm to 41.4 mm. Longer term records (1917–2016) show average annual air temperature at the time of observation and average annual precipitation at 8.6 °C was 810.9 mm, respectively [73].

2.2. Soil Strutural Properties

Soil structural properties were determined using the soil core method [37,74]. The soil core method requires the collection of a known soil volume (i.e., core) by driving a cylindrical sampler into the soil to extract relatively undisturbed samples from a series of profile depths. Each core is weighed and then oven-dried at 105 °C for 24 to 48 h, or until the sample mass remains constant with additional drying time [16,75]. The oven-dried sample mass is cooled to room temperature in a desiccator and weighed [37]. At each study site (n = 8), three 98.17 cm^3 soil cores (n = 179) were collected from depth intervals of 0–5, 25–30, and 45–50 cm at nine equidistant locations across a 2 m × 2 m grid (Figure 1). The grids were designed to be proximal to the piezometers installed along Moore's Run [53]. Depth intervals were selected to characterize the soil structural properties of the upper portion of the unsaturated zone. Soil cores could not be collected at every depth/sampling location due to the presence of large rocks.

After sampling, each core was capped at both ends, stored on ice, and transported to the laboratory within 24 h. Each core was processed and analyzed using the following equations from Hillel [37]. Soil dry bulk density, *bdry*, (g/cm^3) was quantified as

$$bdry = \frac{M_s}{V_t}, \tag{1}$$

where M_s (g) is the mass of the soil core solids and V_t (cm^3) is the soil core total volume. Porosity (unitless) was estimated as

$$Porosity = \frac{V_f}{V_t}, \tag{2}$$

where V_f (cm^3) is the volume of the soil core pore spaces [74,76]. Soil dry bulk density and porosity were quantified because they are structural properties that are known to influence shallow groundwater flow [2,8,38,77] and are easy to measure during routine soil surveys [63]. Two-way analysis of variance (ANOVA) was performed on soil characteristics and tested for significant differences between study site means and independent soil depths [75,78]. After each two-way ANOVA, a Tukey's post hoc multiple-comparison test compared the nominal and measurement variables in all possible combinations [75,78]. All statistical analyses were performed using Origin Pro 2019 (OriginLab Corporation, Northampton, Massachusetts, MA, USA) software.

2.3. Particle Size Fractions and Soil Texture

After the soil cores were dried and weighed, particle size fraction analysis was performed on the 183 samples using a combination of dry sieving and gravimetric filtration [79–81]. Particle size fractions were defined using the NRCS classification system [82]. First, any remaining soil aggregates were carefully broken-up by hand [79–83]. Individual soil core samples were poured into a nest of stacked sieves and separated into three size classes with a mechanical shaker [2,14,81]. The fraction retained on the largest grade sieve (2 mm mesh opening) consisted of coarse gravel, the fraction retained on the fine mesh sieve (53 μm mesh opening) contained sand particles, and the fraction retained in the pan contained fine particles, which consisted of silt and clay [82,84]. After 15 min of sieving, each size fraction was weighed [79,85].

The remaining fine particle fraction was rehydrated with 300 mL of deionized water and vigorously shaken until the particles and the water were well mixed [79,80]. Next, the fine particle mixture was filtered gravimetrically with a vacuum flask [81]. Washed and dried Whatman filter paper (pore size = 2 μm) was used to physically separate the silt and clay size fractions. Filters were oven-dried for one hour at 105 °C, cooled in a desiccator to room temperature, and weighed. The filter drying and weighing process was repeated to confirm that the filtered sample masses agreed within 4% [86]. The mass retained on the filters consisted of the silt fraction [87]. The clay fraction was determined by residual after subtracting the dry-sieved gravel mass, sand mass, and gravimetrically filtered silt mass from the original soil mass [79,81].

The resulting percentages of sand, silt, and clay for each soil core were used to model saturated hydraulic conductivity, as detailed below in Section 2.5. Similar statistical analyses, as detailed in Section 2.2., were conducted on the particle size fraction data. Soil texture was determined for each core using the NRCS Soil Texture Calculator [88].

2.4. Field Saturated Hydraulic Conductivity

A series of falling and rising head slug tests were conducted at the piezometer monitoring locations (n = 8; Figure 1) on 20 September 2019, 20 January 2020, and 6 March 2020, to determine K_{sat} (m/day). Steel drive point piezometers 0.61–1.52 m long with a 3.18 cm inner diameter and a 0.77 m screened bottom segment (i.e., drive point) were driven into the unconfined alluvial aquifer at RMF. Each piezometer was equipped with a Solinst Levelogger Edge [89] pressure transducer to measure water level (M5: error ± 0.003 m; M10: error ± 0.005 m) either every 0.125 s or 0.5 s for the duration of the slug tests. The difference in sampling interval was due to the length of time required for each test and instrument storage capacity. After enough time (i.e., 5–10 min) passed for the water level to equilibrate with the added sensor, a 0.17 cm^3 copper slug was quickly lowered into and removed from each piezometer three times, with time allowed for water levels to equilibrate between each slug insertion and retrieval. Based on earlier field tests, water levels were estimated to equilibrate after a period of at least 5 min. Steady state conditions between replicate slug tests were visually confirmed by identifying asymptotes in the water level data. Average K_{sat} were calculated from the falling and rising water levels using Hvorslev's method [15,17]

$$K_{sat} = \frac{r^2 ln(L_e/R)}{2L_e t_{37}} \tag{3}$$

where r is the radius of the well casing (cm), R is the radius of the well screen (cm), L_e is the length of the well screen (cm), and t_{37} (s) is the time it takes for the water level inside the piezometer to rise or fall 37% of the initial change during a slug test. It should be noted that the average K_{sat} derived from slug tests requires the assumption that the average K_{sat} is representative of the actual field conditions [23].

Falling and rising head slug tests were conducted to assess whether K_{sat} varied during the piezometer emptying and filling as water levels returned to static conditions. Thus, a significant difference in falling- and rising-head-derived K_{sat} demonstrates the presence of a directional dependence on the slug test type that may be attributed to site non-idealities [18]. Slug tests were repeated to increase the statistical power of the results [19]. Differences between the average (n = 3) falling-head-derived K_{sat} and average rising-head-derived K_{sat} (n = 3) were tested with a paired sample T-test [78]. Differences in site-average K_{sat} (n = 6), independent of slug test type, were quantified with a one-way ANOVA and a Tukey's post hoc multiple-comparison test compared pairs of site-average K_{sat} in all possible combinations [78].

2.5. Modeling Saturated Hydraulic Conducitivity

Average site-level K_{sat} was estimated using soil particle size fraction data and five previously published pedotransfer functions [11,27,39–41], with the understanding that K_{sat} predictions reflect neither horizontal nor vertical aquifer properties due to the nature of the soil core sampling procedure [90]. Studies have shown that floodplain soil characteristics converge at depths of approximately 50 cm [75,91], which supports K_{sat} predictions based on data from shallower (i.e., <50 cm) depths. Model results were compared to average measured K_{sat}. Rising and falling head slug tests were combined so that each average measured K_{sat} was based on a larger sample size (n = 6) to increase statistical power and capture unsaturated zone water flow variability. The resulting average K_{sat} for each study site accounted for soil wetting and drying. Model performance was evaluated by comparing predicted K_{sat} to measured K_{sat} at the farm-level.

2.5.1. Puckett et al. Model

Puckett et al. [39] developed a model to predict K_{sat} based on only clay-sized particles. The authors showed that fine sand, sand, and clay percentages were highly correlated with K_{sat}, surface area, and volumetric water content at specific pressure heads [39,82]. The model for estimating K_{sat} based on clay fractions is as follows

$$K_{sat}(p)_i = 4.66 * 10^{-3} exp(-0.1975clay) \tag{4}$$

where K_{sat} $(p)_i$ is the predicted soil saturated hydraulic conductivity (cm/s) at each study site (i) and *clay* represents the average dimensionless clay fraction.

2.5.2. Jabro Model

Jabro [27] proposed a model that used *bdry* and grain size as predictive variables of K_{sat}

$$log\left[K_{sat}(p)_i\right] = 9.59 - 0.81log(silt) - 1.09log(\text{clay}) - 4.64(bdry) \tag{5}$$

where K_{sat} $(p)_i$ is the predicted soil saturated hydraulic conductivity (cm/h) from each site (i); *silt* and *clay* represent site-average dimensionless fractions of silt and clay, respectively; and *bdry* is the site-average dry bulk density (g/cm^3).

2.5.3. Campbell Model

Campbell [40] published a model to predict K_{sat} from existing soil texture data

$$K_{sat}(p)_i = C * exp[-0.025 - 3.63(silt) - 6.9(clay)] \tag{6}$$

in which K_{sat} $(p)_i$ is the predicted soil saturated hydraulic conductivity (mm/h) from each site (i). *Silt* and *clay* represent site average dimensionless fractions of silt and clay, respectively. The constant C is equal to 144 and was derived from previously published studies by Hall et al. [92] and Smettem and Bristow [41].

2.5.4. Smettem and Bristow Model

Smettem and Bristow [41] developed a model to predict K_{sat} from soil clay content using a variety of agricultural topsoil samples [13]. The two-equation K_{sat} model is as follows

$$h_b = 43.5/[-0.25log(clay) + 0.5] \tag{7}$$

$$K_{sat}(p)_i = 2500 * C * {h_b}^{-2} \tag{8}$$

where h_b is the bubbling pressure (mm), *clay* represents the average dimensionless fraction of clay, and K_{sat} $(p)_i$ is the predicted soil saturated hydraulic conductivity (mm/h) at each study site (i). The constant, C, in Equation (8) is the same constant as Equation (6).

2.5.5. Saxton et al. Model

Saxton et al. [11] studied the relationships between soil texture and soil moisture content at saturation (Equation (12)) and soil texture and K_{sat} (Equation (9)). The relationships between these parameters and K_{sat} are

$$K_{sat}(p)_i = 2.778 * 10^{-3}\{exp[A + (B/\theta_s)]\} \tag{9}$$

$$A = 12.012 - 0.0755(sand) \tag{10}$$

$$B = -3.8950 + 0.03671(sand) - 0.1103(clay) + 8.7546 * 10^{-4}(clay)^2 \tag{11}$$

$$\theta_s = 0.332 - 7.251 * 10^{-4}(sand) + 0.1276log(clay) \tag{12}$$

where K_{sat} $(p)_i$ is the predicted soil saturated hydraulic conductivity (mm/s) at each study site (i); *sand* and *clay* represent the average dimensionless fraction of sand and clay, respectively; and θ_s is the soil moisture content at saturation (m^3/m^3).

2.5.6. Statistical Analysis

Farm-level predicted and measured K_{sat} were compared for each K_{sat} model using a statistical analysis outlined in Duan et al. [13]. The mean error (ME), the sum of squared error (SSE), and the root of the mean-square error (RMSE) were quantified for each model. The mean difference between the average predicted and average measured values was determined for the ME with

$$ME = \sum_{i=1}^{n} \frac{K_{sat}(p)_i - K_{sat}(m)_i}{n} \tag{13}$$

where K_{sat} $(m)_i$ is the measured soil-saturated hydraulic conductivity (m/day) from each of study site (i); K_{sat} $(p)_i$ is the predicted soil-saturated hydraulic conductivity (m/day) from each of study site (i); and n is the number of sites included the farm-level metrics (n = 8). The SSE and RMSE were determined using the following equations

$$SSE = \sum_{i=1}^{n}\left[K_{sat}(p)_i - K_{sat}(m)_i\right]^2 \tag{14}$$

$$RMSE = \sqrt{\frac{\sum_{i=1}^{n}\left[K_{sat}(p)_i - K_{sat}(m)_i\right]^2}{n}}. \tag{15}$$

3. Results and Discussion

3.1. Soil Structural Properties

Soils cores were extracted at depths of 0–5, 25–30, and 45–50 cm within the 2 m × 2 m study grid (Figure 1c), for a total sample size ranging from 14 to 27 at each study site (n = 179, total core number). The average soil core results for *bdry* and porosity over the total depth (50 cm) were 1.11, 1.25, and 1.29 g/cm^3 and 0.58, 0.53, and 0.51, respectively (Figure 2). Mean *bdry* ranged from 1.03 g/cm^3 at RMF7 to 1.30 g/cm^3 at RMF5 and RMF8, with an eight site mean of 1.21 g/cm^3. Mean porosity ranged from 0.51 at RMF4, RMF5, and RMF8 to 0.61 at RMF7, with an eight site mean of 0.54. Average *bdry* was below the NRCS range for the region (i.e., 1.34–1.54 g/cm^3) but within the range expected for sandy soils [63,93]. The average porosity values were within the expected soil range of 0.3–0.7 [94]. The differences in *bdry* were likely due to differences in sampling locations for the NRCS study.

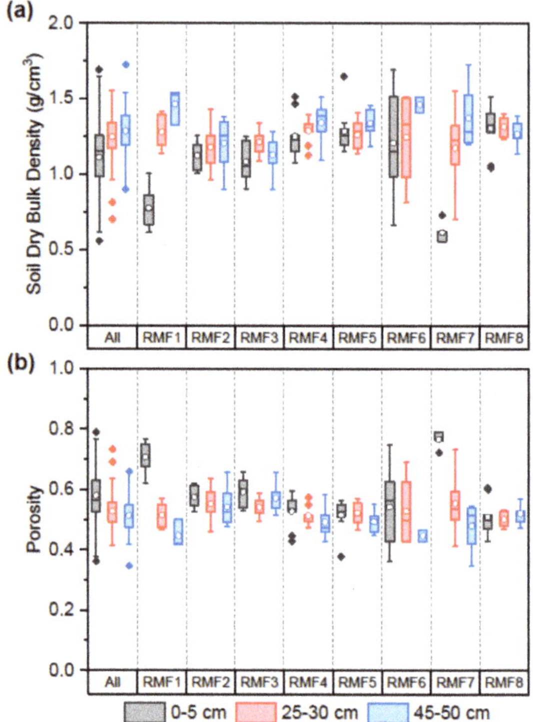

Figure 2. Soil dry bulk density (g/cm^3) (**a**) and porosity (**b**) for all sites, by depth throughout the Moore's Run Watershed, Wardensville, West Virginia, USA. Boxes define the interquartile range (IQR). Vertical lines show the range within 1.5 IQR. Midlines indicate the median. Circles within the boxes denote the mean. Filled-in points indicate potential outliers.

Based on two-way ANOVA results, sites differed significantly in *bdry* ($n = 179$; $p < 0.05$). A comparison of the sites sampled at all depths, using Tukey's post hoc multiple comparison, showed that *bdry* was significantly lower at the 0–5 cm depth ($n = 14$ to $n = 27$ each site; $p < 0.05$) [78]. Porosity was also significantly different among study sites ($n = 179$; $p < 0.05$) and was significantly higher ($n = 14$ to $n = 27$ each site, $p < 0.05$) at the 0–5 cm depth. *bdry* and porosity were statistically similar ($p > 0.05$) at 25–30 and 45–50 cm, confirming that below the surface (i.e., 0–5 cm), RMF alluvial soils are homogeneous [75,91]. However, significant differences ($p < 0.05$) in RMF *bdry* and porosity are evidence of inter-site heterogeneity between RMF study sites.

3.2. Particle Size and Soil Texture

When particle size classes (i.e., sand, silt, and clay) were averaged over the total depth (50 cm), sand was consistently the dominant particle size class, followed by silt, and then clay. Average ($n = 179$) sand, silt, and clay percentages were 92%, 7%, and <1%, respectively (Figure 3). Based on the results of two-way ANOVA tests, the sites differed significantly in all particle size classes ($n = 179$; $p < 0.05$) [78]. The soil core textures were sand ($n = 155$) with a few instances of loamy sand ($n = 24$) [63]. A comparison of the particle size fractions at all sampled depths with Tukey's post hoc multiple comparison tests showed that the sites did not differ significantly in average particle size class percentages between soil depths ($p > 0.05$) [78]. When comparing these results to NRCS-mapped soils from the Web Soil Survey, which included higher percentages of silt and clay fractions, RMF soils had higher sand percentages, and smaller silt and clay percentages [63].

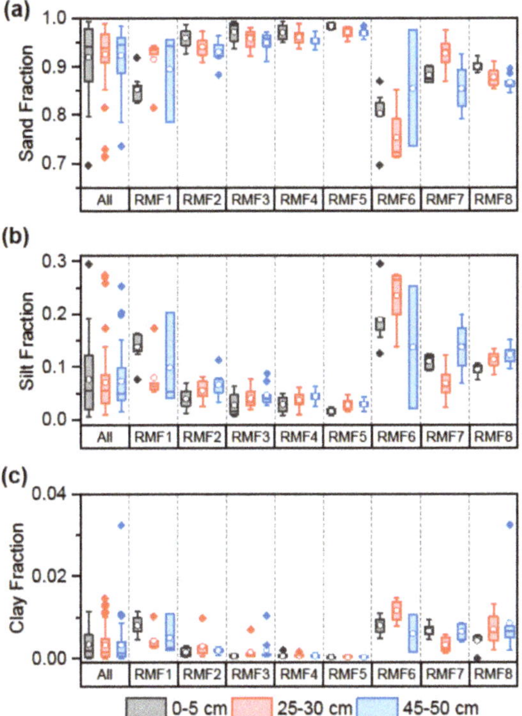

Figure 3. Particle size fractions including Sand (**a**), Silt (**b**), and Clay (**c**) for all sites, by depth throughout the Moore's Run Watershed, Wardensville, West Virginia, USA. Note that the axis ranges differ for each particle size fraction. Boxes define the interquartile range (IQR). Vertical lines show the range within 1.5 IQR. Midlines indicate the median. Circles within the boxes denote the mean. Filled-in points indicate potential outliers. Note the differences in y-axis ranges.

Results confirm that site soil textures were mostly comprised of sand but differed from NRCS findings, which include larger silt and clay fractions (Figure 3) [63,64]. This could be due to the proximity of the sampling grids to Moore's Run and is evidence of inter-site particle size heterogeneity across RMF study sites and the NRCS sampling locations. Differences in particle size fraction percentages with increasing depth were not significant ($p > 0.05$), which supports the use of near-surface (i.e., ≤50 cm) particle size fraction data to model K_{sat} values in the current work.

3.3. Field Saturated Hydraulic Conductivity

Site-level comparisons of slug-test-derived K_{sat} are presented in Figure 4 and separated by slug test type (i.e., FH or RH). FH K_{sat} were typically higher than RH K_{sat}, except for RMF6, where FH and RH K_{sat} were similar. The differences in K_{sat} between the slug test type can be attributed to non-idealities such as well-skin effects [18,19], while the similar K_{sat} values at RMF6 may be explained by differing non-idealities at this site. However, these differences may be attributed to the unique soil properties and site features adjacent to the RMF6 piezometer. A Tukey's post hoc multiple comparison of piezometer adjacent soils revealed that particle size fractions at RMF6 were significantly different ($p < 0.05$) when compared to RMF4 and RMF5, while the RMF6 clay fraction was significantly different ($p < 0.05$) from RMF3 [78]. Additionally, the piezometer at RMF6 was proximal to a buried culvert pipe that may have created an artificial hydraulic boundary, resulting in FH and RH K_{sat} estimates that were different from the other sites.

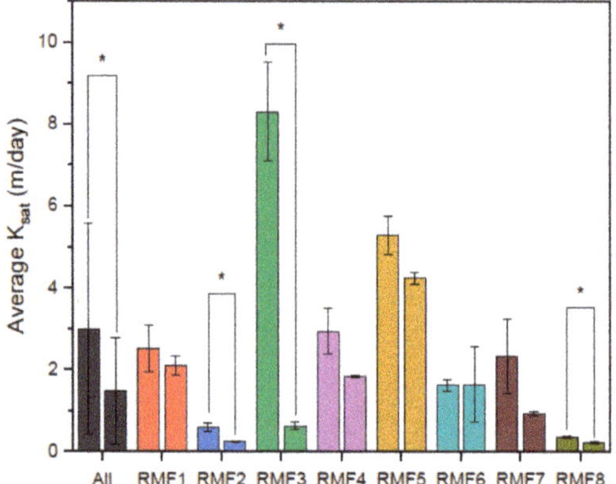

Figure 4. Bar plots of average soil saturated hydraulic conductivity (K_{sat}) (m/day) at eight study sites derived from falling head (left hand bar) and rising head (right hand bar) slug tests during the study period (September 2019–March 2020), Wardensville, West Virginia, USA. Box heights denote the mean value. Vertical lines define the standard deviation. Asterisks denote significant differences ($p < 0.05$) between the falling head and rising head derived average K_{sat} values.

Observed K_{sat} values across all study sites ranged from 0.35–9.33 m/day and 0.21–4.37 m/day for FH and RH slug tests, respectively. The larger range in FH K_{sat} may be attributed to more variable flow resistance during the FH slug tests [18,95]. When compared across all sites ($n = 24$ per slug test type), FH and RH K_{sat} values were significantly different ($p < 0.05$) (Figure 4) [78]. When examined by site ($n = 3$ per slug test type), RMF2, RMF3 and RMF8 had significantly different FH and RH K_{sat} ($p < 0.05$), likely due to differences in flow resistance near the piezometer screened interval [18,19,95,96]. When compared to the expected average K_{sat} for the region, five average FH K_{sat} were higher than K_{sat} range reported by the NRCS of 0.19–1.99 m/day, while only two average RH K_{sat} were higher than the expected range [63,64]. These differences in K_{sat} range may be due to methodology, as NRCS uses double ring infiltrometers and constant head well permeameters for field K_{sat} [97], while soil core *bdry* and particle size distributions are used for laboratory K_{sat} estimates [27].

Site-level average K_{sat} ranged from 0.29 m/day at RMF8 to 4.76 m/day at RMF5, while the average farm-level K_{sat} was 2.24 m/day (Table 2). The results of a one-way ANOVA indicated that the site-level average K_{sat} were significantly different across all study sites ($n = 8$; $p < 0.05$) [78]. Tukey's post-hoc multiple comparison test showed that average measured K_{sat} at RMF 3 and RMF5 were significantly higher than RMF2 and average measured K_{sat} at RMF5 was significantly higher than RMF6, RMF7, and RMF8 ($p < 0.05$). These significant differences highlight the inter-site heterogeneity between RMF study sites [78]. Since site-level average K_{sat} include FH and RH slug test variability, and thus better capture heterogeneity, they were used to assess K_{sat} model performance. Site-level average K_{sat} coefficient of variation (CV) ranged from 0.19 at RMF1 to 0.96 at RMF3, indicating greater variation between slug tests at RMF3. The greater variability in measured K_{sat} at RMF3 likely contributed to the high farm-level CV (Table 2) when all measure K_{sat} were averaged ($n = 48$).

Table 2. Site-level soil saturated hydraulic conductivity (m/day) comparisons derived from the falling head and rising head slug tests. The coefficient of variation (CV) is presented as a unitless ratio of the standard deviation to the mean.

Site	*n*	Mean	Med	Std. Dev.	CV
All	48	2.24	1.80	2.15	0.96
RMF1	6	2.31	2.23	0.45	0.19
RMF2	6	0.43	0.39	0.20	0.46
RMF3	6	4.46	3.85	4.27	0.96
RMF4	6	2.39	2.22	0.70	0.29
RMF5	6	4.76	4.56	0.65	0.14
RMF6	6	1.63	1.54	0.58	0.36
RMF7	6	1.63	1.14	0.96	0.59
RMF8	6	0.29	0.30	0.08	0.26

3.4. Modeled Saturated Hydraulic Conductivity

Predicted site-level K_{sat} and descriptive statistics for farm-level K_{sat} are presented in Table 3 for the Puckett et al. [39], Jabro [27], Campbell [40], Smettem and Bristow [41], and Saxton et al. [11] models. Average farm-level K_{sat} predicted by five tested models ranged from 1.94 m/day with the Saxton et al. [11] model to 39.07 m/day with the Jabro [26] model (Table 3). When compared to the observed farm-level K_{sat} (Table 2), the Puckett et al. [39], Campbell [40], and Saxton et al. [11] models resulted in estimated K_{sat} within one standard deviation of the average farm-level K_{sat} and were similar in magnitude. The Smettem and Bristow [41] model resulted in an estimated K_{sat} within two standard deviations of the farm-level K_{sat}. The Jabro [27] model, which was the only model that included *bdry* as a model parameter, resulted in an unrealistically high K_{sat} that was 178% higher than the observed result. Differences between model results may be attributable to the soil texture range of samples used in the previous studies or the topographic position of the sampling sites. Additionally, the inclusion of additional model parameters did not always result in a lower CV, as evidenced by the increase in CV, apart from the Campbell [40] model which had a slightly improved CV when compared to the Smettem and Bristow [41] model.

Table 3. Site-level soil saturated hydraulic conductivity (m/day) comparisons estimated using five K_{sat} models [11,27,39–41], the resulting farm-level descriptive statistics, and additional model information including data source location(s), type of model parameters(s), and the number of estimated model parameters. The coefficient of variation (CV) is presented as a unitless ratio of the standard deviation to the mean.

Site	Puckett et al.	Jabro	Campbell	Smettem and Bristow	Saxton et al.
RMF1	3.55	6.24	2.14	5.03	2.52
RMF2	3.86	20.72	2.74	6.22	2.52
RMF3	3.92	48.68	2.91	6.73	2.27
RMF4	3.97	54.17	2.94	7.61	1.07
RMF5	4.00	168.07	3.08	8.60	0.29
RMF6	3.39	1.04	1.56	4.70	1.77
RMF7	3.58	11.84	2.14	5.09	2.45
RMF8	3.53	1.84	2.15	4.98	2.60
Mean	3.72	39.07	2.46	6.12	1.94
Std. Dev.	0.24	56.00	0.53	1.43	0.84
CV	0.06	1.43	0.22	0.23	0.44
Location(s)	AL, USA	PA, USA	England/Wales	Australia	USA
Parameter(s)	clay	silt, clay, *bdry*	silt, clay	clay	sand, clay
Parameter #	1	3	2	1	2

When compared to K_{sat} estimates derived from the NRCS Web Soil Survey, all predicted K_{sat} values, except for the Saxton et al. [11], were higher than the NRCS estimated K_{sat} range of 0.19–1.99 m/day [63].

These NRCS underestimates may be attributable to spatial heterogeneity, where the soils were collected, and/or which method NRCS used to estimate K_{sat} [27,82,97]. The results from four models were characterized by CV scores of less than 0.5 (Table 3), apart from the Jabro [27] model, indicating relatively low spatial variability between farm-scale K_{sat} predictions [78].

3.5. Model Performance

Model performance is shown in Figure 5 with average predicted K_{sat} versus averaged measured K_{sat} at each study site. Four out of five models performed similarly well, as predicted average K_{sat} were within the same order of magnitude [11,39–41]. Predicted K_{sat} from the Puckett et al. [39] (Figure 5a) and the Smettem and Bristow [41] (Figure 5d) models scattered away from the 1:1 line with most or all values, respectively, falling between the positive y-axis and 1:1 line, representing an overestimation by the models. In contrast, half of the predicted K_{sat} from the Campbell [40] (Figure 5c) and Saxton et al. [11] models (Figure 5e) scattered along the 1:1 line, indicating smaller relative error and overall better model fits. The Jabro [27] model (Figure 5b) performed poorly, as evidenced by the range of K_{sat} predictions and distance from the 1:1 line, indicating it was not valid for the RMF soil data. Notably, the Jabro [27] model had the most model parameters (i.e., three) and was the only model to incorporate *bdry*, indicating that this pedotransfer function was not suited for RMF soils.

The quantified errors for each model are shown in Table 4. Farm-level ME for the four validated models ranged from 0.26 m/day with the Saxton et al. [11] model to 4.44 m/day with the Smettem and Bristow [41] model, while the Jabro [27] model ME was 37.11 m/day, which was an order of magnitude larger than the other models. Positive MEs indicate that the models generally overestimated RMF K_{sat}, but smaller ME values (<1 m/day) for the Saxton et al. [11] and Campbell [40] model confirms better fits to the observed data (Table 4). Error ranges were similar in magnitude (<10) for four of the five models, while the Jabro [27] model error range was two orders of magnitude larger. Similar model error ranges can be partially explained by the variability in the measured K_{sat}. Although most of the model error ranges were similar, the smaller error ranges for Puckett et al. [39] and Campbell [40] models confirms that they were better fits for the measured data.

Table 4. Performance of five site-level soil saturated hydraulic conductivity (K_{sat}) (m/day) models [11,27,39–41] for RMF soil by site and farm-level.

	Puckett et al.		Jabro		Campbell		Smettem and Bristow		Saxton et al.	
Site	**Error**	**Squared Error**	**Error**	**Squared Error**	**Error**	**Squared Error**	**Error**	**Squared Error**	**Error**	**Squared Error**
RMF1	1.25	1.56	3.93	15.45	−0.17	0.03	2.73	7.43	0.21	0.04
RMF2	3.43	11.79	20.29	411.80	2.31	5.36	5.80	33.59	2.09	4.39
RMF3	−0.55	0.30	46.40	2152.97	2.91	8.44	6.73	45.26	2.27	5.18
RMF4	1.59	2.52	51.78	2681.17	0.55	0.30	5.23	27.30	−1.31	1.73
RMF5	−0.76	0.58	163.31	26670.27	−1.69	2.85	3.83	14.69	−4.48	20.04
RMF6	1.75	3.07	−0.60	0.36	−0.07	0.01	3.06	9.38	0.14	0.02
RMF7	1.94	3.78	10.21	104.20	0.51	0.26	3.46	11.96	0.82	0.67
RMF8	3.24	10.47	1.55	2.40	1.86	3.45	4.68	21.93	2.30	5.29
ME		1.49		37.11		0.78		4.44		0.26
SSE		34.07		32038.62		20.68		171.55		37.36
RMSE		2.06		63.28		1.61		4.63		2.16

Note: ME = mean error (m/day); SSE = sum of squared error (m^2/day^2); RMSE = root of the mean-square error (m/day); unit of error = (m/day); unit of squared error = (m^2/day^2).

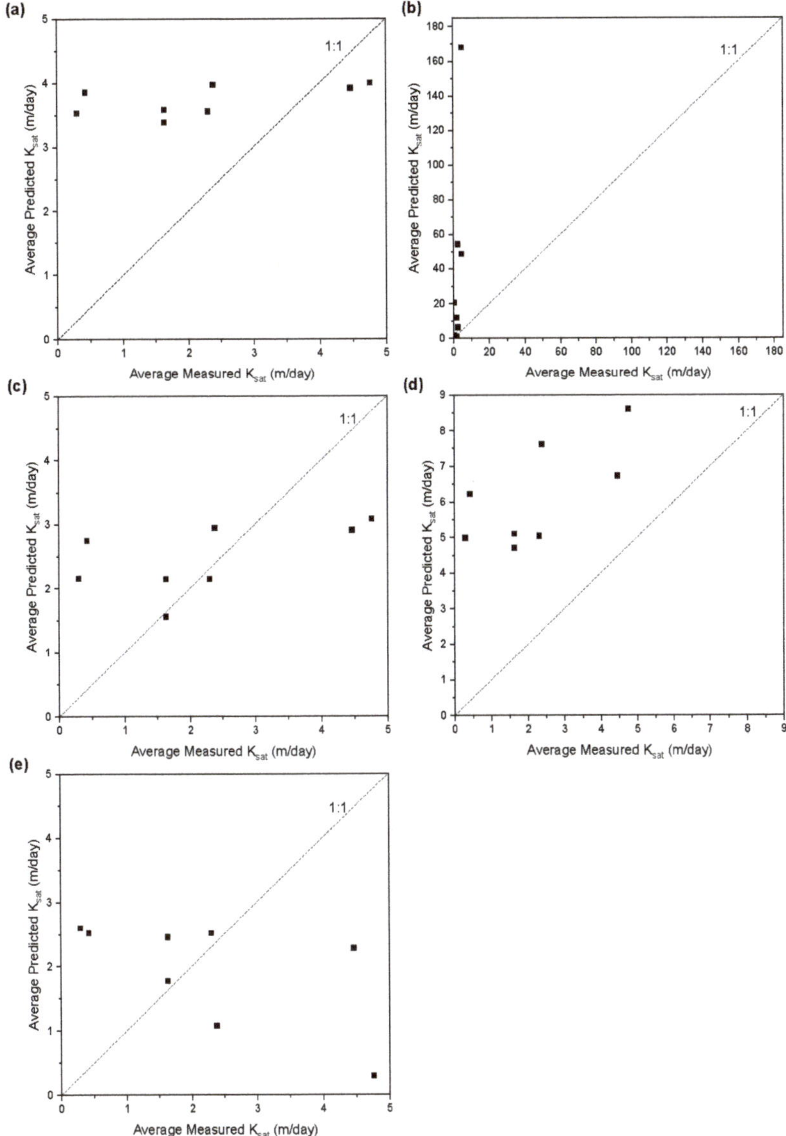

Figure 5. (**a**) Puckett et al. [39] model, (**b**) Jabro [27] model, (**c**) Campbell [40], (**d**) Smettem and Bristow [41], and (**e**) Saxton et al. [11] model predicted soil saturated hydraulic conductivity (K_{sat}) (m/day) versus measured values at Reymann Memorial Farm.

3.6. Model Comparison

Similar magnitude farm-level SSE and RMSE for the Puckett et al. [39], Campbell [40], and Saxton et al. [11] models indicates that these models were similar and better fits for the RMF data when compared to the Jabro [27] and Smettem and Bristow [41] models, the estimates from which were characterized by larger SSE and RMSE values (Table 4). SSE ranged from 20.68 m^2/m^2 for the Campbell [40] model to 32038.62 m^2/m^2 for the Jabro [27] model, while RMSE ranged from 1.61 m/day with the Campbell [40] model to 63.28 m/day with the Jabro [27] model (Table 4). The Campbell [40] model was the best fit for the measured data, as it had the smallest farm-level SSE and RMSE, while the Jabro [27] model did not result in a reasonable fit. Additionally, the Campbell [40] model seemed to perform better when the average site-level K_{sat} were between 1.5 and 2.5 m/day, whereas the other models, did not seem to have a range for better model predictions. The presence of an ideal K_{sat} range demonstrates that soil-property-based K_{sat} models could be characterized by an ideal soil texture range, outside of which they become unsuitable. For example, the Saxton et al. [11] model was a better fit for the sandy RMF soils, as it was developed with more sand-dominant textures, while the Jabro [27] model utilized silt-dominant textures and was not accurate for RMF soils. Additionally, the range of soil textures used to develop each model likely influenced the individual model outcomes.

3.7. Model Results Implications

One of the most important implications of the current work is that four out of the five evaluated pedotransfer functions provide a valid, alternative approach to direct K_{sat} measurements, as evidenced by low ME values (i.e., <4.5 m/day) (Table 4). The similarity in model performance between the Campbell [40] and Saxton et al. [11] models demonstrates their applicability in sand textured soils with similar order of magnitude K_{sat}. Model predictions were improved with the inclusion of a second particle size parameter, as RMSE for the Campbell [40] and Saxton et al. [11] models decreased, or was similar to the one particle size parameter models (i.e., Puckett et al. [39] and Smettem and Bristow [41]). Although slightly more complicated, two parameter models may provide increased accuracy in K_{sat} estimates, as in the current work, and therefore may justify additional complexity [98]. The Jabro [27] model, which utilized three model parameters, including *bdry*, was not a good fit for the measured data, likely due to the RMF soils' high sand content (average 92%). As such, this applied model comparison shows that particle size based K_{sat} models provide a good option for practitioners to predict K_{sat} because (1) gathering the required soil particle size data are relatively simple [36–38] and (2) four particle-size-based K_{sat} models adequately and accurately predicted farm-level K_{sat}, as evidenced by RMSE < 0.5.

The implications of the current work extend beyond the presentation of measured K_{sat} and particle size based K_{sat} model validation. The results increase confidence in K_{sat} models, especially when observed the data are not readily available. Further application of soil property K_{sat} models can improve K_{sat} predictions, and thus understanding of soil–water dynamics, in hydrologic studies throughout the northeast region, Chesapeake Bay Watershed, and areas with readily available particle size data. This model validation work provides practitioners and water resource managers with relatively simple alternatives to easily predict a parameter that is essential for governing shallow groundwater flow, which in turn, can increase K_{sat} estimate availability and transferability.

4. Conclusions

Soil properties and saturated hydraulic conductivity (K_{sat}) were measured in a small forested drainage area within the Northeast USA to validate five soil property K_{sat} models. Soil cores from three sampling depths were collected ($n = 179$) in a grid design to determine soil characteristics and a series of slug tests were performed to quantify K_{sat} at eight study sites. Mean dry bulk density (*bdry*) and porosity ranged from 1.03 to 1.30 g/cm^3 and 0.51 to 0.54, respectively. *bdry* and porosity varied significantly ($n = 179$; $p < 0.05$) at the 0–5 cm depth and with study site location, while *bdry* and porosity were statistically similar at 25–30 and 45–50 cm, indicating that the site soils were homogenous below the surface. On average, soil cores were 92% sand and soil textures were sand or sandy loam. Particle size fractions sites did not differ significantly in average particle size class percentages between soil depths ($n = 179$; $p < 0.05$) but did differ significantly by site ($n = 8$; $p < 0.05$).

Measured K_{sat} were 0.35–9.33 m/day and 0.21–4.37 m/day for FH and RH slug tests, respectively, and varied significantly ($p < 0.05$) with slug test type at three of the eight study sites. Average site K_{sat} ranged from 0.29 to 4.76 m/day and varied significantly with site ($n = 8$; $p < 0.05$). Five pedotransfer functions that predicted K_{sat} from soil property data were tested. Four models performed well and resulted in low, spatial variability between farm-scale estimated K_{sat} (CV < 0.5). The models that relied on particle size parameters performed better (RMSE < 4.64 m/day) than the models that relied on particle sizes and *bdry* (RMSE = 63.26 m/day). Improved K_{sat} estimates justify using a two-parameter particle size model [40] to predict K_{sat} at the farm-level for RMF soils.

Study results provide soil property characteristics and demonstrate that using affordable and readily available soil characteristic data can accurately and efficiently predict K_{sat}. This comparison study validates and supports the use of soil-property-based models to predict K_{sat} in sandy soils. These results are particularly relevant for understanding regional soil-water dynamics but are also informative for hydrologic studies in landscapes with similar soil properties. The broader impacts of this work extend to providing practitioners with an assessment of K_{sat} modeling that can used to effectively inform water resource management decisions and increase the applicability of K_{sat} estimates at locations with available particle size data.

Author Contributions: For the current work, author contributions were as follows: conceptualization, J.A.H., K.S.G., and E.K.; methodology, J.A.H.; formal analysis, K.S.G., E.K., and J.A.H.; investigation, K.S.G., E.K., and J.A.H.; resources, J.A.H.; data curation, K.S.G.; writing—original draft preparation, K.S.G. and J.A.H.; writing—review and editing, J.A.H., E.K., and K.S.G.; visualization, K.S.G. and J.A.H.; supervision, J.A.H.; project administration, J.A.H.; funding acquisition, J.A.H. All authors have read and agreed to the published version of the manuscript.

Funding: This work was supported by the National Science Foundation under Award Number OIA-1458952, the USDA National Institute of Food and Agriculture, Hatch project accession number 1011536, and the West Virginia Agricultural and Forestry Experiment Station. Additional funding was provided by the USDA Natural Resources Conservation Service, Soil and Water conservation, Environmental Quality Incentives Program No: 68-3D47-18-005. Results presented may not reflect the views of the sponsors and no official endorsement should be inferred. The funders had no role in study design, data collection and analysis, decision to publish, or preparation of the manuscript.

Acknowledgments: The authors would like to thank Zackary Heck, Susan Hickman, the Institute of Water Security and Science (https://iwss.wvu.edu/), members of the Interdisciplinary Hydrology Laboratory (https://www.researchgate.net/lab/The-Interdisciplinary-Hydrology-Laboratory-Jason-A-Hubbart), Reymann Memorial Farm, and West Virginia University.

Conflicts of Interest: The authors declare no conflict of interest. The funders had no role in the design of the study; in the collection, analyses, or interpretation of data; in the writing of the manuscript, or in the decision to publish the results.

References

1. Qi, S.; Wen, Z.; Lu, C.; Shu, L.; Shao, J.; Huang, Y.; Zhang, S.; Huang, Y. A new empirical model for estimating the hydraulic conductivity of low permeability media. *IAHS-AISH Proc. Rep.* **2015**, *368*, 478–483. [CrossRef]
2. Hwang, H.T.; Jeen, S.W.; Suleiman, A.A.; Lee, K.K. Comparison of saturated hydraulic conductivity estimated by three different methods. *Water (Switzerland)* **2017**, *9*, 942. [CrossRef]
3. Freeze, R.A.; Cherry, J.A. *Groundwater*; Prentice-Hall, Inc.: Englewood Cliffs, NJ, USA, 1979; ISBN 01-336-53129.
4. Chapuis, R.P. Predicting the saturated hydraulic conductivity of soils: A review. *Bull. Eng. Geol. Environ.* **2012**, *71*, 401–434. [CrossRef]
5. Darcy, H. *The Public Fountains of the City of Dijon*; Dalmont: Paris, France, 1856; ISBN 3663537137.
6. Klute, A. Laboratory Measurement of Hydraulic Conductivity of Unsaturated Soil. In *Methods of Soil Analysis: Part 1-Physical and Mineralogical Properties, Including Statistics of Measurement and Sampling*; Black, C.A., Evans, D.D., White, J.L., Ensminger, L.E., Clark, F.E., Dinauer, R.C., Eds.; American Society of Argronomy: Madison, WI, USA, 1965; pp. 210–221.
7. Boersma, L. Field Measurement of Hydraulic Conductivity Above a Water Table. In *Methods of Soil Analysis: Part 1-Physical and Mineralogical Properties, Including Statistics of Measurement and Sampling*; Black, C.A., Evans, D.D., White, J.L., Ensminger, L.E., Clark, F.E., Dinauer, R.C., Eds.; American Society of Argronomy: Madison, WI, USA, 1965; pp. 222–233.
8. Zhang, Y.; Schaap, M.G. Estimation of saturated hydraulic conductivity with pedotransfer functions: A review. *J. Hydrol.* **2019**, *575*, 1011–1030. [CrossRef]

9. Montzka, C.; Herbst, M.; Weihermüller, L.; Verhoef, A.; Vereecken, H. A global data set of soil hydraulic properties and sub-grid variability of soil water retention and hydraulic conductivity curves. *Earth Syst. Sci. Data Discuss.* **2017**, 1–25. [CrossRef]
10. Vereecken, H.; Schnepf, A.; Hopmans, J.W.; Javaux, M.; Or, D.; Roose, T.; Vanderborght, J.; Young, M.H.; Amelung, W.; Aitkenhead, M.; et al. Modeling Soil Processes: Review, Key Challenges, and New Perspectives. *Vadose Zone J.* **2016**, *15*, vzj2015.09.0131. [CrossRef]
11. Saxton, K.E.; Rawls, W.J.; Romberger, J.S.; Papendick, R. Estimating general soil-water characteristics from texture. *Soil Sci. Soc. Am. J.* **1986**, *5*, 1031–1036. [CrossRef]
12. Kennedy, C.D.; Murdoch, L.C.; Genereux, D.P.; Corbett, D.R.; Stone, K.; Pham, P.; Mitasova, H. Comparison of Darcian flux calculations and seepage meter measurements in a sandy streambed in North Carolina, United States. *Water Resour. Res.* **2010**, *46*, 1–11. [CrossRef]
13. Duan, R.; Fedler, C.B.; Borrelli, J. Comparison of methods to estimate saturated hydraulic conductivity in texas soils with grass. *J. Irrig. Drain. Eng.* **2012**, *138*, 322–327. [CrossRef]
14. Vienken, T.; Dietrich, P. Field evaluation of methods for determining hydraulic conductivity from grain size data. *J. Hydrol.* **2011**, *400*, 58–71. [CrossRef]
15. Fetter, C.W. *Applied Hydrogeology*, 4th ed.; Lynch, P., Ed.; Prentice-Hall, Inc.: Upper Saddle River, NJ, USA, 2001; ISBN 9780130882394.
16. Reynolds, W.D.; Bowman, B.T.; Brunke, R.R.; Drury, C.F.; Tan, C.S. Comparison of tension infiltrometer, pressure infiltrometer, and soil core estimates of saturated hydraulic conductivity. *Soil Sci. Soc. Am. J.* **2000**, *64*, 478–484. [CrossRef]
17. Hvorslev, M.J. Time Lag and Soil Permeability in Ground-Water Observations. *Waterw. Exp. Station Bull.* **1951**, *36*, 1–55.
18. Butler, J.J., Jr. *The Design, Performance, and Analysis of Slug Tests*; CRC Press LLC: Boca Raton, FL, USA, 1998; ISBN 1-56670-230-5.
19. Butler, J.J., Jr.; McElwee, C.D.; Liu, W. Improving the Quality of Parameter Estimates Obtained from Slug Tests. *Ground Water* **1996**, *34*, 480–490. [CrossRef]
20. Dorsey, J.D.; Ward, A.D.; Fausey, N.R.; Bair, E.S. Comparison of four field methods for measuring saturated hydraulic conductivity. *Trans. Am. Soc. Agric. Eng.* **1990**, *33*, 1925–1931. [CrossRef]
21. Black, J.H. The practical reasons why slug tests (including falling and rising head tests) often yield the wrong value of hydraulic conductivity. *Q. J. Eng. Geol. Hydrogeol.* **2010**, *43*, 345–358. [CrossRef]
22. Strayer, D.L.; Beighley, R.E.; Thompson, L.C.; Brooks, S.; Nilsson, C.; Pinay, G.; Naiman, R.J. Effects of land cover on stream ecosystems: Roles of empirical models and scaling issues. *Ecosystems* **2003**, *6*, 407–423. [CrossRef]
23. Kellner, E.; Hubbart, J.A. Continuous and event-based time series analysis of observed floodplain groundwater flow under contrasting land-use types. *Sci. Total Environ.* **2016**, *566–567*, 436–445. [CrossRef]
24. Anderson, M.P. *Characterization of Geological Heterogeneity*; Dagan, G., Neuman, S., Eds.; Cambridge University Press: Cambridge, UK, 1997; ISBN 9780511600081.
25. Chen, X.; Song, J.; Wang, W. Spatial variability of specific yield and vertical hydraulic conductivity in a highly permeable alluvial aquifer. *J. Hydrol.* **2010**, *388*, 379–388. [CrossRef]
26. Van Looy, K.; Bouma, J.; Herbst, M.; Koestel, J.; Minasny, B.; Mishra, U.; Montzka, C.; Nemes, A.; Pachepsky, Y.A.; Padarian, J.; et al. Pedotransfer Functions in Earth System Science: Challenges and Perspectives. *Rev. Geophys.* **2017**, *55*, 1199–1256. [CrossRef]
27. Jabro, J.D. Estimation of saturated hydraulic conductivity of soils from particle size distribution and bulk density data. *Trans. Am. Soc. Agric. Eng.* **1992**, *35*, 557–560. [CrossRef]
28. Suleiman, A.A.; Ritchie, J.T. Estimating saturated hydraulic conductivity from soil porosity. *Trans. Am. Soc. Agric. Eng.* **2001**, *44*, 235–239. [CrossRef]
29. Chapuis, R.P. Predicting the saturated hydraulic conductivity of sand and gravel using effective diameter and void ratio. *Can. Geotech. J.* **2004**, *41*, 787–795. [CrossRef]
30. Urumović, K. The referential grain size and effective porosity in the Kozeny-Carman model. *Hydrol. Earth Syst. Sci.* **2016**, *20*, 1669–1680. [CrossRef]
31. Salem, H.S. Application of the Kozeny-Carman equation to permeability determination for a glacial outwash aquifer, using grain-size analysis. *Energy Sources* **2001**, *23*, 461–473. [CrossRef]

32. Carrier, W.D., III. Goodbye, Hazen; Hello, Kozeny-Carman. *J. Geotech. Geoenviron. Eng.* **2003**, *129*, 1054–1056. [CrossRef]
33. Carman, P.C. Permeability of saturated sands, soils and clays. *J. Agric. Sci.* **1939**, *29*, 262–273. [CrossRef]
34. Hazen, A. Some physical properties of sands and gravels, with special reference to their use in filtration. *24th Annu. Rep.* **1892**, 539–556.
35. Kozeny, J. Ueber kapillare Leitung des Wassers im Boden. *Wien. Akad. Wiss.* **1927**, *136*, 271.
36. Saxton, K.E.; Rawls, W.J. Soil Water Characteristic Estimates by Texture and Organic Matter for Hydrologic Solutions. *Soil Sci. Soc. Am. J.* **2006**, *70*, 1569–1578. [CrossRef]
37. Hillel, D. *Introduction to Environmental Soil Physics*; Academic Press: San Diego, CA, USA, 2004; ISBN 0-12-348655-6.
38. Vereecken, H.; Maes, J.; Feyen, J. Estimating unsaturated hydraulic conductivity from easily measured soil properties. *Soil Sci.* **1990**, *149*, 1–12. [CrossRef]
39. Puckett, W.E.; Dane, J.H.; Hajek, B. Physical and Mineralogical Data to Determine Soil Hydraulic Properties†. *Soil Sci. Soc. Am. J.* **1985**, *49*, 831–836. [CrossRef]
40. Campbell, G.S. *Soil Physics with BASIC: Transport Models for Soil-Plant Systems*; Elsevier Scientific: New York, NY, USA, 1985.
41. Smettem, K.R.J.; Bristow, K.L. Obtaining soil hydraulic properties for water balance and leaching models from survey data. 2. Hydraulic conductivity. *Aust. J. Agric. Res.* **1999**, *50*, 1259–1262. [CrossRef]
42. Schaap, M.G.; Leij, F.J.; van Genuchten, M.T. ROSETTA: A computer program for estimating soil hydraulic parameters with hierarchical pedotransfer functions. *J. Hydrol.* **2002**, *251*, 163–176. [CrossRef]
43. Decharme, B.; Douville, H.; Boone, A.; Habets, F.; Noilhan, J. Impact of an exponential profile of saturated hydraulic conductivity within the ISBA LSM: Simulations over the Rhône basin. *J. Hydrometeorol.* **2006**, *7*, 61–80. [CrossRef]
44. Maest, A.; Kuipers, J.R.; Travers, C.L.; Atkins, D.A. *Predicting Water Quality at Hardrock Mines: Methods and Models, Uncertainties, and State-of-the Art*; Kuipers & Associates: Butte, MT, USA, 2005.
45. United States Army Corps of Engineers. Hydrologic Modeling System HEC-HMS Technical Reference Manual. 2000. Available online: https://www.hec.usace.army.mil/software/hec-hms/documentation/HEC-HMS_Technical%20Reference%20Manual_(CPD-74B).pdf (accessed on 2 April 2020).
46. Chapuis, R.P.; Dallaire, V.; Marcotte, D.; Chouteau, M.; Acevedo, N.; Gagnon, F. Evaluating the hydraulic conductivity at three different scales within an unconfined sand aquifer at Lachenaie, Quebec. *Can. Geotech. J.* **2005**, *42*, 1212–1220. [CrossRef]
47. Harbaugh Arlen, W. MODFLOW-2005, The U.S. Geological Survey Modular Ground-Water Model—The Ground-Water Flow Process. *USGS* **2005**. [CrossRef]
48. Roberts, T.; Lazarovitch, N.; Warrick, A.W.; Thompson, T.L. Modeling Salt Accumulation with Subsurface Drip Irrigation Using HYDRUS-2D. *Soil Sci. Soc. Am. J.* **2009**, *73*, 233–240. [CrossRef]
49. Ortoleva, P.; Merino, E.; Moore, C.; Chadam, J. Geochemical self-organization in reaction-transport feedbacks and modelling approach. *Am. J. Sci.* **1987**, *287*, 979–1007. [CrossRef]
50. Li, Y.; Zhang, Q.; Lu, J.; Yao, J.; Tan, Z. Assessing surface water–groundwater interactions in a complex river-floodplain wetland-isolated lake system. *River Res. Appl.* **2019**, *35*, 25–36. [CrossRef]
51. Lin, Y.-C.; Medina, M.A., Jr. Incorporating transient storage in conjunctive stream–aquifer modeling. *Adv. Water Resour.* **2003**, *26*, 1001–1019. [CrossRef]
52. Hubbart, J.A.; Holmes, J.; Bowman, G. TMDLs: Improving Stakeholder Acceptance with Science-based Allocations. *Watershed Sci. Bull.* **2010**, *1*, 19–24.
53. Hubbart, J.A.; Kellner, E.; Zeiger, S.J. A case-study application of the experimental watershed study design to advance adaptive management of contemporary watersheds. *Water* **2019**, *11*, 2355. [CrossRef]
54. Nichols, J.; Hubbart, J.A.; Poulton, B.C. Using macroinvertebrate assemblages and multiple stressors to infer urban stream system condition: A case study in the central US. *Urban. Ecosyst.* **2016**, *19*, 679–704. [CrossRef]
55. Zeiger, S.J.; Hubbart, J.A. Quantifying suspended sediment flux in a mixed-land-use urbanizing watershed using a nested-scale study design. *Sci. Total Environ.* **2016**, *542*, 315–323. [CrossRef] [PubMed]
56. Tetzlaff, D.; Carey, S.K.; Mcnamara, J.P.; Laudon, H.; Soulsby, C. The essential value of long-term experimental data for hydrology and water management. *Water Resour. Res.* **2017**, *53*, 2598–2604. [CrossRef]

57. Zell, C.; Kellner, E.; Hubbart, J.A. Forested and agricultural land use impacts on subsurface floodplain storage capacity using coupled vadose zone-saturated zone modeling. *Environ. Earth Sci.* **2015**, *74*, 7215–7228. [CrossRef]
58. Gellis, A.C.; Hupp, C.R.; Pavich, M.J.; Landwehr, J.M.; Banks, W.S.L.; Hubbard, B.E.; Landland, M.J.; Ritchie, J.C.; Reuter, J.M. Sources, Transport, and Storage of Sediment at Selected Sites in the Chesapeake Bay Watershed. *USGS* **2009**. [CrossRef]
59. Zhang, Q.; Blomquist, J.D. Watershed export of fine sediment, organic carbon, and chlorophyll-a to Chesapeake Bay: Spatial and temporal patterns in 1984–2016. *Sci. Total Environ.* **2018**, *619–620*, 1066–1078. [CrossRef]
60. Boesch, D.F.; Greer, J. *Chesapeake Futures: Choices for the 21st Century*; Scientific and Technical Advisory Committee: Edgewater, MD, USA, 2003; p. 160.
61. Natural Resource Analysis Center at West Virginia University. WV Land Use Land Cover (NAIP 2016). Available online: http://wvgis.wvu.edu/data/dataset.php?ID=489 (accessed on 15 February 2020).
62. WV GIS Technical Center. Digital Elevation Model 1- to 3-Meter Statewide Mosaic. Available online: http://wvgis.wvu.edu/data/dataset.php?ID=477 (accessed on 2 September 2019).
63. Soil Survey Staff, Natural Resources Conservation Service, United States Department of Agriculture. Web Soil Survey. Available online: https://websoilsurvey.sc.egov.usda.gov/ (accessed on 28 February 2020).
64. Knight, H.G. Reymann Memorial Farms. *W. Va. Agric. For. Exp. Station Bull.* **1925**, *194*, 1–20.
65. Estepp, R. *Soil Survey of Grant and Hardy Counties West Virginia*; United States Department of Agriculture: Washington, DC, USA, 1989.
66. Smith, R.M.; Pohlman, G.G.; Browning, D.R. Some Soil Properties which Influence the Use of Land in West Virginia. *W. Va. Agric. For. Exp. Station Bull.* **1945**, *321*, 1–71.
67. Natural Resources Conservation Service (NRCS) Web Site for Official Soil Series Descriptions and Series Classification. Available online: https://soilseries.sc.egov.usda.gov/ (accessed on 13 February 2020).
68. Dean, S.L.; Lessing, P.; Kulander, B.R. *Geology of the Rio Quadrangle, Hampshire and Hardy Counties*; OF-0004; West Virginia Geological and Economic Survey Open File Publication: Morgantown, WV, USA, 1999.
69. Horton, J.D.; San Juan, C.A.; Stoeser, D.B. The State Geologic Map Compilation (SGMC) geodatabase of the conterminous United States (ver. 1.1, August 2017). In *U.S. Geological Survey Data Series 1052*; U.S. Geological Survey: Reston, VA, USA, 2017; p. 46.
70. Kellner, E.; Hubbart, J. Agricultural and forested land use impacts on floodplain shallow groundwater temperature regime. *Hydrol. Process.* **2015**, *30*, 625–636. [CrossRef]
71. Zeiger, S.J.; Hubbart, J.A. Urban stormwater temperature surges: A central US watershed study. *Hydrology* **2015**, *2*, 193–209. [CrossRef]
72. Hubbart, J.A.; Link, T.E.; Gravelle, J.A.; Elliot, W.J. Timber harvest impacts on water yield in the continental/maritime hydroclimatic region of the United States. *For. Sci.* **2007**, *53*, 169–180. [CrossRef]
73. National Oceanic and Atmospheric Administration (NOAA) Climate Data Online Search. Available online: https://www.ncdc.noaa.gov/cdo-web/search (accessed on 20 April 2020).
74. Blake, G.R. Bulk Density. In *Methods of Soil Analysis: Part 1-Physical and Mineralogical Properties, Including Statistics of Measurement and Sampling*; Black, C.A., Evans, D.D., White, J.L., Ensminger, L.E., Clark, F.E., Dinauer, R.C., Eds.; American Society of Agronomy: Madison, WI, USA, 1965; pp. 374–390.
75. Hubbart, J.A.; Muzika, R.-M.; Dandan, H.; Robinson, A. Bottomland Hardwood Forest Influence on Soil Water Consumption in an Urban floodplain: Potential To Improve Flood Storage Capacity and Reduce Stormwater Runoff. *Watershed Sci. Bull.* **2011**, *2*, 34–43.
76. Vomocil, J.A. Porosity. In *Methods of Soil Analysis: Part 1-Physical and Mineralogical Properties, Including Statistics of Measurement and Sampling*; Black, C.A., Evans, D.D., White, J.L., Ensminger, L.E., Clark, F.E., Dinauer, R.C., Eds.; American Society of Agronomy: Madison, WI, USA, 1965; pp. 299–314.
77. Hoffmann, C.C.; Berg, P.; Dahl, M.; Larsen, S.E.; Andersen, H.E.; Andersen, B. Groundwater flow and transport of nutrients through a riparian meadow—Field data and modelling. *J. Hydrol.* **2006**, *331*, 315–335. [CrossRef]
78. Davis, J.C. *Statistics and Data Analysis in Geology*, 3rd ed.; J. Wiley: New York, NY, USA, 2002; ISBN 0471172758 9780471172758.
79. Gordon, N.D.; McMahon, T.A.; Finlayson, B.L.; Gippel, C.J.; Nathan, R.J. *Stream Hydrology: An Introduction for Ecologists*, 2nd ed.; J. Wiley: New York, NY, USA, 2004; ISBN 978-0-470-84358-1.

80. Skinner, J. A study of the methods used in measurement and analysis of sediment loads in reservoirs. In *Report NN*; Federal Interagency Sedimentation Project: Vicksburg, MS, USA, 2000.
81. Gee, G.W.; Or, D. Particle-Size Analysis. Methods of soil analysis. Part 4. In *Methods of Soil Analysis, Part 4, Phyiscal Methods*; Dane, J.H., Topp, G.C., Eds.; ACSESS: Madison, WI, USA, 2002; pp. 255–293, ISBN 978-0-89118-893-3.
82. Soil Survey Staff. *Soil Taxonomy: A Basic System of Soil Classification for Making and Interpreting Soil Surveys*, 2nd ed.; Natural Resources Conservation Service: Washington, DC, USA, 1999.
83. Nemes, A.; Rawls, W.J. Soil texture and particle-size distribution as input to estimate soil hydraulic properties. In *Developments in Soil Science*; Pachepsky, Y., Rawls, W.J., Eds.; Elsevier Science: Amsterdam, The Netherlands, 2004; Volume 30, pp. 47–70.
84. Song, J.; Chen, X.; Cheng, C.; Wang, D.; Lackey, S.; Xu, Z. Feasibility of grain-size analysis methods for determination of vertical hydraulic conductivity of streambeds. *J. Hydrol.* **2009**, *375*, 428–437. [CrossRef]
85. Day, P.R. Particle Fractionation and Particle-Size Analysis. In *Methods of Soil Analysis: Part 1 Physical and Mineralogical Properties, Including Statistics of Measurement and Sampling*; Black, C.A., Evans, D.D., White, J.L., Ensminger, L.E., Clark, F.E., Dinauer, R.C., Eds.; American Society of Agronomy: Madison, WI, USA, 1965; pp. 545–567.
86. 2540 Solids (2017). Standard Methods for the Examination of Water and Wastewater, 23rd. Available online: https://www.standardmethods.org/doi/abs/10.2105/SMWW.2882.030 (accessed on 2 March 2020).
87. Kettler, T.A.; Doran, J.W.; Gilbert, T.L. Simplified Method for Soil Particle-Size Determination to Accompany Soil-Quality Analyses. *Soil Biol. Biochem.* **2001**, *65*, 849–852. [CrossRef]
88. Natural Resources Conservation Service (NRCS) Soil Texture Calculator. Available online: http://soils.usda.gov/technical/aids/investigations/texture/%5Cnhttp://www.nrcs.usda.gov/wps/portal/nrcs/detail/soils/survey/?cid=nrcs142p2_054167 (accessed on 15 April 2020).
89. Solinst Canada, Ltd. Available online: https://www.solinst.com/products/dataloggers-and-telemetry/3001-levelogger-series/operating-instructions/user-guide/1-introduction/1-1-1-levelogger-edge.php (accessed on 6 February 2020).
90. Cheong, J.Y.; Hamm, S.Y.; Kim, H.S.; Ko, E.J.; Yang, K.; Lee, J.H. Estimating hydraulic conductivity using grain-size analyses, aquifer tests, and numerical modeling in a riverside alluvial system in South Korea. *Hydrogeol. J.* **2008**, *16*, 1129–1143. [CrossRef]
91. Barbour, M.G.; Burk, J.H.; Pitts, W.D.; Gilliam, F.S.; Schwartz, M.W. *Terrestrial Plant Ecology*, 3rd ed.; Addison Wesley Longman: Menlo Park, CA, USA, 1999; ISBN 080530004X.
92. Hall, D.G.M.; Reeve, M.J.; Thomasson, A.J.; Wright, V.F. *Water Retention, Porosity and Density of Field Soils*; Rothamsted Experimental Station: Harpenden, UK, 1977.
93. Rai, R.K.; Singh, V.P.; Upadhyay, A. Soil Analysis. In *Planning and Evaluation of Irrigation Projects: Methods and Implementation*; Maragioglio, N., Ed.; Academic Press: London, UK, 2017; p. 678, ISBN 9780128117484.
94. Nimmo, J.R. Porosity and Pore-Size Distribution. *Encycl. Soils Environ.* **2004**, *4*, 295–303. [CrossRef]
95. McElwee, C.D. Improving the analysis of slug tests. *J. Hydrol.* **2002**, *269*, 122–133. [CrossRef]
96. Stanford, K.L.; McElwee, C.D. Analyzing slug tests in wells screened across the watertable: A field assessment. *Nat. Resour. Res.* **2000**, *9*, 111–124. [CrossRef]
97. Soil Survey Staff. *Soil Survey Field and Laboratory Methods Manual. Soils Survey Investigations Report No. 51, Version 2.0*; Burt, R., Soil Survey Staff, Eds.; United States Department of Agriculture, Natural Resources Conservation Service: Washington, DC, USA, 2014; ISBN 978-0359573516.
98. Lees, M.J. Data-based mechanistic modelling and forecasting of hydrological systems. *J. Hydroinf.* **2000**, *2*, 15–34. [CrossRef]

Case Report

Quantifying *Escherichia coli* and Suspended Particulate Matter Concentrations in a Mixed-Land Use Appalachian Watershed

Fritz Petersen [1] and Jason A. Hubbart [1,2,*]

[1] Division of Plant and Soil Sciences, Davis College of Agriculture, Natural Resources and Design, West Virginia University, Agricultural Sciences Building, Morgantown, WV 26506, USA; fp0008@mix.wvu.edu

[2] Institute of Water Security and Science, Schools of Agriculture and Food, and Natural Resources, Davis College of Agriculture, Natural Resources and Design, West Virginia University, 3109 Agricultural Sciences Building, Morgantown, WV 26506, USA

* Correspondence: Jason.hubbart@mail.wvu.edu; Tel.: +1-304-294-2472

Received: 2 January 2020; Accepted: 12 February 2020; Published: 14 February 2020

Abstract: The relationships between *Escherichia (E) coli* concentration, suspended particulate matter (SPM) particle size class, and land use practices are important in reducing the bacterium's persistence and health risks. However, surprisingly few studies have been performed that quantify these relationships. Conceivably, such information would advance mitigation strategies for practices that address specific SPM size classes and, by proxy, *E. coli* concentration. To advance this needed area of research, stream water was sampled from varying dominant land use practices in West Run Watershed, a representative mixed-land use Appalachian watershed of West Virginia in the eastern USA. Water samples were filtered into three SPM intervals (<5 μm; 5 μm to 60 μm; and >60 μm) and the *E. coli* concentration (colony forming units, CFU) and SPM of each interval was quantified. Statistically significant relationships were identified between *E. coli* concentrations and size intervals ($\alpha < 0.0001$), and SPM ($\alpha = 0.05$). The results show a predominance (90% of total) of *E. coli* CFUs in the <5 μm SPM interval. The results show that land use practices impact the relationships between SPM and *E. coli* concentrations. Future work should include additional combined factors that influence bacterial CFUs and SPM, including hydrology, climate, geochemistry and nutrients.

Keywords: *Escherichia coli*; Suspended particulate matter; Water quality; Land use practices; Watershed management

---

1. Introduction

Globally, there is a need to investigate factors that influence risks facilitated by pathogenic microbes in water sources [1]. The need for research is stimulated by the common and widespread occurrence of fecal pollution and pathogenic water contamination in many parts of the globe [2–5]. The World Health Organization provided an estimate of 2.2 million deaths that occur annually due to waterborne diseases, making it the leading cause of deaths in the developing world [6]. Between 2013 and 2014, pathogen-contaminated drinking water resulted in 1006 cases, 42 outbreaks of disease, 124 hospitalizations, and 13 deaths in the United States of America alone [7]. Conceivably, advancing the understanding of variables that influence the persistence and risk of exposure to potentially harmful fecal microbes (e.g., *E. coli, enterococci* or fecal coliform), including the association of bacteria with suspended particulate matter (SPM), will better inform policy makers and contribute to decreased morbidity and mortality caused by pathogenic microbes.

SPM, defined as heterogeneous aggregates of organic matter, mineral fragments and microbiological fractions [8] in aquatic ecosystems, can facilitate the growth and persistence (survivability) of fecal and pathogenic bacteria [9,10]. For example, Jeng et al. [11] reported that suspended particles extended the survival of fecal indicator organisms for several days through the physical and chemical protection of microbes from biotic and abiotic stresses [12]. Other studies showed that SPM-associated bacteria occur in conjunction with, or adsorbed to, suspended organic or inorganic matter via physical, electrostatic or chemical binding [8,13], and benefit from the increased nutrient and organic matter availability (comparable to biofilms), and optimal light exposure [14,15]. Increased nutrients, organic matter and light that SPM-associated bacteria are exposed to can increase the growth rate of microbes by up to 50% relative to free-floating microbes. SPM-associated bacteria are typically larger, more abundant and feature increased cellular uptake of sugars and amino acids relative to free-floating microbes [16,17]. In addition, the proximity of SPM-associated microbes to each other facilitates horizontal transfer and potential proliferation of resistance genes [18,19]. Advancing the understanding of variables that influence bacterial association with SPM is important to better manage the growth and persistence of microbes in the environment.

Previous work showed that microbes attach preferentially to small particles (e.g., fine clay particles; <2 μm) due to the increased surface area to volume and electrostatic charge [1,12]. Conversely, fecal bacteria (e.g., *E. coli*) associated with large suspended particles will settle to the stream bed more readily than smaller particles [1], potentially decreasing microbial water quality impairment once re-suspension occurs. The potential risk of exposure to *E. coli* through its association with SPM is important given that *E. coli* can be pathogenic [20], while simultaneously indicating the presence of other harmful fecal microbes (e.g., *E. coli* serves as an indicator organism) [2]. This is important because association can increase the risk posed by *E. coli* or by the other fecal microbes that *E. coli* is an indicator for. Thus, an improved understanding of the relationships between *E. coli* concentration and SPM could aid in determining the influence of SPM size on other potentially harmful fecal bacteria, commonly occurring with *E. coli*, such as *Enterococcus* [21]. Pathogenic strains of *E. coli* can cause gastroenteritis, diarrhea, urinary tract infections hemolytic uremic syndrome, and meningitis [22]. Therefore, the increased risk of exposure, through association with SPM, can potentially increase instances of disease outbreaks and subsequent morbidity and mortality [2,3]. In short, advanced understanding of the correlation between *E. coli* concentrations and SPM size class will reduce human health risks by advancing the understanding of exposure and informing effective remediation.

Many knowledge gaps exist regarding *E. coli* (fecal matter) concentration relative to SPM size distribution. For example, previous investigations have been limited by laboratory simulations [1], few sampling locations [11], and/or low to no land use variability [23,24]. The latter is important given land use and land use change has been shown to alter particle size characteristics of SPM in aquatic systems [25,26]. For example, in a mixed-land use watershed (Hinkson Creek, Columbia, Missouri, USA), Hubbart [25] and Kellner and Hubbart [27] reported that suspended sediment displayed a decreasing trend in particle size from agricultural headwaters to urban areas and a subsequent increase in particle size to suburban areas in the lower watershed. Kellner et al. [28] reported a disproportionate contribution of fine sediment from urban areas, relative to receiving waters comprising different land uses (e.g., rural and agriculture) of Flat Branch Creek (a tributary to Hinkson) and Hinkson Creek. The increased fine sediment from urban areas was attributed to preexisting compaction in these areas, increased impervious surface cover relative to other land use types, and in-stream weathering of sediment in conjunction with preferential deposition [25,28]. Ultimately, the results of previous studies have stimulated questions regarding the influence of land use practices, particle size class distribution and *E. coli* (fecal matter) concentration [29].

The Appalachian region of the United States of America (USA) is well-suited for investigations that will resolve existing knowledge gaps regarding the relationship between *E. coli* concentration and SPM size. This region of the USA comprises diverse physiography and widespread, frequent, and problematic fecal pollution [30]. On the basis of physiography alone, Appalachia is comparable to other regions globally. For example, the Central Appalachian region, encompassing a temperate climate, distinct winter and summer periods, and a precipitation regime that is nearly evenly distributed throughout the year is comparable to similarly temperate locations comprising year round precipitation (e.g., Northern Honshu in Japan [31,32]). In rural areas of Appalachia, thousands of residents are exposed to water security issues, in particular microbial contamination [33]. The vulnerabilities of rural Appalachia to decreased water quality due to microbial contamination is exacerbated by some of the greatest levels of poverty, isolation, rough geographical terrain, and inadequate septic treatment systems in the USA [33]. Therefore, water quality is a primary concern and studies investigating fecal pollutants (e.g., *E. coli*) and factors increasing the risk posed by fecal pollution (e.g., by increasing risk of exposure) in this region are greatly needed.

The overarching objective of the current investigation was to quantitatively characterize *E. coli* concentration relative to SPM size distribution from multiple sites in a mixed-land use watershed of Appalachia. A sub-objective was to evaluate the influence of varying land use practices on the relationship between *E. coli* concentration and SPM particle size distribution. This work was also intended to serve as a valuable springboard for future investigations on the alteration of exposure and subsequent health risks of *E. coli* facilitated by SPM of various sizes.

2. Methods

2.1. Study Site Description

This investigation took place in West Run Watershed (WRW), a 3rd order tributary of the Monongahela River, located in Morgantown, West Virginia, USA. The WRW is 23 km^2 in area and is a mixed-land use urbanizing watershed comprising many land use practices, including agriculture, urban and forested areas. At the time of this investigation, forested land use accounted for 50.1%, and agricultural and developed (urban and commercial areas) land use practices accounted for 22.6% and 19%, respectively, of the land use of WRW. The primary stream of WRW, West Run Creek, is typically narrow with small floodplains and is considered to be a moderately entrenched stream [34,35]. Elevation ranges from 420 m above mean sea level, at the headwaters, to 240 m above mean sea level at the confluence of the Monongahela River [35]. The physiography of the watershed comprises relatively rugged terrain, featuring numerous rock outcroppings dating to the Paleozoic era [35]. The oldest recorded geological formation in the watershed is the Upper Kittanning coal, while the most recent formation is the Monongahela series (located in the headwaters) [35]. The water quality in WRW, specifically in the headwaters, has been negatively impacted by historic mining of the Pittsburg coal seam [36].

The climate of West Virginia ranges from temperate and humid with hot summers to cold and humid with warm summers [37]. The climate of Morgantown, WV, located in Monongalia County (and including the WRW), is characterized by warm to hot summers (mean monthly temperature >22 °C), cold winters (mean monthly temperature <0 °C), and no dry season [37]. The average annual precipitation in Morgantown between 1981 and 2010 was approximately 1060 mm. During this time period, July (typically warmest and wettest month) had an average daily temperature of approximately 23 °C and average monthly precipitation of 117 mm. Conversely, the coldest (January) and driest (February) months have an average daily temperature of −0.4 °C and average monthly precipitation of 66 mm, respectively [38].

For the current investigation, a study design including four monitoring sites (gauged sampling locations) was implemented. The sites included varying land use practices (Table 1) and (numbered in downstream order) consisted of 1st and 2nd order confluence tributaries of the West Run Creek (Figure 1). A combination of geographic information system (GIS) data and field surveys were implemented to identify study sites and associated sub-catchments. At the time of this investigation, Site #1 comprised developed and forested lands in the upper sub-catchment, and actively grazed pasture in the lower sub-catchment. The primarily urban site (Site #2) drained a commercial area located on the southern side of the Watershed. Site #3 drained a local farm which includes dairy cattle grazing pastures, holding pens, and livestock manure stacks. Site #4 comprised predominantly (82.4%) forested land use and served as a reference sub-catchment (control) for the current work.

Table 1. Land use/land cover characteristics (% cover) and total drainage area (km^2) at four monitoring sites in West Run Watershed (WRW), West Virginia, USA. Note: land use percentages may not sum to 100%, as not every category is included (i.e., wetland, open water, etc.) and some categories are combinations of others (e.g., developed = urban + residential), or independent (e.g., impervious). Final row indicates total values for the entire watershed.

Site Number and General Description	Forest (%)	Agriculture (%)	Developed (%)	Drainage Area (km^2)
#1 (54.0% Forested)	54.0	31.8	7.6	3.3
#2 (40.6% Developed)	27.6	20.8	40.6	1.0
#3 (49.1% Forested)	49.1	42.1	6.6	0.2
#4 (82.4% Forested)	82.4	13.3	0.8	0.7
WRW (Total)	50.1	22.6	19.0	23.3

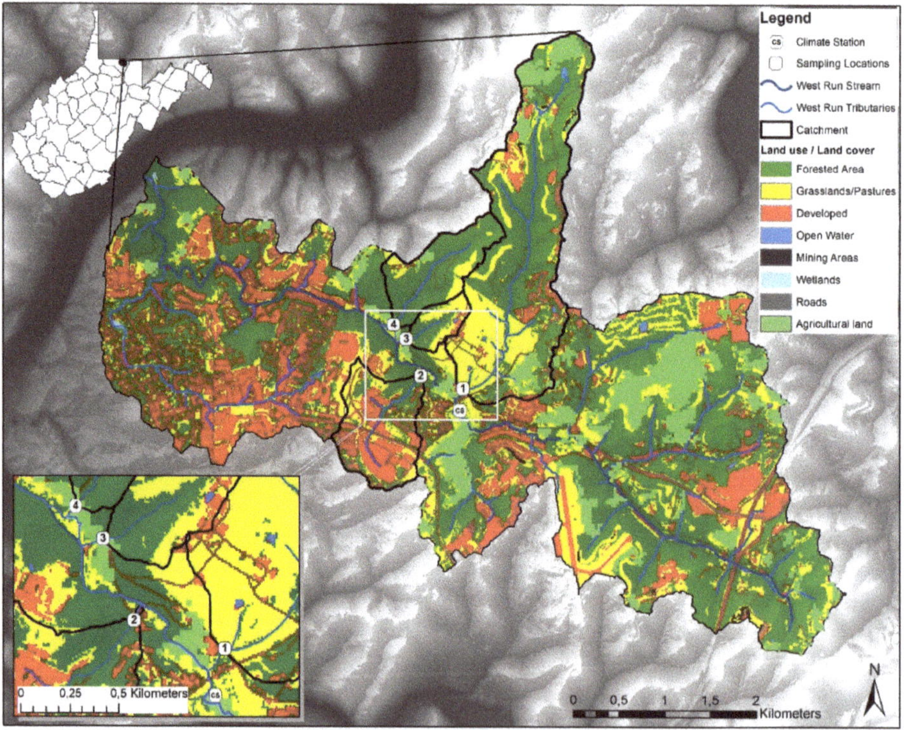

Figure 1. Monitoring/sampling locations for the current investigation, with land use/land cover, in West Run Watershed, Morgantown, West Virginia, USA.

2.2. Data Collection

During the study period (20 July 2018–27 October 2018), climate data were recorded using research-grade climate instrumentation located within approximately 100 m of Site #1 (Figure 1). The climate variables (recorded at a height of 3 m) included precipitation (TE525 Tipping Bucket Rain Gauge), average air temperature and relative humidity (Campbell Scientific HC2S3 Temperature and Relative Humidity Probe), and average wind speed (Met One 034B Wind Set instrument).

Stream water grab-samples were collected following the USGS methods described in the National Field Manual for the Collection of Water-Quality Data [39] and as per Petersen et al. [35], Hubbart et al. [40], Kellner and Hubbart [41], and Zeiger and Hubbart [42,43] from each monitoring site (stream order ≤3). The sampling regime was determined by stream stage (based on the streamflow descriptions by Zeiger and Hubbart [42]) at the sites to ensure sample extraction occurred during low (approximately 25% bankfull), medium (approximately 40–60% bankfull) and high(er) (approximately >60% bankfull) stages. This sampling regime facilitated a distributed SPM concentration data set, providing for robust investigation of the general relationship between SPM and *E. coli* concentration. The stream stage-based sampling regime resulted in irregular sample collection. For example, multiple samples were collected on some days during runoff events (leading and receding limbs of hydrograph, etc.) and other consequent samples extracted multiple days apart. During the course of the investigation, 32 samples were collected from each of the sampling locations. Following collection, the samples were transported to the Interdisciplinary Hydrology Laboratory (https://www.researchgate.net/lab/The-Interdisciplinary-Hydrology-Laboratory-Jason-A-Hubbart), located in the Davis College of Agriculture, Natural Resources and Design at West Virginia University, for analyses. Previous investigations have implemented one filter size (size varies between studies) to separate particle-attached from free-living microbes [13,44–46], although, as noted above, no widely established method exists. During the current investigation, the extracted water grab-samples were subdivided into three water sub-samples. One subsample was processed as per the standard Colilert test (see below) procedures and incubated without filtration (i.e., normal sample processing), while the other two subsamples were filtered using different filter matrices (60 μm and 5 μm), processed (as below) and then incubated. Hydrophilic, nylon net, Merck Millipore filters were used for filtration. The filtration of the samples resulted in water samples containing SPM and *E. coli* of sizes <60 μm and <5 μm, respectively. Therefore, after incubating the samples (process described below), the *E. coli* concentrations of the total sample and the <60 μm and <5 μm sizes were known. Subsequently, subtracting the *E. coli* concentration in the <60 μm filtered sample from the *E. coli* concentration of the total sample thus yielded the *E. coli* concentration in the respective SPM size class interval. Additionally, the *E. coli* concentrations in the <5 μm size were subtracted from the <60 μm size, resulting in *E. coli* concentration data for the intermediate (5 μm < interval < 60 μm) interval. The selection of the filter aperture was determined by approximate soil particle size classifications (i.e., larger than 60 μm = sand; smaller than 60 μm and larger than 4 μm = silt, and smaller than 4 μm = fine silt and clay) [47].

The U.S. Environmental Protection Agency (EPA)-approved Colilert test [29], developed by IDEXX Laboratories Inc., was used to quantify *E. coli* colony forming units (CFU) in the filtered and unfiltered samples. The test, included in Standard Methods for Examination of Water and Wastewater, was developed to estimate fecal concentration in water samples without requiring sample dilution [48,49]. The chances of obtaining inaccurate results with the test is low (chance of reporting false positives ±10%) due to a combination of Colilert's Defined Substrate Technology (DST) nutrient-indicator (ONPG), and a selectively suppressing formulated matrix. ONPG was formulated to render the majority of non-target organisms unable to grow or interfere as they lack the enzyme to metabolize the provided carbon source [48]. The few non-target organisms that can metabolize ONPG were suppressed by the selectively suppressing formulated matrix [48].

The Colilert (ONPG) substrate was added to 100 mL of sampled water and sealed in the Quanti-Tray, prior to the 24-h incubation at 35 °C [29]. Following incubation, the number of fluorescing (positive for *E. coli*) wells was converted (with a 95% confidence interval) into a concentration of *E. coli* CFU per the 100 mL of filtered and unfiltered sample water using a Most Probable Number (MPN) table. The MPN table facilitated an estimation of the concentration value, as it yielded a number of CFU per 100 mL. The Quanti-Tray/MPN table method allowed for the estimation of *E. coli* concentration in the range <1 to 1011.2 CFU. The upper limit of the estimation range from the Quanti-Tray/MPN table method presented a challenge for the current investigation, as *E. coli* concentrations in excess of 1011 CFU per 100 mL could not be accurately estimated, as the test becomes effectively saturated. Due to the fiscal cost associated with the dilution and subsequent incubation of extracted water samples, serial dilutions were not feasible for the current work, as it would have significantly increased the number of samples to be analyzed. Regardless, the method as applied allowed for the detection of small changes in *E. coli* concentration, which are important for small SPM sizes (a focus of the current work), which the literature has shown to be more closely associated with pathogen persistence.

2.3. Data Analysis

Descriptive statistics were generated for *E. coli* and SPM concentrations aggregated to the study period level. The total *E. coli* concentrations (i.e., colony forming unit; CFU) from each filtration interval (<5 μm; 5 μm to 60 μm; and >60) were divided by the total unfiltered *E. coli* concentration. This facilitated the estimation of the average percentage *E. coli* concentration resultant from each respective SPM interval. This process was completed for all four sites for the duration of the study period (20 July 2018–27 October 2018). The average SPM and average *E. coli* concentration from each interval (n = 32) (and totals) were compared to each other to determine the average percentage difference between the sampling locations. Statistical analyses were conducted using Origin Pro 2018 (OriginLab Corporation, Northampton, MA, USA). Normality testing was completed using the Anderson Darling Test, which tests whether a sample of data is drawn from a given probability distribution (normal distribution for the current work) [50]. Ordinal logistic regression was used to explore the effect of SPM size concentrations on *E. coli* concentrations in the same size intervals, while also accounting for the effect of varying land use practices at the four sampling locations. This regression is used to predict an ordinal dependent variable given one or more independent variables and requires that the continuous *E. coli* concentration and SPM concentration data be converted to ordinal data. Therefore, the *E. coli* concentration and SPM concentration data were divided into tertiles in which data in the lowest tertile were assigned a value of 1, data in the second tertile were assigned a value of 2, and data in the upper tertiles were assigned a value of 3. Following the conversion to ordinal data, the regression was calculated using JMP software (JMP®, Version Pro 12.2, SAS Institute Inc., Cary, NC, USA, Copyright ©2015), following the method presented by Stokes et al. [51]. The significance threshold for all statistical tests was set at $\alpha = 0.05$. Principal component analysis (PCA) was used to investigate the relationships between *E. coli* concentrations, SPM and land use practices for both the smallest interval (<5 μm) and the total values (presented in biplots) across all four sampling locations.

3. Results

3.1. Climate during Study

During the study period (20 July 2018–27 October 2018), the total precipitation was 449 mm, the average air temperature was 19 °C, the relative humidity was 82.7%, and the average wind speed was 0.89 m/s. Historic records since 2007 [52] indicated that the study period comprised normal temperatures (average 19 °C), but received 99 mm more precipitation than average (350 mm). A monthly analysis of climate data showed that the month of July included normal average monthly temperatures (difference between recorded values and average was less than 1 °C) for the time period (22 °C). However, there was approximately 20 mm less precipitation than average [52]. Conversely, August had average temperatures that were 2 °C cooler than average and the month was approximately 39 mm wetter than average [52]. Precipitation during September (186mm) was more than double the long-term average (80mm), but the average temperature was consistent with the long-term average (18 °C). October also exhibited long-term average temperatures (12 °C); however, it was drier than average, receiving approximately 25mm less precipitation.

The climate during the period of study (20 July 2018–27 October 2018) was predictably variable and consistent with historic trends (Figure 2). The weather was characteristically humid and warm at the start of the investigation, with temperatures decreasing with transition to winter. As previously noted, stream stage was used to determine the timing of sampling events given that streamflow (depth and velocity) is directly related to SPM transport processes and has been reported to influence *E. coli* concentrations with elevated concentrations typically occurring during high flows [53]. This approach facilitated the opportunity to investigate the relationship between *E. coli* and SPM size under a variety of hydrologic conditions. Thus, climatic conditions that could directly influence stream stage, such as precipitation, were particularly important for this work. During the investigation large precipitation events (e.g., 7/25 and 9/9) resulted in maximum stream stages, while drier periods (e.g., 10/12) resulted in low stream stages, as is common in late summer in the region (Figure 2) [35].

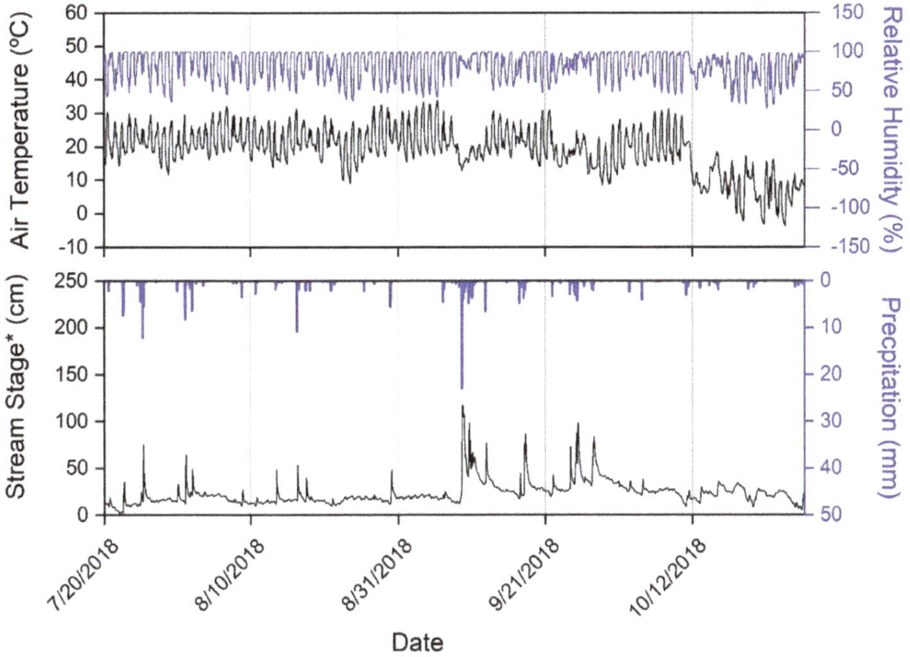

Figure 2. Thirty-minute time series of climate variables (recorded at the climate station) during study period (20 July 2018–27 October 2018) in West Run Watershed, West Virginia, USA. * Note: Stream stage was monitored in the primary stream of WRW, West Run Creek, within approximately 100m of the confluence of Site #1 and West Run Creek (39°40′3.20′′ N 79°55′48.99′′ W).

3.2. Suspended Particulate Matter Concentrations and *Escherichia coli*

The average total SPM concentration (38.52 mg L^{-1}) was the highest at Site #3, which included 49.1% forested area, the highest percentage of agricultural land-use (42.1%) and the highest maximum SPM (402 mg L^{-1}). Site #3 also had the largest standard deviation of SPM (83.91 mg L^{-1}), almost double the standard deviation of the second highest standard deviation (Site #4: 43.28 mg L^{-1}). Site #1 had the highest median (10.83 mg L^{-1}) and minimum (0.67 mg L^{-1}) SPM values. Conversely, Site #2 had the lowest average SPM (12.25 mg L^{-1}), maximum (70 mg L^{-1}), median SPM (2.67 mg L^{-1}), and minimum (0 mg L^{-1}) among the sampled sites.

Site #2 (land use: 40.6% developed, Table 1) had the lowest average *E. coli* concentration (596 CFU per 100 mL) and the lowest median (629 CFU per 100 mL) among the sites. The lowest minimum *E. coli* concentration (38 CFU per 100 mL) was recorded at Site #4 (82.4% forested). Conversely, the highest mean value for *E. coli* concentration (708 CFU per 100 mL), the highest median value for *E. coli* concentration (961 CFU per 100 mL), and the lowest standard deviation (351 CFU per 100 mL) were recorded at Site #3 (42.1% Agricultural, and 49.1% Forested). Site #1, which had the second highest percentage of agricultural land use (31.8%) among the sub-catchments, had the second highest mean value for *E. coli* concentration (676 CFU per 100 mL), the second highest median value for *E. coli* concentration (813 CFU per 100 mL), and second lowest standard deviation (355 CFU per 100 mL).

Box and whisker plots reflecting descriptive statistics of SPM (mg L^{-1}) and *E. coli* concentration (CFU per 100 mL) are provided in Figure 3. In the >60 μm interval, Site #2 (40.6 % developed) had both the smallest SPM concentration range (0–2.5 mg L^{-1}) and the largest *E. coli* concentration range (0–262 CFU per 100 mL). However, in the 5 to 60 μm particle size class, Site #4 (82.4% forested) had the largest SPM concentration range (0–171.5 mg L^{-1}), whereas Site #3 (49.1% forested and 42.1% agriculture) had the largest *E. coli* concentration range (0–276 CFU per 100 mL). In the <5 μm interval, Site #3 had the highest SPM (0–181 mg L^{-1}) and *E. coli* concentration values (0–1011 CFU per 100 mL), while Site #1 (54% forested) had the lowest SPM concentration range (0–30.5 mg $L^{-1)}$ and Site #4 (82.4% forested) generally had the lowest *E. coli* concentration values. The raw (unfiltered) samples had a similar distribution to the <5 μm interval, with Site #3 again featuring the highest SPM (0–402 mg L^{-1}) and *E. coli* concentration values (0–1011 CFU per 100 mL). Similarly, Site #1 (54% forested) comprised the lowest SPM concentration range (0–216 mg L^{-1}); however, Site #2 generally had the lowest *E. coli* concentration values (Figure 3).

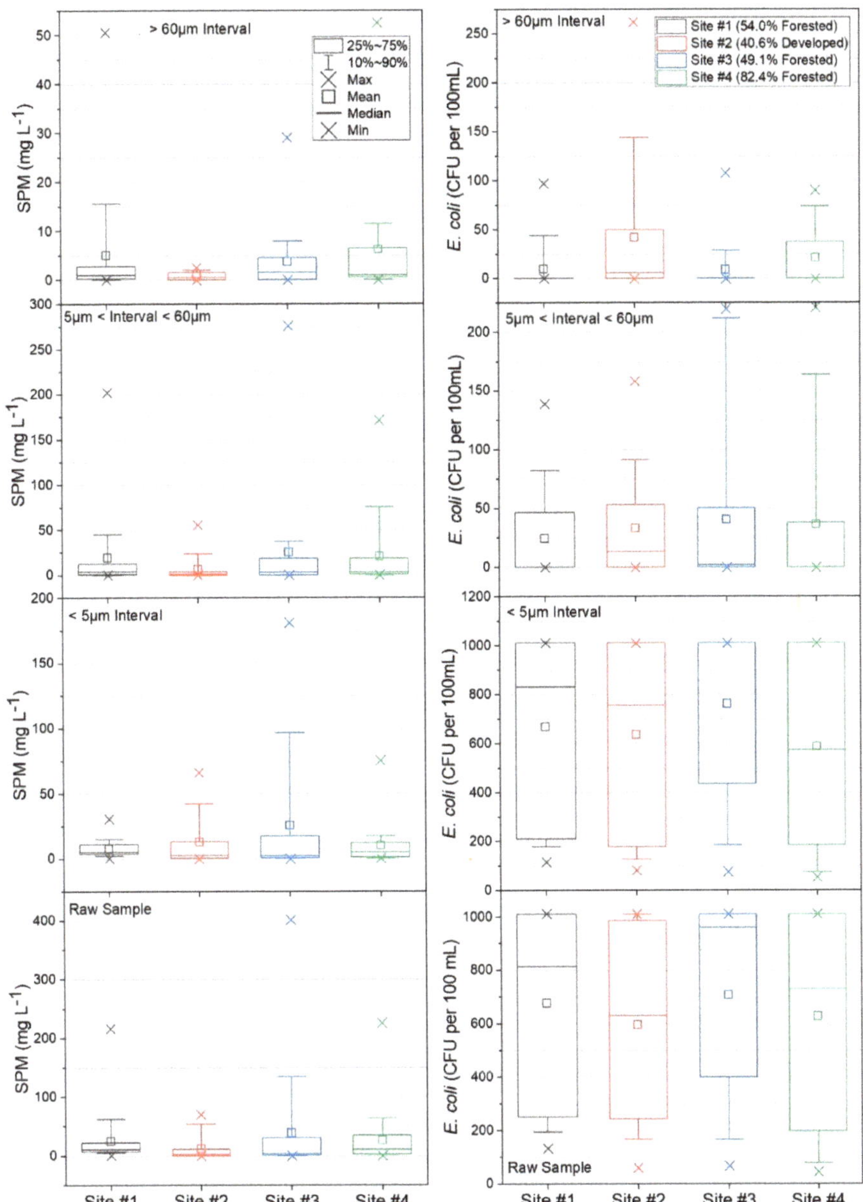

Figure 3. Box and whisker plot of suspended particulate matter (SPM) (mg L^{-1}) and *E. coli* concentration (CFU per 100 mL) at each sampling location (n = 4) during the study period (20 July 2018–27 October 2018) in West Run Watershed, Morgantown, West Virginia, USA. Box delineates 25th and 75th percentiles; line denotes median; square shows mean; whisker describes 10th and 90th percentiles; x shows maximum and minimum when above and below, respectively.

3.3. Non-Parametric Statistical Analysis

Normality test results indicated that *E. coli* concentration data were non-normally distributed, thus a non-parametric statistical method was used for further analysis. Due to the detection range (including an upper limit) of the Colilert method, ordinal logistic regression was deemed the most suitable form of regression to analyze the data [51]. *E. coli* concentration data for all sites were used as the dependent variable with the corresponding size interval, SPM and land use percentages (forest %, agriculture %, and developed %) data from all sites used as independent (explanatory) variables. Across the sites, the results indicate that size interval ($\alpha < 0.0001$), and SPM ($\alpha = 0.05$) showed statistically significant relationships with *E. coli* concentration. Percentage of land use practices across all sites did not display any statistically significant correlations with *E. coli* concentrations, with the percentage of forested land, percentage of agricultural land and percentage of developed land having significance levels (α) of 0.8425, 0.8478, and 0.9031, respectively.

The results from the initial PCA identify two principal components with eigenvalues exceeding 1, an accepted threshold of importance [54] for the smallest interval (<5 µm) (Table 2). *E. coli* concentrations (Eigenvalue = 1.89) and SPM concentrations (Eigenvalue = 1.35) explained approximately 65% of the cumulative variance of the smallest interval data set. The percentage of agricultural land comprised an Eigenvalue of 0.91 and its inclusion accounted for 83% of the cumulative variance of the smallest interval data set. For the total data set (defined as all he SPM and *E. coli* data >0.7 µm) PCA, three principal components were identified with eigenvalues exceeding 1; *E. coli* concentration (1.90), SPM concentration (1.42) and the percentage of agricultural land (1.01) and accounted for 87% of the cumulative variance of the total data set (Table 2). For both PCAs, the percentage of forested land and percentage of developed land accounted for relatively small percentages of variance in the data. Specifically, in the smallest interval, 17% and 0.1% of the variance of the data were accounted for by the percentage of forested land and the percentage of developed land, respectively. Similarly, for the total data set, forested land use accounted for 13% of the data variance and the percentage of developed land accounted for 0.04%.

Table 2. Results of principal component analysis comprising 5 components (*E. coli* concentration, SPM concentration, percentage of agricultural land use, percentage of forested land use and percentage of developed land use) displaying eigenvalues, percentage of variance and cumulative variance during the study period (20 July 2018–27 October 2018) for the smallest interval (<5 µm) and the total data set across the four monitoring sites in West Run Watershed, West Virginia, USA. Note: bold numbers indicate eigenvalues exceeding 1 (representing importance).

Variable	Eigenvalue	Percentage of Variance	Cumulative Variance
	Smallest Interval (<5 µm)		
E. coli concentration	**1.89**	38%	38%
SPM concentration	**1.35**	27%	65%
% Agriculture	0.91	18%	83%
% Forested	0.84	17%	100%
% Developed	0	0	100%
	Total Data Set		
E. coli concentration	**1.90**	38%	38%
SPM concentration	**1.41**	28%	66%
% Agriculture	**1.01**	20%	86%
% Forested	0.67	14%	100%
% Developed	0	0%	100%

4. Discussion

4.1. E. coli and SPM Concentrations

At the time of this study, Site #3 (Table 1) had the highest percentage of agricultural land use among the study sites (42.1%) and comprised the greatest cumulative SPM (1232 mg L^{-1}) (Figure 4).

The cumulative SPM recorded at Site #3 was more than three times the cumulative SPM recorded at Site #2 (40.6% developed), which had the lowest cumulative SPM (391 mg L^{-1}). Previous studies investigating land use practices and SPM also recorded elevated levels of SPM in agricultural land use areas, thereby supporting the results from Site #3 [55]. The flattened cumulative curves (Figure 4) from 28 September 2018 to 19 October 2018 reflect a lack of sample collection during a period of negligible rainfall (and thus runoff; see Figure 2). Cumulative *E. coli* was the highest at Site # 1 (21632 CFU per 100 mL) and Site #3 (22641 CFU per 100 mL) during the sampling period (20 July 2018–27 October 2018) (Figure 4). These two sites drained the highest (42.1%; Site 3) and second highest (31.8%; Site #1) area of agricultural land use practices among the sampled sites (Table 1). The results are supported by previous work that showed increased fecal matter content with agricultural land use practices [53,56]. Conversely, Site #2 and Site #4 had the lowest cumulative *E. coli* CFU (19082 CFU per 100 mL and 20080 CFU per 100 mL, respectively) during the study period. Site #4's low cumulative *E. coli* CFU was anticipated, as this site consisted primarily of forested land use (82.4%) (Table 1) and previous work showed negative correlations between fecal concentration and forest land cover [56]. Furthermore, Site #4 lacked artificial sources of *E. coli* (e.g., livestock manure stacks present at Site #3), which could have increased the *E. coli* concentration in its receiving waters. The low cumulative *E. coli* CFU recorded at Site #2 is contrary to results from previous work linking urban land use practices to increased *E. coli* concentrations in receiving waters [57–59]. Given the differing study design (sampling regime) relative to other studies, these differences are not confounding. It is, however, worth noting that in cases where the current study agreed with previous studies, those agreements are important, given the different sampling regime and yet agreement(s) in results.

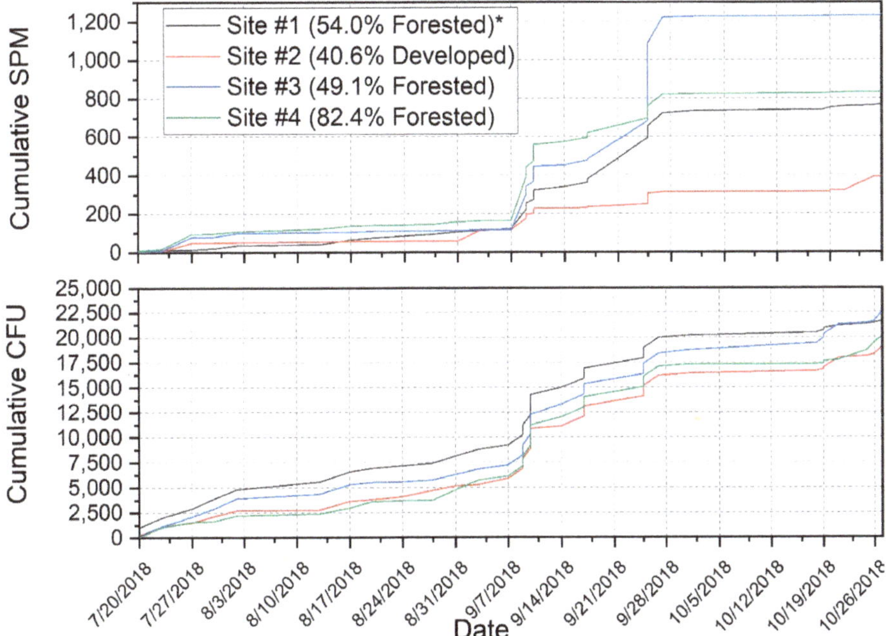

Figure 4. Top: cumulative total SPM (mg L^{-1}). Bottom: cumulative total *E. coli* concentration (CFU per 100 mL) at four monitoring sites during study period (20 July 2018–27 October 2018) in West Run Watershed, Morgantown, West Virginia, USA. * Only predominant land use percentage indicated, for full land use information refer to Table 1.

The average percentage *E. coli* concentration in the ≤5 μm interval exceeded 90 % of the total *E. coli* at all four sites (Figure 5). This finding was supported by ordinal logistic regression results, which showed that the <5 μm size interval had the strongest relationship ($\alpha < 0.0001$) with *E. coli* concentration. The results for Site #2 (40.6% developed land use) differ from the other sites as it had a higher *E. coli* concentration in the >60 μm interval than in the intermediate (5 μm< interval < 60 μm) interval (Figure 6). This result differs from the negative correlations ($p < 0.05$) between particle size and bacterial association typically reported in the literature [60,61]. In the current work, *E. coli* concentrations generally increased with decreasing SPM interval size, with lowest *E. coli* concentration associated with the >60 μm interval, while the highest concentration of *E. coli* were coincident with the <5 μm interval (Figure 5). This result is supported by previous work that reported negative correlations ($p < 0.05$) between particle size and bacterial association [60,61]. Of importance, and as noted earlier, greater concentrations in the smallest interval increase the likelihood that the bacteria will remain buoyant for longer time periods, thereby increasing the downstream extent of decreased microbial water quality [1,12].

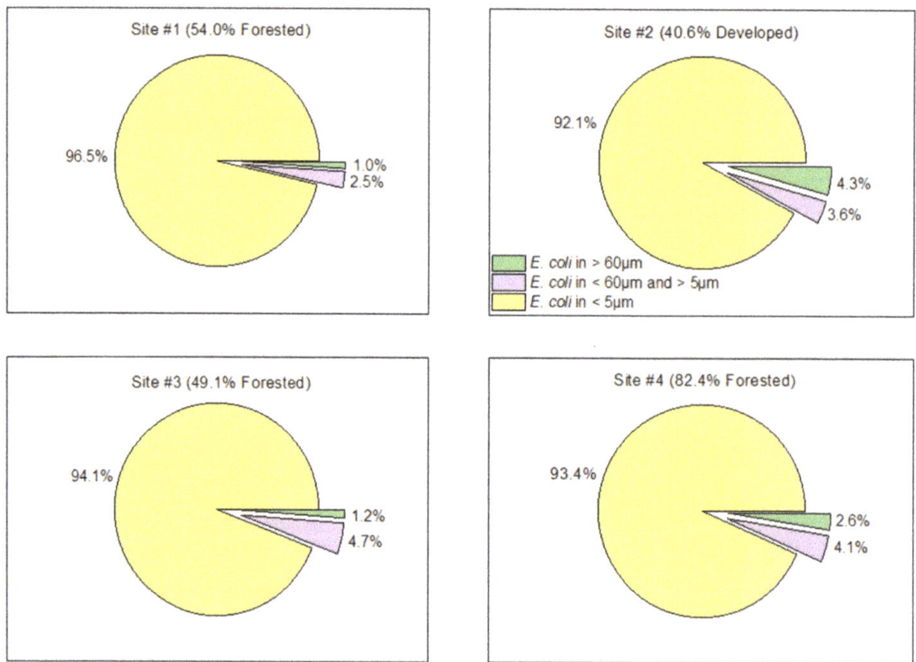

Figure 5. Fractions of *E. coli* concentration (CFU) in three filtered size intervals, at four monitoring sites during the study period (20 July 2018–27 October 2018) in West Run Watershed, Morgantown, West Virginia, USA.

Site #1 (54% forested) had the lowest SPM in the <5 μm interval among the sites (Figure 6). This is consistent with previous studies that showed that developed land use practices typically produce greater quantities of finer (<5 μm) particles relative to other land uses [27], and forested areas originate fewer suspended materials [62–64]. Site #3 (49.1% forested) generally had increased SPM and *E. coli* concentrations in the intermediate (5 μm < interval < 60 μm), <5 μm, and total intervals (Figure 7). Although this site was predominantly forested, it included agricultural land use (42.1%) (including livestock manure stacks), which may account for the elevated SPM and *E. coli* levels [36,65] given the percentage of agricultural land use has been reported to be significantly correlated ($p < 0.04$) with *E. coli* concentrations in receiving waters [35]. Site #4 (82.4% forested; control site) had the lowest average *E. coli* concentration in the < 5 μm interval and the second lowest average total *E. coli* concentration. The low *E. coli* concentrations at Site #4 were expected, given that previous studies reported negative correlations ($p < 0.01$) between *E. coli* concentration and forested land use practices [56]. Decreased *E. coli* in receiving waters from forested locations is usually attributed to decreased endotherm population density relative to agricultural areas [58], lack of artificial sources of *E. coli* (e.g., water infrastructure in urban areas) [65], and decreased run-off compared to urban areas [65]. The decreased *E. coli* and SPM (Figures 4 and 6) from forested areas indicates that the association of *E. coli* with SPM might not be as concerning in forested areas relative to other land use types.

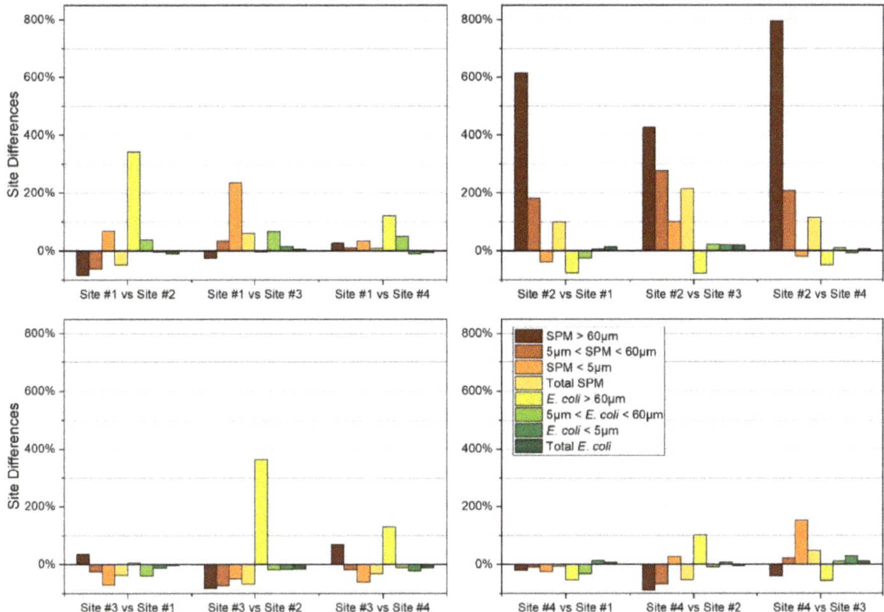

Figure 6. Suspended particulate matter (SPM) and *E. coli* concentration percent difference, separated into filtration intervals (<5 μm; 5 μm < interval < 60 μm; and >60); and total, at four monitoring sites during study period (20 July 2018–27 October 2018) in West Run Watershed, Morgantown, West Virginia, USA.

At Site #2, in the >60 μm interval, there was an increase in *E. coli* concentration and a simultaneous decrease in SPM concentration relative to the other sites (Figure 6). This contradicts the negative correlation between *E. coli* concentration and SPM size reported by previous work [60,61]. However, these results support Figure 5, which shows the apparent preferential association of *E. coli* to SPM in the >60 μm interval relative to the 5 μm to 60 μm interval in developed areas. It is conceivable that there may be an unknown variable that altered the SPM size interval that the *E. coli* at Site #2 preferentially associated with in the current work. This result must be interpreted with caution, however, given differences in sampling regime between this and other studies (as noted earlier). Additionally, given the lack of previous studies, further investigation is needed to validate (or refute) these findings [1].

4.2. Non-Parametric Statistical Analysis

The significant relationship between *E. coli* and SPM discovered using ordinal logistic regression corresponds well with the results from previous work [10,11], which also reported strong relationships between *E. coli* concentrations and SPM. The relationship between *E. coli* and SPM has been shown to be a function of physical processes including runoff events that can influence respective concentrations or it can indicate a potential preferential association of *E. coli* to SPM. However, differentiating between these two possibilities was beyond the scope of the current investigation. The results from ordinal logistic regression indicate no statistically significant (95% confidence interval) relationship between land use practices and *E. coli* concentrations. Therefore, land use practices alone could not be used to predict *E. coli* concentrations in receiving waters in the current investigation. The results from the current work emphasize that additional factors that influence *E. coli* concentrations (i.e., water temperature, pH, and geochemistry) should be addressed in future work.

Increasing SPM concentrations in the intermediate (5 µm < interval < 60 µm) or large (>60 µm) intervals did not result in similar increases to the corresponding *E. coli* concentration (Figure 7). However, increased SPM in the smallest interval (<5 µm) had corresponding increases in the *E. coli* concentration in this interval. This result was attributed to bacteria becoming predominantly attached to, and subsequently transported with, SPM in the smallest interval (<5 µm), a relationship shown preliminarily in previous investigations [66–68]. This relationship may also be a function of the similar transport physics of *E. coli* and SPM particles <5 µm in size, due to similar sizes and buoyancy [69]. In general, this relationship supports the likelihood of *E. coli* remaining in suspension for longer time periods, thereby increasing the stream areas affected by fecal contamination.

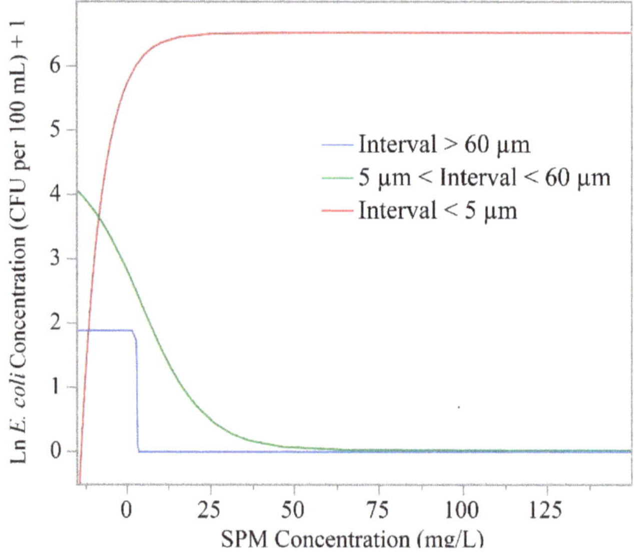

Figure 7. Ln of *E. coli* concentration (CFU per 100 mL) relative to SPM in three filtration size intervals (<5 µm; 5 µm < interval < 60 µm; >60 µm), at four monitoring sites during study period (20 July 2018–27 October 2018) in West Run Watershed, West Virginia, USA. Note all *E. coli* concentration values are +1 to avoid ln values being undefined where *E. coli* concentrations were zero. Additionally, interval size was dictated by the filer sizes (5 µm and 60 µm) that we used during the filtration of the extracted water samples.

Principle component analysis (PCA) showed that the components that account for the maximal variance within a given data set can be identified through the computation of multiple principal components and their respective eigenvalues [70]. Eigenvalues represent the variance of the data in a given direction, therefore components with the highest eigenvalues are principal components [70]. However, most data cannot be well described by a single principal component [70]. Therefore, multiple principal components are typically computed and ranked based on their eigenvalues and displayed visually with biplots [70]. In the current work, principle component biplots showed a distinct spatial distribution of study sites along principal components 1 and 2 for both the smallest interval (<5 µm) and the total data set (defined as all he SPM and *E. coli* data >0.7 µm) (Figure 8). Both biplots highlight the grouping of each of the sites within the idealized vector space defined by principal components 1 and 2. Given the similarity between the sites in terms of geology, topography, and climate, and their close proximity to each other, it can be concluded that the varying land use practices are the principal factors influencing the grouping of the data observed in both biplots [34]. The most obvious patterns are the strong correlation between *E. coli* concentration and SPM concentration in both the smallest interval and total data set (Figure 8). Ultimately, the percentage of agriculture land use was most closely associated with both *E. coli* and SPM concentrations, particularly in the smallest interval, attributable to reasons presented earlier [53,55,56]. This result further emphasizes that agricultural land use practices are strongly correlated with not only the occurrence of fecal contamination [35], but potentially the persistence of fecal microbes in receiving water. Conversely, forested and developed land uses were not as closely correlated with *E. coli* or SPM concentrations (Figure 8). Ultimately, PCA analysis (Table 2) and biplot results (Figure 8) effectively illustrate spatially the relationships between *E. coli* concentration, SPM and land use practices.

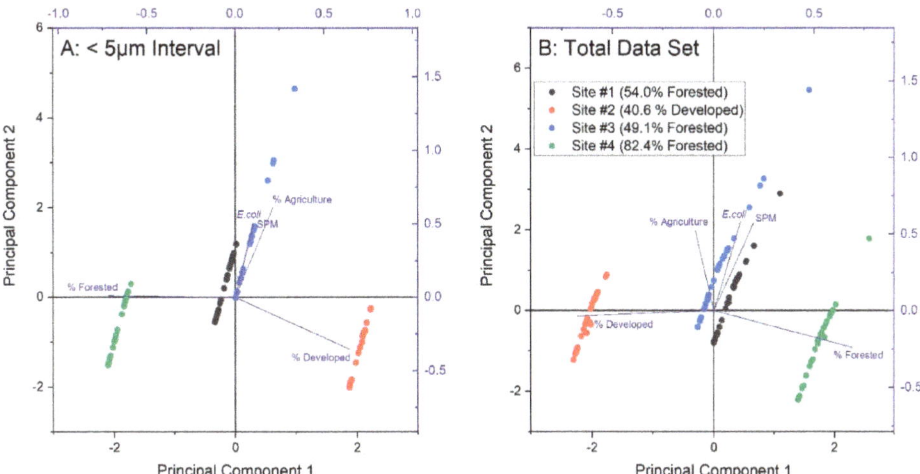

Figure 8. Results of principal components analysis, including biplots A) showing the data in the smallest interval (*E. coli* and SPM) and B) the total data set (defined as all he SPM and *E. coli* data >0.7 µm), for extracted principal components of *E. coli* concentration and SPM concentration at four monitoring sites during study period (20 July 2018–27 October 2018) in West Run Watershed, West Virginia, USA.

4.3. Study Considerations and Future Directions

It is acknowledged that other variables likely affect the lifecycle of *E. coli* (e.g., hydrology and climate) or the association of *E. coli* to SPM (e.g., aquatic geochemistry) [71,72]. Therefore, future work should expand on the results of this investigation by identifying and accounting for these variables. Regression analyses could be a useful tool in establishing the effect of the identified additional variables (e.g., hydrology, climate and aquatic geochemistry) [73]. Future work should also include the use of serial dilutions to avoid sample saturation during the quantification of *E. coli* concentrations. This would improve the accuracy of the results, specifically in areas comprising frequent elevated fecal pollution.

The current work used a study design and sampling regime that was focused on SPM size class and *E. coli*, which was dissimilar to previous work. The study design created challenges in comparing the effects of varying land use practices on *E. coli* or SPM, respectively. Therefore, combining the filtration methodology of the current work with a more traditional sampling regime during future work (i.e., regular temporal and spatial sampling) may facilitate more robust analysis of the influence of different land use practices on the relationship between *E. coli* and SPM of various size intervals. This is particularly important given the strong relationships between *E. coli* concentrations and land use practices identified in previous work [35] and the influence of land use practices on both *E. coli* and SPM concentrations identified in the current investigation. Additionally, the study design applied in the current work can also be used to investigate the association of other constituents in receiving waters with SPM. For example, investigating the association of various microplastics to SPM, and perhaps *E. coli*, in receiving waters, may enhance the understanding of this relatively novel freshwater pollutant. Given the current general lack of process understanding of *E. coli* and emergent co- or in-dependent human population-induced pollutants, such as microplastics, in freshwater sources [74–76], and the potentially harmful effects of microplastics [74,77,78], improving scientific understanding is critical from a water quality perspective.

5. Conclusions

The paucity of field-based research investigating the relationship between *E. coli* (fecal matter) concentration and SPM size distribution and the importance of this relationship, particularly for human health, policy makers, and water resource managers [23,79], provided the impetus for the current work. Similarly, the need to characterize the relationship between SPM size distribution, *E. coli* (fecal matter) and land use practices provided further motivation. A four-site study design (comprising 1st and 2nd order streams) was implemented to advance this understanding in a contemporary representative, mixed-use, urbanizing watershed in the Appalachian region of the eastern United States. The key results show a statistically significant correlation of SPM ($\alpha = 0.05$ with *E. coli* concentration. Moreover, the importance of SPM in the <5 μm interval was also highlighted as 1) more than 90% of *E. coli* data were found in this smallest interval at all four sampling locations, 2) this interval also featured the strongest correlation with *E. coli* concentration data ($\alpha < 0.0001$), and 3) increasing SPM concentrations in the ≤5 μm interval showed corresponding increases in relative *E. coli* concentration. These results generally imply that *E. coli* principally remains free floating or attaches to particles <5 μm in size in receiving waters. Principle component analysis results highlighted the influence of agricultural land use practices on both *E. coli* and SPM concentrations, thereby providing evidence for the potential influence of land use practices on bacterial association with SPM. The work elucidates the effects of SPM and land use practices on *E. coli* concentrations in receiving waters and provides a valuable steppingstone for future research into microbial water quality and fecal pollution. The results from this work better inform policy makers and water resource managers concerned with microbial and fecal pollution in receiving waters, thereby aiding in decision making and the effective management of freshwater resources.

Author Contributions: For the current work author contributions were as follows: conceptualization J.A.H.; methodology J.A.H.; formal analysis F.P. and J.A.H.; investigation J.A.H. and F.P.; resources, J.A.H.; data curation J.A.H.; writing—original draft preparation J.A.H. and F.P.; writing—review and editing J.A.H. and F.P.; visualization, J.A.H. and Petersen; supervision, J.A.H.; project administration, J.A.H.; funding acquisition, J.A.H. All authors have read and agreed to the published version of the manuscript.

Funding: This work was supported by the National Science Foundation under Award Number OIA-1458952, the USDA National Institute of Food and Agriculture, Hatch project accession number 1011536, and the West Virginia Agricultural and Forestry Experiment Station. Results presented may not reflect the views of the sponsors and no official endorsement should be inferred. The funders had no role in study design, data collection and analysis, decision to publish, or preparation of the manuscript.

Acknowledgments: Special thanks are due to many scientists of the Interdisciplinary Hydrology Laboratory (https://www.researchgate.net/lab/The-Interdisciplinary-Hydrology-Laboratory-Jason-A-Hubbart), and the Institute of Water Security and Science (https://iwss.wvu.edu/). The authors extend great appreciation to Ida Holásková of the

Davis College of Agriculture, Natural Resources and Design office of statistics for her expertise and guidance. The authors also appreciate the feedback of anonymous reviewers whose constructive comments improved the article.

Conflicts of Interest: The authors declare no conflict of interest for the current work.

References

1. Oliver, D.M.; Clegg, C.D.; Heathwaite, A.L.; Haygarth, P.M. Preferential attachment of Escherichia coli to different particle size fractions of an agricultural grassland soil. *Water Air Soil Pollut.* **2007**, *185*, 369–375. [CrossRef]
2. Samie, A. *Escherichia coli—Recent Advances on Physiology, Pathogenesis and Biotechnological Applications*; IntechOpen Seattle: Washington, WA, USA, 2017.
3. Mahmoud, M.A.; Abdelsalam, M.; Mahdy, O.A.; El Miniawy, H.M.F.; Ahmed, Z.A.M.; Osman, A.H.; Mohamed, H.M.H.; Khattab, A.M.; Zaki Ewiss, M.A. Infectious bacterial ogens, parasites and pathological correlations of sewage pollution as an important threat to farmed fishes in Egypt. *Environ. Pollut.* **2016**, *219*, 939–948. [CrossRef]
4. Bain, R.; Cronk, R.; Hossain, R.; Bonjour, S.; Onda, K.; Wright, J.; Yang, H.; Slaymaker, T.; Hunter, P.; Prüss-Ustün, A.; et al. Global assessment of exposure to faecal contamination through drinking water based on a systematic review. *Trop. Med. Int. Health* **2014**, *19*, 917–927. [CrossRef]
5. Bain, R.; Cronk, R.; Wright, J.; Yang, H.; Slaymaker, T.; Bartram, J. Fecal contamination of drinking-water in low- and middle-income countries: A systematic review and meta-analysis. *PLoS Med.* **2014**, *11*. [CrossRef]
6. WHO World Water Day Report. Available online: https://www.who.int/water_sanitation_health/takingcharge.html (accessed on 12 December 2019).
7. Benedict, K.M. Surveillance for waterborne disease outbreaks associated with drinking water—United States, 2013–2014. *MMWR Morb. Mortal. Wkly. Rep.* **2017**, *66*, 1216. [CrossRef]
8. Olsen, C.R.; Cutshall, N.H.; Larsen, I.L. Pollutant—Particle associations and dynamics in coastal marine environments: A review. *Mar. Chem.* **1982**, *11*, 501–533. [CrossRef]
9. Jamieson, R.; Joy, D.M.; Lee, H.; Kostaschuk, R.; Gordon, R. Transport and deposition of sediment-associated Escherichia coli in natural streams. *Water Res.* **2005**, *39*, 2665–2675. [CrossRef]
10. Amalfitano, S.; Corno, G.; Eckert, E.; Fazi, S.; Ninio, S.; Callieri, C.; Grossart, H.-P.; Eckert, W. Tracing particulate matter and associated microorganisms in freshwaters. *Hydrobiologia* **2017**, *800*, 145–154. [CrossRef]
11. Jeng, H.C.; England, A.J.; Bradford, H.B. Indicator organisms associated with stormwater suspended particles and estuarine sediment. *J. Environ. Sci. Health* **2005**, *40*, 779–791. [CrossRef]
12. Hassard, F.; Gwyther, C.L.; Farkas, K.; Andrews, A.; Jones, V.; Cox, B.; Brett, H.; Jones, D.L.; McDonald, J.E.; Malham, S.K. Abundance and distribution of enteric bacteria and viruses in coastal and estuarine sediments—A review. *Front. Microbiol.* **2016**, *7*, 1692. [CrossRef]
13. Rieck, A.; Herlemann, D.P.; Jürgens, K.; Grossart, H.-P. Particle-associated differ from free-living bacteria in surface waters of the Baltic Sea. *Front. Microbiol.* **2015**, *6*, 1297. [CrossRef]
14. Grossart, H.-P. Ecological consequences of bacterioplankton lifestyles: Changes in concepts are needed. *Environ. Microbiol. Rep.* **2010**, *2*, 706–714. [CrossRef]
15. Drummond, J.D.; Davies-Colley, R.J.; Stott, R.; Sukias, J.P.; Nagels, J.W.; Sharp, A.; Packman, A.I. Microbial transport, retention, and inactivation in streams: A combined experimental and stochastic modeling approach. *Environ. Sci. Technol.* **2015**, *49*, 7825–7833. [CrossRef]
16. Simon, M.; Grossart, H.-P.; Schweitzer, B.; Ploug, H. Microbial ecology of organic aggregates in aquatic ecosystems. *Aquat. Microb. Ecol.* **2002**, *28*, 175–211. [CrossRef]
17. Bidle, K.D.; Flecthcer, M. Comparison of free-living and particle-associated bacterial communities in the Chesapeake Bay by stable low-molecular-weight Rna analysis. *Appl. Environ. Microbiol.* **1995**, *61*, 944–952. [CrossRef]
18. Allen, H.K.; Donato, J.; Wang, H.H.; Cloud-Hansen, K.A.; Davies, J.; Handelsman, J. Call of the wild: Antibiotic resistance genes in natural environments. *Nat. Rev. Microbiol.* **2010**, *8*, 251. [CrossRef]
19. Corno, G.; Coci, M.; Giardina, M.; Plechuk, S.; Campanile, F.; Stefani, S. Antibiotics promote aggregation within aquatic bacterial communities. *Front. Microbiol.* **2014**, *5*, 297. [CrossRef]

20. Madoux-Humery, A.-S.; Dorner, S.; Sauvé, S.; Aboulfadl, K.; Galarneau, M.; Servais, P.; Prévost, M. The effects of combined sewer overflow events on riverine sources of drinking water. *Water Res.* **2016**, *92*, 218–227. [CrossRef]
21. Jett, B.D.; Huycke, M.M.; Gilmore, M.S. Virulence of enterococci. *Clin. Microbiol. Rev.* **1994**, *7*, 462–478. [CrossRef]
22. Anastasi, E.M.; Matthews, B.; Gundogdu, A.; Vollmerhausen, T.L.; Ramos, N.L.; Stratton, H.; Ahmed, W.; Katouli, M. Prevalence and persistence of Escherichia coli strains with uropathogenic virulence characteristics in sewage treatment plants. *Appl. Environ. Microbiol.* **2010**, *76*, 5882–5886. [CrossRef]
23. Abia, A.L.K.; Ubomba-Jaswa, E.; Genthe, B.; Momba, M.N.B. Quantitative microbial risk assessment (QMRA) shows increased public health risk associated with exposure to river water under conditions of riverbed sediment resuspension. *Sci. Total Environ.* **2016**, *566*, 1143–1151. [CrossRef]
24. Characklis, G.W.; Dilts, M.J.; Simmons III, O.D.; Likirdopulos, C.A.; Krometis, L.-A.H.; Sobsey, M.D. Microbial partitioning to settleable particles in stormwater. *Water Res.* **2005**, *39*, 1773–1782. [CrossRef]
25. Hubbart, J.A. Using sediment particle size class analysis to better understand urban land-use effects. *Int. J. Appl.* **2012**, *2*, 12–27.
26. Kellner, E.; Hubbart, J.A. Improving understanding of mixed-land-use watershed suspended sediment regimes: Mechanistic progress through high-frequency sampling. *Sci. Total Environ.* **2017**, *598*, 228–238. [CrossRef]
27. Kellner, E.; Hubbart, J.A. Flow class analyses of suspended sediment concentration and particle size in a mixed-land-use watershed. *Sci. Total Environ.* **2019**, *648*, 973–983. [CrossRef]
28. Kellner, E.; Hubbart, J.A. Quantifying urban land-use impacts on suspended sediment particle size class distribution: A method and case study. *Stormwater J.* **2014**, *15*, 40–50.
29. Chen, H.J.; Chang, H. Response of discharge, TSS, and *E. coli* to rainfall events in urban, suburban, and rural watersheds. *Environ. Sci. Process. Impacts* **2014**, *16*, 2313–2324. [CrossRef]
30. Cantor, J.; Krometis, L.-A.; Sarver, E.; Cook, N.; Badgley, B. Tracking the downstream impacts of inadequate sanitation in central Appalachia. *J. Water Health* **2017**, *15*, 580–590. [CrossRef]
31. Koppen, W.D. Das geographische system der klimate. *Handb. Klimatol.* **1936**, *46*, 1–44.
32. Central Appalachian Broadleaf Forest—Coniferous Forest—Meadow Province. Available online: https://www.fs.fed.us/land/ecosysmgmt/colorimagemap/images/m221.html (accessed on 12 December 2019).
33. Arcipowski, E.; Schwartz, J.; Davenport, L.; Hayes, M.; Nolan, T. Clean water, clean life: Promoting healthier, accessible water in rural Appalachia. *J. Contemp. Water Res. Educ.* **2017**, *161*, 1–18. [CrossRef]
34. Kellner, E.; Hubbart, J.; Stephan, K.; Morrissey, E.; Freedman, Z.; Kutta, E.; Kelly, C. Characterization of sub-watershed-scale stream chemistry regimes in an Appalachian mixed-land-use watershed. *Environ. Monit. Assess.* **2018**, *190*, 586. [CrossRef]
35. Petersen, F.; Hubbart, J.A.; Kellner, E.; Kutta, E. Land-use-mediated Escherichia coli concentrations in a contemporary Appalachian watershed. *Environ. Earth Sci.* **2018**, *77*, 754. [CrossRef]
36. The West Virginia Water Research Institute, The West Run Watershed Association. *(WVWRI) Watershed Based Plan for West Run of the Monongahela River 2008*; The West Virginia Water Research Institute, The West Run Watershed Association: Morgantown, WV, USA, 2008.
37. Peel, M.C.; Finlayson, B.L.; McMahon, T.A. Updated world map of the Köppen-Geiger climate classification. *Hydrol. Earth Syst. Sci. Discuss.* **2007**, *4*, 439–473. [CrossRef]
38. Arguez, A.; Durre, I.; Applequist, S.; Squires, M.; Vose, R.; Yin, X.; Bilotta, R. NOAA's US climate normals (1981–2010). *NOAA Natl. Cent. Environ. Inf.* **2010**, *10*, V5PN93JP.
39. Myers, M.D. National field manual for the collection of water-quality data. *US Geol. Surv. Tech. Water Resour. Investig. Book* **2006**, *9*. [CrossRef]
40. Hubbart, J.A.; Kellner, E.; Hooper, L.W.; Zeiger, S. Quantifying loading, toxic concentrations, and systemic persistence of chloride in a contemporary mixed-land-use watershed using an experimental watershed approach. *Sci. Total Environ.* **2017**, *581*, 822–832. [CrossRef]
41. Kellner, E.; Hubbart, J. Advancing understanding of the surface water quality regime of contemporary mixed-land-use watersheds: An application of the experimental watershed method. *Hydrology* **2017**, *4*, 31. [CrossRef]
42. Zeiger, S.J.; Hubbart, J.A. Quantifying flow interval–pollutant loading relationships in a rapidly urbanizing mixed-land-use watershed of the Central USA. *Environ. Earth Sci.* **2017**, *76*, 484. [CrossRef]

43. Zeiger, S.J.; Hubbart, J.A. Nested-scale nutrient flux in a mixed-land-use urbanizing watershed. *Hydrol. Process.* **2016**, *30*, 1475–1490. [CrossRef]
44. Crump, B.C.; Armbrust, E.V.; Baross, J.A. Phylogenetic analysis of particle-attached and free-living bacterial communities in the Columbia River, its estuary, and the adjacent coastal ocean. *Appl. Environ. Microbiol.* **1999**, *65*, 3192–3204. [CrossRef]
45. Riemann, L.; Winding, A. Community dynamics of free-living and particle-associated bacterial assemblages during a freshwater phytoplankton bloom. *Microb. Ecol.* **2001**, *42*, 274–285. [CrossRef]
46. Ortega-Retuerta, E.; Joux, F.; Jeffrey, W.H.; Ghiglione, J.-F. Spatial variability of particle-attached and free-living bacterial diversity in surface waters from the Mackenzie River to the Beaufort Sea (Canadian Arctic). *Biogeosciences* **2013**, *10*, 2747–2759. [CrossRef]
47. Wentworth, C.K. A scale of grade and class terms for clastic sediments. *J. Geol.* **1922**, *30*, 377–392. [CrossRef]
48. IDEXX Laboratories Colilert Procedure Manual. Available online: https://www.idexx.com/files/colilert-procedure-en.pdf (accessed on 4 April 2019).
49. Cummings, D. The Fecal Coliform Test Compared to Specific Tests For Escherichia Coli. Available online: https://www.idexx.com/resource-library/water/water-reg-article9B.pdf (accessed on 24 September 2019).
50. Yazici, B.; Yolacan, S. A comparison of various tests of normality. *J. Stat. Comput. Simul.* **2007**, *77*, 175–183. [CrossRef]
51. Stokes, M.E.; Davis, C.S.; Koch, G.G. *Categorical Data Analysis Using SAS*; SAS institute: Cary, NC, USA, 2012.
52. United States Climate Data (USCD). Available online: https://www.usclimatedata.com/climate/morgantown/west-virginia/united-states/uswv0507/2012/7 (accessed on 28 September 2019).
53. Stein, E.D.; Tiefenthaler, L.; Schiff, K. Comparison of stromwater pollutant loading by land use type. *Proc. Water Environ. Fed.* **2007**, *2007*, 700–722. [CrossRef]
54. Bro, R.; Smilde, A.K. Principal component analysis. *Anal. Methods* **2014**, *6*, 2812–2831. [CrossRef]
55. Jeloudar, F.T.; Sepanlou, M.G.; Emadi, S.M. Impact of land use change on soil erodibility. *Glob. J. Environ. Sci. Manag.* **2018**, *4*, 59–70.
56. Tong, S.T.; Chen, W. Modeling the relationship between land use and surface water quality. *J. Environ. Manag.* **2002**, *66*, 377–393. [CrossRef]
57. Wilson, C.; Weng, Q. Assessing surface water quality and its relation with urban land cover changes in the Lake Calumet Area, Greater Chicago. *Environ. Manag.* **2010**, *45*, 1096–1111. [CrossRef]
58. Causse, J.; Billen, G.; Garnier, J.; Henri-des-Tureaux, T.; Olasa, X.; Thammahacksa, C.; Latsachak, K.O.; Soulileuth, B.; Sengtaheuanghoung, O.; Rochelle-Newall, E. Field and modelling studies of Escherichia coli loads in tropical streams of montane agro-ecosystems. *J. Hydro Environ. Res.* **2015**, *9*, 496–507. [CrossRef]
59. Rochelle-Newall, E.J.; Ribolzi, O.; Viguier, M.; Thammahacksa, C.; Silvera, N.; Latsachack, K.; Dinh, R.P.; Naporn, P.; Sy, H.T.; Soulileuth, B. Effect of land use and hydrological processes on Escherichia coli concentrations in streams of tropical, humid headwater catchments. *Sci. Rep.* **2016**, *6*, 32974. [CrossRef]
60. Dong, H.; Onstott, T.C.; DeFlaun, M.F.; Fuller, M.E.; Scheibe, T.D.; Streger, S.H.; Rothmel, R.K.; Mailloux, B.J. Relative dominance of physical versus chemical effects on the transport of adhesion-deficient bacteria in intact cores from South Oyster, Virginia. Environ. *Sci. Technol.* **2002**, *36*, 891–900. [CrossRef]
61. Levy, J.; Sun, K.; Findlay, R.H.; Farruggia, F.T.; Porter, J.; Mumy, K.L.; Tomaras, J.; Tomaras, A. Transport of Escherichia coli bacteria through laboratory columns of glacial-outwash sediments: Estimating model parameter values based on sediment characteristics. *J. Contam. Hydrol.* **2007**, *89*, 71–106. [CrossRef]
62. Mallin, M.A.; Johnson, V.L.; Ensign, S.H. Comparative impacts of stormwater runoff on water quality of an urban, a suburban, and a rural stream. *Environ. Monit. Assess.* **2009**, *159*, 475–491. [CrossRef]
63. Gilley, J.E.; Bartelt-Hunt, S.L.; Eskridge, K.M.; Li, X.; Schmidt, A.M.; Snow, D.D. Setback distance requirements for removal of swine slurry constituents in runoff. *Trans. ASABE* **2017**, *60*, 1885–1894. [CrossRef]
64. Sutherland, D.G.; Ball, M.H.; Hilton, S.J.; Lisle, T.E. Evolution of a landslide-induced sediment wave in the Navarro River, California. *GSA Bull.* **2002**, *114*, 1036–1048. [CrossRef]
65. Gotkowska-Plachta, A.; Golaś, I.; Korzeniewska, E.; Koc, J.; Rochwerger, A.; Solarski, K. Evaluation of the distribution of fecal indicator bacteria in a river system depending on different types of land use in the southern watershed of the Baltic Sea. *Environ. Sci. Pollut. Res.* **2016**, *23*, 4073–4085. [CrossRef]
66. Davies, C.M.; Bavor, H.J. The fate of stormwater-associated bacteria in constructed wetland and water pollution control pond systems. *J. Appl. Microbiol.* **2000**, *89*, 349–360. [CrossRef]

67. Muirhead, R.W.; Collins, R.P.; Bremer, P.J. Interaction of Escherichia coli and soil particles in runoff. *Appl. Environ. Microbiol.* **2006**, *72*, 3406–3411. [CrossRef]
68. Guber, A.K.; Pachepsky, Y.A.; Shelton, D.R.; Yu, O. Effect of bovine manure on fecal coliform attachment to soil and soil particles of different sizes. *Appl. Environ. Microbiol.* **2007**, *73*, 3363–3370. [CrossRef]
69. Southard, J. *Introduction to Fluid Motions, Sediment Transport, and Current-Generated Sedimentary Structures Course Textbook*; MIT OpenCourseWare, Massachusetts Institute of Technology: Cambridge, MA, USA, 2006.
70. Multivariate Analysis: Principal Component Analysis: Biplots-9.3. Available online: http://support.sas.com/documentation/cdl/en/imlsug/64254/HTML/default/viewer.htm#imlsug_ugmultpca_sect003.htm (accessed on 23 December 2019).
71. Presser, K.A.; Ratkowsky, D.A.; Ross, T. Modelling the growth rate of Escherichia coli as a function of pH and lactic acid concentration. *Appl. Environ. Microbiol.* **1997**, *63*, 2355–2360. [CrossRef]
72. Noble, R.T.; Lee, I.M.; Schiff, K.C. Inactivation of indicator micro-organisms from various sources of faecal contamination in seawater and freshwater. *J. Appl. Microbiol.* **2004**, *96*, 464–472. [CrossRef]
73. Helsel, D.R.; Hirsch, R.M. *Statistical Methods in Water Resources*, 1st ed.; Elsevier: Amsterdam, The Netherlands, 1992; Volume 49, ISBN 978-0-08-087508-8.
74. Wagner, M.; Lambert, S. *Freshwater Microplastics: Emerging Environmental Contaminants*; The Handbook of Environmental Chemistry; Barcelo, D., Kostianoy, A.G., Eds.; Springer International Publishing: Cham, Switzerland, 2018; ISBN 978-3-319-61614-8.
75. Rillig, M.C. Microplastic in terrestrial ecosystems and the soil? *Environ. Sci. Technol.* **2012**, *46*, 6453–6454. [CrossRef]
76. Barnes, D.K.A.; Galgani, F.; Thompson, R.C.; Barlaz, M. Accumulation and fragmentation of plastic debris in global environments. *Philos. Trans. R. Soc. B Biol. Sci.* **2009**, *364*, 1985–1998. [CrossRef]
77. Lithner, D.; Damberg, J.; Dave, G.; Larsson, Å. Leachates from plastic consumer products—Screening for toxicity with Daphnia magna. *Chemosphere* **2009**, *74*, 1195–1200. [CrossRef]
78. Bejgarn, S.; MacLeod, M.; Bogdal, C.; Breitholtz, M. Toxicity of leachate from weathering plastics: An exploratory screening study with Nitocra spinipes. *Chemosphere* **2015**, *132*, 114–119. [CrossRef]
79. Pandey, P.K.; Soupir, M.L. Assessing linkages between *E. coli* levels in streambed sediment and overlying water in an agricultural watershed in Iowa during the first heavy rain event of the season. *Trans. ASABE* **2014**, *57*, 1571.

Article

Advancing Understanding of Land Use and Physicochemical Impacts on Fecal Contamination in Mixed-Land-Use Watersheds

Fritz Petersen [1] and Jason A. Hubbart [1,2,*]

1 Division of Plant and Soil Sciences, Davis College of Agriculture, Natural Resources and Design, West Virginia University, Agricultural Sciences Building, Morgantown, WV 26506, USA; fp0008@mix.wvu.edu

2 Institute of Water Security and Science, Schools of Agriculture and Food, and Natural Resources, Davis College of Agriculture, Natural Resources and Design, West Virginia University, 3109 Agricultural Sciences Building, Morgantown, WV 26506, USA

* Correspondence: jason.hubbart@mail.wvu.edu; Tel.: +1-304-293-2472

Received: 31 March 2020; Accepted: 10 April 2020; Published: 12 April 2020

Abstract: Understanding mixed-land-use practices and physicochemical influences on *Escherichia (E.) coli* concentrations is necessary to improve water quality management and human health. Weekly stream water samples and physicochemical data were collected from 22 stream gauging sites representing varying land use practices in a contemporary Appalachian watershed of the eastern USA. Over the period of one annual year, *Escherichia (E.) coli* colony forming units (CFU) per 100 mL were compared to physicochemical parameters and land use practices. Annual average *E. coli* concentration increased by approximately 112% from acid mine drainage (AMD) impacted headwaters to the lower reaches of the watershed (approximate averages of 177 CFU per 100 mL vs. 376 CFU per 100 mL, respectively). Significant Spearman's correlations ($p < 0.05$) were identified from analyses of pH and *E. coli* concentration data representing 77% of sample sites; thus highlighting legacy effects of historic mining (AMD) on microbial water quality. A tipping point of 25–30% mixed development was identified as leading to significant ($p < 0.05$) negative correlations between chloride and *E. coli* concentrations. Study results advance understanding of land use and physicochemical impacts on fecal contamination in mixed-land-use watersheds, aiding in the implementation of effective water quality management practices and policies.

Keywords: *Escherichia coli*; physicochemistry; water quality; land use practices; experimental watershed

---

1. Introduction

Fecal microbes (e.g., *Escherichia (E.) coli*) are sources of waterborne pathogens and water contamination, causing substantial mortality and morbidity among human populations globally [1,2]. Outbreaks of diarrhea, urinary tract infections, respiratory illness, and pneumonia have been traced to increased fecal microbes (e.g., *E. coli*) in freshwater systems [3,4]. In 2018, the World Health Organization (WHO) reported that waterborne diarrheal diseases are the leading cause of mortality in the developing world, causing 2.2 million human deaths annually [5]. Obviously, understanding factors conducive to elevated fecal microbe (e.g., *E. coli*) concentrations in receiving waters is important from water quality and human health perspective. Improved understanding of factors favorable to fecal microbes can be used to inform land use managers in terms of how to effectively reduce fecal contamination, thereby improving water quality, fresh water security, and human health [6].

Fecal microbes in receiving waters can be variously impacted by physicochemical parameters. Previous investigations reported negative correlations between temperature, salinity, oxygen content, pH, and fecal microbe concentrations [7]. However, information on fecal microbes and physicochemical parameters in freshwater systems is limited as most previous investigations occurred in saline environments [7,8]. The available freshwater investigations comprised notable shortcomings including: short sampling periods and/or few collected samples [9], limited sampling locations (in similar land use types) [9], long periods of time (weeks or months) between sample collection [10], sample collection solely during baseflow conditions [10], or failure to account for constituents such as chloride (Cl^-), which has been shown to influence microbe concentrations in laboratory settings [10,11]. Studies from freshwater systems affirmed relationships between temperature, pH, and fecal microbe concentrations [9,10,12]. However, investigations, including many physicochemical (e.g., chloride) relationships with fecal contamination, are greatly lacking. There is, therefore, need for investigations that include high spatio-temporal sampling regimes, occurring at different stage (streamflow and hydro-climate) conditions, in areas comprising different land use practices, over longer time periods. Information gained from these investigations could aid in the implementation of effective strategies to reduce fecal microbe concentrations in receiving waters.

Land use practice impacts on fecal contamination is relatively well-documented [6,13–18]. Both agricultural and urban land use practices have been shown to generally increase fecal concentrations [6,17], attributed to livestock husbandry [14], manure application [16], or poorly maintained wastewater infrastructure [17], and increased impervious surfaces [19,20]. Previous work reported negative correlations between forested areas and fecal concentrations in receiving waters [21]. Shortcomings of previous work include that many investigations occurred in areas comprising similar land use practices (i.e., lacking variability) [22,23] or included few sampling locations [24]. Therefore, information regarding the influence of land use practices on fecal microbe concentrations in mixed-land-use watersheds (comprising the majority of watersheds globally) are quite limited. Furthermore, the predominant focus on storm events [25], limits information regarding the influence of land use practices on *E. coli* concentrations during other flow and transport conditions. Simultaneous combined analysis of fecal pollution (*E. coli* concentration), land use practices, and physicochemical parameters relationships is also lacking in previous investigations. Such advanced integrated understanding will provide land use managers with more detailed information regarding land use impacts on fecal pollution; thereby, aiding in freshwater quality management decisions.

Multiple study designs and sampling regimes have been implemented to investigate fecal pollution in receiving waters, including laboratory and field based designs. Laboratory studies usually included simulations [26], whereas field based designs comprise event based sampling [27], periodic sampling [19], stochastic sampling [28] and, in limited number, scale-nested experimental watersheds [6]. The scale-nested experimental watershed study design has been particularly effective for quantifying factors (e.g., physicochemical parameters, land use practices) influencing response variables of interest (e.g., *E. coli* concentration) in receiving waters, particularly in mixed-land-use watersheds [29–36]. To achieve this, nested watershed study designs divide larger watersheds into a series of sub-catchments, each with a monitoring (gauging) site at its drainage terminus [30,32,36–38]. Hydrologic characteristics and land us practices can then be isolated though sub-catchment delineation [36]. Quantification of the influencing processes recorded at the sub-catchment scale allows for the identification of the influence and cumulative effect of various land use practices on the response variable of interest [39]. Given its use in numerous peer reviewed publications over multiple decades, the scale-nested and paired [36]) experimental watershed study design is the optimal study design for investigating current knowledge gaps regarding fecal contamination, physicochemical parameters, and land use practices.

The widespread, frequent, and persistent fecal pollution characteristics of the Appalachian region of the USA are similar (representative) to numerous locations globally. The region is, therefore, well suited for (transferrable) research into factors affecting fecal contamination in receiving waters [40]. Water quality and security is a primary concern in rural Appalachia, as fecal contamination poses substantial risk to residents [41]. The risk is elevated by inadequate wastewater treatment infrastructure, geographical isolation, inaccessible terrain, and poverty [41]. Fecal contamination is, therefore, a primary concern, and improved understanding of factors increasing fecal contamination can inform better management practices and improve water quality in the region. Furthermore, the Appalachian region is also physiographically diverse, consisting of distinct geographic, climatological, and ecological areas, typically divided into distinct Northern, Central and Southern regions [42]. The diverse physiography makes results from investigations in these distinct regions comparable and transferable to similar areas globally. For example, the central Appalachian region is comparable to areas such as Hokkaido or Northern Honshu in Japan as these areas comprise temperate climates and well-distributed year-round rainfall [43,44].

The overarching objective of the current work was to quantitatively compare relationships between fecal concentration (*E. coli* colony forming units), physicochemical parameters, and land use practices. Sub-objectives included (1) identifying the dominant factors that influence fecal microbe concentrations in receiving waters, and (2) investigating the effects of seasonal variation on factors (e.g., physicochemical parameters, land use practices) influencing fecal microbes (*E. coli*) in freshwater streams. Study outcomes provide new quantitative insight to these issues thereby providing land use managers with advanced science-based information to improve management practices and policies in freshwater systems.

2. Materials and Methods

2.1. Study Site Description

The location for the current work was a 3rd order tributary of the Monongahela River, the 23 km^2 mixed–land use urbanizing West Run Watershed (WRW). The WRW is located in Morgantown, WV, USA, and contains many different land use practices ranging from various mixed development (e.g., urban and commercial), agriculture, and forested practices. Based on 2016 National Agriculture Imagery Program (NAIP) land use and land cover data, at the time of this investigation WRW consisted of 42.7% forested, 37.7% mixed development and 19.4% agricultural land use (Table 1, site #22). West Run Creek, the primary drainage of WRW includes small floodplains and is a narrow, moderately entrenched stream [6,45]. The elevation in WRW ranges between 420 to 240 m above mean sea level from the headwaters to the confluence of the Monongahela River [6]. The geology of WRW comprises numerous Paleozoic era rock outcroppings and the Monongahela series located in the headwaters [6]. Historic mining of the two coal seams in WRW, the Upper Kittanning, and more specifically, the Pittsburg coal seam resulted in pervasive water quality problems in the watershed (particularly in the headwaters) [6,46].

The climate regime in West Virginia and the city of Morgantown, residing in part, in the WRW and in Monongalia County WV, has a climate lacking a dry season and warm summers (average monthly temperature > 22 °C) and cold winters (average monthly temperature < 0 °C) [47]. Between 1981 and 2010 Morgantown received approximately 1060 mm of average annual precipitation, with the wettest and warmest month (July) comprising an average daily temperature and monthly precipitation of 23 °C and 117 mm, respectively [48]. Conversely, the coldest (January) month included an average daily temperature of −0.4 °C and the driest (February) month an average monthly precipitation of 66 mm [48].

A twenty-two study site (i.e., n = 22 stream gauging sites) scale-nested and paired experimental watershed study design [30,39,49–51] was implemented in WRW in 2016. Field surveys and GIS were used to identify study site locations and associated sub-catchments. The sampling sites (numbered in downstream order were located on the 1st and 2nd order confluence tributaries of WRW (#1, #2, #5, #7, #8, #9, # 11, #12, #14, #15, #16, #17, and #20) and along West Run Creek (#3, #4, #6, #10, #13, #18, #19, #21, and #22) and included many land use practices (Table 1; Figure 1). Forested land use was the predominant land use practice in WRW during the time of the investigation, accounting for 42.7% of the total land use practices in the watershed (Table 1, site #22). All the sub-catchments except #1, #11,

#15, #16, and #20 were by majority forested. Sub-catchments #11 and #16 were primarily agricultural, whereas sub-catchments #1, #15 and #20 were primarily mixed development (Table 1).

Table 1. Land use/land cover characteristics (% cover) of 22 monitoring sites in West Run Watershed (WRW), West Virginia, USA, including total drainage area (km^2).

Site	Mixed Development (%)	Agriculture (%)	Forested (%)	Drainage Area (km^2)
1	53.23%	38.70%	8.07%	0.30
2	13.58%	12.20%	74.21%	0.29
3	22.35%	16.17%	61.32%	1.87
4	25.88%	14.91%	59.00%	2.48
5	23.35%	25.51%	51.14%	0.38
6	23.91%	17.25%	58.70%	3.72
7	16.33%	28.60%	54.91%	0.78
8	30.78%	16.47%	52.35%	1.55
9	27.57%	19.33%	52.84%	2.29
10	24.92%	18.40%	56.49%	6.18
11	18.15%	41.87%	39.16%	1.75
12	31.77%	33.72%	34.51%	1.75
13	26.83%	25.77%	47.15%	10.53
14	16.19%	26.43%	56.92%	3.36
15	70.28%	10.31%	19.42%	0.98
16	5.38%	58.72%	35.16%	0.25
17	4.78%	9.38%	85.84%	0.75
18	25.98%	24.88%	48.86%	16.41
19	29.45%	22.45%	47.85%	18.88
20	89.16%	4.19%	6.61%	3.42
21	38.10%	19.46%	42.23%	22.93
22	37.71%	19.38%	42.66%	23.24

Note: due to the omission of certain categories (i.e., wetland, open water, etc.) and certain categories comprising combinations other others (e.g., mixed development = urban and residential) land use percentages may not sum to 100%. Final row (site #22) indicates total values for the entire watershed.

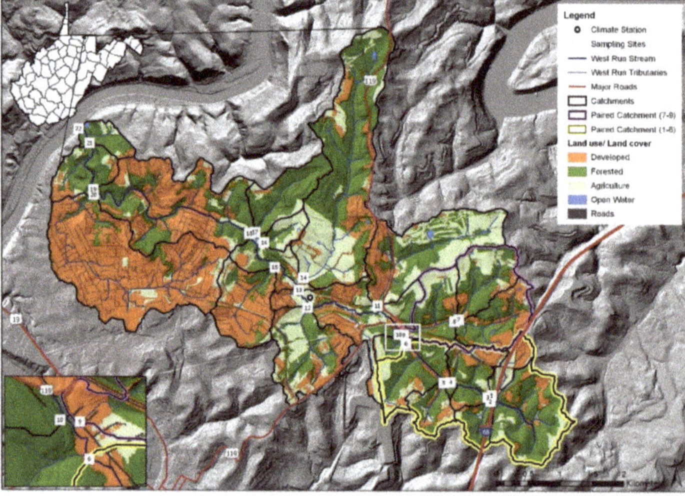

Figure 1. Land use/ land cover of West Run Watershed, Morgantown, WV, USA, including monitoring/ sampling locations for the current investigation.

The predominance of forested cover among the sites should not be taken to imply that catchments were equivalent in terms of land use. For example, the reference sub-catchment (control) for the current work, sub-catchment #17 comprised 85.84% forested land use 9.4% agricultural and 4.8% mixed development land use practices. This sub-catchment comprised the largest percentage of forested land use practices among the sub-catchments (hence, selection as reference sub-catchment). Conversely, sub-catchment #12 comprised 34.5% forested, 33.7% agriculture, and 31.7% agriculture land use practices and had the lowest percentage forested land use among the primarily forested sites (Table 1). Sub-catchment #17 is therefore distinct in terms of land use practices relative to sub-catchment #12, despite both being predominantly forested. In general, at the time of the investigation, mixed development comprised the second largest percentage of land use practices (37.7%) and agricultural land use practices accounted for the lowest percentage of land use practices (19.4%) in WRW.

2.2. Data Collection

Climate data, including precipitation (Campbell Scientific TE525 Tipping Bucket Rain Gauge), average air temperature, relative humidity (Campbell Scientific HC2S3 Temperature and Relative Humidity Probe), and average wind speed (Campbell Scientific Met One 034B Wind Set instrument), were recorded at a three-meter height within approximately 30 m of site #13 (Figure 1), for the duration of the study period (2 January 2018–1 January 2019).

For the current work, weekly water grab-samples were collected as per Petersen et al. [6], Hubbart et al. [52], Kellner and Hubbart [38], and Zeiger and Hubbart [37,53] from each monitoring site (stream order ≤ 3). Water sample collection proceeded through numerical order of sites starting at 09:00 at site #1. To reduce overall sampling time (increasing sample representativeness during processing), sites #9 and #10 were sampled before sites #7 and #8, due to their proximity relative to site #6 (Figure 1). The calendar year duration of the sampling period (1 February 2018–1 January 2019) was longer than previous work on fecal contamination, allowing for assessment of seasonal variability of *E. coli* concentration, physicochemical, and hydro-climate data [54,55]. This resulted in a more comprehensive quantification of fecal contamination (*E. coli*) regimes at sub-catchment mixed-land-use scales than offered through most, if not all previous studies in the published literature. A total of 1166 spatio-temporally delineated fecal contamination (*E. coli*) concentration values were obtained during this high-resolution study.

Once collected, water samples were transported to the Interdisciplinary Hydrology Laboratory, located in the Davis College of Agriculture, Natural Resources and Design at West Virginia University, for analyses. *Escherichia (E) coli* was used as an indicator organism to quantify fecal contamination as per previous work investigating fecal contamination [2,6]. The Colilert test, developed by IDEXX Laboratories Inc. and approved by the U.S. Environmental Protection Agency (EPA) [56], was used to quantify *E. coli* colony forming units (CFU). The test was designed to eliminate the need for sample dilution when evaluating fecal concentration in water samples and is included in the Standard Methods for Examination of Water and Wastewater [56,57]. The likelihood of reporting inaccurate results (i.e., false positives ± 10%) when using the test is low due to Colilert's Defined Substrate Technology nutrient-indicator (ONPG), and a selectively suppressing formulated matrix. The ONPG is a carbon source that most non-target organisms lack the enzyme to utilize, rendering them unable to grow or interfere [56]. The few non-target organism that can use ONPG as a carbon source is suppressed by a selective matrix [56]. As per Colilert instructions, the Colilert (ONPG) substrate was added to 100 mL of sampled water, sealed in the Quanti-Tray, and incubated at 35 °C for 24 h. The Quanti-Tray system comprises 96 total wells: 48 large wells (49, including an overflow well) and 48 small wells. Results are reported in CFU per 100 mL [56]. Following incubation, fluorescing (positive for *E. coli*) wells were enumerated using a UV light (6 watt, 365 nm wavelength) and compared to the Quanti-Tray Most Probable Number (MPN) table. The MPN table is used to convert the number of positive wells to an *E. coli* concentration value (CFU per 100mL), with a 95% confidence interval. The applied method thus used a MPN approach to estimate *E. coli* CFU concentration; therefore, *E. coli* concentration data was referred to as CFU not MPN during the investigation. The *E. coli* concentration range that could be detected using this method was <1 to 1011.2 CFU. *E. coli* concentrations exceeding 1011 CFU per 100 mL; therefore, could not be estimated accurately. However, this shortcoming was deemed acceptable due to the accuracy at detecting low *E. coli* concentrations provided by the method, which is important when sampling for *E. coli* outside of storm events or in land use areas less prone to fecal contamination.

Concurrent with the collection of water grab samples, five physicochemical variables were collected using a handheld multi-parameter water quality sonde (YSI Inc./Xylem Inc.) fitted with an Ion Selective Electrode (ISE) multi-probe [58]. Variables included water temperature (°C), dissolved oxygen (DO), specific conductance (SPC), pH, and Chloride ion. The ISE probe sensed water temperatures ranging from −5 °C to 70 °C with an accuracy of ±0.2 °C, SPC ranging from 0 to 200 mS cm^{-1} with an accuracy of ±0.5% of reading or ±.001 mS cm^{-1}, whichever is greater, (for readings 0–100 mS cm^{-1}) or ±1.0% of reading (for readings 100–200 mS cm^{-1}), DO ranging from 0 to 50 mg L^{-1}, with an accuracy of ±1% (for readings 0–20 mg L^{-1}) or ±8% (for readings 20–50 mg L^{-1}), pH ranging from 0 to 14 units, with an accuracy of ±0.2 units, and Chloride ranging from 0 to 1000 mg/L (at water temperatures from 0 to 40 °C) with an accuracy ±15% of reading or 5 mg/L (whichever is greater) [58].

2.3. Data Analysis

Descriptive statistics for *E. coli*, physicochemical variables and hydro-climate data aggregated to the study period were calculated. Statistical analyses were conducted using OriginPro 2019 (OriginLab Corporation). The Anderson Darling Test was used for normality testing [59]. Land use practices were reclassified into three major categories prior to analysis including mixed development, agriculture, and forested land use [6]. Mixed development included roads, impervious surfaces, mixed development, and barren areas. Agriculture comprised low vegetation, hay pasture, and cultivated crops. Forested land use constituted mine grass, forest, mixed mesophytic forest, dry mesic oak forest, dry oak (pine) forest, and small stream riparian habitats. Seasonal variation was analyzed by dividing annual data into four quarter data subsets, comprising all weekly samples collected in three-month blocks starting 1 January 2018. Consequently, quarter one included 2 January 2018–27 March 2018 (winter); quarter two included 3 April 2018–26 June 2018 (Spring); quarter three included 3 July 2018–25 September 2018 (summer); and quarter four included 2 October, 2018–1 January, 2019 (fall). Spearman correlation tests for both annual and quarterly datasets, with a significance threshold of $\alpha = 0.05$ [60], were used to analyze the relationship between *E. coli* concentration, physicochemical parameters, and land use practices at all twenty-two sites. Finally, the annual data and quarterly data subsets comprising *E. coli* concentrations, physicochemical parameters, and land use practices were analyzed using principal component analysis (PCA) (presented in biplots) across all 22 sampling locations.

3. Results and Discussion

Given the scope of this research and large subsequent dataset, the authors have included the most salient tables and figures in text to facilitate presentation of results and discussion. For the reader wishing to learn more, comprehensive descriptive statistics tables are provided in Appendix A (referenced throughout).

3.1. Climate during Study

West Run Watershed received approximately 20% more total precipitation in 2018 (1378 mm) than annual averages dating back to 2007 (1096 mm) [61]. October was the driest (47 mm) and September was the wettest (186 mm) months during the study period (Figure 2), with September receiving more than double the historic average precipitation (80 mm) [61]. Consequently, September included approximately 14% of the precipitation received during 2018. During 2018, average annual air temperature (12 °C) was close to the historic annual average (11 °C) [61]. The coldest average monthly temperature in WRW was recorded in January (−4 °C), whereas July had the warmest average monthly temperature (22 °C) [61]. Average annual relative humidity was 76% during 2018 and was, therefore, characteristically high (Figure 2), as is common in the region [61]. Ultimately, the climate in WRW during 2018 was predictably variable and consistent with historic trends. As is characteristic of WRW, and the region, 2018 did not include a dry season; however, quarters two and three (spring and summer) received more precipitation than quarters one and four (winter and fall). This was the result of large precipitation events during quarters two/spring (e.g., 6 May; 24 mm) and three/ summer (e.g., 9 September; 60 mm) (Figure 2) [61].

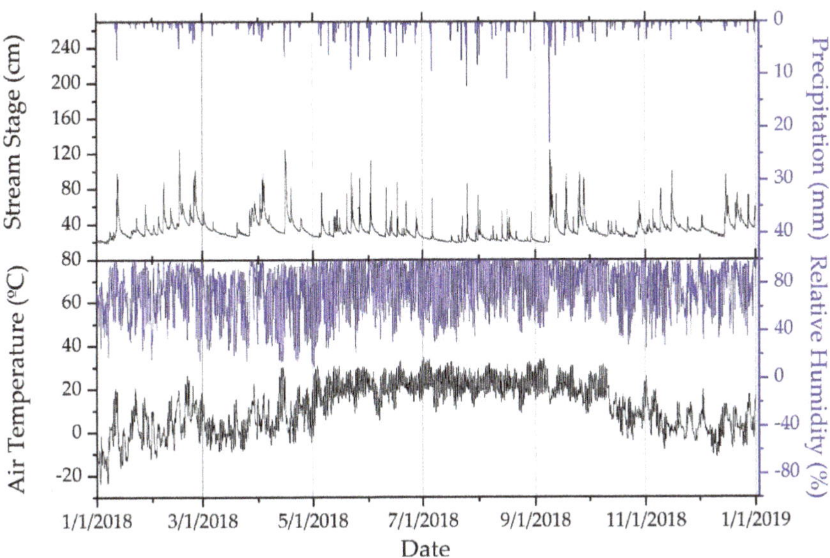

Figure 2. Thirty-minute time series of climate variables during study period (2 January 2018–1 January 2019) in West Run Watershed, West Virginia, USA. Note: Stream stage was monitored in the primary stream of WRW, West Run Creek, at site #13 (Figure 1).

3.2. *E. Coli Concentrations*

Sub-catchment #16, comprising predominantly agricultural land use practices (59%) had the highest *E. coli* concentration (560 CFU per 100 mL) during the current investigation, similar to previous inventions in WRW [6,18]. This sub-catchment also had the highest median concentration (575 CFU per 100 mL) (Figure 3; Appendix A Table A1). These results are supported by previous work that reported significant correlations ($p < 0.04$) between agricultural land use and *E. coli* concentrations [6] and increased fecal contamination in agricultural areas [21,27]. Conversely, forested land use areas (site #2; 74% Forested and site #5; 51% Forested) comprised the lowest *E. coli* concentrations, including the lowest median (3 CFU per 100 mL), and average (34 CFU per 100 mL), respectively (Figure 3; Appendix A Table A1). The low *E. coli* concentrations recorded at these two sites, and others including sites #7, #8, and #9 located in the headwaters of WRW (Figure 1) were impacted by acid mine drainage (AMD), which lowers the pH of receiving waters [62], potentially killing fecal bacteria [63]. For a more comprehensive discussion of pH, the reader is referred to the physicochemical parameters Section 3.3. Under approximately average pH conditions, forested land use areas had lower *E. coli* concentrations (e.g., site # 17: 86% forested average (avg.) 206 CFU per 100 mL) relative to areas comprising either agriculture (e.g., site # 16: 59% agriculture, avg. 560 CFU per 100 mL) or mixed development (e.g., site # 20: 89% mixed development, avg. 415 CFU per 100 mL) (Figure 3). In previous investigations, decreased fecal contamination was recorded in forested areas [21], and attributed to increased receiving water quality [64], thereby supporting the results of the current investigation. The motivation for sampling between storm events, is strengthened by the low average *E. coli* concentrations (specifically in the headwaters) recorded during the study period (2 January 2018—1 January 2019). Results support previous investigations that showed increased *E. coli* concentrations in agricultural areas (e.g., site # 16) and decreased *E. coli* concentrations in forested areas (e.g., site # 17). However, no published investigations included sampling over a full annual year using such a high spatial and temporal resolution. Results here are therefore unique, comprehensive, and may increase confidence in previous study outcomes.

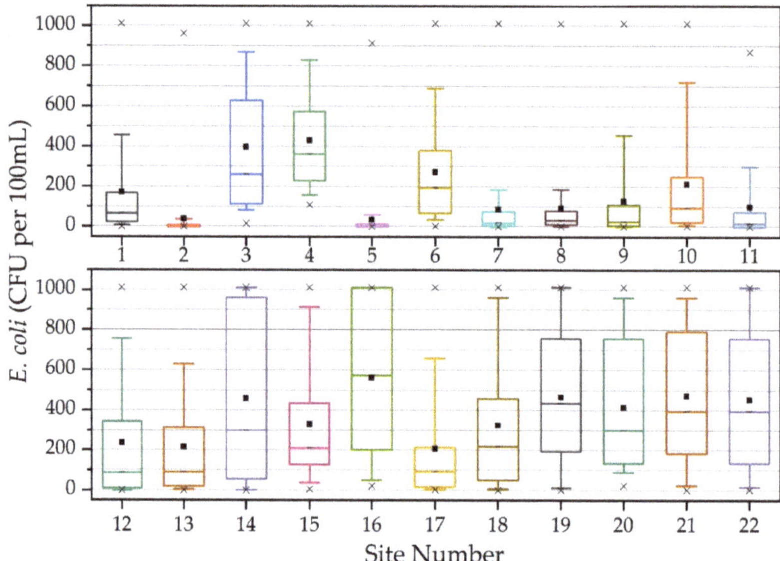

Figure 3. Box and whisker plot of *E. coli* concentration (CFU per 100 mL) descriptive statistics at 22 sampling locations during study period (2 January 2018–1 January 2019) in West Run Watershed, Morgantown, WV, USA. Box delineates 25th and 75th percentiles; line denotes median; square shows mean; whisker describes 10th and 90th percentiles; x shows maximum and minimum when above and below, respectively.

3.3. Physicochemical Parameters

Mixed development land use (e.g., site #20: 89% mixed development) had elevated annual average water temperatures (avg. 12.88 °C) relative to agriculture (e.g., site #16: 59% agriculture, avg. 12.00 °C) and forested land use practices (e.g., site #17: 86% forested, avg. 11.67 °C) (Figure 4; Appendix A Table A2). Previous investigations linking elevated receiving water temperature to mixed development areas support the results of the current study [65–68]. Elevated water temperatures in mixed development areas are typically attributed to decreased vegetation (reduced stream shading) and warmer impervious surfaces (e.g., road and building surfaces) and increased surface runoff (volume and temperature) during storm events [65–68]. During the current investigation water temperatures were lower in forested areas (e.g., site #17: 86% forested), including a lower average (11.67 °C), median (10.40 °C), and maximum (21.70 °C) relative to other land uses (Figure 4; Appendix A Table A2), thereby supporting the results from previous work [69]. The study design and high spatial and temporal resolution-sampling regime of the current investigation resulted in an extensive water temperature and *E. coli* concentration dataset. This is important, given temperature has been identified as the primary factor influencing *E. coli* survival in the environment, accounting for up to 61% of the variance (based on inactivation rates) of *E. coli* populations [70]. Therefore, this analysis (see non-parametric results Sections 3.4 and 3.5) expands current understanding of land use practice and water temperature impacts on *E. coli* concentrations.

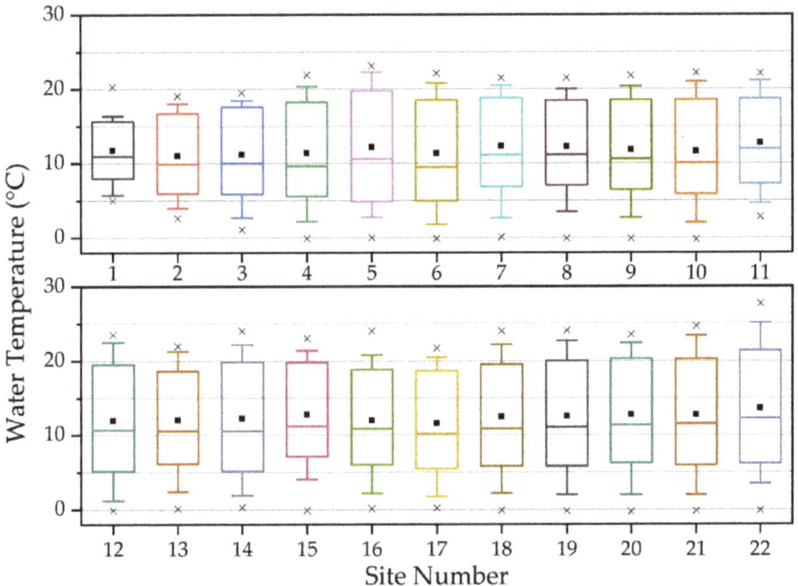

Figure 4. Box and whisker plot of water temperature (°C) descriptive statistics at 22 sampling locations during study period (2 January 2018–1 January 2019) in West Run Watershed, Morgantown, WV, USA. Box delineates 25th and 75th percentiles; line denotes median; square shows mean; whisker describes 10th and 90th percentiles; x shows maximum and minimum when above and below, respectively.

Previous work reported decreased pH in WRW, particularly in the headwaters, attributed to historic mining activities and subsequent AMD [6,18,46]. In the current work, approximately 55% of monitoring sites located in the headwaters had average pH values below six (Figure 5; Appendix A Table A3). The lowest pH was recorded in at site #8 with an annual average pH of 4.37 and median of 4.23. Monitoring sites #7, #8, and #9, (comprising one of the paired watersheds) had the lowest pH values among the sites (annual averages 5.03; 4.37; 5.08 and medians 5.04; 4.23; 4.93, respectively). Sampling locations in lower WRW displayed increased pH values, indicating a dilution of the AMD that was prevalent in the headwaters (no historic mining in lower reaches). For example, when comparing the annual average pH of two sites located on West Run Creek (site #13: 6.10 and #18: 7.18) there was a notable increase in pH (1.08) from acidic to neutral (Figure 5; Appendix A Table A3). Ultimately, the pH data recorded during the current investigation provides high sampling density evidence for legacy effects of historic mining practices. Notably, while mining activities in WRW ceased by 1977 [46]. Long-term effects of those practices were still impacting receiving waters at the time of the current study.

Acid mine drainage (AMD) impacted sites, including the paired watershed comprising sites #7, #8 and #9, generally included the highest specific conductance (SPC) data recorded during the study period (2 January 2018–1 January 2019). For example, site #8, which was the most heavily impacted by AMD (see preceding section) comprised the highest average (1661.13 µS/cm) and median (1480.00 µS/cm) SPC values (Figure 6; Appendix A Table A4). These findings are supported by previous work that reported elevated SPC in AMD impacted locations [71], including previous work investigating the influence of coal mining on conductivity of waters in Appalachia [72]. Increased SPC in AMD impacted areas can be attributed to increased iron, sulfate, copper, cadmium, arsenic, and/or other constituents in the water increasing ion availability [73]. Mixed development sub-catchments (sites #15: 70% mixed development and #20: 89% mixed development) also displayed elevated SPC data (averages of 1392.64 µS/cm and 1463.58 µS/cm, respectively). Moreover, these mixed development areas included the highest recorded maximum SPC (6631.00 µS/cm and 6106.00 µS/cm, respectively) (Figure 6; Appendix A Table A4). These results are supported by previous investigations reporting increased SPC in mixed development locations [74,75], attributed to increased ions in receiving water originating from diverse sources, including transportation, sewage treatment, and infrastructure development [75]. Moreover, the low SPC recorded at forested sites (e.g., site #17: 86% forested; average SPC 249.11 µS/cm), in the lower portion of WRW, during the current work is also supported by previous work that reported negative associations between SPC and forested land cover [76].

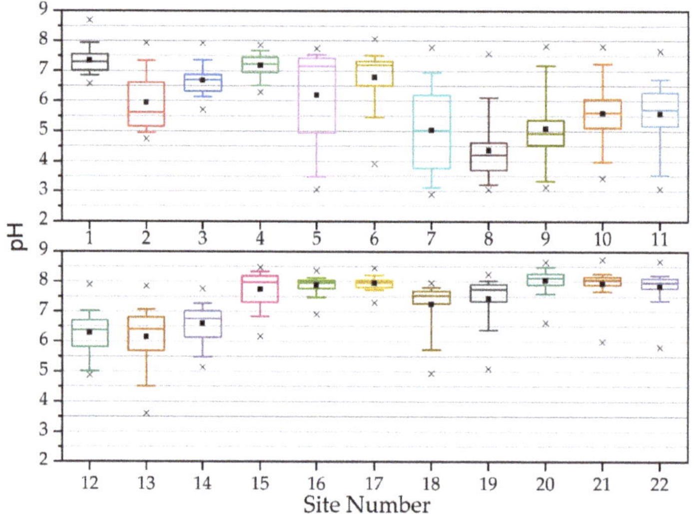

Figure 5. Box and whisker plot of pH descriptive statistics at 22 sampling locations during study period (2 January 2018–1 January 2019) in West Run Watershed, Morgantown, WV, USA. Box delineates 25th and 75th percentiles; line denotes median; square shows mean; whisker describes 10th and 90th percentiles; x shows maximum and minimum when above and below, respectively.

Dissolved oxygen (DO) was lowest at site #1, comprising an annual average of 85.93% and median of 85.10%, and the highest at site #20, with an average of 104.63% and median of 102.40% (Figure 7; Appendix A Table A5). Statistical analysis did not reveal a significant (CI = 0.05) relationship between DO and land use practices, attributable to unmeasured DO influencing factors. Variables beyond the scope of the current investigation included ground water depth and antecedent soil water conditions) [77], aquatic macrophytes [78], aquatic plant photosynthesis [79], and aquatic chemical, physical, and biochemical activities [80]. These factors can influence DO independent from changes to land use practices [77–80], thereby, obscuring the influence of land use changes on DO in receiving waters. However, increased DO variability was recorded at mixed development sub-catchments (site #15: 70% mixed development and site #20: 89% mixed development) in the lower portion of WRW (Figure 7). DO in streams can be impacted by urbanization through increased primary production or decomposition of organic matter [81]. Therefore, increased DO variability in mixed development areas can alter microbial community structures [82] of associated receiving waters, potentially indirectly affecting facultative anaerobic bacteria (e.g., *E. coli*) through inter specific competition [83].

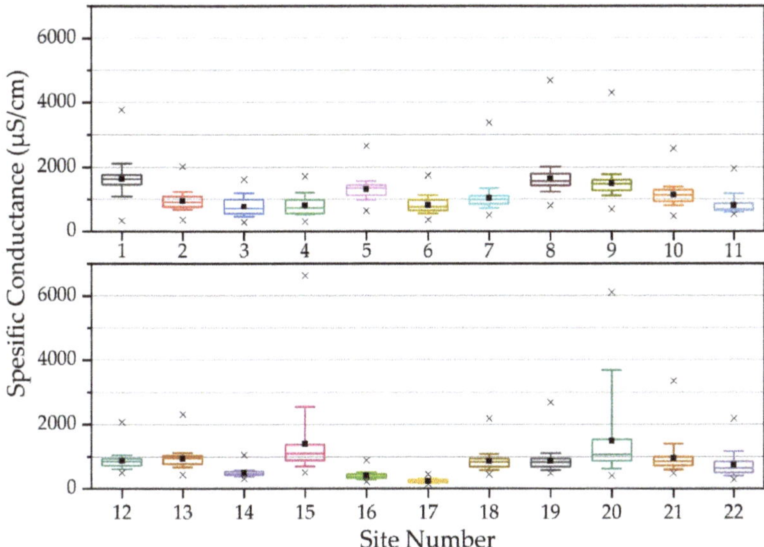

Figure 6. Box and whisker plot of specific conductance (µS/cm) descriptive statistics at 22 sampling locations during study period (2 January 2018–1 January 2019) in West Run Watershed, Morgantown, WV, USA. Box delineates 25th and 75th percentiles; line denotes median; square shows mean; whisker describes 10th and 90th percentiles; x shows maximum and minimum when above and below, respectively.

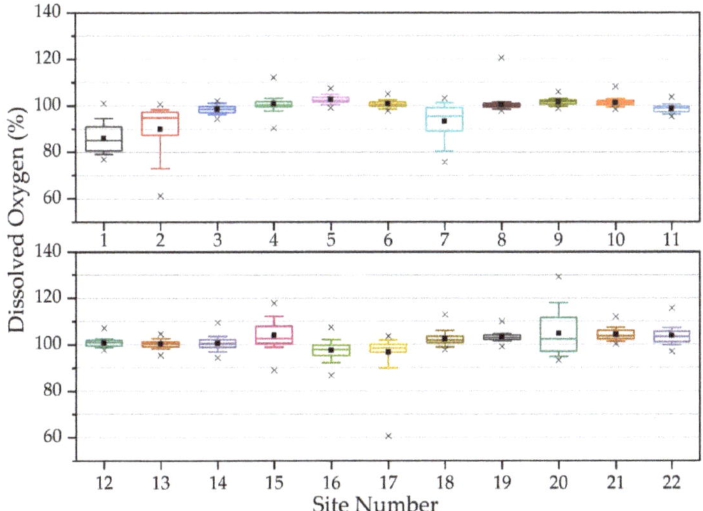

Figure 7. Box and whisker plot of dissolved oxygen (%) descriptive statistics at 22 sampling locations during study period (2 January 2018–1 January 2019) in West Run Watershed, Morgantown, WV, USA. Box delineates 25th and 75th percentiles; line denotes median; square shows mean; whisker describes 10th and 90th percentiles; x shows maximum and minimum when above and below, respectively.

Mixed development sub-catchments showed increased chloride ion (Cl^-) concentrations relative to other land use practices. For example, mixed development at site #1 (53%), site #15 (70%), and site #20 (89%) accounted for the highest chloride concentrations among the sampled locations (average concentrations 272.11 mg/L; 220.17 mg/L and 282.87 mg/L, respectively) (Figure 8; Appendix A Table A6). Previous work, using a similar study design, also reported increased chloride relative to increased mixed development land use practices [52]. The application of road salts in mixed development areas has been presented as a contributor to elevated chloride levels in these land use areas [52,84]. Forested land use areas (site #17: 86% forested) comprised the lowest chloride concentrations in WRW, including the lowest average (13.34 mg/L), median (11.79 mg/L), minimum (6.85 mg/L), maximum (39.82 mg/L), and lowest standard deviation (6.13 mg/L) (Figure 8; Appendix A Table A6). Previous work reported similar low(er) chloride concentrations in forested areas relative to other land use types, thereby validating results from the current investigation [52,84,85]. Given the study design and sampling regime, this work shows convincingly that the impact of chloride on microbe concentrations, including fecal microbes, will be increased in mixed development areas.

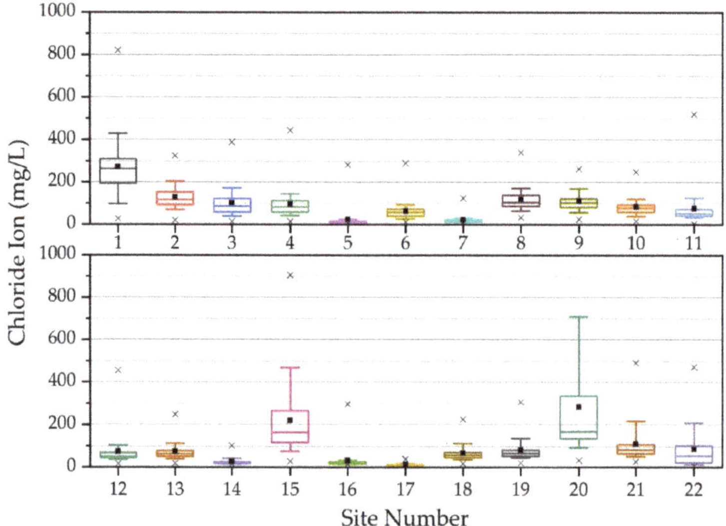

Figure 8. Box and whisker plot of chloride ion (mg/L) descriptive statistics at 22 sampling locations during study period (2 January 2018–1 January 2019) in West Run Watershed, Morgantown, WV, USA. Box delineates 25th and 75th percentiles; line denotes median; square shows mean; whisker describes 10th and 90th percentiles; x shows maximum and minimum when above and below, respectively.

3.4. Annual Non-Parametric Statistical Results

Normality testing showed that annual *E. coli* and physicochemical data were non-normally distributed. Therefore, Spearman's Correlations tests, the non-parametric equivalent of the Pearson's correlations tests [86], were used to investigate annual relationships between *E. coli* concentrations and physicochemical parameters at all 22 sampling locations (Table 2). Results showed that water temperature was significantly ($p < 0.05$) positively correlated with *E. coli* concentrations at 14 (64%) of the 22 sampling locations (Table 2). Six of eight West Run Creek sampling locations (75%) included significant correlations between *E. coli* concentration and water temperature. Despite previous work also reporting spatial and site specific variation regarding water temperature and *E. coli* concentrations [87], water temperature is historically regarded as the primary environmental variable influencing *E. coli* survival in the environment [70]. However, in WRW, pH was significantly negatively correlated ($p < 0.05$) to *E. coli* concentrations at 77% (17 of 22) of sampling locations. Therefore, given the presence of AMD in WRW [6,18,46], which can lower the pH of receiving waters [62] killing (or inactivating) fecal bacteria [63], pH exceeded the influence of water temperatures on *E. coli* concentrations. This is an important finding as it challenges traditional beliefs that temperature is the primary factor influencing the environmental survival of *E. coli*. Moreover, pH values displayed a tipping point (threshold) of between 7.68–7.76, with pH values below this range including significant correlations ($p < 0.05$) with decreased *E. coli* concentrations. Two West Run Creek sites (site #21 and #22) that had insignificant relationships between pH and *E. coli* concentrations were located near the terminus (confluence with the Monongahela River) of the watershed. At these sites, AMD was diluted to levels not influencing *E. coli* survival. Therefore, results also provide evidence for the dilution of AMD impacted waters and subsequent decreased impact on fecal microbe viability. Subsequently, *E. coli* concentrations could potentially be used to assess the freshwater health for aquatic organisms' sensitive to decreased pH and AMD, essentially serving as a bioindicator. Notably, SPC, which is known to be impacted by pH and AMD, displayed significant correlations ($p < 0.05$) with *E. coli* at three sites (#7, #8, and #9) which were particularly heavily impacted by AMD. Generally, SPC did not show consistent correlations with *E. coli* concentrations across the sampling locations, as only 55% (12 of the 22) sites displayed significant relationships. Therefore, based on the Spearman's Correlations tests from the current investigation, SPC was poorly related to *E. coli* CFU's in the current study.

Table 2. Results of Spearman's Correlation test, including annual *E. coli* concentration (colony forming units (CFU) per 100mL) water temperature (°C), pH, specific conductance (SPC; µS/cm), dissolved oxygen (DO; %) and chloride ion (Cl^-; mg/L) at each sampling location (n = 22) during study period (2 January 2018—1 January 2019) in West Run Watershed, WV, USA.

		Site Number										
		#1	**#2**	**#3**	**#4**	**#5**	**#6**	**#7**	**#8**	**#9**	**#10**	**#11**
Water Temp.	SCC	**0.35**	**0.59**	0.20	0.19	**0.75**	**0.37**	0.12	0.26	0.20	**0.50**	**0.52**
	p-value	**0.01**	**0.00**	0.18	0.19	**0.00**	**0.01**	0.43	0.08	0.18	**0.00**	**0.00**
pH	SCC	**0.36**	**0.47**	**0.54**	**0.31**	**0.69**	**0.44**	**0.55**	**0.46**	**0.55**	**0.70**	**0.81**
	p-value	**0.01**	**0.00**	**0.00**	**0.03**	**0.00**	**0.00**	**0.00**	**0.00**	**0.00**	**0.00**	**0.00**
SPC	SCC	**−0.48**	−0.23	−0.17	−0.18	−0.16	−0.16	**−0.80**	**−0.51**	**−0.55**	−0.14	**−0.41**
	p-value	**0.00**	0.11	0.24	0.21	0.27	0.27	**0.00**	**0.00**	**0.00**	0.33	**0.00**
DO	SCC	**0.31**	−0.14	−0.18	−0.24	−0.01	**−0.39**	−0.16	−0.15	**−0.37**	**−0.40**	**−0.49**
	p-value	**0.03**	0.35	0.21	0.10	0.93	**0.01**	0.27	0.32	**0.01**	**0.00**	**0.00**
Cl^-	SCC	−0.14	−0.20	−0.24	−0.26	−0.16	−0.21	**−0.35**	**−0.45**	−0.28	−0.12	**−0.48**
	p-value	0.34	0.18	0.09	0.07	0.28	0.16	**0.01**	**0.00**	0.05	0.43	**0.00**

		Site Number										
		#12	**#13**	**#14**	**#15**	**#16**	**#17**	**#18**	**#19**	**#20**	**#21**	**#22**
Water Temp.	SCC	**0.71**	**0.66**	**0.71**	**0.37**	0.13	**0.69**	**0.56**	**0.46**	0.05	**0.33**	0.24
	p-value	**0.00**	**0.00**	**0.00**	**0.01**	0.37	**0.00**	**0.00**	**0.00**	0.73	**0.02**	0.10
pH	SCC	**0.62**	**0.74**	**0.69**	**−0.35**	−0.14	0.16	**0.56**	**0.57**	−0.08	0.22	0.25
	p-value	**0.00**	**0.00**	**0.00**	**0.02**	0.36	0.30	**0.00**	**0.00**	0.61	0.14	0.09
SPC	SCC	−0.16	**−0.38**	−0.06	**−0.70**	0.21	**0.30**	**−0.30**	−0.12	**−0.50**	**−0.29**	**−0.33**
	p-value	0.28	**0.01**	0.67	**0.00**	0.16	**0.04**	**0.04**	0.43	**0.00**	**0.05**	**0.03**
DO	SCC	**−0.41**	**−0.56**	0.00	**−0.73**	**−0.47**	**−0.58**	−0.22	−0.20	−0.25	−0.06	**−0.31**
	p-value	**0.00**	**0.00**	0.97	**0.00**	**0.00**	**0.00**	0.14	0.17	0.09	0.68	**0.03**
Cl^-	SCC	**−0.57**	**−0.41**	−0.23	**−0.52**	0.16	−0.15	**−0.44**	−0.27	**−0.37**	**−0.41**	**−0.30**
	p-value	**0.00**	**0.00**	0.13	**0.00**	0.27	0.33	**0.00**	0.07	**0.01**	**0.00**	**0.04**

Note: bold values indicate significant correlations ($p < 0.05$).

Dissolved oxygen (DO) lacked consistent correlations with *E. coli* concentrations, as only 55% (12 out of 22) of sites comprised significant correlations (Table 2). Furthermore, no significant relationship between land use practices and *E. coli* concentrations and DO were found. Previous investigations showed that facultative anaerobic characteristics of *E. coli* decrease its dependence on oxygen for survival [88], which may account for these results (e.g., fish) [89]. Spearman's correlation results between *E. coli* and chloride concentrations showed consistent (with the exception of site #1) significant negative correlations in sub-catchments comprising mixed development land uses in excess of 25% to 30% (Figure 1; Table 2). Analyses also identified that if mixed development land use is less than 25% to 30%, chloride is less likely to influence *E. coli* concentrations in receiving waters. Notably, this tipping point should not be interpreted to imply that lower concentrations are ecologically benign. The insignificant correlation at site #1 may be a function of the relatively small drainage of this sub-catchment (0.30 km^2) and shorter stream distance, relative to other larger catchments with mixed development land use practices in excess of 25–30%. Previous work investigating the influence of mixed development, specifically urban, land use on chloride concentrations reported tipping points approaching 25%; thus, supporting the results from the current investigation [84]. Of importance, the distinct data set of the current work facilitated advanced understanding of the influence of land use practices on *E. coli* concentrations by Cl^- (possible attribute of winter road salting) that may suppress (inactivate or kill) fecal bacteria in receiving waters.

Principal component analysis (PCA) can be implemented to determine which explanatory variables account for the maximal variance in a data set, through the computation of multiple principal components and their respective Eigenvalues [90]. A principal component is defined as a linear function of original data set variables, which maximize variance and is uncorrelated with other principal components [91]. Eigenvalues are used to identify principal components based on the assumption that components comprising the highest Eigenvalues will constitute principal components as Eigenvalues symbolize the variance of the data in that direction [90]. Given most data cannot be accurately described by a single principal component, numerous principal components are typically calculated and ranked based on their Eigenvalues [90]. For the current work, annual PCA results displayed three principal components with Eigenvalues exceeding 1 (an accepted threshold of importance [18,92]). The three principal components comprised Eigenvalues of 2.64, 1.99, and 1.33, respectively, and combined accounted for 66.27% of the variance on the annual data (Table 3). Consequently, the remaining six principal components accounted for only 33.73% of the variance of the data, of which principal component four accounted for 10.91% of the variance. Notably, principal component 4 comprised an Eigenvalue of 0.98, very close to the threshold of importance. For more comprehensive list of the coefficients of the variables comprising the three identified important principal components of the annual PCA please see Appendix A Table A7).

Table 3. Results of principal component analysis comprising 9 variables (*E. coli* concentration, water temperature, pH, SPC, DO, chloride, percentage of agricultural land use, percentage of forested land use, and percentage of developed land use) used to define 9 principal components, displaying eigenvalues, percentage of variance, and cumulative variance during the study period (2 January 2018–1 January 2019) across the 22 monitoring sites in West Run Watershed, West Virginia, USA.

Principal Component	Eigenvalue	Percentage of Variance	Cumulative Variance
1	**2.64**	29.34%	29.34%
2	**1.99**	22.14%	51.48%
3	**1.33**	14.79%	66.27%
4	0.98	10.91%	77.18%
5	0.79	8.80%	85.98%
6	0.61	6.73%	92.71%
7	0.44	4.94%	97.65%
8	0.21	2.35%	100.00%
9	0.00	0.00%	100.00%

Note: bold numbers indicate eigenvalues exceeding 1 (representing importance).

The Annual PCA biplot (Figure 9) compliments the Spearman' correlation results for pH and *E. coli* concentration. Water temperature was also closely correlated to *E. coli* concentrations based on annual PCA results. Both water temperature and pH has been reported to be closely correlated with fecal bacteria concentrations in previous work [9,10,12], thereby supporting results of the current investigation. Historic land use in WRW, specifically mining, influences the pH in the watershed, which is closely correlated with *E. coli* concentrations (Table 2; Figure 9). Moreover, water temperature, which is influenced by land use practices [65–68] is also closely correlated with *E. coli* concentrations. Therefore, annual PCA biplot results emphasize the influence of both historic and contemporary land use practices on fecal bacteria in receiving waters. The biplots of annual land use practices, physicochemical parameters, and *E. coli* concentration relationships facilitates visual assessment of land use impacts on physicochemical parameters, which influences fecal microbes. For example, forested land use showed a negative correlation with water temperature and *E. coli* concentrations (Figure 9) by reduced solar radiation reaching the stream [93]. The decreased water temperatures may suppress *E. coli* concentrations, as the microbe is sensitive to temperature changes [94]. The influence of other physicochemical parameters (e.g., chloride) on *E. coli* concentrations were overshadowed by temperature and pH (Figure 9). Therefore, study results (based on annual average values) highlight temperature and pH as priority factors influencing *E. coli* in the receiving waters of WRW. Notably in watersheds with more neutral pH values and decreased legacy land use impacts (e.g., mining), the influence of physicochemical parameters on *E. coli* concentrations may be different [95]. In the current work, biplot results did not display a strong negative correlation between chloride and *E. coli* (Figure 9). Ultimately, results from the current study highlight legacy land use impacts (mining and AMD) as important considerations regarding microbial water quality management.

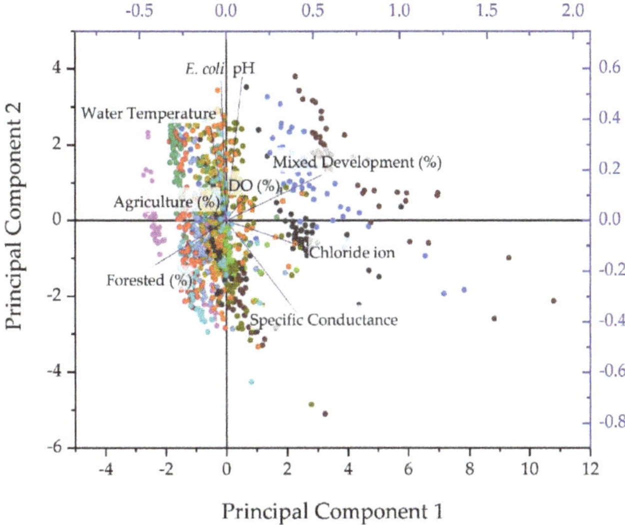

Figure 9. Results of principal components analysis, including biplots, for extracted principal components of annual *E. coli* concentration (CFU per 100 mL), water temperature (°C), pH, specific conductance (μS/cm), dissolved oxygen (DO; %), and chloride ion (mg/L) at 22 monitoring sites (indicated by the different colors) during study period (2 January 2018–1 January 2019) in West Run Watershed, West Virginia, USA.

3.5. Quarterly Non-Parametric Statistical Results

Quarterly PCA results displayed varying Eigenvalues between quarters. Quarter one comprised three principal components, which accounted for 70% of the cumulative data variance (Table 4). Conversely, quarters two, three, and four included four principal components that accounted for 81%, 84%, and 80% for the cumulative variance, respectively (Table 4). The remaining six principal components of quarter one and five, principal components of quarter two to four, did not comprise eigenvalues denoting importance, accounted for only 30%, 19%, 16%, and 20% of the data variance in their respective quarters (Table 4). Appendix A (Tables A8–A11) includes a more thorough presentation of the coefficients comprising the principal components of the PCA for all four quarters.

Table 4. Results of principal component analysis comprising 9 variables (*E. coli* concentration, water temperature, pH, SPC, DO, chloride, percentage of agricultural land use, percentage of forested land use and percentage of developed land use) used to define 9 principal components, displaying eigenvalues, percentage of variance and cumulative variance during quarter one (winter: 2 January 2018–27 March 2018); quarter two (spring: 3 April 2018–26 June 2018); quarter three (summer: 3 July, 2018–25 September, 2018); and quarter four (fall: 2 October 2018–1 January 2019) across the 22 monitoring sites in West Run Watershed, West Virginia, USA.

Principal Component	Eigenvalue	Percentage of Variance	Cumulative Variance	Eigenvalue	Percentage of Variance	Cumulative Variance
	Quarter 1				Quarter 2	
1	**3.36**	37.31%	37.31%	**2.81**	31.22%	31.22%
2	**1.53**	16.95%	54.26%	**1.98**	21.97%	53.19%
3	**1.44**	16.00%	70.25%	**1.41**	15.63%	68.82%
4	0.92	10.22%	80.48%	**1.06**	11.74%	80.56%
5	0.73	8.13%	88.60%	0.85	9.39%	89.95%
6	0.53	5.94%	94.54%	0.48	5.36%	95.32%
7	0.26	2.94%	97.48%	0.30	3.38%	98.70%
8	0.23	2.52%	100.00%	0.12	1.30%	100.00%
9	0.00	0.00%	100.00%	0.00	0.00%	100.00%
	Quarter 3				Quarter 4	
1	**2.89**	32.07%	32.07%	**2.74**	30.39%	30.39%
2	**2.14**	23.80%	55.88%	**1.87**	20.73%	51.13%
3	**1.34**	14.91%	70.79%	**1.47**	16.30%	67.43%
4	**1.16**	12.92%	83.71%	**1.09**	12.14%	79.57%
5	0.58	6.44%	90.15%	0.79	8.83%	88.40%
6	0.39	4.29%	94.44%	0.57	6.38%	94.78%
7	0.31	3.49%	97.93%	0.29	3.22%	97.99%
8	0.19	2.07%	100.00%	0.18	2.01%	100.00%
9	0.00	0.00%	100.00%	0.00	0.00%	100.00%

Note: bold numbers indicate eigenvalues exceeding 1 (representing importance).

Quarterly PCA biplots (Figure 10) show the predominant influence of pH (presumably AMD driven) on fecal bacteria concentrations in WRW, as pH was closely correlated to *E. coli* concentrations during all four quarters of the study period (2 January 2018–1 January 2019). Conversely, the relationships between *E. coli* concentration and other physicochemical parameters varied between quarters. For example, water temperature was closely correlated with *E. coli* concentrations during quarters two and three, but not during quarter's one and two (Figure 10). The changing relationship between *E. coli* and water temperature is attributable to seasonal changes in air temperature and water temperatures [96]. Colder temperatures are known to suppress *E. coli* concentrations in receiving waters [94]. Thus, temporal and seasonal fluctuations of physicochemical parameters constitute important considerations regarding fecal concentration regimes (and mitigation practices) in receiving waters. The relationships between *E. coli* concentrations and land use practices also lacked consistency (Figure 10). This may have been due to the consistent close correlation between *E. coli* concentrations and pH. Legacy land use practices could be further confounding the influence of contemporary land use practices. Additionally, seasonal variation in land use practices (e.g., application of road salts during winter months) may contribute to the lack of consistent relationships. The high spatial temporal sampling regime and experimental watershed study design in conjunction with quarterly PCA biplots provided new (high-resolution) insight, emphasizing the complexity of temporal changes in *E. coli* concentration, physicochemical parameters, and land use practices. Results can be used by land use managers to inform water quality management strategies during different quarters (seasons), thereby improving efficacy. For example, focusing management strategies, including limiting livestock stream crossings through temporary fencing [97] and irrigation management [98] in quarter one in agricultural areas may be more effective in reducing *E. coli* concentrations than focusing on quarter four in some geographic locations.

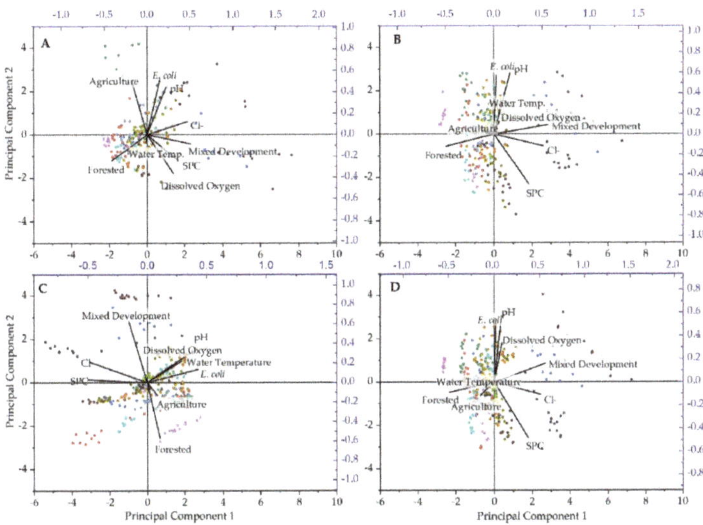

Figure 10. Results of principal components analysis, including biplots, for extracted principal components of quarterly *E. coli* concentration (CFU per 100mL), water temperature (°C), pH, specific conductance (μS/cm), dissolved oxygen (%) and chloride ion (mg/L) at 22 monitoring sites (indicated by the different colors) during study period (2 January 2018–1 January 2019) in West Run Watershed, West Virginia, USA. Note: (**A**) represents quarter one (winter: 2 January 2018–27 March 2018); (**B**) represents quarter two (spring: 3 April 2018–26 June 2018); (**C**) represents quarter three (summer: 3 July 2018–25 September 2018); (**D**) represents quarter four (fall: 2 October 2018–1 January 2019).

3.6. Study Implications and Future Directions

The scale nested experimental watershed study design, calendar-year sampling period, and high spatial and temporal sampling regime used in this work allowed for the identification of legacy land use impacts on *E. coli* concentrations in receiving waters. Legacy land use impacts, specifically mining and subsequent AMD, were identified major influencers of *E. coli* concentrations (and by extension microbial water quality) in receiving waters, even exceeding the influence of water temperature, commonly regarded as the primary factor influencing the environmental survival of *E. coli*. Additionally, the 7.68–7.76 pH tipping point identified regarding the significant correlations between pH and *E. coli* concentrations may indicate the use of *E. coli* as a potential bioindicator species for assessing the freshwater health, specifically in AMD impacted streams. Study results clearly identify legacy land use impacts of mining activity as a major influencer of microbial water quality. A threshold (tipping point) of 25–30% was identified regarding mixed development land use practices and significant ($p < 0.05$) negative correlations between *E. coli* and chloride concentrations. Increased chloride in the receiving waters of mixed development areas has been attributed to road salting [52,84]. This work shows that road salting in mixed development areas exceeding 25–30% total area may impact microbial water quality, through the suppression of *E. coli* concentrations. Future work should include the implementation of similar study design in physiographically dissimilar areas, including simultaneous implementation across different watersheds. This would allow for the comparison of data from climatically distinct regions, further improving understanding regarding physicochemical and land use impacts on *E. coli* regimes. Replication of the experimental watershed study design in areas not effected by AMD could potentially result in the identification of other tipping points regarding land use practices and physicochemical and *E. coli* relationships. For example, the influence of pH in the current work may have obscured the influence of other physicochemical parameters. Future investigations should be undertaken that includes multi-year sampling to better account for annual climate variability.

4. Conclusions

Fecal bacteria concentrations were investigated in a mixed-land-use watershed in the Appalachian region of the eastern United States, using a 22-site nested scale experimental watershed study design. Specific focus was given to the relationships between *E. coli* concentrations, physicochemical parameters (water temperature, pH, SPC, DO, chloride) and land use practices. In the study watershed, there was an approximate 112% increase in *E. coli* concentrations from the AMD impacted headwaters (avg. 177 CFU per 100mL) to the lower portion of the watershed (avg. 376 CFU per 100mL), an approximate 7 Km stream distance. Study results highlight the legacy impacts of historic mining (acid mine drainage) on *E. coli* concentrations, as Spearman correlation test results showed significant correlation ($p < 0.05$) between pH and *E. coli* concentrations at 77% of sample sites. Moreover, a pH tipping point (threshold) in the range of 7.68–7.76 was identified in the current investigation, with pH values below this range including significant correlations ($p < 0.05$) with *E. coli* concentrations. Consequently, pH values in receiving waters below the 7.68–7.76 tipping point will start significantly impacting (decreasing) active *E. coli* concentrations. Furthermore, a land cover tipping point of 25–30% was identified for mixed development land use practices and significant ($p < 0.05$) negative correlations between *E. coli* and chloride concentrations. Therefore, study results indicate that in areas comprising mixed development in excess of 25% to 30%, the application of road salts may suppress fecal bacteria in receiving waters. The importance of seasonal variability on fecal concentrations in receiving waters was illustrated by temporal variability in quarterly PCA biplots of *E. coli* concentrations, physicochemical parameters, and land use practices. The current work advances understanding of land use practice (both historic and current) and physicochemical parameter influences on *E. coli* concentrations in contemporary mixed-land-use watersheds. Results will aid policy makers and land use managers in effective water quality management, in watersheds with fecal contamination challenges.

Author Contributions: For the current work author contributions were as follows: conceptualization, J.A.H.; methodology, J.A.H.; formal analysis, F.P. and J.A.H.; investigation, F.P. and J.A.H.; resources, J.A.H.; data curation, J.A.H.; writing—original draft preparation, F.P. and J.A.H.; writing—review and editing, J.A.H. and F.P.; visualization, F.P. and J.A.H.; supervision, J.A.H.; project administration, J.A.H.; funding acquisition, J.A.H. All authors have read and agreed to the published version of the manuscript.

Funding: This work was supported by the National Science Foundation under Award Number OIA-1458952, the USDA National Institute of Food and Agriculture, Hatch project accession number 1011536, and the West Virginia Agricultural and Forestry Experiment Station. Results presented may not reflect the views of the sponsors and no official endorsement should be inferred. The funders had no role in study design, data collection and analysis, decision to publish, or preparation of the manuscript.

Acknowledgments: Special thanks are due to many scientists of the Interdisciplinary Hydrology Laboratory (https://www.researchgate.net/lab/The-Interdisciplinary-Hydrology-Laboratory-Jason-A-Hubbart). The authors also appreciate the feedback of anonymous reviewers whose constructive comments improved the article.

Conflicts of Interest: The authors declare no conflict of interest for the current work.

Appendix A

Table A1. Descriptive statistics of *E. coli* concentration (CFU per 100 mL) at each sampling location (n = 22) during study period (2 January 2018–1 January 2019) in West Run Watershed, WV, USA. Avg. = average, Med = median, Min. = minimum, Max. = maximum and Std. Dev. = standard deviation.

	Site Number										
	#1	#2	#3	#4	#5	#6	#7	#8	#9	#10	#11
Avg.	170	38	397	429	34	269	84	89	127	210	98
Med.	66	3	260	361	4	194	20	32	25	93	16
Min.	0	0	15	107	0	2	0	0	0	3	0
Max.	1011	961	1011	1011	914	1011	1011	1011	1011	1011	870
Std. Dev.	251	139	315	249	129	276	179	180	241	273	202
	Site Number										
	#12	#13	#14	#15	#16	#17	#18	#19	#20	#21	#22
Avg.	234	215	457	330	560	206	324	466	415	471	452
Med.	88	91	299	211	575	93	218	436	299	397	397
Min.	0	0	0	5	22	3	0	1	23	2	3
Max.	1011	1011	1011	1011	1011	1011	1011	1011	1011	1011	1011
Std. Dev.	305	266	406	293	373	288	342	339	340	342	345

Table A2. Descriptive statistics of water temperature (°C) at each sampling location (n = 22) during study period (2 January 2018–1 January 2019) in West Run Watershed, WV, USA. Avg. = average, Med = median, Min. = minimum, Max. = maximum and Std. Dev. = standard deviation.

	Site Number										
	#1	#2	#3	#4	#5	#6	#7	#8	#9	#10	#11
Avg.	11.77	11.09	11.27	11.48	12.28	11.44	12.36	12.32	11.88	11.72	12.84
Med.	11.00	9.90	10.20	9.80	10.85	9.70	11.40	11.30	10.60	10.10	12.80
Min.	5.00	2.60	1.10	−0.10	0.00	−0.10	0.10	−0.10	−0.10	−0.20	2.80
Max.	20.30	19.10	19.50	21.90	23.10	22.10	21.50	21.50	21.80	22.20	22.10
Std. Dev.	4.20	5.51	6.15	6.95	7.73	7.21	6.65	6.28	6.72	7.02	6.22
	Site Number										
	#12	#13	#14	#15	#16	#17	#18	#19	#20	#21	#22
Avg.	12.10	12.09	12.34	12.86	12.00	11.67	12.53	12.64	12.88	12.87	13.74
Med.	10.70	10.70	10.60	11.40	11.10	10.40	10.90	11.10	11.80	11.60	13.05
Min.	−0.10	0.10	0.30	−0.10	0.10	0.20	−0.10	−0.20	−0.30	−0.20	−0.10
Max.	23.50	22.00	24.00	23.00	24.00	21.70	24.00	24.10	23.50	24.60	27.70
Std. Dev.	7.87	6.91	7.66	6.89	6.99	6.99	7.44	7.72	7.76	7.88	8.61

Table A3. Descriptive statistics of pH at each sampling location (n = 22) during study period (2 January 2018–1 January 2019) in West Run Watershed, WV, USA. Avg. = average, Med = median, Min. = minimum, Max. = maximum and Std. Dev. = standard deviation.

	Site Number										
	#1	#2	#3	#4	#5	#6	#7	#8	#9	#10	#11
Avg.	7.33	5.94	6.68	7.17	6.19	6.79	5.03	4.37	5.08	5.60	5.56
Med.	7.30	5.62	6.70	7.24	7.17	7.19	5.04	4.23	4.93	5.62	5.70
Min.	6.57	4.74	5.71	6.29	3.05	3.92	2.89	3.05	3.13	3.43	3.08
Max.	8.69	7.93	7.92	7.85	7.74	8.06	7.78	7.58	7.83	7.81	7.66
Std. Dev.	0.44	0.93	0.51	0.42	1.63	0.90	1.43	1.04	1.24	1.08	1.08
	Site Number										
	#12	#13	#14	#15	#16	#17	#18	#19	#20	#21	#22
Avg.	6.13	6.10	6.54	7.68	7.80	7.86	7.18	7.38	7.97	7.86	7.76
Med.	6.38	6.38	6.75	7.97	7.96	7.96	7.51	7.72	8.11	8.06	7.96
Min.	0.25	3.61	4.01	4.30	4.23	4.00	3.99	4.14	4.27	4.32	4.27
Max.	7.89	7.85	7.76	8.47	8.36	8.45	7.95	8.23	8.65	8.73	8.67
Std. Dev.	1.13	1.07	0.77	0.76	0.58	0.59	0.91	0.92	0.66	0.70	0.72

Table A4. Descriptive statistics of water specific conductance (μS/cm) at each sampling location (n = 22) during study period (2 January 2018–1 January 2019) in West Run Watershed, WV, USA. Avg. = average, Med = median, Min. = minimum, Max. = maximum and Std. Dev. = standard deviation.

	Site Number										
	#1	#2	#3	#4	#5	#6	#7	#8	#9	#10	#11
Avg.	1639.28	958.51	778.70	810.22	1320.75	827.24	1053.28	1661.13	1497.19	1143.77	812.42
Med.	1643.00	918.00	718.00	727.00	1353.00	769.00	999.00	1562.00	1480.00	1129.00	695.00
Min.	331.60	358.10	286.20	301.80	633.00	354.30	503.00	813.00	688.00	476.60	530.00
Max.	3778.00	2024.00	1620.00	1709.00	2654.00	1745.00	3378.00	4692.00	4302.00	2572.00	1955.00
Std. Dev.	484.90	301.93	305.86	303.74	307.58	263.52	404.69	555.53	496.47	349.67	284.72
	Site Number										
	#12	#13	#14	#15	#16	#17	#18	#19	#20	#21	#22
Avg.	885.06	961.58	503.58	1392.64	413.31	249.11	868.99	877.08	1463.58	945.33	738.12
Med.	861.00	958.00	480.40	1116.00	385.00	226.40	845.00	831.00	1045.00	850.00	651.00
Min.	516.00	449.50	321.50	510.00	231.60	144.40	451.60	494.00	408.90	486.50	297.60
Max.	2072.00	2315.00	1052.00	6631.00	890.00	449.90	2184.00	2690.00	6106.00	3345.00	2185.00
Std. Dev.	279.25	292.86	111.90	1041.80	120.58	68.43	306.89	355.26	1193.23	458.84	359.11

Table A5. Descriptive statistics of dissolved oxygen (%) at each sampling location (n = 22) during study period (2 January 2018–1 January 2019) in West Run Watershed, WV, USA. Avg. = average, Med = median, Min. = minimum, Max. = maximum and Std. Dev. = standard deviation.

	Site Number										
	#1	#2	#3	#4	#5	#6	#7	#8	#9	#10	#11
Avg.	85.93	89.95	98.47	100.78	102.52	100.62	93.20	100.38	101.42	101.16	98.50
Med.	85.10	94.90	98.50	100.50	102.25	100.60	95.00	100.10	101.70	101.30	98.90
Min.	76.90	61.30	93.20	90.40	97.10	95.20	75.70	94.60	94.80	95.30	92.70
Max.	101.10	100.70	102.00	112.10	107.40	105.00	103.20	120.50	106.00	108.10	103.60
Std. Dev.	6.02	10.04	1.97	3.48	1.92	1.68	7.64	3.14	1.60	1.73	1.89
	Site Number										
	#12	#13	#14	#15	#16	#17	#18	#19	#20	#21	#22
Avg.	100.76	100.31	100.39	103.78	97.40	96.68	102.30	103.09	104.63	104.22	103.68
Med.	100.90	100.50	100.50	102.70	97.70	98.50	102.00	103.10	102.40	103.80	103.50
Min.	94.00	92.90	93.40	89.10	86.80	60.70	93.80	96.90	93.40	96.70	94.80
Max.	107.20	104.60	109.50	117.80	107.40	103.70	112.90	110.00	129.30	112.00	115.60
Std. Dev.	2.08	2.09	3.05	5.56	3.97	6.60	3.13	1.78	9.06	2.74	3.80

Table A6. Descriptive statistics of chloride ion (mg/L) at each sampling location (n = 22) during study period (2 January 2018–1 January 2019) in West Run Watershed, WV, USA. Avg. = average, Med = median, Min. = minimum, Max. = maximum and Std. Dev. = standard deviation.

	Site Number										
	#1	#2	#3	#4	#5	#6	#7	#8	#9	#10	#11
Avg.	272.11	128.83	101.63	95.87	23.41	62.49	22.53	119.36	109.73	84.04	76.27
Med.	263.40	118.25	86.62	83.73	12.89	57.47	18.25	105.28	102.46	78.75	52.46
Min.	27.08	19.78	13.76	14.31	4.97	12.14	6.38	32.65	24.29	16.35	3.39
Max.	821.26	325.24	388.96	444.44	282.40	289.87	124.07	340.28	263.38	249.22	520.25
Std. Dev.	146.30	55.88	68.56	68.34	45.50	41.00	18.04	55.72	47.16	40.79	84.18
	Site Number										
	#12	#13	#14	#15	#16	#17	#18	#19	#20	#21	#22
Avg.	73.42	74.23	27.94	220.17	32.16	13.34	67.52	80.97	282.87	111.21	87.82
Med.	51.01	63.76	22.94	164.83	21.56	11.79	57.96	66.06	168.78	84.08	56.97
Min.	16.48	17.88	15.26	29.23	14.13	6.85	17.69	21.87	33.96	27.24	7.71
Max.	455.64	248.34	101.48	905.31	297.32	39.82	225.82	307.39	1242.68	491.53	472.78
Std. Dev.	74.83	45.41	17.91	176.53	44.33	6.13	40.02	54.18	260.03	95.67	99.41

Table A7. Coefficients of annual principal components comprising 9 variables (*E. coli* concentration, water temperature, DO, SPC, pH, chloride, percentage of mixed developed land use, percentage of agricultural land use and percentage of forested land use) used to define 9 principal components, during study period (2 January 2018–1 January 2019) in West Run Watershed, WV, USA.

Variables	Coefficients of PC1	Coefficients of PC2	Coefficients of PC3
E. coli	−0.03	0.55	0.00
Water Temp	−0.09	0.37	−0.14
DO	0.12	0.11	0.53
SPC	0.41	−0.37	−0.08
pH	0.09	0.57	0.06
Cl-	0.50	−0.12	−0.03
Mixed Development	0.55	0.18	0.08
Agriculture	−0.15	0.02	−0.73
Forested	−0.47	−0.20	0.40

Table A8. Coefficients of annual principal components comprising 9 variables (*E. coli* concentration, water temperature, DO, SPC, pH, chloride, percentage of mixed developed land use, percentage of agricultural land use and percentage of forested land use) used to define 9 principal components, during quarter one (winter: 2 January 2018–27 March 2018) in West Run Watershed, WV, USA.

Variables	Coefficients of PC1	Coefficients of PC2	Coefficients of PC3
E. coli	−0.03	0.55	0.00
Water Temp	−0.09	0.37	−0.14
DO	0.12	0.11	0.53
SPC	0.41	−0.37	−0.08
pH	0.09	0.57	0.06
Cl-	0.50	−0.12	−0.03
Mixed Development	0.55	0.18	0.08
Agriculture	−0.15	0.02	−0.73
Forested	−0.47	−0.20	0.40

Table A9. Coefficients of annual principal components comprising 9 variables (*E. coli* concentration, water temperature, DO, SPC, pH, chloride, percentage of mixed developed land use, percentage of agricultural land use and percentage of forested land use) used to define 9 principal components, during quarter two (spring: 3 April 2018–26 June 2018) in West Run Watershed, WV, USA.

Variables	Coefficients of PC1	Coefficients of PC2	Coefficients of PC3	Coefficients of PC4
E. coli	0.03	0.58	−0.09	−0.09
Water Temp	0.05	0.12	−0.12	0.91
DO	0.04	0.09	0.59	−0.18
SPC	0.36	−0.49	−0.08	0.17
pH	0.17	0.60	0.04	0.09
Cl^-	0.51	−0.12	−0.01	−0.11
Mixed Development	0.56	0.09	0.15	-0.01
Agriculture	−0.12	0.03	−0.71	−0.23
Forested	−0.50	−0.12	0.31	0.16

Table A10. Coefficients of annual principal components comprising 9 variables (*E. coli* concentration, water temperature, DO, SPC, pH, chloride, percentage of mixed developed land use, percentage of agricultural land use and percentage of forested land use) used to define 9 principal components during quarter three (summer: 3 July 2018–25 September 2018) in West Run Watershed, WV, USA.

Variables	Coefficients of PC1	Coefficients of PC2	Coefficients of PC3	Coefficients of PC4
E. coli	0.43	0.14	−0.29	−0.20
Water Temp	0.21	0.18	0.58	0.33
DO	0.30	0.22	0.48	0.16
SPC	−0.50	0.03	0.24	0.14
pH	0.39	0.32	−0.20	−0.18
Cl^-	−0.48	0.21	−0.13	−0.12
Mixed Development	−0.16	0.63	0.01	−0.15
Agriculture	0.09	−0.13	−0.41	0.78
Forested	0.11	−0.57	0.26	−0.36

Table A11. Coefficients of annual principal components comprising 9 variables (*E. coli* concentration, water temperature, DO, SPC, pH, chloride, percentage of mixed developed land use, percentage of agricultural land use and percentage of forested land use) used to define 9 principal components during quarter four (fall: 2 October 2018–1 January 2019) in West Run Watershed, WV, USA.

Variables	Coefficients of PC1	Coefficients of PC2	Coefficients of PC3	Coefficients of PC4
E. coli	−0.03	0.55	0.00	0.21
Water Temp	−0.09	0.37	−0.14	0.79
DO	0.12	0.11	0.53	−0.01
SPC	0.41	−0.37	−0.08	0.19
pH	0.09	0.57	0.06	−0.14
Cl^-	0.50	−0.12	−0.03	−0.17
Mixed Development	0.55	0.18	0.08	0.09
Agriculture	−0.15	0.02	−0.73	−0.44
Forested	−0.47	−0.20	0.40	0.20

References

1. Mahmoud, M.A.; Abdelsalam, M.; Mahdy, O.A.; El Miniawy, H.M.F.; Ahmed, Z.A.M.; Osman, A.H.; Mohamed, H.M.H.; Khattab, A.M.; Zaki Ewiss, M.A. Infectious bacterial pathogens, parasites and pathological correlations of sewage pollution as an important threat to farmed fishes in Egypt. *Environ. Pollut.* **2016**, *219*, 939–948. [CrossRef] [PubMed]
2. Price, R.G.; Wildeboer, D.E. coli as an Indicator of Contamination and Health Risk in Environmental Waters. In *Escherichia coli—Recent Advances on Physiology, Pathogenesis and Biotechnological Applications*; InTech: Vienna, Austria, 2017.
3. Madoux-Humery, A.-S.; Dorner, S.; Sauvé, S.; Aboulfadl, K.; Galarneau, M.; Servais, P.; Prévost, M. The effects of combined sewer overflow events on riverine sources of drinking water. *Water Res.* **2016**, *92*, 218–227. [CrossRef] [PubMed]
4. *E. coli (Escherichia coli)|E. coli|*CDC. Available online: https://www.cdc.gov/ecoli/index.html (accessed on 8 January 2020).
5. WHO. World Water Day Report. Available online: https://www.who.int/water_sanitation_health/takingcharge.html (accessed on 12 December 2019).
6. Petersen, F.; Hubbart, J.A.; Kellner, E.; Kutta, E. Land-use-mediated Escherichia coli concentrations in a contemporary Appalachian watershed. *Environ. Earth Sci.* **2018**, *77*, 754. [CrossRef]

7. Hirn, J.; Viljamaa, H.; Raevuori, M. The effect of physicochemical, phytoplankton and seasonal factors on faecal indicator bacteria in northern brackish water. *Water Res.* **1980**, *14*, 279–285. [CrossRef]
8. Goyal, S.M.; Gerba, C.P.; Melnick, J.L. Occurrence and distribution of bacterial indicators and pathogens in canal communities along the Texas coast. *Appl. Environ. Microbiol.* **1977**, *34*, 139–149. [CrossRef]
9. Islam, M.M.M.; Hofstra, N.; Islam, M.A. The Impact of Environmental Variables on Faecal Indicator Bacteria in the Betna River Basin, Bangladesh. *Environ. Process.* **2017**, *4*, 319–332. [CrossRef]
10. Verhougstraete, M.P.; Martin, S.L.; Kendall, A.D.; Hyndman, D.W.; Rose, J.B. Linking fecal bacteria in rivers to landscape, geochemical, and hydrologic factors and sources at the basin scale. *Proc. Natl. Acad. Sci. USA* **2015**, *112*, 10419–10424. [CrossRef]
11. Chou, C.-C.; Cheng, S.-J. Recovery of low-temperature stressed *E. coli* O157:H7 and its susceptibility to crystal violet, bile salt, sodium chloride and ethanol. *Int. J. Food Microbiol.* **2000**, *61*, 127–136. [CrossRef]
12. Hong, H.; Qiu, J.; Liang, Y. Environmental factors influencing the distribution of total and fecal coliform bacteria in six water storage reservoirs in the Pearl River Delta Region, China. *J. Environ. Sci.* **2010**, *22*, 663–668. [CrossRef]
13. Gotkowska-Plachta, A.; Golaś, I.; Korzeniewska, E.; Koc, J.; Rochwerger, A.; Solarski, K. Evaluation of the distribution of fecal indicator bacteria in a river system depending on different types of land use in the southern watershed of the Baltic Sea. *Environ. Sci. Pollut. Res.* **2016**, *23*, 4073–4085. [CrossRef]
14. Rwego, I.B.; Gillspie, T.R.; Isabirye-Basuta, G.; Goldberg, T.L. High Rates of Escherichia coli Transmission between Livestock and Humans in Rural Uganda. *J. Clin. Microbiol.* **2008**, *46*, 3187–3191. [CrossRef] [PubMed]
15. Causse, J.; Billen, G.; Garnier, J.; Henri-des-Tureaux, T.; Olasa, X.; Thammahacksa, C.; Latsachak, K.O.; Soulileuth, B.; Sengtaheuanghoung, O.; Rochelle-Newall, E. Field and modelling studies of Escherichia coli loads in tropical streams of montane agro-ecosystems. *J. Hydro-Environ. Res.* **2015**, *9*, 496–507. [CrossRef]
16. Jamieson, R.; Joy, D.M.; Lee, H.; Kostaschuk, R.; Gordon, R. Transport and deposition of sediment-associated Escherichia coli in natural streams. *Water Res.* **2005**, *39*, 2665–2675. [CrossRef] [PubMed]
17. Wu, J.; Yunus, M.; Islam, M.S.; Emch, M. Influence of climate extremes and land use on fecal contamination of shallow tubewells in Bangladesh. *Environ. Sci. Technol.* **2016**, *50*, 2669–2676. [CrossRef]
18. Petersen, F.; Hubbart, J.A. Quantifying Escherichia coli and Suspended Particulate Matter Concentrations in a Mixed-Land Use Appalachian Watershed. *Water* **2020**, *12*, 532. [CrossRef]
19. Rochelle-Newall, E.J.; Ribolzi, O.; Viguier, M.; Thammahacksa, C.; Silvera, N.; Latsachack, K.; Dinh, R.P.; Naporn, P.; Sy, H.T.; Soulileuth, B. Effect of land use and hydrological processes on Escherichia coli concentrations in streams of tropical, humid headwater catchments. *Sci. Rep.* **2016**, *6*, 32974. [CrossRef]
20. Wilson, C.; Weng, Q. Assessing surface water quality and its relation with urban land cover changes in the Lake Calumet Area, Greater Chicago. *Environ. Manag.* **2010**, *45*, 1096–1111. [CrossRef]
21. Tong, S.T.; Chen, W. Modeling the relationship between land use and surface water quality. *J. Environ. Manag.* **2002**, *66*, 377–393. [CrossRef]
22. Abia, A.L.K.; Ubomba-Jaswa, E.; Genthe, B.; Momba, M.N.B. Quantitative microbial risk assessment (QMRA) shows increased public health risk associated with exposure to river water under conditions of riverbed sediment resuspension. *Sci. Total Environ.* **2016**, *566–567*, 1143–1151. [CrossRef]
23. Characklis, G.W.; Dilts, M.J.; Simmons, O.D., III; Likirdopulos, C.A.; Krometis, L.-A.H.; Sobsey, M.D. Microbial partitioning to settleable particles in stormwater. *Water Res.* **2005**, *39*, 1773–1782. [CrossRef]
24. Jeng, H.C.; England, A.J.; Bradford, H.B. Indicator organisms associated with stormwater suspended particles and estuarine sediment. *J. Environ. Sci. Health* **2005**, *40*, 779–791. [CrossRef] [PubMed]
25. Kändler, M.; Blechinger, K.; Seidler, C.; Pavlů, V.; Šanda, M.; Dostál, T.; Krása, J.; Vitvar, T.; Štich, M. Impact of land use on water quality in the upper Nisa catchment in the Czech Republic and in Germany. *Sci. Total Environ.* **2017**, *586*, 1316–1325. [CrossRef] [PubMed]
26. Paster, E.; Ryu, W.S. The thermal impulse response of Escherichia coli. *Proc. Natl. Acad. Sci. USA* **2008**, *105*, 5373–5377. [CrossRef] [PubMed]
27. Stein, E.D.; Tiefenthaler, L.; Schiff, K. Comparison of stormwater pollutant loading by land use type. *Proc. Water Environ. Fed.* **2007**, *2007*, 700–722. [CrossRef]

28. Widgren, S.; Engblom, S.; Emanuelson, U.; Lindberg, A. Spatio-temporal modelling of verotoxigenic Escherichia coli O157 in cattle in Sweden: Exploring options for control. *Vet. Res.* **2018**, *49*, 78. [CrossRef] [PubMed]
29. Tetzlaff, D.; Carey, S.K.; McNamara, J.P.; Laudon, H.; Soulsby, C. The essential value of long-term experimental data for hydrology and water management. *Water Resour. Res.* **2017**, *53*, 2598–2604. [CrossRef]
30. Zeiger, S.; Hubbart, J.A. Quantifying suspended sediment flux in a mixed-land-use urbanizing watershed using a nested-scale study design. *Sci. Total Environ.* **2016**, *542*, 315–323. [CrossRef]
31. Leopold, L.B. *Hydrologic Research on Instrumented Watersheds*; International Association of Scientific Hydrology: Wellington, New Zealand, 1970; pp. 135–150.
32. Hewlett, J.D.; Lull, H.W.; Reinhart, K.G. In Defense of Experimental Watersheds. *Water Resour. Res.* **1969**, *5*, 306–316. [CrossRef]
33. Bosch, J.M.; Hewlett, J.D. A review of catchment experiments to determine the effect of vegetation changes on water yield and evapotranspiration. *J. Hydrol.* **1982**, *55*, 3–23. [CrossRef]
34. Nichols, J.; Hubbart, J.A. Using macroinvertebrate assemblages and multiple stressors to infer urban stream system condition: A case study in the central US. *Urban Ecosyst.* **2016**, *19*, 679–704. [CrossRef]
35. Zeiger, S.; Hubbart, J.A.; Anderson, S.A.; Stambaugh, M.C. Quantifying and modelling urban stream temperature: A central US watershed study. *Hydrol. Process.* **2015**, *30*, 503–514. [CrossRef]
36. Hubbart, J.A.; Kellner, E.; Zeiger, S. A Case-Study Application of the Experimental Watershed Study Design to Advance Adaptive Management of Contemporary Watersheds. *Water* **2019**, *11*, 2355. [CrossRef]
37. Zeiger, S.J.; Hubbart, J.A. Nested-scale nutrient flux in a mixed-land-use urbanizing watershed. *Hydrol. Process.* **2016**, *30*, 1475–1490. [CrossRef]
38. Kellner, E.; Hubbart, J. Advancing understanding of the surface water quality regime of contemporary mixed-land-use watersheds: An application of the experimental watershed method. *Hydrology* **2017**, *4*, 31. [CrossRef]
39. Hubbart, J.A. Urban Floodplain Management: Understanding Consumptive Water-Use Potential in Urban Forested Floodplains. *Stormwater J.* **2011**, *12*, 56–63.
40. Cantor, J.; Krometis, L.-A.; Sarver, E.; Cook, N.; Badgley, B. Tracking the downstream impacts of inadequate sanitation in central Appalachia. *J. Water Health* **2017**, *15*, 580–590. [CrossRef]
41. Arcipowski, E.; Schwartz, J.; Davenport, L.; Hayes, M.; Nolan, T. Clean water, clean life: Promoting healthier, accessible water in rural Appalachia. *J. Contemp. Water Res. Educ.* **2017**, *161*, 1–18. [CrossRef]
42. Dykeman, W. Appalachian Mountains. Available online: https://www.britannica.com/place/Appalachian-Mountains (accessed on 20 November 2019).
43. Köppen, W. Das geographische System der Klimate. *Handb. Klimatol.* **1936**, *35*, 44.
44. Central Appalachian Broadleaf Forest—Coniferous Forest—Meadow Province. Available online: https://www.fs.fed.us/land/ecosysmgmt/colorimagemap/images/m221.html (accessed on 12 December 2019).
45. Kellner, E.; Hubbart, J.; Stephan, K.; Morrissey, E.; Freedman, Z.; Kutta, E.; Kelly, C. Characterization of sub-watershed-scale stream chemistry regimes in an Appalachian mixed-land-use watershed. *Environ. Monit. Assess.* **2018**, *190*, 586. [CrossRef]
46. The West Virginia Water Research Institute, The West Run Watershed Association. *(WVWRI) Watershed Based Plan for West Run of the Monongahela River*; The West Virginia Water Research Institute, The West Run Watershed Association: Morgantown, WV, USA, 2008.
47. Peel, M.C.; Finlayson, B.L.; McMahon, T.A. Updated world map of the Köppen-Geiger climate classification. *Hydrol. Earth Syst. Sci. Discuss.* **2007**, *4*, 439–473. [CrossRef]
48. Arguez, A.; Durre, I.; Applequist, S.; Squires, M.; Vose, R.; Yin, X.; Bilotta, R. NOAA's US climate normals (1981–2010). *NOAA Natl. Cent. Environ. Inf.* **2010**, *10*, V5PN93JP.
49. Hubbart, J.A.; Muzika, R.-M.; Huang, D.; Robinson, A. Improving Quantitative Understanding of Bottomland Hardwood Forest Influence on Soil Water Consumption in an Urban Floodplain. *Watershed Sci. Bull.* **2011**, *3*, 34–43.
50. Hubbart, J.A. Measuring and Modeling Hydrologic Responses to Timber Harvest in a Continental/Maritime Mountainous Environment. Ph.D. Dissertation, University of Idaho, Moscow, ID, USA, 2007.
51. Wei, L.; Hubbart, J.A.; Zhou, H. Variable Streamflow Contributions in Nested Subwatersheds of a US Midwestern Urban Watershed. *Water Resour. Manag.* **2018**, *32*, 213–228. [CrossRef]

52. Hubbart, J.A.; Kellner, E.; Hooper, L.W.; Zeiger, S. Quantifying loading, toxic concentrations, and systemic persistence of chloride in a contemporary mixed-land-use watershed using an experimental watershed approach. *Sci. Total Environ.* **2017**, *581*, 822–832. [CrossRef] [PubMed]
53. Zeiger, S.J.; Hubbart, J.A. Quantifying flow interval–pollutant loading relationships in a rapidly urbanizing mixed-land-use watershed of the Central USA. *Environ. Earth Sci.* **2017**, *76*, 484. [CrossRef]
54. Dusek, N.; Hewitt, A.J.; Schmidt, K.N.; Bergholz, P.W. Landscape-Scale Factors Affecting the Prevalence of Escherichia coli in Surface Soil Include Land Cover Type, Edge Interactions, and Soil pH. *Appl. Environ. Microbiol.* **2018**, *84*, e02714-17. [CrossRef] [PubMed]
55. Desai, M.A.; Rifai, H.S. Variability of Escherichia coli Concentrations in an Urban Watershed in Texas. *J. Environ. Eng.* **2010**, *136*, 1347–1359. [CrossRef]
56. IDEXX. Laboratories Colilert Procedure Manual. Available online: https://www.idexx.com/files/colilert-procedure-en.pdf (accessed on 4 April 2019).
57. Cummings, D. The Fecal Coliform Test Compared to Specific Tests for Escherichia coli. IDEXX. Available online: https://www.idexx.com/resource-library/water/water-reg-article9B.pdf (accessed on 24 September 2019).
58. Yellow Springs Instruments. ProDSS User Manual. Available online: https://www.ysi.com/File%20Library/Documents/Manuals/YSI-ProDSS-110714-Rev-B-626973-User-Manual.pdf (accessed on 8 January 2020).
59. Yazici, B.; Yolacan, S. A comparison of various tests of normality. *J. Stat. Comput. Simul.* **2007**, *77*, 175–183. [CrossRef]
60. Fisher, R.A. Statistical Methods for Research Workers. In *Breakthroughs in Statistics: Methodology and Distribution*; Kotz, S., Johnson, N.L., Eds.; Springer Series in Statistics; Springer: New York, NY, USA, 1992; pp. 66–70. ISBN 978-1-4612-4380-9.
61. United States Climate Data (USCD). Available online: https://www.usclimatedata.com/climate/morgantown/west-virginia/united-states/uswv0507/2012/7 (accessed on 28 September 2019).
62. RoyChowdhury, A.; Sarkar, D.; Datta, R. Remediation of Acid Mine Drainage-Impacted Water. *Curr. Pollut. Rep.* **2015**, *1*, 131–141. [CrossRef]
63. Méndez-García, C.; Peláez, A.I.; Mesa, V.; Sánchez, J.; Golyshina, O.V.; Ferrer, M. Microbial diversity and metabolic networks in acid mine drainage habitats. *Front. Microbiol.* **2015**, *6*, 475.
64. Brown, T.C.; Binkley, D.; Brown, D. Water quality on forest lands|Rocky Mountain Research Station. Available online: /rmrs/projects/water-quality-forest-lands (accessed on 11 March 2020).
65. Nelson, K.C.; Palmer, M.A. Stream temperature surges under urbanization and climate change: Data, models and responses. *J. Am. Water Resour. Assoc.* **2007**, *43*, 440–452. [CrossRef]
66. Wang, L.; Kanehl, P. Influences of Watershed Urbanization and Instream Habitat on Macroinvertebrates in Cold Water Streams1. *JAWRA J. Am. Water Resour. Assoc.* **2003**, *39*, 1181–1196. [CrossRef]
67. Wang, L.; Lyons, J.; Kanehl, P. Impacts of Urban Land Cover on Trout Streams in Wisconsin and Minnesota. *Trans. Am. Fish. Soc.* **2003**, *132*, 825–839. [CrossRef]
68. Imberger, S.J.; Walsh, C.J.; Grace, M.R. More microbial activity, not abrasive flow or shredder abundance, accelerates breakdown of labile leaf litter in urban streams. *JNBS* **2008**, *27*, 549–561. [CrossRef]
69. Mustafa, M.; Barnhart, B.; Babber-Sebens, M.; Ficklin, D. Modeling landscape change effects on stream temperature using the soil and water assessment tool. *Water* **2018**, *10*, 1143. [CrossRef]
70. McQuestin, O.J.; Shadbolt, C.T.; Ross, T. Quantification of the Relative Effects of Temperature, pH, and Water Activity on Inactivation of Escherichia coli in Fermented Meat by Meta-Analysis. *Appl. Environ. Microbiol.* **2009**, *75*, 6963–6972. [CrossRef]
71. Lyew, D.; Sheppard, J. Use of conductivity to monitor the treatment of acid mine drainage by sulphate-reducing bacteria. *Water Res.* **2001**, *35*, 2081–2086. [CrossRef]
72. Cormier, S.M.; Wilkes, S.P.; Zheng, L. Relationship of land use and elevated ionic strength in Appalachian watersheds. *Environ. Toxicol. Chem.* **2013**, *32*, 296–303. [CrossRef]
73. Gray, N.F. Field assessment of acid mine drainage contamination in surface and ground water. *Geo* **1996**, *27*, 358–361. [CrossRef]
74. Zampella, R.A.; Procopio, N.A.; Lathrop, R.G.; Dow, C.L. Relationship of Land-Use/Land-Cover Patterns and Surface-Water Quality in The Mullica River Basin1. *JAWRA J. Am. Water Resour. Assoc.* **2007**, *43*, 594–604. [CrossRef]

75. Baker, M.E.; Schley, M.L.; Sexton, J.O. Impacts of Expanding Impervious Surface on Specific Conductance in Urbanizing Streams. *Water Resour. Res.* **2019**, *55*, 6482–6498. [CrossRef]
76. Ding, J.; Jiang, Y.; Fu, L.; Liu, Q.; Peng, Q.; Kang, M. Impacts of land use on surface water quality in a subtropical river basin: A case study of the dongjian river basin, southeastern China. *Water* **2015**, *7*, 4427–4445. [CrossRef]
77. Abdul-Aziz, O.I.; Ahmed, S. Relative linkages of stream water quality and environmental health with the land use and hydrologic drivers in the coastal-urban watersheds of southeast Florida. *GeoHealth* **2017**, *1*, 180–195. [CrossRef] [PubMed]
78. Guenther, P.M.; Hubert, W.A. Factors influencing dissolved oxygen concentrations during winter in small Wyoming reservoirs. *Great Basin Nat.* **1991**, *51*, 282–285.
79. Wetzel, R.G.; Likens, G.E. Dissolved Oxygen. In *Limnological Analyses*; Wetzel, R.G., Likens, G.E., Eds.; Springer: New York, NY, USA, 2000; pp. 73–84. ISBN 978-1-4757-3250-4.
80. De Vivo, B.; Belkin, H.E.; Lima, A. *Environmental Geochemistry: Site Characterization, Data Analysis and Case Histories*, 2nd ed.; Elsevier: Amsterdam, The Netherlands, 2018; ISBN 978-0-444-63763-5.
81. Grese, M.M.; Kaushal, S.S.; Newcomer, T.A.; Findlay, S.E.G.; Groffman, P.M. Effects of urbanization on variability in temperature and diurnal oxygen patterns in streams. In Proceedings of the Global Warming Esa, Pittsburg, CA, USA, 22 September 2010; The David L. Lawrence Convention Center: Pittsburg, CA, USA, 2010.
82. Niu, T.; Zhou, Z.; Shen, X.; Qiao, W.; Jiang, L.-M.; Pan, W.; Zhou, J. Effects of dissolved oxygen on performance and microbial community structure in a micro-aerobic hydrolysis sludge in situ reduction process. *Water Res.* **2016**, *90*, 369–377. [CrossRef] [PubMed]
83. Hibbing, M.E.; Fuqua, C.; Parsek, M.R.; Peterson, S.B. Bacterial competition: Surviving and thriving in the microbial jungle. *Nat. Rev. Microbiol.* **2010**, *8*, 15–25. [CrossRef] [PubMed]
84. Lax, S.M.; Peterson, E.W.; Van der Hoven, S.J. Stream chloride concentrations as a function of land use: A comparison of an agricultural watershed to an urban agricultural watershed. *Environ. Earth Sci.* **2017**, *76*, 708. [CrossRef]
85. Kaushal, S.S.; Groffman, P.M.; Likens, G.E.; Belt, K.T.; Stack, W.P.; Kelly, V.R.; Band, L.E.; Fisher, G.T. Increased salinization of fresh water in the northeastern United States. *Proc. Natl. Acad. Sci. USA* **2005**, *102*, 13517–13520. [CrossRef]
86. Hauke, J.; Kossowski, T. Comparison of values of Pearson's and Spearman's correlation coefficients on the same sets of data. *Quaest. Geogr.* **2011**, *30*, 87–93. [CrossRef]
87. Blaustein, R.A.; Pachepsky, Y.; Hill, R.L.; Shelton, D.R.; Whelan, G. Escherichia coli survival in waters: Temperature dependence. *Water Res.* **2013**, *47*, 569–578. [CrossRef]
88. Ishii, S.; Sadowsky, M.J. Escherichia coli in the Environment: Implications for Water Quality and Human Health. *Microbes Environ.* **2008**, *23*, 101–108. [CrossRef] [PubMed]
89. Spietz, R.L.; Williams, C.M.; Rocap, G.; Horner-Devine, M.C. A Dissolved Oxygen Threshold for Shifts in Bacterial Community Structure in a Seasonally Hypoxic Estuary. *PLoS ONE* **2015**, *10*, e0135731. [CrossRef] [PubMed]
90. Multivariate Analysis: Principal Component Analysis: Biplots-9.3. Available online: http://support.sas.com/documentation/cdl/en/imlsug/64254/HTML/default/viewer.htm#imlsug_ugmultpca_sect003.htm (accessed on 23 December 2019).
91. Jolliffe, I.T.; Cadima, J. Principal component analysis: A review and recent developments. *Philos Trans. A Math. Phys. Eng. Sci.* **2016**, *374*, 20150202. [CrossRef] [PubMed]
92. Bro, R.; Smilde, K.A. Principal component analysis. *Anal. Methods* **2014**, *6*, 2812–2831. [CrossRef]
93. Dingman, S.L. *Phyiscal Hydrology*, 2nd ed.; Prentice Hall: Englewood Cliffs, NJ, USA, 2002; ISBN 0-13-099695-5.
94. Farewell, A.; Neidhardt, F.C. Effect of Temperature on In Vivo Protein Synthetic Capacity in Escherichia coli. *J. Bacteriol.* **1998**, *180*, 4704–4710. [CrossRef] [PubMed]
95. Hajmeer, M.; Ceylan, E.; Marsden, J.L.; Fung, D.Y.C. Impact of sodium chloride on Escherichia coli O157:H7 and Staphylococcus aureus analysed using transmission electron microscopy. *Food Microbiol.* **2006**, *23*, 446–452. [CrossRef]
96. Harvey, R.; Lye, L.; Khan, A.; Paterson, R. The Influence of Air Temperature on Water Temperature and the Concentration of Dissolved Oxygen in Newfoundland Rivers. *Can. Water Resour. J.* **2011**, *36*, 171–192. [CrossRef]

97. Davies-Colley, R.J.; Nagels, J.W.; Smith, R.A.; Young, R.G.; Phillips, C.J. Water quality impact of a dairy cow herd crossing a stream. *N. Z. J. Mar. Freshw. Res.* **2004**, *38*, 569–576. [CrossRef]
98. Collins, R.; Mcleod, M.; Hedley, M.; Donnison, A.; Close, M.; Hanly, J.; Horne, D.; Ross, C.; Davies-Colley, R.; Bagshaw, C.; et al. Best management practices to mitigate faecal contamination by livestock of New Zealand waters. *N. Z. J. Agric. Res.* **2007**, *50*, 267–278. [CrossRef]

Article

Spatial and Temporal Characterization of *Escherichia coli*, Suspended Particulate Matter and Land Use Practice Relationships in a Mixed-Land Use Contemporary Watershed

Fritz Petersen [1,*] and Jason A. Hubbart [1,2]

1 Division of Plant and Soil Sciences, Davis College of Agriculture, Natural Resources and Design, West Virginia University, Agricultural Sciences Building, Morgantown, WV 26506, USA; Jason.hubbart@mail.wvu.edu
2 Institute of Water Security and Science, Schools of Agriculture and Food, and Natural Resources, Davis College of Agriculture, Natural Resources and Design, West Virginia University, 3109 Agricultural Sciences Building, Morgantown, WV 26506, USA
* Correspondence: fp0008@mix.wvu.edu; Tel.: +1-304-413-5449

Received: 12 March 2020; Accepted: 23 April 2020; Published: 25 April 2020

Abstract: Understanding land use practice induced increases in *Escherichia (E.) coli* and suspended particulate matter (SPM) concentrations is necessary to improve water quality. Weekly stream water samples were collected from 22 stream gauging sites with varying land use practices in a representative contemporary mixed-land use watershed of the eastern USA. Over the period of one annual year, *Escherichia (E.) coli* colony forming units (CFU per 100 mL) were compared to suspended particulate matter (SPM) concentrations (mg/L) and land use practices. Agricultural land use sub-catchments comprised elevated *E. coli* concentrations (avg. 560 CFU per 100 mL) compared to proximate mixed development (avg. 330 CFU per 100 mL) and forested (avg. 206 CFU per 100 mL) sub-catchments. Additionally, agricultural land use showed statistically significant relationships ($p < 0.01$) between annual *E. coli* and SPM concentration data. Quarterly PCA biplots displayed temporal variability in land use impacts on *E. coli* and SPM concentrations, with agricultural land use being closely correlated with both pollutants during Quarters 2 and 3 but not Quarters 1 and 4. The data collected during this investigation advance the understanding of land use impacts on fecal contamination in receiving waters, thereby informing land use managers on the best management practices to reduce exposure risks.

Keywords: *Escherichia coli*; suspended particulate matter; water quality; land use practices; experimental watershed

---

1. Introduction

Fecal pollution is the greatest contributor to water borne disease human morbidity and mortality rates globally [1]. Freshwater fecal pollution and subsequent increases in pathogenic bacteria (e.g., *Escherichia (E)* coli), cause disease outbreaks, including diarrhea, urinary tract infections, respiratory illness and pneumonia [2,3]. The World Health Organization reported that 2.2 million deaths are caused by diarrhea annually, due to the consumption of fecal contaminated water [1]. An improved understanding of factors leading to increased fecal contamination in receiving waters will be useful in reducing outbreaks of waterborne disease and improving water quality. Furthermore, understanding the factors impacting the health and exposure risks of fecal pollution can be used to decrease the threat posed by fecal organisms. For example, the environmental persistence of fecal microbes can be extended when occurring with suspended particulate matter (SPM) [4,5]. Therefore, monitoring SPM in conjunction with fecal pollution can provide greater insight into water quality, through more accurate assessments of the persistence of fecal microbes.

Suspended particulate matter (SPM), defined as heterogeneous aggregates of mineral fragments, organic matter and microbial fractions, comprises the greatest water pollutant by volume globally [6]. Excess SPM in freshwater can impact water quality by decreasing the amount of transmitted light, thereby restricting or eliminating the photosynthesis of aquatic plants and dramatically influencing the aquatic food chain [7,8]. Therefore, understanding the factors influencing fluxes in SPM (e.g., land use practices) is important from an ecosystem management perspective. Additionally, increases in SPM can clog the gills of fish, thus lowering resistance to disease and decreasing developmental growth rates [7] while also elevating water temperatures, thereby disrupting the metabolic processes of various aquatic biota [8]. Thus, changes in SPM concentrations in receiving waters can entail serious consequences for various aquatic organisms. SPM can also act as a conveyance system for other pollutants including heavy metals, chemicals and pathogens, including fecal microbes (as discussed above) [8–12]. Insofar as excess sediment can be harmful to aquatic ecosystems, too little sediment can also be harmful, leading to the scouring of river channels, erosion and reduced nutrient inputs [8,13]. Consequently, understanding the factors leading to increases or decreases of SPM in receiving waters is important from a water quality perspective [14].

Previous work investigating fecal contamination and SPM reported statically significant, Pearson's product moment, correlations ($r > 0.9$) between the two pollutants [15]. Moreover, the strength of the relationships between fecal pollution and SPM reported in previous work [15,16] has led to speculation that SPM concentrations (and the turbidity caused by SPM) can potentially serve as a proxy for fecal contamination [16]; however, this is yet to be verified. The relationship between fecal contamination and SPM has been attributed to similar transport processes influencing both pollutants during run-off events [17], including similar in-stream transport physics [18], and the sorption of fecal microbes to SPM [4,5]. Additionally, certain land use practices can simultaneously elevate both SPM and fecal microbe concentrations in receiving waters [19,20]. For example, agricultural land use practices are commonly associated with increased fecal (e.g., *E. coli*) and SPM concentrations relative to other land use types [19–22]. This is often related to the presence of livestock [23], with livestock population density being correlated to fecal indicator organism concentration [24]. Manure application in agricultural areas has also been linked to increased concentrations of fecal microbes in receiving waters [4]. Conversely, agricultural practices such as soil tillage and soil exposure yield increased SPM concentrations in the receiving waters of agricultural areas [19,21]. Differing land use practices in a given area (watershed) can therefore account for differing *E. coli* and SPM concentrations in receiving waters and should thus be accounted for when monitoring these pollutants.

Despite the progress of previous research, knowledge gaps regarding the relationship between fecal contamination and SPM remain. For example, few studies investigated the relationships between SPM and fecal contamination in mixed land use settings. Furthermore, the majority of studies included limited sampling locations [20,25] and tended to occur in areas of similar land use types [16,26], or were controlled laboratory simulations [27]. Additionally, previous work on fecal contamination typically focused on storm events and therefore report disproportionately elevated fecal concentrations in receiving waters [28]. Clearly, knowledge regarding fecal concentrations outside of storm events is lacking, creating challenges for proper management practices. Similarly, previous work investigating fecal concentrations in receiving waters typically utilized shorter (weeks or months) sampling periods [20], which fail to account for the seasonal variability in land use practices. For example, seasonal variability in agricultural land use practices can lead to changes in fecal microbe concentrations in receiving water, particularly in areas where manure is applied [29]. Manure is typically only applied in specific seasons, thereby leading to corresponding increases in the fecal contamination of associated receiving waters during these seasons [29]. Therefore, a study design capable of distinguishing the effects of different land uses and seasonality would be useful for investigating fecal contamination and SPM in receiving waters and for better informing water resource managers.

Previous studies have used many different study design methods, including different sampling regimes, to advance the understanding of *E. coli* regimes. Study designs have included laboratory based designs comprising simulations [30] and field based designs comprising event based sampling [31], periodic sampling [32], stochastic sampling [33] and nested-scale experimental watersheds [20]. The nested-scale and paired experimental watershed study design is a method that has been successfully used to quantify the effects of land use practices on receiving waters in mixed land use settings [34–40]. Nested watershed study designs divide a larger watershed into a series of sub-catchments to investigate the influence of land use practices on the environmental variables of interest [35,37,41–43]. Sub-catchment delineation isolates different land use practices and hydrologic characteristics [41]. Paired watersheds comprise at least two physiographically similar watersheds (control and treatment) from which data are collected [41]. The study design enables the identification of the influence and cumulative effect of various land use practices on the response variable of interest through the quantification of the influencing processes observed at the sub-catchment scale [44]. Therefore, the approach allows for the effective disentanglement of factors (e.g., land use practices and SPM) that influence a given response variable of interest (e.g., fecal microbe concentration), thus providing quantitative information regarding hydrologic and water quality regimes related to specific land-uses [41]. Given its proven application in over a century of studies, the nested-scale and paired experimental watershed study design is an accepted optimal study design for investigating current knowledge gaps regarding fecal contamination, SPM and land use practices.

The Appalachian region of the USA is well-suited for researching knowledge gaps concerning fecal contamination, SPM and land use practices. The region is representative of many locations globally given that it suffers from widespread, frequent, and problematic fecal pollution [45]. Additionally, Appalachia is physiographically diverse, encompassing distinct Northern, Central and Southern regions, consisting of dissimilar geographic, climatological, and ecological characteristics [46]. For example, the temperate climate and well-distributed year-round rainfall characteristics of Central Appalachia [47] are similar to those of areas such as Uruguay or Southern Brazil [48], and many other locations. Conceivably, other temperate areas comprising year-round precipitation (e.g., Uruguay) will benefit from research conducted in the Central Appalachian region as the results will be comparable and transferable. Furthermore, water quality is a primary concern in rural Appalachia as thousands of residents are exposed to water quality problems, specifically regarding microbial contamination [49]. Water quality problems in rural areas are exacerbated by inadequate wastewater treatment infrastructure, isolation due to geographically inaccessible terrain, and poverty [49]. Consequently, water quality is a primary concern, and insight into both SPM and fecal pollution (e.g., *E. coli* concentrations) is necessary to effectively inform policy makers and water resource managers regarding water quality to make the best management practice decisions in Appalachia and physiographically similar locations globally.

The overarching objective of the current investigation was to quantify fecal contamination (*E. coli* concentration) and SPM concentrations in receiving waters relative to differing land use practices from numerous sites in a mixed-land use contemporary watershed of Appalachia. Sub-objectives included (1) investigating the relationship between fecal microbe concentration (*E. coli* colony forming units) and SPM concentrations in receiving waters, and (2) determining the influence of quarterly (seasonal) changes on the relationship between *E. coli* and SPM. The study outcomes were to improve the understanding of the influence of land use practices on both fecal contamination and SPM pollution, providing land use managers with insight into factors influencing water quality in receiving waters.

2. Methods

2.1. Study Site Description

This research took place in West Run Watershed (WRW) a 23 km^2 mixed-land use urbanizing watershed located in Morgantown, West Virginia, USA. West Virginia's climate varies between cold and humid with warm summers, to temperate and humid with hot summers [50]. In Morgantown, WV, located in Monongalia County (and including the WRW), the climate is characterized by the lack of a dry season, cold winters (mean monthly temperature < 0 °C) and warm-to-hot summers (mean monthly temperature > 22 °C) [50]. Historically (1981–2010), Morgantown received approximately 1060 mm of average annual precipitation, with the coldest (January) and driest (February) months having an average daily temperature of −0.4 °C and an average monthly precipitation of 66 mm, respectively [51]. Conversely, the warmest and wettest month (July) comprised an average daily temperature of approximately 23 °C and an average monthly precipitation of 117 mm [51].

West Run Creek, the primary drainage of WRW, is a third order tributary of the Monongahela River, and includes many land use practices including agriculture, urban and forested areas [22]. Based on the 2016 National Agriculture Imagery Program (NAIP) land use and land cover data, WRW includes 42.7% forested land use, 37.7% mixed development (urban and commercial areas) land use and 19.4% agricultural land use practices. West Run Creek is a narrow, moderately entrenched stream with multiple small floodplains [20,52]. The elevation of the headwaters of WRW is 420 m above mean sea level [22]. Conversely, the elevation of the confluence of WRW with the Monongahela River is 240 m above mean sea level [20]. The watershed includes relatively rugged terrain, featuring numerous Paleozoic era rock outcroppings [20]. The headwaters of WRW contain the most recent geological formation (Monongahela series) [20]. Two coal formations are also present in the watershed, namely the Upper Kittanning coal and the Pittsburg coal seam [20]. Historic mining of the Pittsburg coal seam negatively impacted water quality in WRW, particularly in the headwaters [53].

A nested-scale and paired experimental watershed study design [35,44,54–56] comprising twenty-two study sites (i.e., gauge sites) was implemented in 2017. Sampling sites (numbered in downstream order) were located in West Run Creek (#3, #4, #6, #10, #13, #18, #19, #21 and #22) and its first and second order confluence tributaries (#1, #2, #5, #7, #8, #9, # 11, #12, #14, #15, #16, #17 and #20) and included varying land use practices (Table 1; Figure 1). Both field surveys and GIS were used to identify the study sites and related sub-catchments. At the time of this investigation, forested land use was the predominant land use in WRW, accounting for 42.7% of the total land use practices in the watershed. Additionally, forested land use was the predominant land use type in all sub-catchments except #1, #11, #15, #16 and #20. Sub-catchments #1, #15 and #20 were primarily mixed development, whereas sub-catchments #11 and #16 where primarily agricultural (Table 1). Conversely, 85.84% of sub-catchment #17 was forested land use, the highest among the sub-catchments. This sub-catchment, which served as a reference sub-catchment (control) for the current work, also comprised 9.4% agricultural and 4.8% mixed development land use practices. Sub-catchment #17 is, therefore, considerably different to sub-catchment #12 (34.5% forested, 33.7% agriculture and 31.7% agriculture) despite both comprising predominantly forested land use practices. In general, at the time of the investigation, mixed development comprised the second largest percentage of land use practices (37.7%) and agricultural land use practices accounted for the lowest percentage of land use practices (19.4%) in WRW.

Table 1. Land use/land cover characteristics (% cover) and total drainage area (km^2) at 22 monitoring sites in West Run Watershed (WRW), West Virginia, USA. Note: land use percentages may not sum to 100%, as not every category is included (i.e., wetland, open water, etc.) and some categories are combinations of others (e.g., mixed development = urban + residential). Final row (Site #22) indicates the total values for the entire watershed.

Site	Mixed Development (%)	Agriculture (%)	Forested (%)	Drainage Area (km^2)
1	53.23%	38.70%	8.07%	0.30
2	13.58%	12.20%	74.21%	0.29
3	22.35%	16.17%	61.32%	1.87
4	25.88%	14.91%	59.00%	2.48
5	23.35%	25.51%	51.14%	0.38
6	23.91%	17.25%	58.70%	3.72
7	16.33%	28.60%	54.91%	0.78
8	30.78%	16.47%	52.35%	1.55
9	27.57%	19.33%	52.84%	2.29
10	24.92%	18.40%	56.49%	6.18
11	18.15%	41.87%	39.16%	1.75
12	31.77%	33.72%	34.51%	1.75
13	26.83%	25.77%	47.15%	10.53
14	16.19%	26.43%	56.92%	3.36
15	70.28%	10.31%	19.42%	0.98
16	5.38%	58.72%	35.16%	0.25
17	4.78%	9.38%	85.84%	0.75
18	25.98%	24.88%	48.86%	16.41
19	29.45%	22.45%	47.85%	18.88
20	89.16%	4.19%	6.61%	3.42
21	38.10%	19.46%	42.23%	22.93
22	37.71%	19.38%	42.66%	23.24

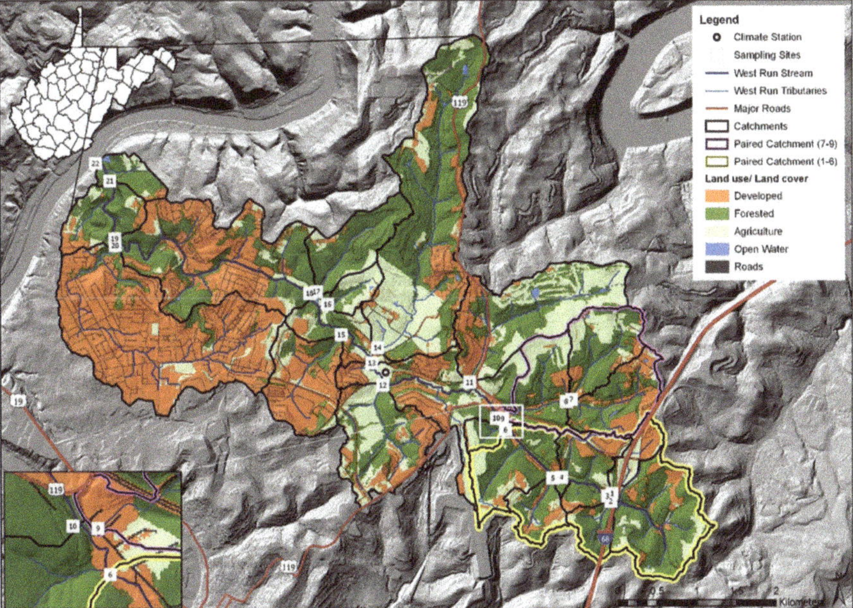

Figure 1. Monitoring/sampling locations for the current investigation, with land use/land cover, in West Run Watershed, Morgantown, West Virginia, USA.

2.2. Data Collection

Climate data collected for the current work included precipitation (Campbell Scientific TE525 Tipping Bucket Rain Gage), average air temperature, relative humidity (Campbell Scientific HC2S3 Temperature and Relative Humidity Probe), and average wind speed (Campbell Scientific Met One 034B Wind Set instrument). Data were recorded at a 3 m height during the study period (2 January 2018–1 January 2019) by a climate station located within approximately 30 m of Site #13 (Figure 1).

For the current work, weekly water grab-samples were collected as per Petersen et al. [20], Hubbart et al. [57], Kellner and Hubbart [43], and Zeiger and Hubbart [42,58] from each monitoring site (stream order ≤ 3). Water sample collection was initiated at 09:00 at Site #1 and continued in numerical order of sites. Sites #9 and #10 were exceptions, as they were sampled before Sites #7 and #8, due to their location relative to other sites (Figure 1). The proximity of Sites #9 and #10 to Site #6 meant that overall sampling time was reduced by sampling them after Site #6, increasing the comparability of the samples during sample processing. The sampling period for the study (2 January 2018–1 January 2019, thus 53 weeks) was one calendar year to account for seasonal variability in the *E. coli* concentration and SPM data. Notably, the sampling period was longer than in typical studies on fecal contamination [59,60], allowing for a comprehensive quantification of fecal contamination (*E. coli*) regimes at sub-catchment mixed-land-use scales. The high-resolution study design resulted in a total of 1166 spatio-temporally delineated fecal contamination (*E. coli*) concentration and SPM concentration values.

Following collection, the samples were transported to the Interdisciplinary Hydrology Laboratory, located in the Davis College of Agriculture, Natural Resources and Design at West Virginia University, for analyses. In the laboratory, water samples were refrigerated (at 3.3 °C), and gravimetric analyses (vacuum filtration) were conducted as per the American Society for Testing and Materials, test number D 3977-97, [61] within a few days of collection to determine the mass of suspended particulate matter (SPM). Additionally, fecal contamination was quantified immediately upon arrival at the laboratory using *Escherichia (E) coli* as an indicator organism, as per previous work [20,62]. *E. coli* coliform forming units (CFU) were enumerated using the U.S. Environmental Protection Agency (EPA) approved Colilert test [63], developed by IDEXX Laboratories Inc. The applied method used an MPN approach to estimate the *E. coli* CFU concentration; therefore, *E. coli* concentration data were referred to as CFU, not MPN, during the investigation. The test, included in Standard Methods for Examination of Water and Wastewater was developed to estimate fecal concentrations in water samples without requiring sample dilution [63,64]. A combination of Colilert's Defined Substrate Technology nutrient-indicator (ONPG), and a selectively suppressing formulated matrix created low chances of recording inaccurate results (chance of reporting false positives ±10%). With this test, most non-target organisms are unable to grow given that they lack the enzyme to metabolize the provided carbon source (ONPG) [63]. The formulated matrix selectively suppresses the few non-target organisms that can metabolize ONPG [63]. The number of *E. coli* colony-forming units (CFU) per 100 mL of sampled water was estimated using the Quanti-Tray system, comprising 96 total wells: 48 large wells (49, including the overflow well) and 48 small wells [63]. The Colilert (ONPG) substrate was added to 100 mL of sampled water, sealed in the Quanti-Tray, and incubated at 35 °C for 24 hours, as per Colilert's instructions [29]. Following incubation, fluorescing (positive for *E. coli*) wells were quantified using a UV light and converted, with a 95% confidence interval, into a concentration of *E. coli* (CFU per 100 mL) using the Quanti-Tray Most Probable Number (MPN) table. The *E. coli* concentration range resultant from the Quanti-Tray/MPN table method was <1 to 1011.2 CFU. Therefore, *E. coli* concentrations in excess of 1011 CFU per 100 mL could not be accurately estimated. This limitation was an allowable shortcoming of the current work given the focus on the weekly detection of small *E. coli* concentrations occurring between storm events.

2.3. Data Analysis

Descriptive statistics were generated for *E. coli* and SPM concentrations and aggregated for the study period. Average percentage differences between sites were determined by comparing the average SPMs and average *E. coli* concentrations between sites. Statistical analyses were conducted using Origin Academic 2018 (OriginLab Corporation). Normality testing was completed using the Anderson Darling Test [65]. Land use practices were reclassified (lumped) into three major categories prior to analysis, namely mixed development, agriculture, and forested [20]. Mixed development constituted roads, impervious surfaces, mixed developments and barren areas. Agriculture included low vegetation, hay pasture and cultivated crops. Forested land use included mine grass, forest, mixed mesophytic forest, dry mesic oak forest, dry oak (pine) forest and small stream riparian habitats. Annual data were also analyzed in four quarter data subsets, comprising all weekly samples collected in three month blocks starting on January 1st, 2018, to analyze seasonal variation. Thus, Quarter 1 included 2 January 2018–27 March 2018 (winter), Quarter 2 included 3 April 2018–26 June 2018 (Spring), Quarter 3 included 3 July 2018–25 September 2018 (summer), and Quarter 4 included 2 October 2018–1 January 2019 (fall). Spearman correlation tests, with a significance threshold of $\alpha = 0.05$ [66], were used to analyze the relationship between *E. coli* concentration, suspended sediment, and land use practices at all twenty-two sites, as per Petersen et al. [20] for the complete annual data set and the four quarterly data subsets. Finally, principal component analysis (PCA) was used to investigate the relationships between *E. coli* concentrations, SPM and land use practices (presented in biplots) across all 22 sampling locations for the annual data set and the four quarterly data subsets.

3. Results and Discussion

3.1. Climate during Study

Total precipitation was 1378 mm in 2018 in WRW. This was approximately 20% more precipitation than the historic annual average (1096 mm) dating back to 2007 [67]. September (186 mm) and October (47 mm) were the wettest and driest months, respectively, during 2018 (Figure 2). Approximately 14% of the annual precipitation was received in September. This was more than double the historic average precipitation (80 mm) for that month [67]. The average air temperature, during the study period was approximately 12 °C, which is close to the historic average of 11 °C [67]. July (22 °C) and January (−4 °C) comprised the warmest and coldest average monthly temperatures, respectively, in WRW during 2018. Relative humidity was characteristically high during 2018 (Figure 2), comprising a yearly average of 76%. Generally, climate during the period of study (2 January 2018–1 January 2019) was predictably variable and consistent with historic trends (Figure 3), including humid and warm weather during the summer months, with temperatures decreasing over the transition to winter (Figure 3). As is typical of the region, there was no dry season; however, large precipitation events during Quarters 2 (spring; e.g., May 6th: 24 mm) and 3 (summer; e.g., September 9th; 60 mm) resulted in greater quarterly (seasonal) variation in precipitation (Figure 2) [67]. Quarters 2 and 3 (spring and summer; 850 mm) therefore received 67% more precipitation than Quarters 1 and 4 (winter and fall; 510 mm).

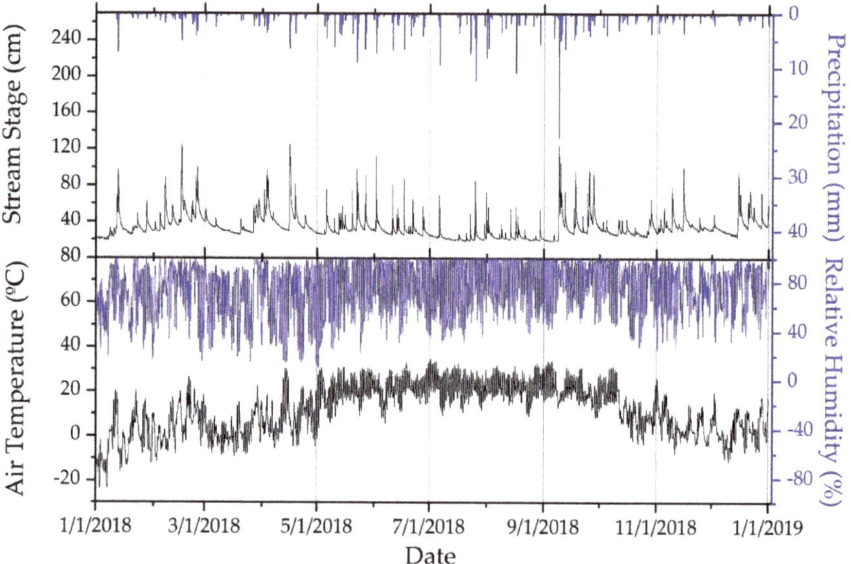

Figure 2. Thirty-minute time series of climate variables during the study period (2 January 2018–1 January 2019) in West Run Watershed, West Virginia, USA. Note: stream stage was measured in the primary stream of WRW, West Run Creek, within approximately 30 m of Site #13 and West Run Creek.

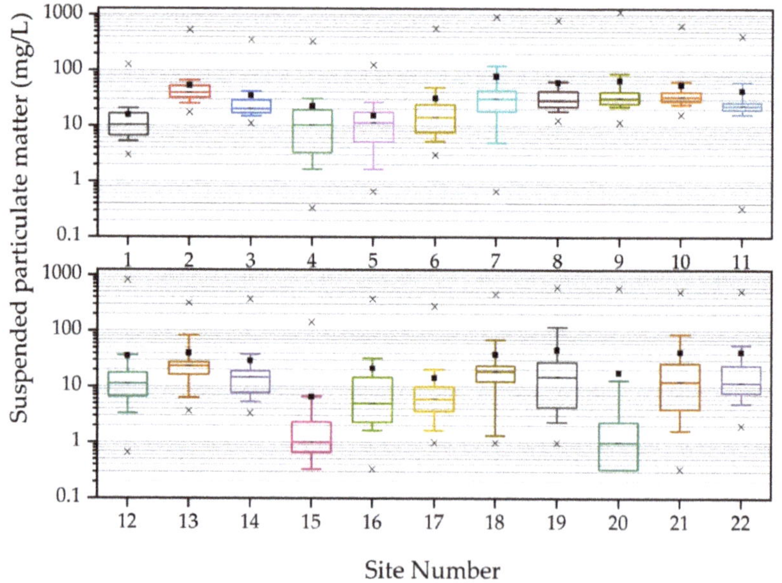

Figure 3. Box and whisker plots of suspended particulate matter (mg L^{-1}; log^{10} scale) at each sampling location (n = 22) during the study period (2 January 2018–1 January 2019) in West Run Watershed, Morgantown, West Virginia, USA. Boxes delineate 25th and 75th percentiles; lines denotes medians; squares shows means; whiskers describe 10th and 90th percentiles; x shows maxima and minima when above and below, respectively. Note: different box colors represent data from different sites.

3.2. Annual Suspended Particulate Matter, E. coli Concentrations and Land Use Practices

The results showed that forested sub-catchments had the highest average (Site #7; 55% forested; 78.4 mg/L), maximum (Site #9; 53% forested; 1140 mg/L) and minimum (Site #8; 52% forested; 12.7 mg/L) SPM concentrations (Table 2; Figure 3). Notably, these sub-catchments constituted one of the paired watersheds of the paired study design and were in close proximity to each other (Figure 1). Consequently, these sub-catchments were subject to similar land use activities and processes leading to elevated SPM in this region of the watershed. For example, the agricultural land use practices in the headwaters of sub-catchments #7 and #8 (Figure 1) could have elevated the SPM in the entire paired catchment (Sites #7, #8 and #9) as previous work has reported increased SPM in agricultural areas [19–22]. SPM concentrations were also elevated in West Run Creek (combined average of sites in West Run Creek, 39 mg/L) relative to sites located in the first and second order confluence tributaries (combined average, 35 mg/L) (Table 2; Figure 3). The increased SPM in West Run Creek was attributable to (1) the greater volumetric streamflow in West Run Creek relative to in its tributaries, as increased streamflow can increase the SPM concentration [68]; and (2) increased SPM sources due to an increased drainage area relative to its tributaries (23 km^2 and 15 km^2 respectively) (Table 1; Figure 1). Conversely, SPM concentrations were decreased in mixed development sub-catchments (Sites #15: 70% mixed development and #20: 89% mixed development) comprising the lowest average (6.5 mg/L), lowest median (1 mg/L), and lowest minimum (0 mg/L) (Table 2; Figure 3). Site #15 also had low SPM concentrations during previous work conducted in the WRW, thereby supporting the results from the current investigation [20]. Mixed development areas can comprise decreased exposed soil surfaces and subsequent reductions in the SPM sources relative to other land use types, which can account for the decreased SPM concentrations [69].

Table 2. Descriptive statistics of suspended particulate matter (mg L^{-1}) at each sampling location (n = 22) during the study period (2 January 2018–1 January 2019) in West Run Watershed, WV, USA. Note: all average values presented in the current work constitute arithmetic means.

	Site Number										
	#1	#2	#3	#4	#5	#6	#7	#8	#9	#10	#11
Avg.	15.6	53.0	34.5	22.2	15.2	31.3	78.4	61.1	66.4	55.7	42.9
Med.	10.3	39.3	20.3	10.3	11.3	14.3	30.7	28.7	31.7	33.3	23.0
Min.	3.0	17.3	11.0	0.3	0.7	3.0	0.7	12.7	11.7	16.0	0.3
Max.	126.7	528.7	357.0	332.3	125.3	569.7	928.3	803.3	1140.0	642.0	417.3
Std. Dev.	20.8	69.8	57.5	49.5	19.0	79.9	176.5	123.5	159.7	93.3	77.2
	Site Number										
	#12	#13	#14	#15	#16	#17	#18	#19	#20	#21	#22
Avg.	35.0	40.1	29.1	6.5	21.1	14.4	38.2	46.1	17.7	41.9	42.0
Med.	11.3	23.3	14.7	1.0	5.0	6.0	18.7	15.0	1.0	12.3	11.8
Min.	0.7	3.7	3.3	0.0	0.3	1.0	1.0	1.0	0.0	0.3	2.0
Max.	819.3	316.0	370.7	144.7	376.0	277.3	456.7	603.3	590.0	502.0	518.0
Std. Dev.	116.4	62.0	59.7	21.5	59.3	38.6	75.3	103.7	83.3	101.1	99.1

The study results showed that *E. coli* concentrations were the highest at sub-catchments comprising the greatest percentage agricultural land use area (Site #16: 59% agricultural). These results are similar to those of previous investigations in WRW reporting increased *E. coli* concentrations in agricultural land use sub-catchments [20]. This predominantly agricultural sub-catchment comprised the highest average (560 CFU per 100 mL) and median (575 CFU per 100 mL) (Table 3; Figure 4) *E. coli* concentrations over the period of investigation. Previous investigations in the USA (California and Ohio) reported increased fecal contamination with agricultural land use practices [31,70], and a significant correlation ($p < 0.04$) between agricultural land use and *E. coli* concentrations [20], thereby supporting the results recorded during the current work. The lowest *E. coli* concentrations were recorded at two forested sites (Site #2: 74% forested and Site #5: 51% forested) comprising the lowest median (3 CFU per

100 mL) and average (34 CFU per 100 mL) amongst the sites respectively (Table 3; Figure 5). These two sites, and the forested sub-catchments comprising one of the paired watersheds (Sites #7, #8 and #9) located in the headwaters of WRW (Figure 1), were heavily impacted by acid mine drainage (AMD) from historic mining activities [20,22,53], which likely, at least in part, explains the low *E. coli* concentrations observed at these sites. This is an important finding given that previous studies showed that AMD lowers the pH of receiving waters [71] and that the current results indicate that AMD may also lower *E. coli* concentrations. Forested sites generally had lower *E. coli* concentrations (e.g., Site #17: 86% forested; average *E. coli* concentration: 206 CFU per 100 mL) during the study period than sites comprising other land use practices (e.g., Site # 20: 89% mixed development; average *E. coli* concentration: 415 CFU per 100 mL) (Figure 4). These results align well with previous studies reporting decreased fecal contamination in forested areas [70] and are attributable to the increased quality of receiving waters in forested areas [72]. Consequently, both forested land use practices and AMD lowered *E. coli* concentrations in WRW during the investigation. Notably, the low average *E. coli* concentrations recorded during the study period (2 January 2018–1 January 2019), specifically in the headwaters, affirms the study objective of analyzing samples collected between storm events that comprise lower *E. coli* concentrations. Additionally, in no other study has there been such high spatial and temporal resolution sampling over a full annual year. This allowed for a more comprehensive analysis of *E. coli* concentration regimes and relationships with SPM and land use, including accounting for seasonality, than is available in the literature surrounding contemporary mixed land use watersheds. The current study therefore lends greatly needed confirmation through high spatial and temporal resolution of previous studies.

Table 3. Descriptive statistics of *E. coli* concentration (CFU per 100 mL) at each sampling location (n = 22) during the study period (2 January 2018–1 January 2019) in West Run Watershed, WV, USA.

	Site Number										
	#1	**#2**	**#3**	**#4**	**#5**	**#6**	**#7**	**#8**	**#9**	**#10**	**#11**
Avg.	170	38	397	429	34	269	84	89	127	210	98
Med.	66	3	260	361	4	194	20	32	25	93	16
Min.	0	0	15	107	0	2	0	0	0	3	0
Max.	1011	961	1011	1011	914	1011	1011	1011	1011	1011	870
Std. Dev.	251	139	315	249	129	276	179	180	241	273	202
	Site Number										
	#12	**#13**	**#14**	**#15**	**#16**	**#17**	**#18**	**#19**	**#20**	**#21**	**#22**
Avg.	234	215	457	330	560	206	324	466	415	471	452
Med.	88	91	299	211	575	93	218	436	299	397	397
Min.	0	0	0	5	22	3	0	1	23	2	3
Max.	1011	1011	1011	1011	1011	1011	1011	1011	1011	1011	1011
Std. Dev.	305	266	406	293	373	288	342	339	340	342	345

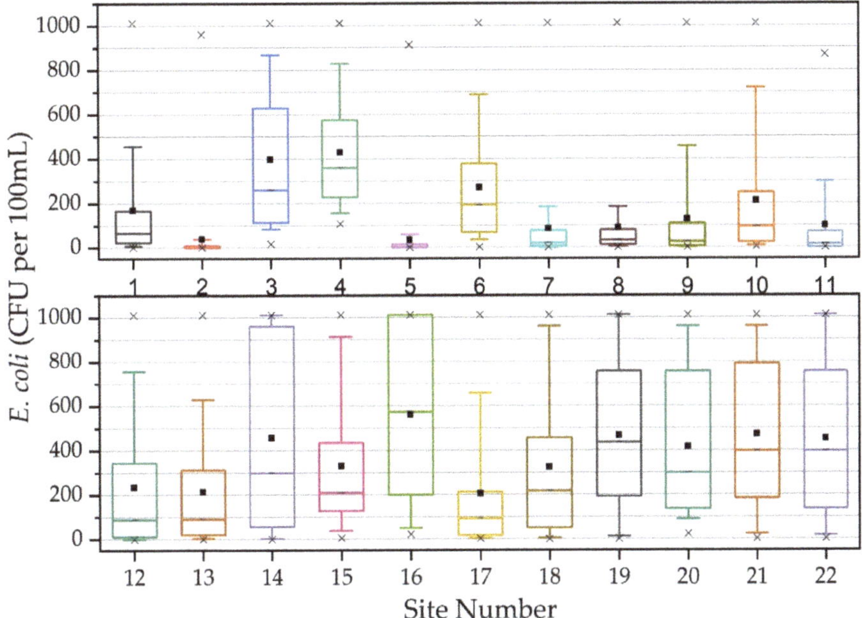

Figure 4. Box and whisker plot of *E. coli* concentration (CFU per 100 mL) at each sampling location (n = 22) during the study period (2 January 2018–1 January 2019) in West Run Watershed, Morgantown, West Virginia, USA. Boxes delineate 25th and 75th percentiles; lines denote medians; squares show means; whiskers describe 10th and 90th percentiles; x shows maxima and minima when above and below, respectively. Note: different box colors represent data from different sites.

E. coli concentrations showed a general increase from the headwaters of WRW to the confluence of the Monongahela River, with larger average concentrations typically being observed in the lower portions of the watershed (Figure 5). In the current work, AMD may account for the lower *E. coli* concentrations in the upper watershed (as discussed above). However, in the lower elevations of WRW, land use practices may be the predominant factor influencing *E. coli* concentrations. For example, in West Run Creek (Sites #13–#21), there was a notable increase in cumulative *E. coli* concentrations and a simultaneous increase in agricultural and mixed development land use practices (Figure 6). Previous work reported on the increased fecal contamination associated with increased agricultural and urban areas [31,70,73,74], commonly attributed to increased sources (livestock and manure) [23] and increased (concentrated flow) run-off during precipitation events, respectively [73], and urban stream syndrome [75], thereby supporting the results from the current investigation. The inter-site relationship between *E. coli* concentrations and SPM was not clearly discernable based on average values (Figure 5) or cumulative values in West Run Creek (Figure 6), as increases in SPM were not always accompanied by similar increases or decreases in *E. coli* between the different sampling locations. A potential explanation for these results may be that SPM and *E. coli* concentrations are affected by different factors at different sites (e.g., geochemistry, land use and antecedent soil water conditions). Thus, in WRW, there may exist a spatial disconnect regarding the factors influencing *E. coli* concentrations and subsequently impacting the relationship between *E. coli* and SPM in the watershed. To account for the spatial disconnect, site specific analysis of *E. coli* and SPM correlations, including separate analysis for different time periods in the year (quarters), was required to improve the current understanding regarding this relationship and to assess the use of SPM as a proxy for fecal contamination.

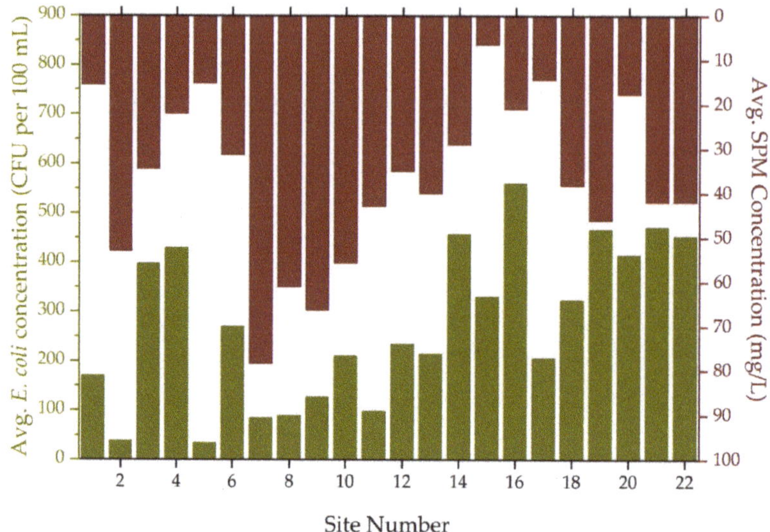

Figure 5. Average suspended particulate matter (mg L^{-1}) and *E. coli* concentration (CFU per 100 mL) at each sampling location (n = 22) during the study period (2 January 2018–1 January 2019) in West Run Watershed, Morgantown, West Virginia, USA.

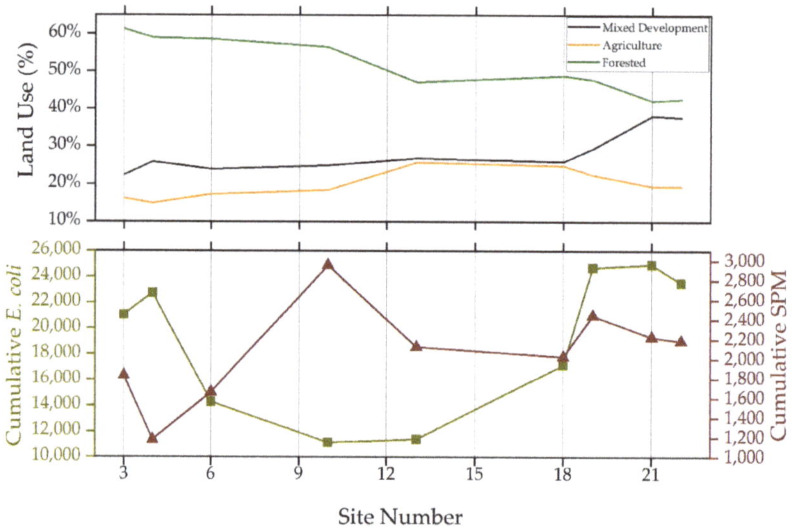

Figure 6. Land use percentage relative to cumulative annual *E. coli* concentration (CFU per 100 mL) and SPM concentration (mg/L) at West Run Creek monitoring sites (n = 9) during the study period (2 January 2018–1 January 2019) in West Run Watershed, Morgantown, West Virginia, USA. Note: West Run Creek included the following site numbers: #3, #4, #6, #10, #13, #18, #19, #21 and #22.

3.3. Quarterly Suspended Particulate Matter, E. coli Concentrations and Land Use Practices

Average *E. coli* and SPM concentration data showed notable temporal variation during 2018, based on quarterly analysis (Figure 7). *E. coli* concentrations were elevated during Quarter 2 (spring; 3 April 2018–26 June 2018) and Quarter 3 (summer; 3 July 2018–25 September 2018) of 2018, coinciding with the warmer spring and summer months (average maximum daily temperatures: Quarters 2 and 3 = 26 °C; Quarters 1 and 4 = 9 °C) of the year and higher stream flows induced by larger and more frequent precipitation events (Figure 3). Previous work noted correlations between elevated air temperatures and *E. coli* concentrations in freshwater [74], thereby supporting the results from the current work reporting greater *E. coli* concentrations during warmer months. Furthermore, the second and third quarters included some of the largest precipitation events of 2018 (Figure 3). These precipitation events could, at least in part, account for the elevated *E. coli* concentrations recorded during this time period, as previous work linked precipitation events to elevated *E. coli* concentrations in receiving waters [32,74]. Conversely, the highest average SPM concentrations were recorded during Quarters 1 (winter) and 2 (spring), with average concentrations decreasing during the second half of the year (Figure 7). The high SPM recorded during this time period may be attributable to decreased vegetation cover throughout WRW, leading to increased exposed soil surfaces, owing to the seasonal changes in vegetation (i.e., many plant species senesce during the cold winter months) [76,77]. Once vegetation throughout WRW increases during Quarter 3 (summer), exposed soil surfaces and sources of SPM become more limited. Quarter 4 (fall) comprised low *E. coli* and SPM concentrations. During this time period, WRW received less precipitation than in the preceding quarters (Figure 2), which could have limited the transport of pollutants (i.e., *E. coli* and SPM) to receiving waters [78]. Additionally, *E. coli* concentrations in the receiving water could have been suppressed by the colder temperatures [79] and decreased nutrient availability owing to the drier antecedent conditions and greater infiltration [77,80].

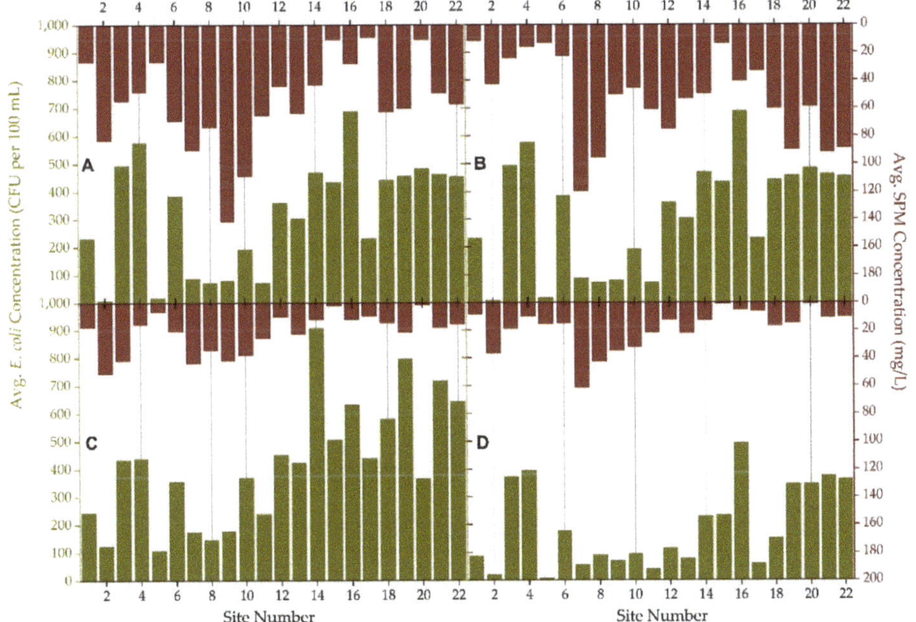

Figure 7. Quarterly average *E. coli* concentration (CFU per 100 mL) and SPM concentration (mg/L) at each sampling location (n = 22) during the study period (1/2/18–1/1/19) in West Run Watershed, Morgantown, West Virginia, USA. Note: (**A**) represents Quarter 1 (winter: 2 January 2018–27 March 2018); (**B**) represents Quarter 2 (spring: 3 April 2018–26 June 2018); (**C**) represents Quarter 3 (summer: 3 July 2018–25 September 2018); (**D**) represents Quarter 4 (fall: 2 October 2018–1 January 2019).

3.4. Non-Parametric Statistical Results

Normality test results indicated that the *E. coli* concentration data were non-normally distributed, thus Spearman correlation coefficients (nonparametric version of the Pearson product moment correlation) were used to quantify the relationships between *E. coli* concentration, SPM concentration, and land use at each site. *E. coli* concentrations and SPM concentrations were not significantly correlated at all sites; however, nine of the 22 sites (Sites #4, #7, #8, #9, #11, #15, #16, #17 and #20) did have significant correlations (Table 4). Notably, Sites #7, #8 and #9, which had the highest SPM during the investigation (Table 2; Figure 3), displayed statistically significant positive correlations ($p < 0.01$ for all three sites) between *E. coli* concentrations and SPM concentrations. Based on these relationships, SPM may serve as a relatively accurate proxy for *E. coli* concentrations in similarly physiographic catchments. Mixed development and forested sub-catchments did not display consistent significant correlations ($p < 0.05$) between *E. coli* and SPM concentrations. For example, Sites #15 (70.3% mixed development) and #17 (85.8% forested) both comprised statistically significant relationships between SPM and *E. coli* concentrations ($p < 0.01$and $p < 0.05$, respectively) despite comprising different predominant land use practices. However, Site #1 (53.2% mixed development) and Site #5 (51.1 % forested) displayed statistically insignificant correlations, despite including approximately similar dominant land use practices to Sites #15 and #17. As discussed in the preceding sections, AMD in the headwaters of WRW could be affecting *E. coli* concentrations, thereby creating inconsistency in the correlations between the *E. coli* and SPM concentrations. Therefore, the *E. coli* and SPM correlation results imply a spatial disconnect in terms of the influence of land use practices—in particular, mixed devolvement and forested areas—on the use of SPM as a proxy for fecal contamination. Notably, it seems likely, based on these results, that legacy effects (AMD) of historic land use practices (coal mining) may be impacting *E. coli* concentrations and affecting the observed relationships between *E. coli* and SPM concentrations. Conversely, both of the predominantly agricultural sites (Site #11 and #16) displayed significant correlations between SPM concentrations and *E. coli* concentrations, supporting previous work that reported elevated *E. coli* and SPM concentrations in the receiving waters of agricultural areas [19,20]. Therefore, in agricultural areas, SPM could be implemented as a proxy for *E. coli* with greater accuracy and less difficulty than in other land use areas.

Table 4. Results of Spearman's correlation test, including annual *E. coli* concentration (CFU per 100 mL) and annual SPM concentration (mg/L) at each sampling location (n = 22) during the study period (2 January 2018–1 January 2019) in West Run Watershed, WV, USA. Note: bold values indicate significant correlations ($p < 0.05$).

	Site Number										
	#1	#2	#3	#4	#5	#6	#7	#8	#9	#10	#11
SCC	0.19	0.18	0.04	**0.30**	−0.11	0.04	**0.46**	**0.52**	**0.73**	0.24	**0.56**
p-value	0.17	0.19	0.77	**0.03**	0.42	0.78	**<0.01**	**<0.01**	**<0.01**	0.09	**<0.01**
	Site Number										
	#12	**#13**	**#14**	**#15**	**#16**	**#17**	**#18**	**#19**	**#20**	**#21**	**#22**
SCC	0.00	0.13	−0.25	**0.42**	**0.64**	**0.27**	0.02	−0.14	**0.55**	0.01	0.17
p-value	0.99	0.34	0.07	**<0.01**	**<0.01**	**0.05**	0.88	0.33	**<0.01**	0.94	0.24

SCC = Spearman correlation coefficient.

Explanatory variables that account for the maximal variance in a data set can be identified via principle component analysis (PCA), through the computation of multiple principal components and their respective Eigenvalues [81]. Components comprising the highest Eigenvalues are assumed principal components, given that Eigenvalues represent the variance of the data in that direction [81]. A principle component is a linear function of the variables in an original data set that successively maximize variance and that are uncorrelated with each other [82]. Multiple principal components are typically calculated and ranked based on their Eigenvalues as most data cannot be well-described by a single principal component [81]. For the current work, the results showed three principal components with Eigenvalues exceeding 1 (an accepted threshold of importance [22,83]), including Principal Component 1 (Eigenvalue = 1.83), Principal Component 2 (Eigenvalue = 1.22) and Principal Component 3 (Eigenvalue = 1.17). These three principle components explained approximately 85%

of the cumulative variance of the data set. Conversely, Principal Components 4 and 5 accounted for approximately 16% and 0% of the variance of the data set. For the current work, the principle component biplots showed distinct spatial distributions for study sites along Principal Components 1 and 2 (Figure 8). The idealized biplot vector space defined by Principal Components 1 and 2 is characterized by the grouping of the sites. Given these results, it can be concluded that land use practices are the primary factors influencing the grouping of the data in the biplot, given the similarity of the sites in terms of geology, topography and climate, and their close proximity to each other [52]. The strongest correlation illustrated by the biplot is between SPM concentration and agricultural land use practices (Figure 8). However, *E. coli* concentration is also closely related to both, attributable to agricultural land use practices (i.e., the rearing of livestock, manure application, soil tillage and increased exposed soil surfaces), as discussed in the preceding sections and reported by previous investigations [19–23]. Ultimately, the PCA results were analogous to the Spearman correlation coefficient results, indicating that SPM could potentially serve as proxy for *E. coli* in agricultural areas, especially during periods with lower levels of fecal contamination between storm events.

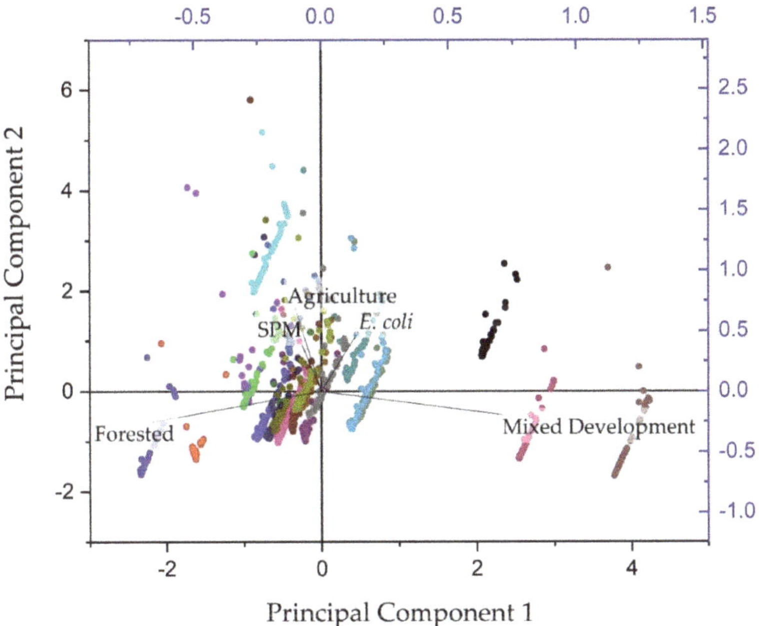

Figure 8. Results of principal component analysis, including biplots, for extracted principal components of annual *E. coli* concentration (CFU per 100 mL) and annual SPM concentration (mg/L) at 22 monitoring sites (indicated by the different colors) during the study period (2 January 2018–1 January 2019) in West Run Watershed, West Virginia, USA.

The Spearman correlation test results based on the quarterly analysis of data indicated predictable temporal variation in the correlation between *E. coli* and SPM concentrations, with only Site #9 showing significant correlations ($p < 0.05$) throughout all four quarters ($p < 0.01$ Quarters 1 to 3; $p = 0.02$ Quarter 4) (Table 5). Quarter 1 displayed the most significant correlations (10; Site #2, #7, #8, #9, #10, #11, #12, #13, #15, and #18), whereas Quarter 4 had the fewest significant correlations (two; Site #9 and #16) (Table 5). The temporal variation can be explained by the different impacts seasonal variation have on *E. coli* and SPM concentrations. For example, as discussed above, SPM concentrations will be influenced by changes in vegetation, with elevated concentrations typically occurring when vegetation cover decreases and decreasing as vegetation cover increases [76,77]. Therefore, elevated

SPM concentrations can be expected during and immediately after the cold winter months, with decreased concentrations during the warmer summer months. Conversely, previous work linked elevated *E. coli* concentrations with warmer water (and air) temperatures [84–86]. Consequently *E. coli* concentrations can be expected to be elevated during the warmer summer months and decrease during the colder winter months. Ultimately, there is a temporal (seasonal) difference between periods of elevated SPM and *E. coli* concentrations. For example, Quarter 3 included the highest average (424 CFU per 100 mL) *E. coli* concentrations across all 22 sampling locations, while Quarter 1 comprised the lowest average (187 CFU per 100 mL) *E. coli* concentration. Thus, between Quarter 1 and 3 there was more than a 100% increase in average *E. coli* concentrations across the 22 sampling sites. Conversely, Quarter 1 had the highest average (55.7 mg/L) SPM across all 22 sampling locations, while Quarter 4 comprised the lowest average (18.2 mg/L) SPM. Consequently, there was more than a 65% decrease in average SPM, across the 22 sampling locations, between Quarter 1 and Quarter 4. These temporal differences, driven by changes in precipitation [78], antecedent conditions [87], seasonal land cover [76] or land use practices [29] may account for the variable *E. coli* and SPM concentrations correlations identified in Table 5.

Table 5. Results of Spearman's correlation test, including quarterly *E. coli* concentration (CFU per 100 mL) and quarterly SPM concentration (mg/L) at each sampling location (n = 22) during the study period (2 January 2018–1 January 2019) in West Run Watershed, WV, USA. Note: Quarter 1 represents 2 January 2018–27 March 2018; Quarter 2 represents 3 April 2018–26 June 2018; Quarter 3 represents 3 July 2018–25 September 2018; Quarter 4 represents 2 October 2018–1 January 2019. Bold values indicate significant correlations ($p < 0.05$).

		Site Number										
		#1	**#2**	**#3**	**#4**	**#5**	**#6**	**#7**	**#8**	**#9**	**#10**	**#11**
Quarter 1	SCC	0.40	**0.60**	0.26	0.48	0.52	0.35	**0.76**	**0.70**	**0.73**	**0.63**	**0.69**
	p-value	0.18	**0.03**	0.39	0.09	0.09	0.23	**<0.01**	**0.01**	**<0.01**	**0.02**	**0.01**
Quarter 2	SCC	0.19	0.18	0.03	0.26	−0.04	0.07	**0.78**	**0.88**	**0.78**	0.51	**0.93**
	p-value	0.53	0.56	0.91	0.38	0.89	0.81	**<0.01**	**<0.01**	**<0.01**	0.08	**<0.01**
Quarter 3	SCC	−0.05	−0.40	0.32	0.50	**0.76**	**0.69**	−0.27	**0.58**	**0.79**	0.25	0.54
	p-value	0.88	0.17	0.29	0.08	**<0.01**	**0.01**	0.37	**0.04**	**<0.01**	0.42	0.06
Quarter 4	SCC	−0.30	−0.32	−0.39	−0.04	−0.02	−0.01	−0.08	−0.09	**0.62**	−0.22	0.51
	p-value	0.31	0.27	0.17	0.89	0.95	0.98	0.78	0.75	**0.02**	0.45	0.06

		Site Number										
		#12	**#13**	**#14**	**#15**	**#16**	**#17**	**#18**	**#19**	**#20**	**#21**	**#22**
Quarter 1	SCC	**0.63**	**0.62**	−0.16	**0.60**	0.36	0.10	**0.68**	0.16	0.35	0.12	0.40
	p-value	**0.02**	**0.02**	0.59	**0.03**	0.22	0.74	**0.01**	0.61	0.24	0.69	0.19
Quarter 2	SCC	0.17	0.44	**0.58**	**0.71**	**0.91**	0.42	0.54	0.24	**0.75**	0.39	0.37
	p-value	0.57	0.13	**0.04**	**0.01**	**<0.01**	0.15	0.06	0.43	**<0.01**	0.19	0.22
Quarter 3	SCC	**0.60**	0.45	0.22	0.29	**0.72**	**0.68**	0.51	0.34	**0.58**	0.38	0.40
	p-value	**0.03**	0.13	0.47	0.33	**0.01**	**0.01**	0.07	0.26	**0.04**	0.19	0.18
Quarter 4	SCC	0.09	−0.16	−0.39	−0.28	**0.56**	0.07	−0.31	−0.18	0.38	0.19	0.28
	p-value	0.75	0.58	0.17	0.33	**0.04**	0.82	0.27	0.53	0.19	0.53	0.34

The varying correlations displayed in Table 5 constitute important results regarding the use of SPM as a proxy for *E. coli*. Temporal changes in correlation significance indicate that SPM cannot be a consistently accurate proxy for fecal contamination throughout all the quarters (seasons) of a year. Even agricultural sub-catchments displayed insignificant correlations during certain quarters (Quarter 3 and 4 for Site #11 and Quarter 1 for Site #16), despite being significantly correlated for the annual time period. These results differ from previous investigations that showed strong correlations between *E. coli* and SPM concentrations [15,16], particularly small SPM particles [22]. However, these studies did not comprise high frequency or sufficiently long sampling regimes that would allow for quarterly (seasonal) analysis. Therefore, the current work is among the first to include seasonal analysis of the relationship (correlation) between *E. coli* and SPM concentrations in receiving waters.

Similarly to in the annual results, Quarter 1 (winter) had three principal components with Eigenvalues exceeding 1 (Eigenvalues = 1.86, 1.35 and 1.20, respectively), which accounted for 88% of the cumulative variance in the data. Quarters 2 and 3 also included three principal components with Eigenvalues exceeding 1 (1.85, 1.29 and 1.15; and 1.88, 1.22 and 1.16, respectively). The principal components explained 86% of the variance in Quarter 2 and 85% of that in Quarter 3. Two principal components, accounting for 63% of the data variance, were identified for Quarter 4, comprising Eigenvalues of 1.95 and 1.21. The quarterly PCA results illustrated the correlation between agricultural land use and *E. coli* and SPM during Quarters 2 (spring) and 3 (summer) (Figure 9). The agricultural land use impacts on *E. coli* and SPM concentrations during these quarters were subjected to the largest precipitation events during the study period (Figure 3). Precipitation has been reported to exacerbate the impact of land use practices on receiving waters [19,74,88]; therefore, land use impacts (particularly agriculture) were elevated during Quarters 2 (spring) and 3 (summer). The elevated impacts of land use practices during precipitation events are attributable to the increased transport of both *E. coli* and SPM during runoff events and subsequent increased *E. coli* and SPM concentrations in the receiving waters [88]. Conversely, during Quarters 1 (winter) and 4 (fall), none of the land use classes were closely correlated with either SPM or *E. coli* concentrations (Figure 9). Small and fewer precipitation events during Quarters 1 (winter) and 4 (fall) (Figure 3) may account for these results as fewer runoff events would lead to decreased concentrations of *E. coli* and SPM in the associated receiving waters [88]. The reduced precipitation during Quarters 1 and 4 would also have led to drier antecedent soil water conditions [89], leading to greater infiltration during subsequent precipitation events [90], further reducing the transport of both *E. coli* and SPM to the receiving waters. These results highlight the varied seasonality of land use impacts on *E. coli* and SPM concentrations and thus advance the science-based understanding of temporal fluctuations in *E. coli* concentration regimes.

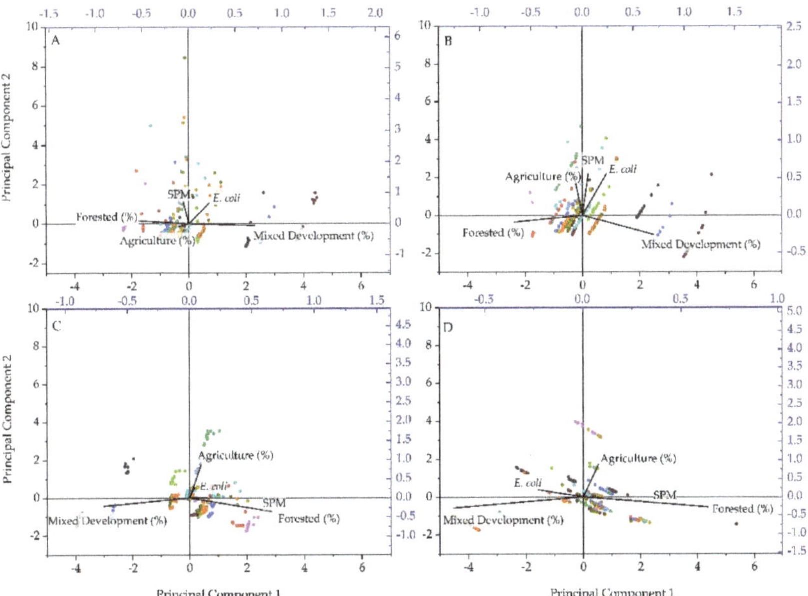

Figure 9. Results of principal component analysis, including biplots, for extracted principal components of quarterly *E. coli* concentration (CFU per 100 mL) and quarterly SPM concentration (mg/L) at each sampling location (n = 22) during the study period (2 January 2018–1 January 2019) in West Run Watershed, WV, USA. Note: (**A**) represents Quarter 1 (winter: 2 January 2018–27 March 2018); (**B**) represents Quarter 2 (spring: 3 April 2018–26 June 2018); (**C**) represents Quarter 3 (summer: 3 July 2018–25 September 2018); (**D**) represents Quarter 4 (fall: 2 October 2018–1 January 2019).

3.5. Study Implications and Future Work

The scale-nested experimental watershed study design and the high spatial and temporal sampling period implemented during the current work allowed for the collection of a unique data set. PCA biplots illustrated the close correlation between agricultural land use practices and both *E. coli* and SPM concentrations, relative to mixed development and forested land use practices. Additionally, spatial and temporal variability in the significant correlations between *E. coli* and SPM concentrations indicated that SPM would not be a suitable proxy for fecal contamination. The recorded lack of consistent *E. coli* and SPM relationships constitutes an important result for the development of accurate predictive fecal pollution models. The investigation emphasized the efficacy of the nested-scale experimental watershed study design to elucidate land use influences on fecal pollution in receiving waters. Future work should expand on the results from the current investigation by attempting to determine the precise tipping points associated with different land use practices' influence on *E. coli* and SPM concentrations. Implementing a similar study design, as in the current work, in mixed land use watersheds not impacted by legacy land use impacts (e.g., AMD) could provide useful information regarding the precise tipping points for various land use practices. Furthermore, due to the results of the current investigation indicating that *E. coli* concentrations are not solely influenced by land use practices and given the previously reported influence of physicochemical parameters (e.g., pH and water temperature) on *E. coli* [84–86], future work should focus on the identification of additional variables (e.g., physicochemical and geochemical) influencing *E. coli* concentrations in receiving waters. The incorporation of a multi-year study period would also allow future work to expand on the results of this study to account for annual variations in climate.

4. Conclusions

A 22-site, nested-scale, experimental watershed study design was implemented to investigate *E. coli* concentrations in a mixed land use watershed in the Appalachian region of the eastern United States. Specific focus was given to the relationship between *E. coli* concentrations, SPM concentrations and land use practices, including an evaluation of the potential use of SPM concentrations as a proxy for *E. coli* concentrations. Agricultural land use sub-catchments comprised elevated *E. coli* concentrations (avg. 560 CFU per 100 mL) compared to adjacent mixed development (avg. 330 CFU per 100 mL) and forested (avg. 206 CFU per 100 mL) sub-catchments. Annual *E. coli* and SPM concentration data displayed a statically significant relationship ($p < 0.01$) in agricultural areas. However, quarterly SCC analysis highlighted fluctuations between significance ($p < 0.05$) and insignificance ($p > 0.05$) in the correlations between *E. coli* and SPM concentrations across all land use classes. Therefore, SPM lacked the consistent significant correlations with *E. coli* concentrations required to be a suitable proxy for fecal contamination. The annual PCA results illustrated the influence of agricultural land use practices on both *E. coli* and SPM concentrations, serving as validation for previous investigations, which typically included less temporally and spatially robust sampling regimes. The quarterly PCA results highlighted the seasonal variability of land use impacts on both *E. coli* and SPM concentrations, with Quarters 2 and 3's biplots displaying greater correlations between agricultural land use practices, *E. coli* and SPM concentrations than Quarters 1 and 2. Combined, Quarters 2 and 3 received 67% more precipitation (850 mm) than Quarters 1 and 2 (510 mm), accounting for the temporal variation in land use impacts depicted by the quarterly biplots. Ultimately, the current investigation advances the understanding of the influence of land use practices on *E. coli* and SPM concentrations, thereby contributing to the current understanding of fecal contamination regimes in contemporary mixed land use watersheds. The results better inform model builders, policy makers and land use managers regarding the factors influencing freshwater fecal contamination, thereby aiding in effective decision making and effective water quality management.

Author Contributions: For the current work author contributions were as follows: conceptualization, J.A.H.; methodology, J.A.H.; formal analysis, F.P. and J.A.H.; investigation, F.P. and J.A.H.; resources, J.A.H.; data curation, J.A.H.; writing—original draft preparation, F.P. and J.A.H.; writing—review and editing, F.P. and J.A.H.; visualization, F.P. and J.A.H.; supervision, J.A.H.; project administration, J.A.H.; funding acquisition, J.A.H. All authors have read and agreed to the published version of the manuscript.

Funding: This work was supported by the National Science Foundation under Award Number OIA-1458952, the USDA National Institute of Food and Agriculture, Hatch project accession number 1011536, and the West Virginia Agricultural and Forestry Experiment Station. Results presented may not reflect the views of the sponsors and no official endorsement should be inferred. The funders had no role in study design, data collection and analysis, decision to publish, or preparation of the manuscript.

Acknowledgments: Special thanks are due to many scientists of the Interdisciplinary Hydrology Laboratory (https://www.researchgate.net/lab/The-Interdisciplinary-Hydrology-Laboratory-Jason-A-Hubbart). The authors also appreciate the feedback of anonymous reviewers whose constructive comments improved the article.

Conflicts of Interest: The authors declare no conflict of interest for the current work.

References

1. WHO. World Water Day Report. Available online: https://www.who.int/water_sanitation_health/takingcharge.html (accessed on 12 December 2019).
2. CDC. *E. coli (Escherichia coli)*. Available online: https://www.cdc.gov/ecoli/index.html (accessed on 8 January 2020).
3. Madoux-Humery, A.-S.; Dorner, S.; Sauvé, S.; Aboulfadl, K.; Galarneau, M.; Servais, P.; Prévost, M. The effects of combined sewer overflow events on riverine sources of drinking water. *Water Res.* **2016**, *92*, 218–227. [CrossRef] [PubMed]
4. Jamieson, R.; Joy, D.M.; Lee, H.; Kostaschuk, R.; Gordon, R. Transport and deposition of sediment-Associated *Escherichia coli* in natural streams. *Water Res.* **2005**, *39*, 2665–2675. [CrossRef] [PubMed]
5. Amalfitano, S.; Corno, G.; Eckert, E.; Fazi, S.; Ninio, S.; Callieri, C.; Grossart, H.-P.; Eckert, W. Tracing particulate matter and associated microorganisms in freshwaters. *Hydrobiologia* **2017**, *800*, 145–154. [CrossRef]
6. Bernard, J.M.; Steffen, L.L.; Iivari, T.A. Has the US sediment pollution problem been solved? In Proceedings of the Sixth Federal Interagency Sedimentation Conference, Sahara Hotel, Las Vegas, NV, USA, 10–14 March 1996; United States Geologic Survey. 1996; pp. 7–13. Available online: https://water.usgs.gov/osw/ressed/references/Bernard-Iivari-6Fisc2-8.pdf (accessed on 24 April 2020).
7. Ongley, E.D. *Control of Water Pollution from Agriculture*; Food and Agriculture Organization of the United Nations: Rome, Italy, 1996; ISBN 92-5-103875-9.
8. Hubbart, J.A.; Kellner, E.; Freeman, G. A case study considering the comparability of mass and volumetric suspended sediment data. *Environ. Earth Sci.* **2014**, *71*, 4051–4060. [CrossRef]
9. Uri, N.D. The environmental implications of soil erosion in the United States. *Environ. Monit. Assess.* **2001**, *66*, 293–312. [CrossRef] [PubMed]
10. Foster, I.D.L.; Charlesworth, S.M. Heavy metals in the hydrological cycle: Trends and explanation. *Hydrol. Process.* **1996**, *10*, 227–261. [CrossRef]
11. Oschwald, W.R. Sediment-water interactions 1. *J. Environ. Qual.* **1972**, *1*, 360–366. [CrossRef]
12. Russell, M.A.; Walling, D.E.; Webb, B.W.; Bearne, R. The composition of nutrient fluxes from contrasting UK river basins. *Hydrol. Process.* **1998**, *12*, 1461–1482. [CrossRef]
13. Walling, D. The changing sediment loads of the world's rivers. *Ann. Wars. Univ. Life Sci. SGGW Land Reclam.* **2008**, *39*, 3–20. [CrossRef]
14. Bilotta, G.S.; Brazier, R.E. Understanding the influence of suspended solids on water quality and aquatic biota. *Water Res.* **2008**, *42*, 2849–2861. [CrossRef]
15. Irvine, K.N.; Pettibone, G.W.; Droppo, I.G. Indicator Bacteria-Sediment Relationships: Implications for Water Quality Modeling and Monitoring. *J. Water Manag. Model.* **1995**, *3*, 205–230. [CrossRef]
16. Abia, A.L.K.; Ubomba-Jaswa, E.; Genthe, B.; Momba, M.N.B. Quantitative microbial risk assessment (QMRA) shows increased public health risk associated with exposure to river water under conditions of riverbed sediment resuspension. *Sci. Total Environ.* **2016**, *566–567*, 1143–1151. [CrossRef] [PubMed]
17. Cho, K.H.; Pachepsky, Y.A.; Oliver, D.M.; Muirhead, R.W.; Park, Y.; Quilliam, R.S.; Shelton, D.R. Modeling fate and transport of fecally-Derived microorganisms at the watershed scale: State of the science and future opportunities. *Water Res.* **2016**, *100*, 38–56. [CrossRef] [PubMed]

18. Southard, J. *Introduction to Fluid Motions, Sediment Transport, and Current-Generated Sedimentary Structures Course Textbook*; Massachusetts Institute of Technology, MIT OpenCourseWare: Cambridge, MA, USA, 2006; pp. 1–536.
19. Jeloudar, F.T.; Sepanlou, M.G.; Emadi, S.M. Impact of land use change on soil erodibility. *Glob. J. Environ. Sci. Manag.* **2018**, *4*, 59–70.
20. Petersen, F.; Hubbart, J.A.; Kellner, E.; Kutta, E. Land-Use-Mediated *Escherichia coli* concentrations in a contemporary Appalachian watershed. *Environ. Earth Sci.* **2018**, *77*, 754. [CrossRef]
21. AL-Kaisi, M. Frequent Tillage and its Impact on Soil Quality | Integrated Crop Management. 2020. Available online: https://crops.extension.iastate.edu/encyclopedia/frequent-tillage-and-its-impact-soil-quality (accessed on 14 February 2020).
22. Petersen, F.; Hubbart, J.A. Quantifying *Escherichia coli* and suspended particulate matter concentrations in a mixed-Land use Appalachian watershed. *Water* **2020**, *12*, 532. [CrossRef]
23. Rwego, I.B.; Gillspie, T.R.; Isabirye-Basuta, G.; Goldberg, T.L. High rates of *Escherichia coli* transmission between livestock and humans in rural Uganda. *J. Clin. Microbiol.* **2008**, *46*, 3187–3191. [CrossRef]
24. Causse, J.; Billen, G.; Garnier, J.; Henri-des-Tureaux, T.; Olasa, X.; Thammahacksa, C.; Latsachak, K.O.; Soulileuth, B.; Sengtaheuanghoung, O.; Rochelle-Newall, E. Field and modelling studies of *Escherichia coli* loads in tropical streams of montane agro-Ecosystems. *J. Hydro Environ. Res.* **2015**, *9*, 496–507. [CrossRef]
25. Jeng, H.C.; England, A.J.; Bradford, H.B. Indicator organisms associated with stormwater suspended particles and estuarine sediment. *J. Environ. Sci. Health* **2005**, *40*, 779–791. [CrossRef]
26. Characklis, G.W.; Dilts, M.J.; Simmons, O.D., III; Likirdopulos, C.A.; Krometis, L.-A.H.; Sobsey, M.D. Microbial partitioning to settleable particles in stormwater. *Water Res.* **2005**, *39*, 1773–1782. [CrossRef]
27. Oliver, D.M.; Clegg, C.D.; Heathwaite, A.L.; Haygarth, P.M. Preferential attachment of *Escherichia coli* to different particle size fractions of an agricultural grassland soil. *Water Air Soil Pollut.* **2007**, *185*, 369–375. [CrossRef]
28. Kändler, M.; Blechinger, K.; Seidler, C.; Pavlů, V.; Šanda, M.; Dostál, T.; Krása, J.; Vitvar, T.; Štich, M. Impact of land use on water quality in the upper Nisa catchment in the Czech Republic and in Germany. *Sci. Total Environ.* **2017**, *586*, 1316–1325. [CrossRef] [PubMed]
29. Thurston-Enriquez, J.A.; Gilley, J.E.; Eghball, B. Microbial quality of runoff following land application of cattle manure and swine slurry. *J. Water Health* **2005**, *3*, 157–171. [CrossRef] [PubMed]
30. Paster, E.; Ryu, W.S. The thermal impulse response of *Escherichia coli*. *Proc. Natl. Acad. Sci. USA* **2008**, *105*, 5373–5377. [CrossRef] [PubMed]
31. Stein, E.D.; Tiefenthaler, L.; Schiff, K. Comparison of storm water pollutant loading by land use type. In *Southern California Coastal Water Research Project 2008 Annual Report*; 3535 Harbor Blvd STE 110: Costa Mesa, CA, USA, 2008; pp. 15–27.
32. Rochelle-Newall, E.J.; Ribolzi, O.; Viguier, M.; Thammahacksa, C.; Silvera, N.; Latsachack, K.; Dinh, R.P.; Naporn, P.; Sy, H.T.; Soulileuth, B. Effect of land use and hydrological processes on *Escherichia coli* concentrations in streams of tropical, humid headwater catchments. *Sci. Rep.* **2016**, *6*, 32974. [CrossRef]
33. Widgren, S.; Engblom, S.; Emanuelson, U.; Lindberg, A. Spatio-Temporal modelling of verotoxigenic *Escherichia coli* O157 in cattle in Sweden: Exploring options for control. *Vet. Res.* **2018**, *49*, 78. [CrossRef]
34. Tetzlaff, D.; Carey, S.K.; McNamara, J.P.; Laudon, H.; Soulsby, C. The essential value of long-term experimental data for hydrology and water management. *Water Resour. Res.* **2017**, *53*, 2598–2604. [CrossRef]
35. Zeiger, S.; Hubbart, J.A. Quantifying suspended sediment flux in a mixed-land-use urbanizing watershed using a nested-scale study design. *Sci. Total Environ.* **2016**, *542*, 315–323. [CrossRef]
36. Leopold, L.B. Hydrologic research on instrumented watersheds. In *International Symposium on the Results of Research on Representative and Experimental Basins*; International Association of Scientific Hydrology: Wallingford, UK, 1970; pp. 135–150.
37. Hewlett, J.D.; Lull, H.W.; Reinhart, K.G. In Defense of experimental watersheds. *Water Resour. Res.* **1969**, *5*, 306–316. [CrossRef]
38. Bosch, J.M.; Hewlett, J.D. A review of catchment experiments to determine the effect of vegetation changes on water yield and evapotranspiration. *J. Hydrol.* **1982**, *55*, 3–23. [CrossRef]
39. Nichols, J.; Hubbart, J.A. Using macroinvertebrate assemblages and multiple stressors to infer urban stream system condition: A case study in the central US. *Urban Ecosyst.* **2016**, *19*, 679–704. [CrossRef]

40. Zeiger, S.; Hubbart, J.A.; Anderson, S.A.; Stambaugh, M.C. Quantifying and modelling urban stream temperature: A central US watershed study. *Hydrol. Process.* **2015**, *30*, 503–514. [CrossRef]
41. Hubbart, J.A.; Kellner, E.; Zeiger, S. A case-study application of the experimental watershed study design to advance adaptive management of contemporary watersheds. *Water* **2019**, *11*, 2355. [CrossRef]
42. Zeiger, S.J.; Hubbart, J.A. Nested-scale nutrient flux in a mixed-land-use urbanizing watershed. *Hydrol. Process.* **2016**, *30*, 1475–1490. [CrossRef]
43. Kellner, E.; Hubbart, J. Advancing understanding of the surface water quality regime of contemporary mixed-land-use watersheds: An application of the experimental watershed method. *Hydrology* **2017**, *4*, 31. [CrossRef]
44. Hubbart, J.A. Urban floodplain management: Understanding consumptive water-use potential in urban forested floodplains. *Stormwater J.* **2011**, *12*, 56–63.
45. Cantor, J.; Krometis, L.-A.; Sarver, E.; Cook, N.; Badgley, B. Tracking the downstream impacts of inadequate sanitation in central Appalachia. *J. Water Health* **2017**, *15*, 580–590. [CrossRef]
46. Dykeman, W. Appalachian Mountains. Available online: https://www.britannica.com/place/Appalachian-Mountains (accessed on 20 November 2019).
47. Central Appalachian Broadleaf Forest—Coniferous Forest—Meadow Province. Available online: https://www.fs.fed.us/land/ecosysmgmt/colorimagemap/images/m221.html (accessed on 12 December 2019).
48. Köppen, W. Das geographische System der Klimate. In *Handb. Der Klimato*; Gebruder Borntraeg: Berlin, Germany, 1936.
49. Arcipowski, E.; Schwartz, J.; Davenport, L.; Hayes, M.; Nolan, T. Clean water, clean life: Promoting healthier, accessible water in rural Appalachia. *J. Contemp. Water Res. Educ.* **2017**, *161*, 1–18. [CrossRef]
50. Peel, M.C.; Finlayson, B.L.; McMahon, T.A. Updated world map of the Köppen-Geiger climate classification. *Hydrol. Earth Syst. Sci. Discuss.* **2007**, *4*, 439–473. [CrossRef]
51. Arguez, A.; Durre, I.; Applequist, S.; Squires, M.; Vose, R.; Yin, X.; Bilotta, R. NOAA's US climate normals (1981–2010). *NOAA Natl. Cent. Environ. Inf.* **2010**, *10*, V5PN93JP.
52. Kellner, E.; Hubbart, J.; Stephan, K.; Morrissey, E.; Freedman, Z.; Kutta, E.; Kelly, C. Characterization of sub-watershed-scale stream chemistry regimes in an Appalachian mixed-land-use watershed. *Environ. Monit. Assess.* **2018**, *190*, 586. [CrossRef] [PubMed]
53. The West Virginia Water Research Institute (WVWRI); The West Run Watershed Association. *Watershed Based Plan for West Run of the Monongahela River*; The West Virginia Water Research Institute, The West RunWatershed Association: Morgantown, WV, USA, 2008; pp. 1–59.
54. Hubbart, J.A.; Muzika, R.-M.; Huang, D.; Robinson, A. Improving quantitative understanding of bottomland hardwood forest influence on soil water consumption in an urban floodplain. *Watershed Sci. Bull.* **2011**, *3*, 34–43.
55. Hubbart, J.A. Measuring and Modeling Hydrologic Responses to Timber Harvest in a Continental/Maritime Mountainous Environment. Ph.D. Thesis, University of Idaho, Moscow, ID, USA, 2007.
56. Wei, L.; Hubbart, J.A.; Zhou, H. Variable streamflow contributions in nested subwatersheds of a US Midwestern urban watershed. *Water Resour. Manag.* **2018**, *32*, 213–228. [CrossRef]
57. Hubbart, J.A.; Kellner, E.; Hooper, L.W.; Zeiger, S. Quantifying loading, toxic concentrations, and systemic persistence of chloride in a contemporary mixed-land-use watershed using an experimental watershed approach. *Sci. Total Environ.* **2017**, *581*, 822–832. [CrossRef] [PubMed]
58. Zeiger, S.J.; Hubbart, J.A. Quantifying flow interval–pollutant loading relationships in a rapidly urbanizing mixed-land-use watershed of the Central USA. *Environ. Earth Sci.* **2017**, *76*, 484. [CrossRef]
59. Dusek, N.; Hewitt, A.J.; Schmidt, K.N.; Bergholz, P.W. Landscape-scale factors affecting the prevalence of *Escherichia coli* in surface soil include land cover type, edge interactions, and soil pH. *Appl. Environ. Microbiol.* **2018**, *84*, 1–19. [CrossRef]
60. Desai, M.A.; Rifai, H.S. Variability of *Escherichia coli* concentrations in an urban watershed in Texas. *J. Environ. Eng.* **2010**, *136*, 1347–1359. [CrossRef]
61. American Society for Testing and Materials. *Standard Methods for Determining Sediment Concentrations in Water Samples*; American Society for Testing and Materials: West Conshohocken, PA, USA, 2007; pp. 1–6.
62. Price, R.G.; Wildeboer, D. *E. coli* as an indicator of contamination and health risk in environmental waters. In *Escherichia coli—Recent Advances on Physiology, Pathogenesis and Biotechnological Applications*; Intech: London, UK, 2017. [CrossRef]

63. IDEXX. Laboratories Colilert Procedure Manual. Available online: https://www.idexx.com/files/colilert-procedure-en.pdf (accessed on 4 April 2019).
64. Cummings, D.; IDEXX. The Fecal Coliform Test Compared to Specific Tests for Escherichia Coli. Available online: https://www.idexx.com/resource-library/water/water-reg-article9B.pdf (accessed on 24 September 2019).
65. Yazici, B.; Yolacan, S. A comparison of various tests of normality. *J. Stat. Comput. Simul.* **2007**, *77*, 175–183. [CrossRef]
66. Fisher, R.A. Statistical Methods for Research Workers. In *Breakthroughs in Statistics: Methodology and Distribution*; Kotz, S., Johnson, N.L., Eds.; Springer Series in Statistics; Springer: New York, NY, USA, 1992; pp. 66–70. ISBN 978-1-4612-4380-9.
67. United States Climate Data (USCD). Available online: https://www.usclimatedata.com/climate/morgantown/west-virginia/united-states/uswv0507/2012/7 (accessed on 28 September 2019).
68. Chapman, D. Rivers. In *Water Quality Assessments—A Guide to Use of Biota, Sediments and Water in Environmental Monitoring*; UNESCO/WHO/UNEP: Paris, France, 1992; ISBN 0 419 21590 5.
69. Lazar, J.A. Urban and Agricultural Land Cover Impacts on Storm Flow and Nutrient Concentrations in SW Ohio Streams. Master's Thesis, Miami University, Oxford, OH, USA, 2018.
70. Tong, S.T.; Chen, W. Modeling the relationship between land use and surface water quality. *J. Environ. Manag.* **2002**, *66*, 377–393. [CrossRef]
71. RoyChowdhury, A.; Sarkar, D.; Datta, R. Remediation of acid mine drainage-impacted water. *Curr. Pollut. Rep.* **2015**, *1*, 131–141. [CrossRef]
72. Brown, T.C.; Binkley, D.; Brown, D. Water Quality on Forest Lands | Rocky Mountain Research Station. Available online: /rmrs/projects/water-quality-forest-lands (accessed on 11 March 2020).
73. Gotkowska-Plachta, A.; Golaś, I.; Korzeniewska, E.; Koc, J.; Rochwerger, A.; Solarski, K. Evaluation of the distribution of fecal indicator bacteria in a river system depending on different types of land use in the southern watershed of the Baltic Sea. *Environ. Sci. Pollut. Res.* **2016**, *23*, 4073–4085. [CrossRef] [PubMed]
74. Wu, J.; Yunus, M.; Islam, M.S.; Emch, M. Influence of climate extremes and land use on fecal contamination of shallow tubewells in Bangladesh. *Environ. Sci. Technol.* **2016**, *50*, 2669–2676. [CrossRef] [PubMed]
75. Booth, D.B.; Roy, A.H.; Smith, B.; Capps, K.A. Global perspectives on the urban stream syndrome. *Freshw. Sci.* **2016**, *35*, 412–420. [CrossRef]
76. Kim, Y.; Wang, G. Modeling seasonal vegetation variation and its validation against Moderate Resolution Imaging Spectroradiometer (MODIS) observations over North America. *J. Geophys. Res. Atmos.* **2005**, *110*, 1–13. [CrossRef]
77. Loch, R.J. Effects of vegetation cover on runoff and erosion under simulated rain and overland flow on a rehabilitated site on the Meandu Mine, Tarong, Queensland. *Soil Res.* **2000**, *38*, 299–312. [CrossRef]
78. Chen, H.J.; Chang, H. Response of discharge, TSS, and *E. coli* to rainfall events in urban, suburban, and rural watersheds. *Environ. Sci. Process. Impacts* **2014**, *16*, 2313–2324. [CrossRef]
79. Farewell, A.; Neidhardt, F.C. Effect of temperature on in vivo protein synthetic capacity in *Escherichia coli*. *J. Bacteriol.* **1998**, *180*, 4704–4710. [CrossRef]
80. Mcmillan, S.K.; Wilson, H.F.; Tague, C.L.; Hanes, D.M.; Inamdar, S.; Karwan, D.L.; Loecke, T.; Morrison, J.; Murphy, S.F.; Vidon, P. Before the storm: Antecedent conditions as regulators of hydrologic and biogeochemical response to extreme climate events. *Biogeochemistry* **2018**, *141*, 487–501. [CrossRef]
81. Multivariate Analysis: Principal Component Analysis: Biplots—9.3. Available online: http://support.sas.com/documentation/cdl/en/imlsug/64254/HTML/default/viewer.htm#imlsug_ugmultpca_sect003.htm (accessed on 23 December 2019).
82. Jolliffe, I.T.; Cadima, J. Principal component analysis: A review and recent developments. *Philos. Trans. A Math. Phys. Eng. Sci.* **2016**, *374*, 1–18. [CrossRef]
83. Bro, R.; Smilde, A.K. Principal component analysis. *Anal. Methods* **2014**, *6*, 2812–2831. [CrossRef]
84. Verhougstraete, M.P.; Martin, S.L.; Kendall, A.D.; Hyndman, D.W.; Rose, J.B. Linking fecal bacteria in rivers to landscape, geochemical, and hydrologic factors and sources at the basin scale. *Proc. Natl. Acad. Sci. USA* **2015**, *112*, 10419–10424. [CrossRef] [PubMed]
85. Hong, H.; Qiu, J.; Liang, Y. Environmental factors influencing the distribution of total and fecal coliform bacteria in six water storage reservoirs in the Pearl River Delta Region, China. *J. Environ. Sci.* **2010**, *22*, 663–668. [CrossRef]

86. Islam, M.M.M.; Hofstra, N.; Islam, M.A. The impact of environmental variables on faecal indicator bacteria in the Betna River Basin, Bangladesh. *Environ. Process.* **2017**, *4*, 319–332. [CrossRef]
87. Biron, P.M.; Roy, A.G.; Courschesne, F.; Hendershot, W.H.; Côté, B.; Fyles, J. The effects of antecedent moisture conditions on the relationship of hydrology to hydrochemistry in a small forested watershed. *Hydrol. Process.* **1999**, *13*, 1541–1555. [CrossRef]
88. Shi, P.; Zhang, Y.; Li, Z.; Li, P.; Xu, G. Influence of land use and land cover patterns on seasonal water quality at multi-spatial scales. *Catena* **2017**, *151*, 182–190. [CrossRef]
89. Ali, S.; Ghosh, N.C.; SIngh, R. Rainfall–Runoff simulation using a normalized antecedent precipitation index. *Hydrol. Sci. J.* **2010**, *55*, 266–274. [CrossRef]
90. Song, S.; Wang, W. Impacts of antecedent soil moisture on the rainfall-runoff transformation process based on high-resolution observations in soil. *Water* **2019**, *11*, 296. [CrossRef]

 *water*

Review

Physical Factors Impacting the Survival and Occurrence of *Escherichia coli* in Secondary Habitats

Fritz Petersen [1,*] and Jason A. Hubbart [1,2]

[1] Division of Plant and Soil Sciences, Davis College of Agriculture, Natural Resources and Design, West Virginia University, Agricultural Sciences Building, Morgantown, WV 26506, USA; jason.hubbart@mail.wvu.edu

[2] Institute of Water Security and Science, Schools of Agriculture and Food, and Natural Resources, Davis College of Agriculture, Natural Resources and Design, West Virginia University, 3109 Agricultural Sciences Building, Morgantown, WV 26506, USA

* Correspondence: fp0008@mix.wvu.edu; Tel.: +1-304-413-5449

Received: 13 May 2020; Accepted: 20 June 2020; Published: 23 June 2020

Abstract: *Escherichia (E.) coli* is a fecal microbe that inhabits the intestines of endotherms (primary habitat) and the natural environment (secondary habitats). Due to prevailing thinking regarding the limited capacity of *E. coli* to survive in the environment, relatively few published investigations exist regarding environmental factors influencing *E. coli's* survival. To help guide future research in this area, an overview of factors known to impact the survival of *E. coli* in the environment is provided. Notably, the lack of historic field-based research holds two important implications: (1) large knowledge gaps regarding environmental factors influencing *E. coli's* survival in the environment exist; and (2) the efficacy of implemented management strategies have rarely been assessed on larger field scales, thus leaving their actual impact(s) largely unknown. Moreover, the persistence of *E. coli* in the environment calls into question its widespread and frequent use as a fecal indicator microorganism. To address these shortcomings, future work should include more field-based studies, occurring in diverse physiographical regions and over larger spatial extents. This information will provide scientists and land-use managers with a new understanding regarding factors influencing *E. coli* concentrations in its secondary habitat, thereby providing insight to address problematic fecal contamination effectively.

Keywords: bacteria; water quality; land-use practices; environmental persistence

---

1. Introduction

Escherichia coli (*E. coli*) is a fecal indicator microbe with a life history that cycles between two principal habitats, intestines of endotherms (primary habitat) and environmental water, sediment, and soils (secondary habitats). These habitats differ markedly with respect to physical conditions (e.g., temperature) and nutrient availability [1]. For example, the temperature remains relatively constant (approximately 37 °C) in the primary habitat but can vary greatly in the secondary habitat where annual average temperatures can range from below freezing (0 °C) to approximately 18 °C, or higher [1]. Additionally, the primary habitat is an anaerobic environment [2], whereas the secondary environment varies between aerobic and anaerobic (e.g., deep soil, sediment, and water resources) [3]. Nutrients in the secondary environment are also typically less abundant, especially in soil and sediment [4], which, with respect to bacterial growth, are in a state of constant nutrient deficiency [5]. In water, nutrients can vary from being abundant (e.g., receiving waters in agricultural areas) to scarce (e.g., open ocean) [6,7]. This contrasts primary habitat (i.e., colon) nutrient conditions, comprising consistently high levels that support rapid bacterial growth [1]. Consequently, the secondary habitat will typically place greater strain on the growth and survival of *E. coli*, as it is not the habitat the *Escherichia* genus predominantly evolved in.

The conditions of the secondary habitat (environment) will, therefore, inhibit the growth and survival of *E. coli* when the microbe's tolerance thresholds are exceeded. The tolerance thresholds of *E. coli* can be used to predict changes in *E. coli* concentrations based on changes in the conditions of the secondary environment. This can be used by scientists and land-use managers concerned with microbial pollution in the environment.

When present in the environment, fecal microbes, such as the O157:H7 serotype of *E. coli*, can pose considerable health risks to endotherms (humans and animals), particularly if ingested [8]. Exposure to these microbes contributes significantly to morbidity and mortality in the global human population (3.4 million annual deaths) [9]. Therefore, managing the abundance of fecal microbes in the environment is important from a human health perspective. Managing fecal microbe concentrations and, by extension, the water quality of receiving waters requires an understanding of factors influencings the lifecycles and concentrations of these microorganisms in their secondary habitat. Therefore, understanding temperature and solar insolation influences, hydrologic requirements, chemistry, and nutrient availability, and land-use impacts on *E. coli* populations are critical for the proper implementation of effective management strategies.

Historically, *E. coli* was thought to be poorly adapted to survival in the environment and believed to comprise an average half-life of two days [1]. Additionally, it was believed that *E. coli* cells could only originate in the intestines of endotherms before excretion into the secondary habitat. The transfer from primary to secondary habitats was represented by the following two relationships [1]:

$$\frac{dP}{dt} = \gamma P - \beta P \tag{1}$$

and

$$\frac{dS}{dt} = \beta P - \delta S \tag{2}$$

where P represents *E. coli* populations in the primary habitat, S represents *E. coli* populations in the secondary habitat, and γ, β, and δ represent effective growth, bulk transfer, and death rate, respectively. Traditional thought surrounding *E. coli*'s limited survival in the secondary habitat held two important implications: (1) deposition of new fecal matter was required to increase *E. coli* populations in the secondary habitat, and (2) the probability of new host colonization was low [1]. Direct deposition of new fecal matter remains an important factor influencing fecal microbe concentrations in the environment, due to the direct proportionality between endotherm population density and fecal contamination (i.e., more animals produce more waste) [10–12]. However, recent investigations have reported on the ability of *E. coli* to persist (survive and reproduce) in the environment for extended periods, thereby increasing the risk of colonizing a new host. This shift in the understanding of the persistence of *E. coli* in the secondary habitat implies that there may be ample time for the microbe to become naturalized into the soil microbiome [13]. Once naturalized, the microbe may become autochthonous and thus, capable of surviving and reproducing in the environment without being directly deposited or replaced by microorganisms from animal feces or water [13]. Given the extended persistence of *E. coli* and its widespread use as a fecal indicator organism [14,15], understanding the relationships between the environment and this microbe will aid land managers and scientists concerned with mitigation of microbial contamination in soil and water resources.

The objective of this review is to provide an overview of factors currently understood to affect *E. coli* lifecycles and concentrations in the secondary habitat (encompassing both soil and water environments), including how those factors can be used to predict *E. coli* population changes quantitatively. Notably, few previous publications have included reviews of quantitative predictive equations used in the estimation of *E. coli* occurrence or survival. A sub-objective was to supply a summary of contemporary management strategies aimed at reducing *E. coli* and fecal microbe concentrations in the environment, specifically management strategies that can be applied by land-use managers and policy-makers at a sub-catchment scale. The review also includes the identification of research needs and future directions regarding the survival of *E. coli* in the secondary habitat.

2. *E. coli* and the Environment

2.1. Temperature

In nutrient-rich environments (e.g., canned meat products), temperature (approximately 0 to 47 °C) is typically considered the primary factor influencing *E. coli* survival, accounting for up to 61% of *E. coli* population variance (centered on inactivation rates) based on an Arrhenius model [16]. Given its relative importance, temperature must be accounted for when assessing local environmental parameters that influence *E. coli* lifecycles and concentrations in the secondary habitat. Thus, the growth limit and tolerance range of the microbe, 7 °C and −20 °C to 66 °C, respectively, provide insight as to how temperature can alter the bacteria's lifecycle [17]. Previous investigations reported bacterial temperature dependencies, based on the first-order inactivation rates k_c, defined by the Chick equation [18,19] as

$$C = C_0 e^{-k_c t} \tag{3}$$

or alternatively

$$\ln C = lnC_0 - k_c t \tag{4}$$

where C represents bacterial concentration, C_0 represents initial bacterial concentration, k_c represents the inactivation rate, and t represents time in days [19]. In the above equations, the temperature dependence of bacterial inactivation (k_c) rates can be expressed using either the Q_{10} or Mancini equations [19]. The Q_{10} equation is defined as follows:

$$\frac{k}{k_*} = Q_{10}{}^{(T_c - T_{c*})/10} \tag{5}$$

where k represents the first-order inactivation at temperature T_c, k_* represents reference temperature first-order inactivation, Q_{10} represents the rate of change in the inactivation rate due to temperature increases in 10 °C increments, *Tc* represents temperature in °C, and T_{c*} represents a reference temperature (usually 20 °C) [19]. Conversely, the Mancini equation, frequently implemented in water quality models [19], can be defined as follows:

$$\frac{k}{k_*} = \theta^{T_c - T_{c*}} \tag{6}$$

where k represents first-order inactivation (i.e., die-off) at temperature T_c, k_* represents reference temperature first-order inactivation, θ represents temperature sensitivity of the microbe, T_c represents temperature in °C, and T_{c*} represents a reference temperature (usually 20 °C). Consequently, Q_{10} can be converted to θ as follows:

$$\theta = Q_{10}{}^{1/10} \tag{7}$$

Additionally, the temperature dependence of first-order reaction rates can be found through the Arrhenius equation:

$$k = Ae^{(-\frac{E_a}{RT})} \tag{8}$$

where k represents the kinetic rate constant at temperature T, A represents the constant prefactor (collision frequency factor), E_a represents activation energy, R represents the gas constant, and T represents temperature in Kelvin.

Notably, environmental factors, such as water purity, can influence *E. coli* inactivation rates, as pristine water usually comprises higher average Q_{10} values but lower first-order inactivation rates at 20 °C. For example, based on a review of 450 *E. coli* survival datasets from 70 peer-reviewed investigations by Blaustein et al. [19], pristine water (defined by the authors as water originating from caves or springs and including fewer impurities) comprised an average Q_{10} of 2.066 ± 0.190 and an average first-order inactivation rate at 20 °C (k_{20}/day^{-1}) of 0.063 ± 0.007. Conversely, groundwater, agricultural waters and wastewater comprised the following Q_{10} and k_{20}/day^{-1} values: 1.783 ± 0.702, 0.504 ± 0.136; 1.548 ± 0.161, 0.388 ± 0.024, and 1.358 ± 0.238, 0.672 ± 0.114, respectively [19]. Consequently, the source of water can be used to approximate inactivation for aquatic *E. coli*. However, the impacts of biological and physical survival factor variations can cause variability in site- and source-specific *E. coli* survival rates at the same temperature [19]. Therefore, site-specific data (e.g., physicochemical parameters) would greatly improve the accuracy of predictions regarding *E. coli* inactivation and survival in water resources.

Based on laboratory investigations, *E. coli* was shown to survive for up to 32 days in soil when incubated at 15 °C [13]. In situ investigations also support the extended survival of *E. coli* in soils, even at colder temperatures. For example, *E. coli* in soil samples extracted from soils surrounding Lake Superior, United States of America (USA), in October 2003 and April 2004 had a 92% DNA fingerprint similarity [13]. This high similarity indicates these *E. coli* strains became naturalized. Naturalized, in this context, indicates the process by which non-native *E. coli* becomes integrated into the secondary habitat and reproduces at a sufficient rate to maintain its population [20]. Through naturalization, *E coli* become autochthonous members of the soil microbial community, capable of enduring the cold winter months (including numerous freeze–thaw cycles) and growing during the warmer summer months [13]. Additionally, the growth and replication of *E. coli* in soils have been verified by laboratory studies. For example, the bacteria grew to high cell densities (4.2×10^5 colony forming units (CFU)/g soil) when incubated at 30 °C to 37 °C in nonsterile soils [13]. Additionally, when 9×10^2 CFU/g soil *E. coli* was inoculated and incubated for 16 days at 15 °C before a temperature increase (37 °C for 8 days), cell density decreased to 1.1×10^2 CFU/g soil (during the 16-day period), before increasing 10-fold to 1.04×10^3 CFU/g soil 4 days after the temperature increase [13]. However, there was a subsequent decrease in cell numbers to 7.7 CFU/g soil 8 days after the temperature increase, attributed to nutrient depletion (see Section 2.6) [13]. Soil *E. coli* population density is also subject to seasonal variation with the highest cell densities, up to 3×10^3 CFU/g soil, reported in warmer months (summer to autumn), and the lowest numbers, ≤1 CFU/g soil, reported in the colder months (winter to spring) [13]. Consequently, *E. coli* growth is directly influenced by soil temperature fluctuations, with rapid growth increases occurring as soil temperature rises from 15 °C (no growth) to 37 °C [13]. Ultimately, given its reported influence on bacterial survival and growth [13,16,21], temperature constitutes a vital environmental factor influencing *E. coli's* survival in the secondary habitat.

2.2. Solar Insolation

Research regarding the impacts of solar insolation on fecal bacteria (e.g., *E. coli*) survival and inactivation, have been predominantly focused on marine waters [22–24]. However, in a study conducted at Lake Michigan, USA, day length and exposure to insolation during sunny days resulted in an exponential decrease in *E. coli* counts [25]. Additionally, diminished *E. coli* inactivation was reported during cloudy days [25]. For example, *E. coli* concentrations frequently exceeded safe swimming criteria (threshold *E. coli* concentration in water at which the bacteria becomes hazardous to human health, approximately >235 CFU 100 mL^{-1}) during partly cloudy or completely cloudy conditions, but rarely exceeded this threshold during sunny conditions [25]. Similarly, results from both a marsh and lagoon in California indicated that first-order *E. coli* decay rate constants varied between 1 to 2 days during low light conditions and 6 days during high light conditions [26]. Furthermore, submersion depth also impacted *E. coli* decay rates. For example, sunny condition decay rates at 45 cm and 90 cm depths were $Y_{45} = 48091e^{-0.4682t}$ and $Y_{90} = 12746e^{-0.4184t}$, respectively, where Y represents *E. coli* concentration (CFU 100 mL^{-1}), and t represents time (hour) [25]. Notably, different components of sunlight can yield different responses in *E. coli* [27]. For example, ultraviolet B-ray (UVB) intensity has been reported to impact first-order decay rates, as the two are highly correlated ($\alpha < 0.05$) [26]. Moreover, short exposure (six hours) to UVB was shown to be sufficient to decrease culturability and reduce the activity of E. coli, thus eliciting similar effects as exposure to sunlight [27]. Conversely, exposing E. coli to ultraviolet A-rays (UVA) or photosynthetically active radiation (PAR) reduces the culturability of the cells to 10%, despite remaining metabolically active [27]. The impact of insolation on E. coli inactivation is also subject to initial bacterial concentrations, with higher concentrations having quicker decay rates [25]. Moreover, lake E. coli density could be more accurately predicted by exposure time (dosage) than insolation [25]. Thus, the impact of extended periods of insolation exceeded the effect of intense insolation over shorter periods. In the Lake Michigan study, insolation was the predominant abiotic factor influencing E. coli inactivation, accounting for 40% of the variance as opposed to 7% by temperature and 8% for relative lake level [25]. Therefore, the results from this study challenge the assumption that temperature is the primary factor influencing E. coli survival, specifically at the surface (upper 90 cm) of freshwater bodies. Consequently, in shallow streams and headwaters, the inactivation of E. coli could be primarily driven by insolation and not temperature.

2.3. Suspended and Settled Solids

The survival of E. coli in water can be influenced by suspended solids concentrations in terms of how readily microbes can attach to those particles [28]. Association can increase nutrient and organic matter availability, particularly when the suspended solids include organic material (e.g., fallen leaf litter), while also providing optimal light exposure [29,30]. In addition, the close proximity of suspended particle-associated microbes to each other can facilitate the horizontal transfer and proliferation of resistance genes [31,32]. The horizontal transfer of genetic material can be expedited when two microbes come into close contact with each other and remain that way until the transfer of genetic material is completed [31,32]. Thus, if two or more microbes associate with the same suspended particle, the likelihood of horizontal genetic transfer increases relative to free-floating microbes. If resistance genes are transferred in this manner, over time, the microbial population may display increased resistance to stressors, such as chemical disinfectants, excessive photosynthetically active radiation (PAR) radiation, ultraviolet (UV) radiation, and predation [33–35]. However, the effect of suspended solids, including sediment, on the inactivation and survival of E. coli in the secondary habitat (the environment) is yet to be quantified as the majority of studies that attempted to quantify this relationship also include temperature fluxes which have a greater impact on E. coli variance [16,36]. Consequently, no equations are available that relate changes in suspended solids to associated changes in E. coli concentrations. Nevertheless, given current understanding, decreased suspended solids in receiving waters will decrease E. coli survivability, thereby decreasing concentrations of this microbe.

The survivability of E. coli in settled sediments has been quantified [37] using an exponential die-off model based on the Chick [18] equation:

$$\ln C = lnC_0 - \mu_t \quad (9)$$

where C represents current concentration; C_0 represents initial concentration; μ represents inactivation rate, and t represents time. Different inactivation rates have been reported by previous investigations. For example, settled sediment from the lakes of eastern United States comprised an E. coli inactivation (i.e., die-off) rate (the authors of this work used inactivation rate synonymously with die-off rate) of approximately 0.54 d^{-1} [37], whereas settled sediments of Southern Ontario creeks included an inactivation rate of 0.15 d^{-1}[38], and the Hillsborough River in Florida comprised an inactivation rate of 0.07 d^{-1} [39]. Like other environmental (secondary habitat) variables (see Section 2.1), temperature influences the die-off of E. coli in sediment, with more rapid die-off occurring at warmer temperatures [37]. Additionally, sediment particle size impacts temperature-driven die-off rates, with survival rates being less sensitive to temperature in finer soil. Garzio-Heardick et al. [40] reported the temperature-driven die-off rates relative to soil types as follows:

$$\text{Sand}: \ \mu = 0.109 * 1.133^{T-20} \quad (10)$$

$$\text{Sandy Loam}: \ \mu = 0.051 * 1.105^{T-20} \tag{11}$$

$$\text{Silt Loam}: \ \mu = 0.046 * 1.054^{T-20} \tag{12}$$

where μ represents the die-off rate, and T represents the temperature in °C. Ultimately, *E. coli*'s survival in settled sediment will vary geographically, due to changes in both temperature and physical soil characteristics. Therefore, to understand *E. coli* survival (including changes in survival due to land-use changes or mitigation strategies) in sediment at a specified location, site-specific information would be required, to avoid broad assumptions potentially leading to prediction inaccuracies.

2.4. Hydrologic Conditions

Intense precipitation and subsequent runoff events can increase pollutant transport, thereby deteriorating surface water quality by increasing turbidity, suspended solid concentrations, organic matter, and fecal contamination during stormwater discharge events [11,12]. Similarly, increased overland and streamflow, during storm events, have been linked to increased *E. coli* concentrations relative to baseflow conditions [41,42]. The magnitude of the *E. coli* concentration increase varies between 15-fold [43] to 1000-fold [12], such that the concentration increase can be represented by the formula below:

$$C_s \geq C_0 I_s \tag{13}$$

where C_s and C_0 represent storm and base flow *E. coli* concentrations, respectively, and I_s represents the coefficient of increase ranging between 15.8283 and 1000. Factors impacting *E. coli* concentrations during storm-generated overland flow include rainfall intensity and duration, upland agricultural manure application, type and age of fecal deposits, and *E. coli* adsorption to soil particles [12]. The coefficient of increase (I_s) is subject to change based on these factors. Moreover, the relationship between streamflow and *E. coli* concentration is not linear, as increases in discharge during stormflow may dilute *E. coli* concentrations. For example, in systems where the contribution of groundwater flow to streamflow is high, storm events may result in a decrease in receiving water *E. coli* concentrations [12]. This is due to groundwater typically comprising low *E. coli* concentrations [44,45]. However, while groundwater can dilute stream water, high groundwater contributions to streamflow can result in increased bed and bank shear stress, increasing resuspension of streambed sediment, and elevating *E. coli* numbers [46]. This resuspension can account for approximately 11% of the total *E. coli* load during storm events [46]. The concentration of *E. coli* that can become resuspended can be calculated as follows [47]:

$$R_0 = C_s \times E_0 \left(\frac{\tau_b - \tau_{cn}}{t_c - t_{cn}}\right)^{n_a} \tag{14}$$

where R_0 represents resuspended *E. coli* (CFU/m^{2}S); E_0 represents erosion rate (cm/s), τ_b represents bottom shear stress caused by water flow (Pa), τ_{cn} represent the critical shear stresses (N/m^2) of non-cohesive sediments, τ_c represents the critical shear stresses (N/m^2) of cohesive sediments, and n_a represents particle size (diameter < 432 μm, n_a =2) [45]. The τ_b can be calculated using the specific gravity of water, γ (N/m^3), hydraulic radius, R (m), and water surface slope, S, (m/m) (τ_b = γ.R.S). Conversely, τ_{cn} can be calculated using particle size, d (m), (τ_{cn} = d.4.14 × 10^{-3}), and τ_c can be calculated using Lick's equation as follows [45]:

$$\tau_c = \tau_{cn}\left(\frac{1 + a e^{b_c p \delta}}{d^2}\right) \tag{15}$$

where a and b_c represent constants of 8.5 × 10^{-16} and 9.07 cm^3/g, respectively, p is water density (M L^{-3}), and d is particle size. Resuspension during high flows can be driven by three resuspension mechanisms [48], (1) a steep-fronted wave (caused by the influx of water entering a given stream during a precipitation event resulting in a flashy leading hydrograph edge), with a wave height in excess of the preceding water depth, can lift microbes from the bottom sediment, holding them in the turbulent wave front [37,48], (2) a less steep front or falling wave can resuspend microbes without maintaining them in the wave overrun [37,48], and (3) high flow turbulence (decrease in laminar flow due to increased kinetic energy) can cause steady-flow stochastic erosion of bed and bank sources, thereby maintaining elevated microbe concentrations relative to periods of lower flow [37,48]. Currently, equations that describe the general relationships between increased *E. coli* concentrations or survival due to storm flow-induced increased suspended solids or increased turbidity are lacking in the literature. This is attributable to the high geographic variation and subsequent site-specificity of this relationship. Subsequently, compiling site-specific data from diverse geophysical environments would be useful for the development of general equations relating to streamflow changes to *E. coli* concentrations and survival.

2.5. Water Chemistry

Few large-scale published investigations are available regarding the influence of water chemistry on *E. coli* in the environment. Therefore, laboratory investigations are most often relied on and extrapolated to determine the growth limits of *E. coli* regarding water chemistry variables. However, the sole impact of water chemistry variables of *E. coli* is obscured by the inclusion of temperature as an independent variable in addition to the chemical aspect being investigated, by the majority of previous investigations [16,49–51]. These studies invariably conclude that ambient temperature greatly influences water chemistry impacts on *E. coli*, as, for example, *E. coli* can tolerate lower pH at higher temperatures [49]. Additionally, Presser et al. 1997, reported the effects of temperature, pH, water activity and lactic acid and concluded that these factors were synergistic in limiting *E. coli* growth [49]. In this investigation water activity (defined as the partial vapor pressure of water in a substance divided by the standard state partial vapor pressure of water) values of 0.985 and 0.975 and temperatures ≥ 25 °C resulted in a minimum *E. coli* growth pH of approximately 4. However, temperature decreases raised the minimum pH slightly [49]. Consequently, the growth rate equations presented below assume a constant temperature as temperature fluctuations could impact water chemistry and *E. coli* relationships and alter growth and survival thresholds.

Previous investigations reported optimal *E. coli* survival between pH 5 and pH 7 with increased acidity or alkalinity resulting in decreased survival [52]. The growth limit of *E. coli* is at approximately a pH of 4, however *E. coli* in the stationary phase (the period during which the number of viable bacteria cells remains the same) can survive in pH 2 to 3 for several hours [53]. The following equation can be used to estimate the growth rate of *E. coli* based on pH, assuming the growth rate is proportional to the amount by which the pH is more than the minimum value which prohibits growth [49]:

$$rate = \left(c \times 10^{-pH_{min}}\right)\left(\frac{10^{-pH_{min}} - 10^{-pH}}{10^{pH_{min}}}\right) \tag{16}$$

where *rate* represents *E. coli* growth rate; *c* represents a constant of proportionality; pH_{min} represents minimum growth tolerable pH, and pH represents ambient pH [49]. Similar equations have been created for other chemistry variables constituting linear relationships with *E. coli* growth. For example, the growth rate of *E. coli* as determined by organic acid concentration can be estimated as follows:

$$rate = c' \times (C_{min} - C) \tag{17}$$

where c' represents a proportionality constant; C_{min} represents the theoretical minimum growth inhibitory concentration of the organic acid, and C is the measured concentration of the organic acid [49]. The equation is predicated on the assumption that the growth rate is proportional to the amount by which the concentration of the organic acid is less than the minimum concentration, which prevents growth. Analogous equations can be created for inorganic acids, water activity chloride, and salinity by applying similar assumptions. The concentrations of inorganic and organic acids capable of preventing the growth of *E. coli* is dependent on the specific chemical under consideration.

However, general minimum growth inhibitory concentrations for water activity, chloride concentration, and salinity concentrations are 0.95 [54], 1.5 mg/L [55], and 20% NaCl (complete die-off in 72 h) in nutrient-rich media and 3.5% NaCl (limit for growth) in nutrient-depleted media [56], respectively.

2.6. Nutrients and Nutrient Availability

Environmental nutrient conditions impact *E. coli* growth and survival in secondary habitats. For example, previous investigations reported *E. coli* populations were three times greater in soils rich in organic matter relative to nutrient-depleted sandy soils, suggesting that soil nutrients and organic matter facilitate the growth of bacteria [57]. Additionally, *E. coli* cell density incubated at 30 °C and 37 °C in soils decreased in the days following rapid initial cell growth [13]. This indicates that the final population of the soil *E. coli* was determined predominantly through either the exhaustion of bioavailable nutrients or predation [13]. Nutrient limitation on *E. coli* growth was also evident in laboratory studies where *E. coli* growth in M9 (minimal growth) medium without C and N was limited to a less than one log increase in CFU [21]. In addition to nutrients, soil water potential can also influence the growth of *E. coli* in soil due to its impact on nutrient availability and bacterial movement [21]. For example, regression analysis from previous work indicated that *E. coli* growth (population doubling time) at 37 °C was significantly related to soil water potential ($r^2 = 0.70$, $p < 0.001$) [21]. Soil water potential can also impact the motility of microbes in soil, as lower water potential (−1.5 or −0.1 MPa), results in negligible bacterial movement, and decreased solute diffusion (half the rate observed under saturated conditions) and limited nutrient supply [21,58]. In aquatic environments, dissolved nutrients (glucose and peptone) have been shown to greatly increase the survival of *E. coli* [59]. Additionally, nutrient availability, specifically glucose, can alter *E. coli*'s response to stressors [60]. For example, *E. coli* displayed increased sensitivity to secondary stressors and short term nutrient availability following a period of starvation (nutrient deprivation) [60]. Ultimately, nutrient abundance and availability (as determined by factors such as soil water potential) constitute important factors impacting the survival and growth of *E. coli* in the environment.

2.7. Land-Use Practices

Previous work linked land-use practices, including agricultural and urban land uses, to increased *E. coli* concentrations in receiving waters [14,28,61]. In agricultural regions, increased *E. coli* concentrations are primarily driven by manure applications [62] and the population density of endotherms (livestock) [10,12,63]. During manure application, the environmental inactivation or die-off of *E. coli* results in decreasing concentrations with time passed since application. This can be estimated using the equation below:

$$C = C_i - C_i(R_d \times t) \tag{18}$$

where C represents the current *E. coli* concentration at the soil surface, C_i represents the initial soil *E. coli* concentration immediately after manure application, R_d represents the die-off rate (or rate of inactivation), and t represents the time since the manure was applied. This equation can be adapted to estimate *E. coli* concentrations in associated receiving waters if the *E. coli* transfer rate between the soil and associated receiving waters is known or assuming that 89% of stream *E. coli* concentrations originate directly from surface runoff, as per Ribolzi et al. [46]. Conversely, in agricultural areas comprised primarily of the rearing of livestock, the animal population numbers can be used to approximate the input (addition) of *E. coli* as follows [12]:

$$I = c \times \frac{\left(N_p \times w_i\right)}{A} \tag{19}$$

where I represents the new input of *E. coli* in a specified area over a specified time period; c represents a constant (which varies based on the type and size of the animals under consideration, assuming larger animals will produce more waste, contributing more *E. coli* to the area), Np represents the population number of the animals (endotherms), *Wi* represents the waste per individual animal, and *A* represents the area which the animals inhabit. This equation only approximates new additions of *E. coli* over a specified period and does not approximate total *E. coli* concentrations, bacterial persistence, bacterial inactivation, or die-off rates. Notably, if the area that an animal population inhabits can be expanded, the concentration of new *E. coli* inputs will decrease, due to the inverse proportionality between land-use area and *E. coli* concentration based on animal waste. Thus, the concentrations of *E. coli* inputs can be reduced without reducing animal population numbers simply by increasing the animal grazing area. The above equation can be adapted to estimate contributions of newly deposited *E. coli* to associated receiving waters if the *E. coli* transferal rate between the soil and associated receiving waters is known or assuming that 89% of stream *E. coli* concentrations originate directly from surface runoff, as per Ribolzi et al. [46].

Predicting *E. coli* population increases from urban areas is more complex as *E. coli* concentrations are elevated due to two primary reasons: (1) leaking wastewater infrastructure [10,11,61] and (2) increased runoff during precipitation events due to increased impervious surfaces [10,64]. Predicting leaks from wastewater infrastructure is difficult, and, therefore, the quantitative effect of leaks in urban areas is rarely accounted for, despite potentially contributing significantly to *E. coli* concentrations in receiving waters [65]. In a simplified form, the contribution of a leak on the *E. coli* population in a region (*C*) will be impacted by the fecal concentrations of the leak (C_l), the specific discharge (q_l), and the removal of bacteria in the soil due to the filtering effect of the soil (calculated by the removal rate divided by the distance flowed) (f_r):

$$C = C_l \times q_l \times f_r \tag{20}$$

Assuming saturated conditions, the specific discharge of the leak can be calculated with Darcy's Law [62]:

$$q_l \equiv \frac{Q_l}{A_l} = -K_{hl}\frac{dh}{dl} \tag{21}$$

where q_l represents the specific discharge of the leak, Q_l represents the volume discharge of the leak, K_{hl} represents the hydraulic conductivity, and *dh/dl* is the gradient of the total hydraulic head [66]. Conversely, spatial data software and models, such as the fate transport advection dispersion overland flow modeling approach, can be used to predict the effect of land cover changes on the transport of pollutants (including *E. coli*) during overland flow from precipitation events [67–69]. These models, also known as mass balance water quality models, utilize annual average export coefficients from land use and land cover data to estimate in-stream pollutant (e.g., *E. coli*) loadings [67–69]. An example of an overland flow model is the St. Venant equation [70]:

$$\frac{\delta h(x,t)}{\delta t} + \frac{5(S_0)^{\frac{1}{2}}}{3n_s}h^{\frac{2}{3}}\frac{\delta h(x,t)}{\delta x} = f(t) - i(t) \tag{22}$$

and

$$q = \frac{(S_0)^{\frac{1}{2}}}{n_s}h^{\frac{5}{3}} \tag{23}$$

where *h(x,t)* represents overland flow depth, *q(x,t)* represents overland flow discharge per unit width, *f(t)* represents rainfall rate, *i(t)* represents infiltration rate, n_s represents Manning's roughness coefficient, and S_0 is the channel slope [70]. Notably, complex hydrological equations can be used to predict vertical flow in soil and flow in unsaturated groundwaters, which can include the transport of *E. coli*.

3. Mitigation Strategies

Current mitigation strategies to control freshwater *E. coli* contamination include minimizing the transport of *E. coli* during overland flow or reducing sources of *E. coli*. Strategies include (1) vegetation management, (2) restricting livestock grazing and movement, (3) altering manure application strategies, and (4) wastewater infrastructure maintenance. The maintenance of adequate vegetation or use of vegetative filter strips (strips of vegetation planted for the sole purpose of reducing pollutant transport during runoff events) can reduce the rate and energy of runoff, thereby reducing the concentration of pollutants transported to receiving waters [12,71]. Given that 89% of stream *E. coli* concentrations result from overland flow [46], a reduction in the transport of *E. coli* from soil surfaces to associated receiving waters will proportionately decrease the concentration of the microbe in the water. Restricting the movement and grazing of cattle, using temporary fencing or active herding will reduce the amount of fecal matter, including *E. coli*, that is deposited in a specified area over a given period [68]. For example, McDowell et al. [72] reported that restricting the grazing time of dairy cows to three hours decreased *E. coli* concentrations in associated receiving water to below water quality guidelines of New Zealand and the United States Environmental Protection Agency (126 CFU per 100 mL). Limiting the use of manure in the growing of crops can also decrease *E. coli* concentrations due to a reduction in the sources of the microbe. Warnemuende and Kanwar [73] investigated the effects of swine manure application on bacterial quality of leachate and reported that "an increase in application rate is more likely to cause greater bacterial contamination". Therefore, limiting the application rate (frequency) of manure can improve microbial water quality and decrease *E. coli* concentrations and population numbers in associated receiving waters. Notably, very few large-scale field-based case studies investigating the effect of varying manure application on *E. coli* or fecal concentrations currently exist in the literature. Thus, the true effectiveness of this form of mitigation remains largely unknown. The same holds true for the precise effect of frequent and proper maintenance of wastewater infrastructure in urban land use areas. Due to the sporadic and unpredictable occurrence of leaks, it is hard to quantify their exact effect on *E. coli* concentrations. However, studies have reported that in developing nations, leaking wastewater infrastructure contributed significantly to *E. coli* concentrations in urban receiving waters, specifically during storm events [10,11,65]. Finally, the creation of artificial wetlands can also reduce secondary habitat *E. coli* populations, as open-water treatment wetlands are effective at reducing fecal indicator organisms present in water, including *E. coli*, due to increased exposure to solar insolation [74].

4. Future Directions

Currently, there exists a general lack of field-based investigations studying the environmental factors impacting *E coli*'s survival in the secondary habitat [75]. Consequently, current understanding is predominantly based on laboratory studies, including many uncertainties and excluding factors that could influence identified relationships in uncontrolled environmental settings. For example, it is known that suspended solids can increase the persistence of *E. coli* in water resources [38]. Yet, no widely accepted equations are available that relate changes in suspended solids to associated changes in *E. coli* concentrations. Additionally, the laboratory or small field scale methods implemented by previous investigations [73] imply that large scale work is needed, especially given the processes governing *E. coli* survival and concentrations could differ by spatial scale. For example, animal population density could be a key consideration in a small agricultural area [59] but could become irrelevant if the surrounding area comprises large cities. In this example, the impact of the physical habitat changes brought about by urbanization may impact the survival of *E. coli* to a greater extent than the few animals still present in the area. Site-specific modeling techniques, including geophysical characteristics, need to be developed and applied to a wide variety of areas under varying climatic conditions, to improve understanding of the dynamics of the different water masses and their associated *E. coli* concentrations [12]. This type of work is specifically needed given the contrasting roles of groundwater outflow on stream *E. coli* concentrations during precipitation events: 1) the dilution of stream *E. coli* concentrations, 2) and the resuspension of streambed *E. coli* [12]. Having more fine-scale data also constitutes a fundamental requirement for improving predictive *E. coli* models [12] and a process-based understanding of *E. coli* concentration fluctuation in the secondary habitat. Finally, the identification of naturalized soil *E. coli* communities calls into question the microbe's use as an indicator of fecal contamination [13,21]. Subsequently, studies are required to investigate the movement of naturalized *E. coli* strains and their relative contribution to stream *E. coli* concentrations. Environmental effects on this relative contribution also warrant further investigation, as changing climatic or secondary habitat conditions could potentially alter the movement of both naturalized and newly deposited *E. coli* to associated receiving waters. *E. coli* strains can differ in terms of their metabolic and physiological characteristics [76]. Therefore, strain-specific response to physicochemical changes in the secondary habitat also warrants further investigation.

5. Conclusions

Given the health risks posed by the consumption of *E. coli* contaminated stream water (a single exposure exceeding 500 colonies 100 mL^{-1} has a 10% chance to result in gastrointestinal illness [77]) and the bacteria's widespread use as a fecal indicator organism, understanding the survival of this microbe in the environment is important from a human health perspective. Based on the limited published investigations regarding the environmental requirements of *E. coli* factors, including temperature [13], solar insolation [25], suspended and settled solids [29,30], hydrologic conditions [42], water chemistry [49], nutrient conditions [57], and land-use practices, impact the survival of *E. coli* in the environment [12,78,79]. With more information, the implementation of effective management strategies should be possible and widely applied, given the widespread occurrence of fecal water contamination [9]. However, the effectiveness of implemented management strategies is rarely assessed on large scales, using field-based methods. Therefore, their usefulness remains largely unknown. Consequently, future *E. coli*-focused work should attempt to expand on the current limited number of field-based published works and investigate both the survival of *E. coli* under different environmental conditions in the secondary habitat and the effectiveness of implemented management strategies, specifically on larger scales. This information will provide scientists and land-use managers with new insight to effectively address problematic fecal contamination, thereby aiding in the reduction in disease outbreaks caused by contaminated water.

Author Contributions: For the current work author contributions were as follows: conceptualization, F.P.; methodology, F.P.; formal analysis, F.P. and J.A.H.; investigation, F.P.; resources, J.A.H.; data curation, J.A.H.; writing—Original draft preparation, F.P.; writing—Review and editing, J.A.H. and F.P.; visualization, F.P. and J.A.H.; supervision, J.A.H.; project administration, J.A.H.; funding acquisition, J.A.H. All authors have read and agreed to the published version of the manuscript.

Funding: This work was supported by the National Science Foundation under Award Number OIA-1458952, the USDA National Institute of Food and Agriculture, Hatch project accession number 1011536, and the West Virginia Agricultural and Forestry Experiment Station. Results presented may not reflect the views of the sponsors, and no official endorsement should be inferred. The funders had no role in study design, data collection and analysis, decision to publish, or preparation of the manuscript.

Acknowledgments: Special thanks are due to many scientists of the Interdisciplinary Hydrology Laboratory (https://www.researchgate.net/lab/The-Interdisciplinary-Hydrology-Laboratory-Jason-A-Hubbart). The authors also appreciate the feedback of anonymous reviewers whose constructive comments improved the article.

Conflicts of Interest: The authors declare no conflict of interest for the current work.

References

1. Savageau, M.A. *Escherichia coli* Habitats, Cell Types, and Molecular Mechanisms of Gene Control. *Am. Nat.* **1983**, *122*, 732–744. [CrossRef]
2. Freter, R. Factors controlling the composition of the intestinal microflor. *Sp. Suppl. Microbiol.* **1976**, *1*, 109–120.
3. Bonde, G.J. Pollution of a marine environment. *J. Water Pollut.* **1967**, *39*, R45–R63.
4. Wetzel, R.G. *Limnology*, 1st ed.; W.B. Saunders Co: Philadelphia, PA, USA; London, UK; Toronto, ON, Canada, 1975.
5. Marshall, K.C. Adsorption of Microorganisms to Soils and Sediment. In *Adsorption of Microorganisms to Surface*, 1st ed.; Wiley: New York, NY, USA, 1980.
6. Bristow, L.A.; Mohr, W.; Ahmerkamp, S.; Kuypers, M.M.M. Nutrients that limit growth in the ocean. *Curr. Biol.* **2017**, *27*, R474–R478. [CrossRef]

7. Chambers, P.A.; Vis, C.; Brua, R.B.; Guy, M.; Culp, J.M.; Benoy, G.A. Eutrophication of agricultural streams: Defining nutrient concentrations to protect ecological condition. *Water Sci. Technol.* **2008**, *58*, 2203–2210. [CrossRef]
8. *E. coli (Escherichia coli)|E. coli|*CDC. Available online: https://www.cdc.gov/ecoli/index.html (accessed on 8 January 2020).
9. WHO. World Water Day Report. Available online: https://www.who.int/water_sanitation_health/takingcharge.html (accessed on 12 December 2019).
10. Causse, J.; Billen, G.; Garnier, J.; Henri-des-Tureaux, T.; Olasa, X.; Thammahacksa, C.; Latsachak, K.O.; Soulileuth, B.; Sengtaheuanghoung, O.; Rochelle-Newall, E. Field and modelling studies of *Escherichia coli* loads in tropical streams of montane agro-ecosystems. *J. Hydro-Environ. Res.* **2015**, *9*, 496–507. [CrossRef]
11. Wu, J.; Yunus, M.; Islam, M.S.; Emch, M. Influence of climate extremes and land use on fecal contamination of shallow tubewells in Bangladesh. *Environ. Sci. Technol.* **2016**, *50*, 2669–2676. [CrossRef]
12. Rochelle-Newall, E.J.; Ribolzi, O.; Viguier, M.; Thammahacksa, C.; Silvera, N.; Latsachack, K.; Dinh, R.P.; Naporn, P.; Sy, H.T.; Soulileuth, B. Effect of land use and hydrological processes on *Escherichia coli* concentrations in streams of tropical, humid headwater catchments. *Sci. Rep.* **2016**, *6*, 32974. [CrossRef]
13. Ishii, S.; Ksoll, W.B.; Hicks, R.E.; Sadowsky, M.J. Presence and Growth of Naturalized *Escherichia coli* in Temperate Soils from Lake Superior Watersheds. *Appl. Environ. Microbiol.* **2006**, *72*, 612–621. [CrossRef] [PubMed]
14. Petersen, F.; Hubbart, J.A.; Kellner, E.; Kutta, E. Land-use-mediated *Escherichia coli* concentrations in a contemporary Appalachian watershed. *Environ. Earth Sci.* **2018**, *77*, 754. [CrossRef]
15. Petersen, F.; Hubbart, J.A. Advancing Understanding of Land Use and Physicochemical Impacts on Fecal Contamination in Mixed-Land-Use Watersheds. *Water* **2020**, *12*, 1094. [CrossRef]
16. McQuestin, O.J.; Shadbolt, C.T.; Ross, T. Quantification of the Relative Effects of Temperature, pH, and Water Activity on Inactivation of *Escherichia coli* in Fermented Meat by Meta-Analysis. *Appl. Environ. Microbiol.* **2009**, *75*, 6963–6972. [CrossRef] [PubMed]
17. Jones, T.; Gill, C.O.; McMullen, L.M. The behaviour of log phase *Escherichia coli* at temperatures that fluctuate about the minimum for growth. *Lett. Appl. Microbiol.* **2004**, *39*, 296–300. [CrossRef] [PubMed]
18. Chick, H. An Investigation of the Laws of Disinfection. *J. Hyg.* **1908**, *8*, 92–158. [CrossRef]
19. Blaustein, R.A.; Pachepsky, Y.; Hill, R.L.; Shelton, D.R.; Whelan, G. *Escherichia coli* survival in waters: Temperature dependence. *Water Res.* **2013**, *47*, 569–578. [CrossRef]
20. Jang, J.; Hur, H.G.; Sadowsky, M.J.; Byappanahalli, M.N.; Yan, T.; Ishii, S. Environmental *Escherichia coli*: Ecology and public health implications—A review. *J. Appl. Microbiol.* **2017**, *123*, 570–581. [CrossRef]
21. Ishii, S.; Yan, T.; Vu, H.; Hansen, D.L.; Hicks, R.E.; Sadowsky, M.J. Factors Controlling Long-Term Survival and Growth of Naturalized *Escherichia coli* Populations in Temperate Field Soils. *Microbes Environ.* **2010**, *25*, 8–14. [CrossRef]
22. Davies-Colley, R.J.; Bell, R.G.; Donnison, A.M. Sunlight inactivation of enterococci and fecal coliforms in sewage effluent diluted in seawater. *Appl. Environ. Microbiol.* **1994**, *60*, 2049–2058. [CrossRef]
23. Fujioka, R.S.; Hashimoto, H.H.; Siwak, E.B.; Young, R.H. Effect of sunlight on survival of indicator bacteria in seawater. *Appl. Environ. Microbiol.* **1981**, *41*, 690–696. [CrossRef]
24. Sinton, L.W.; Finlay, R.K.; Lynch, P.A. Sunlight inactivation of fecal bacteriophages and bacteria in sewage-polluted seawater. *Appl. Environ. Microbiol.* **1999**, *65*, 3605–3613. [CrossRef]
25. Whitman, R.L.; Nevers, M.B.; Korinek, G.C.; Byappanahalli, M.N. Solar and Temporal Effects on *Escherichia coli* Concentration at a Lake Michigan Swimming Beach. *Appl. Environ. Microbiol.* **2004**, *70*, 4276–4285. [CrossRef] [PubMed]
26. Maraccini, P.A.; Mattioli, M.C.C.; Sassoubre, L.M.; Cao, Y.; Griffith, J.F.; Ervin, J.S.; van de Werfhorst, L.C.; Boehm, A.B. Solar Inactivation of Enterococci and *Escherichia coli* in Natural Waters: Effects of Water Absorbance and Depth. *Environ. Sci. Technol.* **2016**, *50*, 5068–5076. [CrossRef] [PubMed]
27. Muela, A.; Garcia-Bringas, J.M.; Arana, I.; Barcina, I. The Effect of Simulated Solar Radiation on *Escherichia coli*: The Relative Roles of UV-B, UV-A, and Photosynthetically Active Radiation. *Microb. Ecol.* **2000**, *39*, 65–71. [CrossRef] [PubMed]
28. Petersen, F.; Hubbart, J.A. Quantifying *Escherichia coli* and Suspended Particulate Matter Concentrations in a Mixed-Land Use Appalachian Watershed. *Water* **2020**, *12*, 532. [CrossRef]

29. Grossart, H.-P. Ecological consequences of bacterioplankton lifestyles: Changes in concepts are needed. *Environ. Microbiol. Rep.* **2010**, *2*, 706–714. [CrossRef]
30. Drummond, J.D.; Davies-Colley, R.J.; Stott, R.; Sukias, J.P.; Nagels, J.W.; Sharp, A.; Packman, A.I. Microbial transport, retention, and inactivation in streams: A combined experimental and stochastic modeling approach. *Environ. Sci. Technol.* **2015**, *49*, 7825–7833. [CrossRef]
31. Allen, H.K.; Donato, J.; Wang, H.H.; Cloud-Hansen, K.A.; Davies, J.; Handelsman, J. Call of the wild: Antibiotic resistance genes in natural environments. *Nat. Rev. Microbiol.* **2010**, *8*, 251. [CrossRef]
32. Corno, G.; Coci, M.; Giardina, M.; Plechuk, S.; Campanile, F.; Stefani, S. Antibiotics promote aggregation within aquatic bacterial communities. *Front. Microbiol.* **2014**, *5*, 297. [CrossRef] [PubMed]
33. Mamane, H. Impact of particles on UV disinfection of water and wastewater effluents: A review. *Rev. Chem. Eng.* **2008**, *24*, 67–157. [CrossRef]
34. Tang, K.W.; Dziallas, C.; Grossart, H.-P. Zooplankton and aggregates as refuge for aquatic bacteria: Protection from UV, heat and ozone stresses used for water treatment. *Environ. Microbiol.* **2011**, *13*, 378–390. [CrossRef]
35. Callieri, C.; Amalfitano, S.; Corno, G.; Bertoni, R. Grazing-induced Synechococcus microcolony formation: Experimental insights from two freshwater phylotypes. *FEMS Microbiol. Ecol.* **2016**, *92*, fiw154.
36. Czajkowski, D.; Witkowska-Gwiazdowska, A.; Sikorska, I.; Boszczyk-Maleszak, H.; Horoch, M. Survival of *Escherichia coli* Serotype O157:H7 in Water and in Bottom-Shore Sediments. *Pol. J. Environ. Stud.* **2005**, *14*, 423–430.
37. Pachepsky, Y.A.; Shelton, D.R. *Escherichia Coli* and Fecal Coliforms in Freshwater and Estuarine Sediments. *Crit. Rev. Environ. Sci. Technol.* **2011**, *41*, 1067–1110. [CrossRef]
38. Jamieson, R.; Joy, D.M.; Lee, H.; Kostaschuk, R.; Gordon, R. Persistence of enteric bacteria in alluvial streams. *J. Environ. Eng. Sci.* **2004**, *3*, 202–213. [CrossRef]
39. Anderson, K.L.; Whitlock, J.E.; Harwood, V.J. Persistence and Differential Survival of Fecal Indicator Bacteria in Subtropical Waters and Sediments. *Appl. Environ. Microbiol.* **2005**, *71*, 3041–3048. [CrossRef]
40. Garzio-Hadzick, A.; Shelton, D.R.; Hill, R.L.; Pachepsky, Y.A.; Guber, A.K.; Rowland, R. Survival of manure-borne *E. coli* in streambed sediment: Effects of temperature and sediment properties. *Water Res.* **2010**, *44*, 2753–2762. [CrossRef]
41. Ribolzi, O.; Cuny, J.; Sengsoulichanh, P.; Mousquès, C.; Soulileuth, B.; Pierret, A.; Huon, S.; Sengtaheuanghoung, O. Land Use and Water Quality Along a Mekong Tributary in Northern Lao P.D.R. *Environ. Manag.* **2011**, *47*, 291–302. [CrossRef] [PubMed]
42. Ekklesia, E.; Shanahan, P.; Chua, L.H.C.; Eikaas, H.S. Temporal variation of faecal indicator bacteria in tropical urban storm drains. *Water Res.* **2015**, *68*, 171–181. [CrossRef]
43. Knierim, K.J.; Hays, P.D.; Bowman, D. Quantifying the variability in *Escherichia coli* (*E. coli*) throughout storm events at a karst spring in northwestern Arkansas, United States. *Environ. Earth Sci.* **2015**, *74*, 4607–4623. [CrossRef]
44. Hunter, C.; McDonald, A.; Beven, K. Input of fecal coliform bacteria to an upland stream channel in the Yorkshire Dales. *Water Resour. Res.* **1992**, *28*, 1869–1876. [CrossRef]
45. Weiskel, P.K.; Howes, B.L.; Heufelder, G.R. Coliform Contamination of a Coastal Embayment: Sources and Transport Pathways. *Environ. Sci. Technol.* **1996**, *30*, 1872–1881. [CrossRef]
46. Ribolzi, O.; Evrard, O.; Huon, S.; Rochelle-Newall, E.; Henri-des-Tureaux, T.; Silvera, N.; Thammahacksac, C.; Sengtaheuanghoung, O. Use of fallout radionuclides (7Be, 210Pb) to estimate resuspension of *Escherichia coli* from streambed sediments during floods in a tropical montane catchment. *Environ. Sci. Pollut. Res.* **2016**, *23*, 3427–3435. [CrossRef] [PubMed]
47. Pandey, P.K.; Soupir, M.L. Assessing the Impacts of *E. coli* Laden Streambed Sediment on *E. coli* Loads over a Range of Flows and Sediment Characteristics. *J. Am. Water Resour. Assoc.* **2013**, *49*, 1261–1269. [CrossRef]
48. Wilkinson, J.; Kay, D.; Wyer, M.; Jenkins, A. Processes driving the episodic flux of faecal indicator organisms in streams impacting on recreational and shellfish harvesting waters. *Water Res.* **2006**, *40*, 153–161. [CrossRef] [PubMed]
49. Presser, K.A.; Ratkowsky, D.A.; Ross, T. Modelling the growth rate of *Escherichia coli* as a function of pH and lactic acid concentration. *Appl. Environ. Microbiol.* **1997**, *63*, 2355–2360. [CrossRef] [PubMed]
50. Deng, Y.; Ryu, J.-H.; Beuchat, L.R. Influence of temperature and pH on survival of *Escherichia coli* O157:H7 in dry foods and growth in reconstituted infant rice cereal. *Int. J. Food Microbiol.* **1998**, *45*, 173–184. [CrossRef]

51. Conner, D.E.; Kotrola, J.S. Growth and survival of *Escherichia coli* O157:H7 under acidic conditions. *Appl. Environ. Microbiol.* **1995**, *61*, 382–385. [CrossRef]
52. De W Blackburn, C.; Curtis, L.M.; Humpheson, L.; Billon, C.; McClure, P.J. Development of thermal inactivation models for Salmonella enteritidis and *Escherichia coli* O157:H7 with temperature, pH and NaCl as controlling factors. *Int. J. Food Microbiol.* **1997**, *38*, 31–44. [CrossRef]
53. Small, P.; Blankenhorn, D.; Welty, D.; Zinser, E.; Slonczewski, J.L. Acid and base resistance in *Escherichia coli* and Shigella flexneri: Role of rpoS and growth pH. *J. Bacteriol.* **1994**, *176*, 1729–1737. [CrossRef]
54. Fontana, A.J. Appendix D: Minimum Water Activity Limits for Growth of Microorganisms. In *Water Activity in Foods*; John Wiley & Sons, Ltd: Hoboken, NJ, USA, 2008; p. 405. ISBN 978-0-470-37645-4.
55. Owoseni, M.C.; Olaniran, A.O.; Okoh, A.I. Chlorine Tolerance and Inactivation of *Escherichia coli* recovered from Wastewater Treatment Plants in the Eastern Cape, South Africa. *Appl. Sci.* **2017**, *7*, 810. [CrossRef]
56. Hrenovic, J.; Ivankovic, T. Survival of *Escherichia coli* and Acinetobacter junii at various concentrations of sodium chloride. *EurAsia J. Biosci.* **2009**, 144–151. [CrossRef]
57. Tate, R.L. Cultural and environmental factors affecting the longevity of *Escherichia coli* in Histosols. *Appl. Environ. Microbiol.* **1978**, *35*, 925–929. [CrossRef] [PubMed]
58. Griffin, D.M. Water Potential as a Selective Factor in the Microbial Ecology of Soils 1. *Water Potential Relat. Soil Microbiol.* **1981**, *9*, 141–151. [CrossRef]
59. Milne, D.P.; Curran, J.C.; Findlay, J.S.; Crowther, J.M.; Bennet, J.; Wood, B.J.B. The Effect of Dissolved Nutrients and Inorganic Suspended Solids on the Survival of *E. coli* in Seawater. *Water Sci. Technol.* **1991**, *24*, 133–136. [CrossRef]
60. Wu, S.Y.; Klein, D.A. Starvation effects on *Escherichia coli* and aquatic bacterial responses to nutrient addition and secondary warming stresses. *Appl. Environ. Microbiol.* **1976**, *31*, 216–220. [CrossRef]
61. Gotkowska-Plachta, A.; Golaś, I.; Korzeniewska, E.; Koc, J.; Rochwerger, A.; Solarski, K. Evaluation of the distribution of fecal indicator bacteria in a river system depending on different types of land use in the southern watershed of the Baltic Sea. *Environ. Sci. Pollut. Res.* **2016**, *23*, 4073–4085. [CrossRef]
62. Jamieson, R.C.; Gordon, R.J.; Sharples, K.E.; Stratton, G.W.; Madani, A. Movement and persistence of fecal bacteria in agricultural soils and subsurface drainage water: A review. *Can. Biosyst. Eng.* **2002**, *44*, 1–9.
63. Rwego, I.B.; Gillspie, T.R.; Isabirye-Basuta, G.; Goldberg, T.L. High Rates of *Escherichia coli* Transmission between Livestock and Humans in Rural. *J. Clin. Mircobiol.* **2008**, *46*, 3187–3191. [CrossRef]
64. Wilson, C.; Weng, Q. Assessing surface water quality and its relation with urban land cover changes in the Lake Calumet Area, Greater Chicago. *Environ. Manag.* **2010**, *45*, 1096–1111. [CrossRef]
65. Fewtrell, L.; Kay, D. Recreational Water and Infection: A Review of Recent Findings. *Curr. Environ. Health Rep.* **2015**, *2*, 85–94. [CrossRef]
66. Dingman, S.L. *Phyiscal Hydrology*, 2nd ed.; Prentice Hall: Upper Saddle River, NJ, USA, 2002; ISBN 0-13-099695-5.
67. Syed, A.U.; Jodoin, R.S. *Estimation of Nonpoint-Source Loads of Total Nitrogen, Total Phosphorous, and Total Suspended Solids in the Black, Belle, and Pine River Basins, Michigan, by Use of the PLOAD Model*; USGS: Reston, VA, USA, 2006.
68. Van der Tak, L.; Edwards, C. *An ArcView GIS Tool to Calculate Nonpoint Sources of Pollution in Watershed and Stormwater Projects*; USEPA: Washington, DC, USA, 2001.
69. Howard, D.M. *Aquatic Life Water Quality Standards Draft Technical Support Document for Total Suspended Solids (Turbidity)*; Minnesota Polution Control Agency: Saint Paul, MN, USA, 2011; pp. 1–50.
70. Massoudieh, A.; Huang, X.; Young, T.M.; Mariño, M.A. Modeling Fate and Transport of Roadside-Applied Herbicides. *J. Environ. Eng.* **2005**, *131*, 1057–1067. [CrossRef]
71. Olilo, C.O.; Muia, A.W.; Moturi, W.N.; Onyando, J.O.; Amber, F.R. The current state of knowledge on the interaction of *Escherichia coli* within vegetative filter strips as a sustainable best management practice to reduce fecal pathogen loading into surface waters. *Energy Ecol. Environ.* **2016**, *1*, 248–266. [CrossRef] [PubMed]
72. McDowell, R.W.; Drewry, J.J.; Muirhead, R.W.; Paton, R.J. Restricting the grazing time of cattle to decrease phosphorus, sediment and *E. coli* losses in overland flow from cropland. *Soil Res.* **2005**, *43*, 61–66. [CrossRef]
73. Warnemuende, E.A.; Kanwar, R.S. Effects of swine manure application on bacterial quality of leachate from intact soil columns. *Am. Soc. Agric. Eng.* **2002**, *45*, 1849–1857. [CrossRef]

74. Bear, S.E.; Nguyen, M.T.; Jasper, J.T.; Nygren, S.; Nelson, K.L.; Sedlak, D.L. Removal of nutrients, trace organic contaminants, and bacterial indicator organisms in a demonstration-scale unit process open-water treatment wetland. *Ecol. Eng.* **2017**, *109*, 76–83. [CrossRef]
75. Soller, J.; Bartrand, T.; Ravenscroft, J.; Molina, M.; Whelan, G.; Schoen, M.; Ashbolt, N. Estimated human health risks from recreational exposures to stormwater runoff containing animal faecal material. *Environ. Model. Softw.* **2015**, *72*, 21–32. [CrossRef]
76. Meier, S.; Jensen, P.R.; Duus, J.O. Direct Observation of Metabolic Differences in Living *Escherichia coli* Strains K-12 and BL21. *ChemBioChem* **2012**, *13*, 308–310. [CrossRef] [PubMed]
77. WHO. Guidelines for Safe Recreational Water Environments. Available online: http://www.who.int/water_sanitation_health/publications/srwe1/en/ (accessed on 20 March 2020).
78. Crowther, J.; Kay, D.; Wyer, M.D. Faecal-indicator concentrations in waters draining lowland pastoral catchments in the UK: Relationships with land use and farming practices. *Water Res.* **2002**, *36*, 1725–1734. [CrossRef]
79. Servais, P.; Garcia-Armisen, T.; George, I.; Billen, G. Fecal bacteria in the rivers of the Seine drainage network (France): Sources, fate and modelling. *Sci. Total Environ.* **2007**, *375*, 152–167. [CrossRef]

Review

Potential Health Risks Linked to Emerging Contaminants in Major Rivers and Treated Waters

James Kessler [1], Diane Dawley [2], Daniel Crow [2], Ramin Garmany [1] and Philippe T. Georgel [1,3,*]

[1] Department of Biological Sciences, Marshall University, Huntington, WV 25755, USA; kessler42@live.marshall.edu (J.K.); garmany.ramin@mayo.edu (R.G.)

[2] Joan C. Edwards School of Medicine, Marshall University, Huntington, WV 25755, USA; dawleyd@live.marshall.edu (D.D.); dtcrow1@gmail.com (D.C.)

[3] Department of Biological Sciences, College of Science, Cell Differentiation and Development Center, Marshall University, Huntington, WV 25755, USA

* Correspondence: georgel@marshall.edu; Tel.: +1-(304)-696-3965; Fax: +1-(304)-696-3766

Received: 30 October 2019; Accepted: 4 December 2019; Published: 11 December 2019

Abstract: The presence of endocrine-disrupting chemicals (EDCs) in our local waterways is becoming an increasing threat to the surrounding population. These compounds and their degradation products (found in pesticides, herbicides, and plastic waste) are known to interfere with a range of biological functions from reproduction to differentiation. To better understand these effects, we used an in silico ontological pathway analysis to identify the genes affected by the most commonly detected EDCs in large river water supplies, which we grouped together based on four common functions: Organismal injuries, cell death, cancer, and behavior. In addition to EDCs, we included the opioid buprenorphine in our study, as this similar ecological threat has become increasingly detected in river water supplies. Through the identification of the pleiotropic biological effects associated with both the acute and chronic exposure to EDCs and opioids in local water supplies, our results highlight a serious health threat worthy of additional investigations with a potential emphasis on the effects linked to increased DNA damage.

Keywords: endocrine disrupting chemical; opioid; pathway analysis; ontology; metabolomics

---

1. Introduction

Outline: endocrine-disruption chemical (EDC) effects are not solely mediated by competing for hormone receptors in cells, but may also be caused by an increase in reactive oxygen species, leading to DNA damage. There is also a potential epigenetic effect linking to changes in levels of DNA methylation

The quality of tap and drinking water is an important issue affecting most countries worldwide. Contamination and its downstream effect on human health can vary widely between regions. In most industrialized countries, a significant portion of the tap water used for daily consumption comes from local waterways. Industrialization results in greater populations which necessitates larger water sources, increasing the likelihood that various chemical contaminants will be present. Concerns about water quality were initially triggered in the 1990s due to the presence of pharmaceuticals such as antibiotics at detectable, biologically significant concentrations [1]. In addition, the requirement for increased food production and the control of insect populations has led to the augmented use of herbicides and pesticides. Simultaneously, urban development has coincided with an increased use of detergents and plastic materials, leading to by-product dissipation in local water supplies. Of specific concern among these emerging contaminants are endocrine disrupting chemicals (EDCs), compounds which interfere with hormone metabolism in the body (National Institutes of Environmental Health Science: Endocrine disruptors fact sheet; for review, see [2] and opioids [3]). These polyphenolic chemical structures found in waste, such as plastic containers, detergents, herbicides, and pesticides, can mimic hormones and cause major disturbances (even at low concentrations) in cellular homeostasis, functionality, and differentiation [4,5]. For example, links between EDCs and elevated cancer risks have clearly been established [6]. In addition, exposure to EDCs has been demonstrated to negatively impact the immune response [7], and compounding health factors, such as obesity, can exacerbate the severity of this reaction [8]. Chronic exposure to EDCs in local waterways can also have environmental consequences, including the feminization of several species of *crustacea*, fish, and other vertebrates [9]. Importantly, EDCs have been recognized to be capable of disrupting neuroendocrine processes, modifying the patterns of expression and production in neurotransmitters, resulting in the alteration of physiological and behavioral responses [10,11]. Another factor which may complicate the analysis of the results of studies involving EDCs' effects on organisms relates to the genetic variability in individuals which, in turn, can lead to different biological responses to environmental exposures [12,13]. Although not yet fully understood, the molecular reactions to EDC exposure have recently been linked with epigenetic changes, potentially affecting cells in a genome-wide manner [14].

To examine the potential biological effects of EDCs contaminating major waterways, we selected commonly used polyphenolic hormone mimetics utilized as herbicides, insecticides, and detergent as our representative examples (these products have been shown to be present in the Ohio River and other major water sources, see Table 1 for detected concentrations and references). Data on EDC concentrations in surface water, rivers, sediments, and tap water have been recorded for extended periods of time. This data set is summarized in Table 1. Note that the range of concentrations reported may be influenced by the location of collection of the samples. The proximity of chemical or pharmaceutical plants, or even hospitals, may significantly affect the values reported (for example, see the values reported for buprenorphine in Table 1). Also, note that the values reported do not reflect toxicity, but simply the concentration of the various compounds investigated.

We selected atrazine as an herbicidal EDC example, as it is arguably the most commonly used agricultural herbicide. It has been detected in tap water at levels above 3 ppb in 19 U.S. states (https://www.ewg.org/tapwater). As for insecticides and pesticides, we decided to focus on Chlorpyrifos (http://www.health.state.mn.us/divs/eh/risk/guidance/gw/chlorpyinfo.pdf. [15,16]) and Endosulfan, two widely-used EDCs which have both been detected in major U.S. waterways [17,18] and tap water (http://md.water.usgs.gov/nawqa; [18]).

Bisphenol A was chosen as a known organic wastewater EDC contaminant detected in various large waterways world-wide, reaching median concentrations ranging from 0.016 to 0.5 μg/L in reported European and U.S. studies [19]. It is also present at detectable levels in drinking water from Asia, Europe and North America [20,21]. To complete our basic investigation, we included a very common EDC detergent, known as Nonyl Phenol, commonly detected in water and accumulating in sediments due to its poor solubility [22]. Despite standard tap water purification procedures, Nonyl Phenol has been detected as a contaminant in tap water [23].

Recently, a significant increase in the use of opioids, as both a therapeutic agent and recreational drug, has raised concerns that they will soon emerge as a major contaminant in local water treatment facilities [24], Among addiction treatment options for opiate-abusing patients, buprenorphine has become a popular alternative to methadone as a maintenance/weaning agent, particularly in the Appalachian region [25]. As a result, renewed efforts to monitor local water treatment plants for buprenorphine have been initiated. Highlighting a need for a more rigorous monitoring system, a recent study in France has revealed the presence of hot spots in water treatment plants containing buprenorphine levels capable of generating the biological responses known to affect brain function and development [26,27].

In order to further understand the pleiotropic levels of disruption caused by EDC and buprenorphine exposure, we undertook a bioinformatics in silico gene ontology investigation focused on the various cellular functions and genetic pathways that are affected by these emerging contaminants detected in a variety of water sources worldwide [1]. Our goal for this analysis was to identify and evaluate the potential biological effects and modes of action of each contaminant investigated. By finding the common trends among these affected functions and pathways, we gained valuable insight on the potential mechanisms of action of these emerging contaminants and outlined the alarming possibility that additive and/or synergistic properties could enhance the deleterious effects resulting from acute or long-term co-exposure from multiple sources.

2. Material and Methods

Ingenuity Pathway Analysis (IPA®, Ingenuity Pathway Analysis, Qiagen Corp, Germantown, MD, USA): The identification of genes affected by the emerging contaminants was accomplished using the Ingenuity Pathway Analysis software. Information related to each individual emerging contaminant was entered into the software, resulting in a list of associated genes generated by the Ingenuity Pathways Knowledge Base. The cellular network and localization of these genes were then algorithmically elucidated using the IPA® mapping tools, allowing for the visualization of the various interactive disease and functions nodes affected by each individual emerging contaminant. Matching tables were compiled using the "BioProfiler" function of the IPA® mapping tool, and the display of our analysis was focused on the gene's symbols, synonyms(s), NCBI Entrez gene names, cellular location, and nature/function (shown as "Family" in the tables). All of the compounds were analyzed for their roles, links, and functions related to the following IPA® selection "Families": Organismal injury, cell death, organismal survival, cancer, and behavior (which were the top five functions outlined using the default ontologic/metabolomic analytical "BioProfiler" as a basis for our specific analysis settings). The tables and figures presented are based on our analysis and display all of the data generated by the above-mentioned algorithm. The absence of data for any specific "Family" indicates a lack of IPA-available information about the particular compound analyzed. All data were obtained without statistical bias or any favorable weight assigned. For each "Family", the tables were compiled to present all the genes affected by a specific compound (see below for a more detailed algorithm and Tables in Supplementary Data).

Enter name of Chemical investigated Select IPA Analysis Tool: Path Designer, Biomarker filters: Path Explorer General Filter: Interactions Direct and Indirect; data Source: All; Species: All; All Tissues and Cell Lines: All; Relation Types: All; Node Types: All; Diseases: All; Biofluids: All; Biomarkers: All; Mutations: All. Select Option: Gens and Chemicals: Enter chemical name, run search. Select Analysis Display in BioProfiler results. Select Display Option: Display by protein cellular location.

3. Results and Discussion

3.1. Atrazine: Standard Herbicide, Potential Carcinogen, and Environmental Immune Disruptor

Atrazine is an herbicide belonging to the triazine family (see Figure 1A for structure) that has been widely used by agriculture workers in the USA for decades and has long been suspected to have multiple deleterious effects on both invertebrates and vertebrates. Based on this suspicion, it was banned in 2004 by the European Union, as the levels in ground water were exceeding the legal limits considered safe for the environment (See European Commission Decision C (2004) 731 [28]). Despite this fact, in 2003, the Environmental Protection Agency (EPA) estimated that "cumulative exposures to these pesticides (atrazine and simazine) through food and drinking water are safe and meet the rigorous human health standards set forth in the Food Quality Protection Act (FQPA)." (http://www.epa.gov/pesticides/reregistration/atrazine/atrazine_update.htm#atrazine). This statement was regarded as being highly controversial, as recent publications have indicated that chronic exposure to the presence of atrazine and/or degradations products is likely to have serious negative effects on human health [29]. To further investigate these potential effects of atrazine exposure, we performed an in silico gene ontology analysis using the Ingenuity Pathway Analysis (IPA) software package, focusing on identifying how atrazine affects cellular functions, such as organismal injuries, cell death, cancer and behavior.

(A) Organismal injury: As was expected, based on its chemical structure, atrazine has an effect on the expression and cellular pathways of genes involved in hormonal responses. The testosterone, estrone, estrogen, and estradiol pathways are all prime targets in atrazine-exposed cells (see Figure 1B and Supplementary Materials Table S1). Increased levels of expression were noted for specific hormone receptors, such as the androgen receptor (AR), glucocorticoid receptor (GR), growth hormone-releasing hormone receptor (GHRHR), and the estrogen receptor (ER). The cytochrome p450 family genes CYP11A1 and CYP19A1, known to be involved in drug metabolism and detoxification, were found to be affected by atrazine exposure as well [30].

(B) Cell death: Many of the pathways and receptors associated with organismal injury are also involved in atrazine-mediated cell death (Figure 1C and Supplementary Materials Table S2), in addition to the growth factor CLEC11A, a member of the C-Type lectin superfamily-3 [31], and known to be involved in apoptosis [32]. Based on our analysis, most of the genes that we identified to be involved in the atrazine exposure response have been shown to be linked with triggering apoptosis in fish [33], amphibians [34], and mice [35,36]. This effect can be mediated by changes in access to membrane receptors, such as the glucocorticoid receptor (GR, [37,38]), a process which has been linked with the gene NR3C1, coding for the glucocorticoid nuclear receptor variant 1 ([39], see Figure 1C). The intra-cellular response involves detoxification through cytochrome p450 genes, such as CYP11A1 and CYP19A.1, Interestingly, these genes are also known as the key regulators in cytotoxicity and apoptosis [30,40].

Table 1. Concentrations of emerging contaminants. Data were collected from various databases, reports, and poster presentations. μg L^{-1} (microgram per liter), μg K^{-1} (microgram per kilogram)1. ORSANCO, Ohio River, Evansville, IN (2016); 2. https://www.ewg.org/tapwater); 3. http://www.health.state.mn.us/divs/eh/risk/guidance/gw/chlorpyinfo.pdf.

EDC	Concentration (ppb)	μg L^{-1}	μg.K^{-1} (in Sediments)	Presence in Tap/Drinking Water
Atrazine (ppb[1])	0.19 to 1.88 (Ohio River)[1]			Detected in 28 U.S. states[2]
Bis-Phenol A		0.016–0.5 [26–29]		Detected in Asia, Europe, North America [26]
Chlorpyrifos	0.24 (ground water, MN)[3]	0–2.828 [30]		Detected in drinking water[3] [30]
Endosulfan		<1 (WHO), 0.020–0.11 [31]		Detected in drinking water [31]
Nonyl Phenol		0.1–0.5 (Ohio Tributary) [32] 0000.1-0.0027 (Tap water, Chongqing China [28])	75–340 (Ohio Tributary) [32]	Detected in drinking water [28,32]
Buprenorphine		0.042–0.195 (sewage water, Paris) France [7]		Detected in waste water [7,33,34]

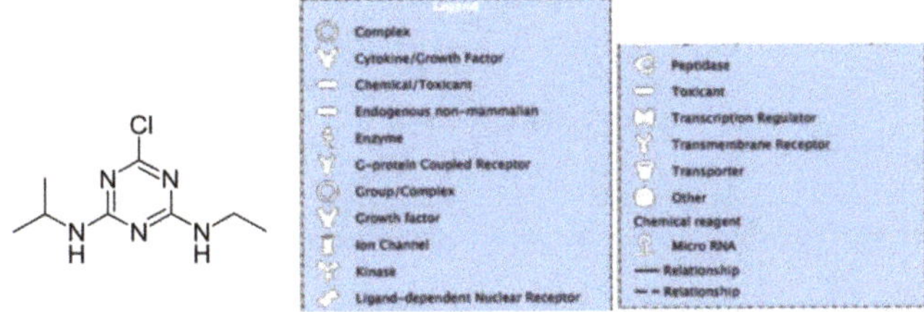

A: Atrazine

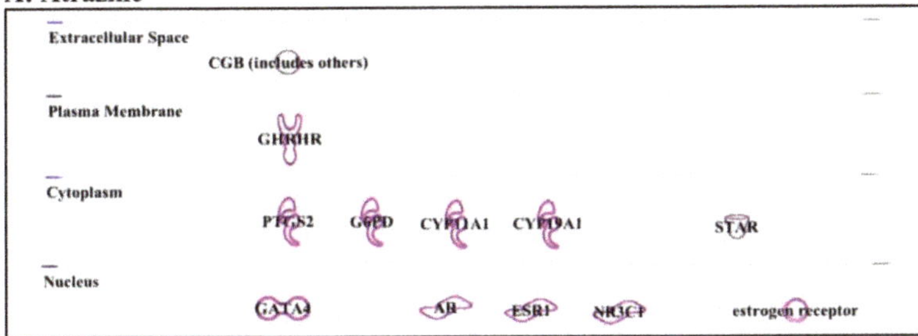

B: Organismal Injury

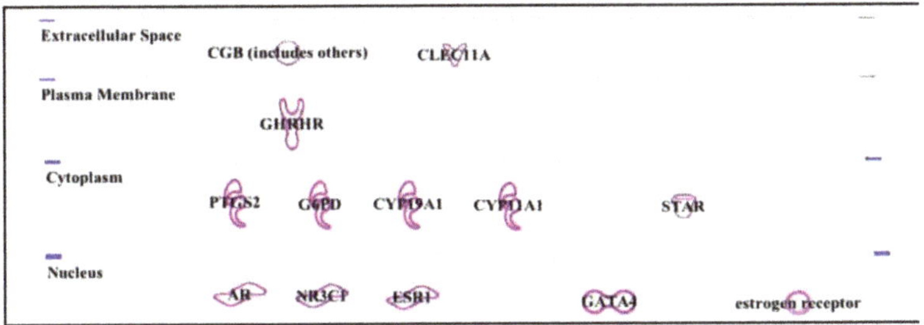

C: Cell Death

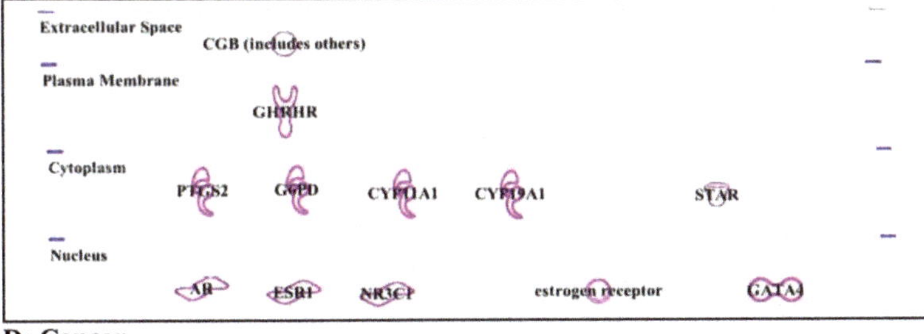

D: Cancer

Figure 1. *Cont.*

E: Behavior

Figure 1. Atrazine. Panel **A**: Structure of atrazine and legend of symbols used for panels **B** to **E** (as provided by Ingenuity Pathway Analysis (IPA) modeling "Bioprofiler" function). Panel **B–E**: Cellular location and names of genes affected by atrazine exposure linked to, respectively, Panel **B**: Cellular response to organismal injury, Panel **C**: Cell death, Panel **D**: Cancer. Panel **E**: Behavior. Note that all symbols and genes names and functions are described in Supplementary Materials Tables S1–S4. The data were generated using human and animal model data. They include information from in vivo and in vitro experiments.

(C) Cancer: Again, the genes affected by atrazine exposure manifest their effects through responses linked to hormone receptor pathways, involving various genes from the cytochrome p450 family genes [30,41] (Figure 1D and Supplementary Materials Table S3). The dysregulation of the aromatase CYP19, mediated by the steroidogenic factor 1 (SF-1), leads to increased risks of developing prostate and breast cancers in human [42,43]. The general mechanism of action of atrazine is considered to involve oxidative stress and the production of reactive oxygen species (ROS) and hydroxyl radicals, processes which are known to contribute to increased DNA damage [41]. An increased DNA damage response in human breast epithelial cells (MCF-10A) has recently been reported in response to atrazine exposure-mediated DNA double-strand breaks [44]. The atrazine-mediated increase in the expression of GATA4, a transcription factor involved in DNA damage response, confirms this connection [45]. GATA4 may also indirectly cause an increase in inflammation [46]. As an increasing amount of evidence suggests a role for atrazine as an oncogenesis trigger in breast cancer, one may envision that atrazine may soon be officially considered as a potential carcinogen, based on its xenoestrogen properties as well as its ability to induce DNA double-strand breaks.

(D) Behavior: The effects attributed to an exposure to atrazine and behavior changes are also mediated by hormone receptors and cytochrome p450 family genes [30,41] (Figure 1E and Supplementary Materials Table S4). An exposure to low doses of atrazine has been shown to affect the behaviors of young male mice [47,48]. The reproductive behaviors of male *Drosophila* have also been shown to be affected by atrazine exposure [49]. Atrazine-mediated mis-regulation of acetylcholinesterase was recently shown to affect defensive behavior in zebrafish [50].

As we investigated the effects of atrazine on various biological systems and functions, our results indicate that this specific herbicide harbors a strong carcinogenic potential, as has been long suspected [51]. As a hormone mimic, the main identified targets of atrazine are hormone receptors, some of which (AR, GR, and ER) are strongly associated with breast [52,53] and/or prostate cancers [29]. These receptors are also involved in the processes of growth and development, therefore long-term exposure should be considered potentially hazardous for developing aquatic and semi-aquatic animals (amphibians, fish, etc.) [54], as well as mammals exposed to contaminated water [55,56]. In addition, atrazine has also been recently associated with disruptions of immune evasion, a mechanism involved in regulating tumor formation, progression, and evasion [57,58], and future studies are likely to uncover additional effects mediated by atrazine exposure involving the immune response and their connection with oncogenesis. Deficiencies in processes such as tumor evasion prevent one's immune system from recognizing and disposing of malignant tumors. Importantly, recent studies have also linked exposure to atrazine with increased DNA damage in fish [59] and mammals [60], including humans [44,61]. As previously mentioned, this aspect of atrazine toxicity is likely to be connected with the increase in the formation of ROS caused by the reduced expression of cytochrome p450 family genes, hence leading to increased DNA damage. This effect on impeding DNA damage might give a clue to the pluripotent effects of atrazine, as the increased presence of ROS will promote DNA single and double-strand breaks, causing mutations and chromosomal instability.

3.2. Chlorpyrifos: Organophosphates and Their Neural and Hepatic Toxicities

The use of organophosphate insecticides is widespread in agriculture, despite being a frequent source of poisoning around the world. In 2002, there were an estimated 3,000,000 cases of organophosphate poisoning globally, which resulted in 300,000 deaths [62]. The United States has had far fewer cases of lethal organophosphate poisoning reported compared to other developing nations. This lower number might be partially due to better access to the main antidotes to chlorpyrifos poisoning—atropine and pralidoxime. Nonetheless, organophosphates such as chlorpyrifos (for structure, see Figure 2A) are still found in commercially available household insecticides, which increases the average individual's risk of exposure to these harmful compounds.

Although the modern use of organophosphates is generally limited to insecticides, these compounds have also been used historically in chemical warfare due to their inherent lethality in humans. While unfortunate, this has resulted in a wealth of anthropocentric research data not available when compared to with many other insecticides. Examples of organophosphates include physostigmine, which is naturally found in the Calabar bean and has been used for centuries in West African witch trials [63], and sarin gas, first used during World War II and a known contributor to the high rate of insecticide-mediated suicides in modern Asia [64]. On a molecular level, these deadly chemicals exert their effects on the brain through the inhibition of the critical enzyme cholinesterase (an enzyme involved in the degradation of the neurotransmitter acetylcholine).

Much of the early research on organophosphate toxicity failed to definitively link exposure to any long-term effects in humans [65]. However, decades of investigation have followed and led to a change of opinion, as several neuropsychiatric conditions have become strongly associated with organophosphate exposure [66]. Broadly speaking, long-term toxicity results in similar symptoms to those seen in acute exposure, and a single severe episode of acute poisoning may lead to chronic effects long after functional cholinesterase levels are restored [67]. Adding to the current knowledge, our study has linked organophosphate exposure to changes in gene expression involved in organismal injury, cell death, cancer, and behavior (Figure 2B–E).

(A) Organismal injury: Chlorpyrifos exposure increases the expression of many genes associated with stress (Figure 2B and Supplementary Materials Table S5). These include *CRH* (corticotropin-releasing hormone), an important regulator in the hypothalamic–pituitary axis; *GAL* (galanin), which may be neuroprotective; and *ABCG2*, a multi-drug resistance transporter gene. The choline acetyltransferase-encoding gene *ChAT* was affected, which is not surprising considering the direct mechanism of action of chlorpyrifos is through the inhibition of the cholinesterase enzyme. Other gene targets of chlorpyrifos exposure include succinate dehydrogenase, one of the key enzymes in the citric acid cycle and electron transport chain and *Sod*, a superoxide dismutase gene which is important in apoptosis, and is linked with DNA damage.

(B) Cell death/survival: The cell death- and survival-associated genes targeted by chlorpyrifos are similar to those involved in organismal injury (Figure 2C and Supplementary Materials Table S6). Examples include *Sod*, *ChAT*, *ABCG2*, *CRH*, *NPY*, and *GA*. In addition, a heme-oxygenase encoding gene known as *HMOX1* is affected by chlorpyrifos exposure. Overall, the profile of cell death and organismal injury targeted by chlorpyrifos activity demonstrates that the toxicity of this chemical affects not only the existing cholinesterase enzyme, but also the genes which are key factors involved in cell protection and metabolism.

(C) Cancer: Only three cancer-related genes were identified as being altered by chlorpyrifos exposure, and all three were common to the other categories delineated in this paper: *ChAT*, *GSR*, and *HMOX 1* (see sections A, B, and D, Figure 2D and Supplementary Materials Table S7). *GSR* encodes glutathione disulfide reductase, a highly conserved gene which is important for preventing oxidative stress in humans. This function is highly relevant in cancer, as the conversion of GSSG (oxidated Glutathion) to GSH (Glutathion) promotes increased DNA synthesis, which aids the growth and development of tumors [68].

(D) Behavior: The main behavior-related genes identified in this ontological study were *ESR1*, *NR3C1*, and genes involved in the conversion of fatty acids or cholesterol to hormones. These include *PTGS2*, *CYP11A1*, and *CYP19A1*, which encodes aromatase (Figure 2E and Supplementary Materials Table S8). These findings suggest that chlorpyrifos exposure is closely associated with the production of sex steroids and the resultant signaling pathways, via the estrogen receptor-encoding genes *ESR1* and *NR3C1*.

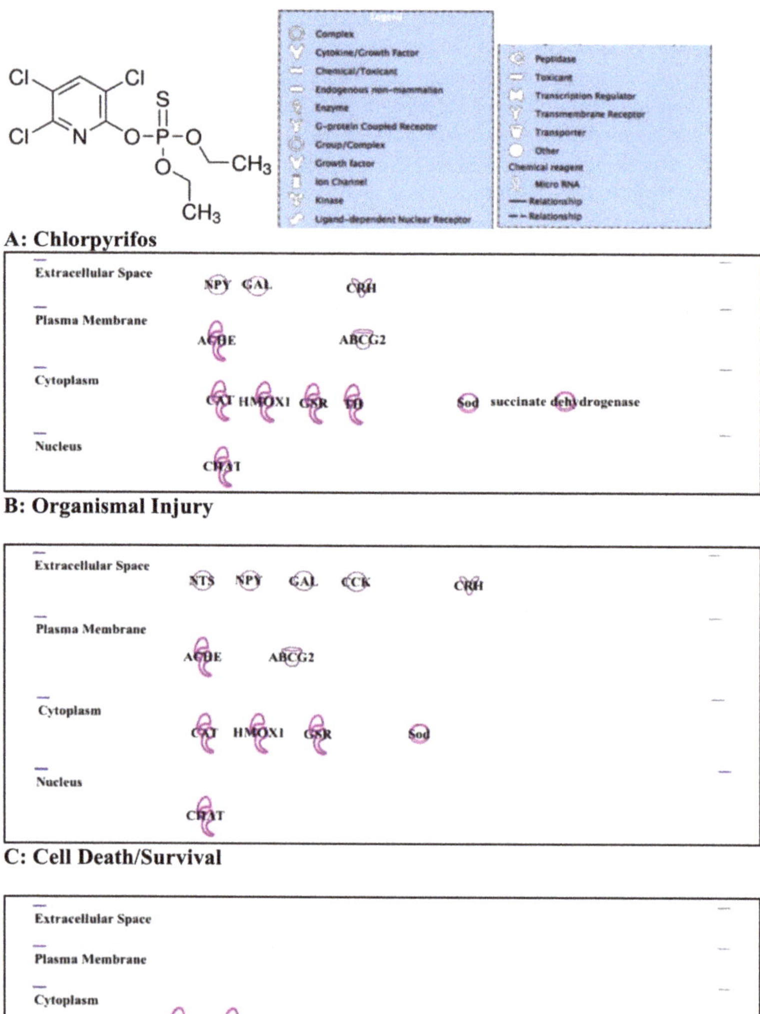

Figure 2. *Cont.*

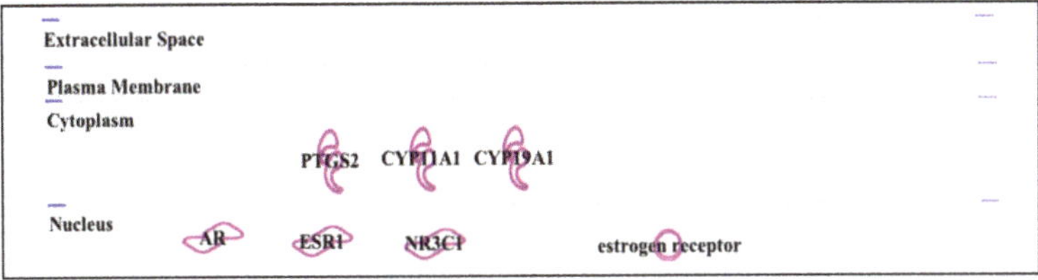

E: Behavior

Figure 2. Chlorpyrifos. Panel **A**: Structure of chlorpyrifos and legend of symbols used for panels **B** to **E** (as provided by IPA modeling "Bioprofiler" function). Cellular location and names of genes affected by chlorpyrifos exposure linked to, respectively, Panel **B**: Cellular response to organismal injury, Panel **C**: Cell death/survival, Panel **D**: Cancer, Panel **E**: Behavior. Note that all symbols and genes names and functions are described in Supplementary Materials Tables S5–S8. The data were generated using human and animal model data. They include information from in vivo and in vitro experiments.

As was expected in the case for atrazine, based on its chemical structure, chlorpyrifos is a strong disruptor of hormonal responses. However, its biological effects are more targeted towards the processes involving cellular detoxification and glutathione (GSH) [68]. Although the majority of the data available were generated using the planktonic crustacean Daphnia as a model system, several reports suggest that its effects follow a similar mode of action in humans [69,70], also potentially mediated by GSH availability. Recent studies of chlorpyrifos exposure also give indications that it may alter proper brain function and development by impairing the cortical axon functions in rats [71], as well as proper differentiation of neural stem cells into neuronal and glial cell phenotypes [72]. The links between chlorpyrifos toxicity with cancer are less obvious than that observed with atrazine, as they are apparently mediated by defective detoxification genes, such as *GSR*. This might be a secondary effect, possibly related more to a potential red-ox imbalance (oxidative stress in general) than a direct carcinogenic effect. As a potential link to this oxidative stress, DNA damage associated with chrlorpyrifos exposure has been reported in various mammalian tissue, most notably in the brain [73].

3.3. *Endosulfan: Pesticide*

Endosulfan is an organochlorine cyclodiene pesticide (see Figure 3A for structure) considered to be highly toxic because of its endocrine effects and high potential for bioaccumulation (EPA toxicity Class I, for additional details see https://www.epa.gov/sites/production/files/2014-07/documents/chapter7_revised_final_0714.pdf). Based on our analysis, all changes in organismal injuries, cell death, cancer and behavior a similar sub-set of genes, which are generally associated with endosulfan's ability to be recognized as a hormone mimetic substance (Figure 3B and Supplementary Table S9). Most reports on the toxicity of endosulfan are based on its biological effects on aquatic organisms [74]. In contrast, less information is available on its effects on humans, although initial reports, from as early as 1982, described its potential bioaccumulation in humans as well as several other non-target species (for more details, see [75]). Aquatic species appear to be more sensitive to the bioaccumulation of endosulfan, therefore experiencing a higher toxicity [74–76]. According to studies performed on various animal models and as seen with the previously-mentioned EDCs and based on its chemical structure, endosulfan can predictably disrupt hormonal responses in both fish and mammals (including humans) [77]. Our ontological study results identified a similar set of genes affected by endosulfan exposure in all the investigated cellular-function scenarios. These deleterious effects can result in the triggering of developmental issues in multiple cell types including, but not limited to, the reproductive tracts [78]. Additionally, two important genes involved in cell proliferation, the estrogen receptor *ESR1* and *ESR2*, are strongly affected by endosulfan exposure [11]. Another report associates endosulfan toxicity with prolactin (PRL) expression [79], an effect which may also be mediated by changes in expression of the nitric oxide synthase genes *NOS1* and *NOS2*, altering the normal functions of the pituitary glands [80]. In addition, exposure to endosulfan has been linked to an increased incidence of various types of cancers, with cell types involved in sexual development, primarily breast, as primary targets [78–82]. Other tissues or cell types are also targeted, such as in colon cancer, in which jun/AP-1 is the main pathway affected by endosulfan exposure known to date [83]. As a possible linkage to explain its activity in cancer cells, endosulfan has been identified as an apoptotic agent [84]. The affected cell types by this mode of action cover a wide spectrum, from human T-cells [84] and peripheral mononuclear cells [85] to umbilical, embryonic, and placental cells [86].

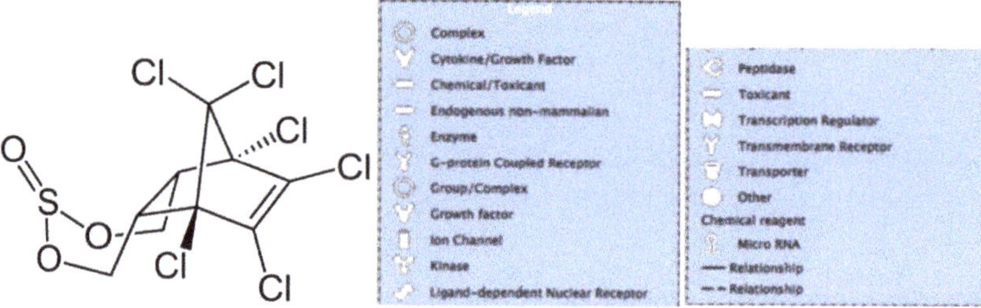

A: Endosulfan

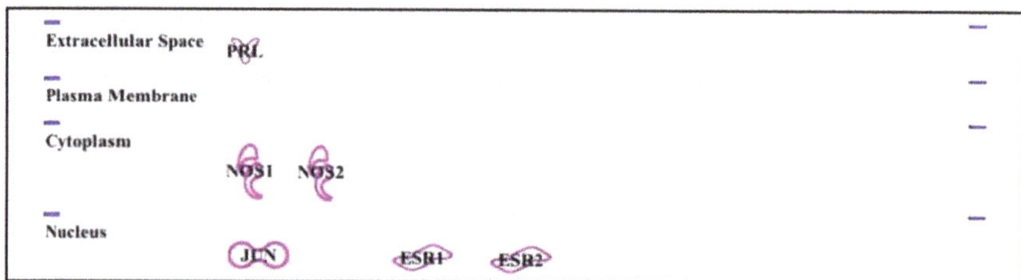

B: Organismal Injury, Cell Death, Cancer, Behavior

Figure 3. Endosulfan. Panel **A**: Structure of endosulfan and legend of symbols used for panel **B** (as provided by IPA modeling "Bioprofiler" function). Panel **B**: Cellular location and names of genes affected by endosulfan exposure linked to cellular response to organismal injury, cell death, cancer, and behavior. Note that all symbols and genes names and functions are described in Supplementary Materials Table S9. The data were generated using human and animal model data. They include information from in vivo and in vitro experiments.

As revealed in our chlorpyrifos analysis, the bulk of evidence for endosulfan toxicity uncovered in our ontologic study was based on experiments performed using aquatic species. Nonetheless, a small number of mammalian (mostly rat) studies on endosulfan exist providing valuable information pertaining to human health. Exposure to endosulfan in young male rats interferes with normal development of the mammary glands [87], an expected result for EDC toxicity. Recent work has also linked endosulfan exposure to deficiencies in the brain and behavioral functions [88]. Aiming to determine its general mechanism of action, additional studies have indicated that endosulfan appears to be associated with an apparent increase in non-sequence-specific DNA damage (for review, see Sebastian and Raghavan, 2017 [89]). This might explain how endosulfan can affect a large number of genes involved in a variety of cellular functions instead of having more defined and specific genetic targets. This potential mechanism of action involving genomic instability raises the possibility that endosulfan displays carcinogenic properties.

3.4. Nonyl Phenol and Nonyl Phenol Ethoxylates: Endocrine Disruptor

Nonyl phenol, or 4-nonyl phenol (NP), is a non-ionic surfactant (see Figure 4A for structure) commonly derivatized to generate Nonyl phenol ethoxylates (NPE) which are used as emulsifiers, detergents, and dispersing agents [90]. Since 2000, NP and NPE have been highly regulated in European Union countries because of their inherent toxicities [91] (PARCOM 92/8, 2000; Directive 2000/60/EC, 2000; [92] Directive 2003/53/EC, 2003).

As was shown in planarians and numerous other non-vertebrates, organismal survival and ability to respond to injury can be affected by NP exposure, similarly to what has been described for other EDCs [93] (for examples, see [94,95]). In zebrafish, long-term exposure was shown to be a factor in survival by reducing the reproduction rate [96]. Also, the survival and injury response may be reduced because of the ability of NP to form adducts with DNA which leads to possible increases in mutation rate [97].

(A) Cell death and organismal injury/survival: Exposure to NP has been reported to induce apoptosis in various cell types, including sexually related cells, neurons, and neural stem cells [98]. As NP acts as a hormone mimic, the effect on sexual organs was expected; however, the apoptotic induction in neurons and neuronal stem cells was more surprising. Nonetheless, the consequences of such exposure during gestation and/or early brain development may be a very serious issue that has not yet received full attention. These observed effects can be linked to hormone receptors mediated by the signal transduction triggered by several types of surface receptors such as the FAS receptor (also referred to as the "apoptosis antigen") and the anion transporter SCL22A6 ([99] (see Figure 4B–D and Supplementary Materials Tables S10–S12).

(B) Cancer: As an EDC, the health risk related to exposure to NP and/or NPE stems from their ability to act as a hormone mimic [100]. Studies have linked NP with breast cancer, as it can mimic 17B-oestradiol and compete for the binding site of oestrogen receptors *ESR1*, *ESR2*, and the related steroid receptor co-activator-1 (SRC-1/NCOA1) [101] (see Figure 4E and Supplementary Materials Table S13). Also, a linkage with androgen receptors has also been reported. This suggests a potential involvement of NP exposure with prostate cancer [102]. This mechanism may not involve a direct interaction, but instead may be mediated by cross-talks with other hormone receptor(s). As there is very little information connecting NP with other types of cancers, it is possible that its effect is exerted in a compound manner with other EDCs or emerging water contaminants [90].

(C) Behavior: In addition to a propensity for promoting breast cancer, NP can also alter the neuroendocrine system [90,103,104]. This interference may be associated, directly or indirectly, with behavioral modifications, such as motility changes (swimming in fish), or social behaviors [103,105,106]. Interestingly, these modifications have been observed over two generations, suggesting a potentially epigenetic mechanism of transmission [107]. As was the case for the cancer-associated effects, hormone receptors (ESR1/2, NCOA1, see Figure 4F) may play a significant role in these behavioral changes.

Overall, as with most EDCs, NP can act on a plethora of genes, resulting in alterations of multiple functions. The effects are most often mediated by changes in hormone receptor activity. Interestingly, and similarly to endosulfan, NP exposure was shown, in human keratinocytes, to be linked with DNA damage. This mechanism involved multiple factors, including ATM (ataxia-telangiectasia mutated), the tumor suppressor p53, and the histone H2A.X (a marker of DNA double-strand breaks). The same study linked NP to apoptosis, mediated by activation of poly(ADP-ribose) polymerase (PARP) and caspase 3 [108]. Although not clearly outlined by our bioinformatics analysis, very recent literature has connected NP with liver toxicity in mammals, including humans, as its pro-inflammatory properties negatively affects liver cells [109].

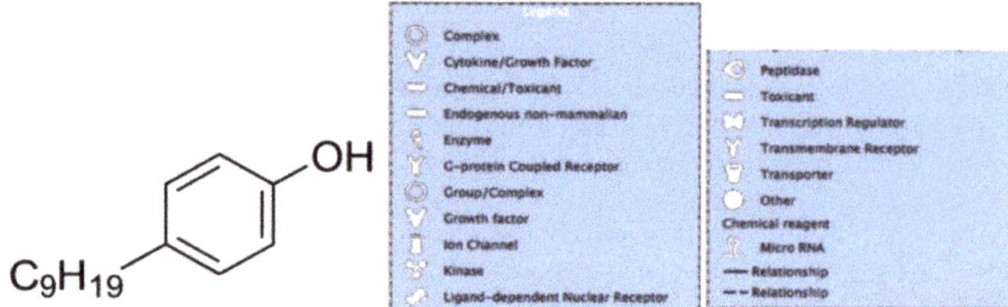

A: Nonyl Phenol

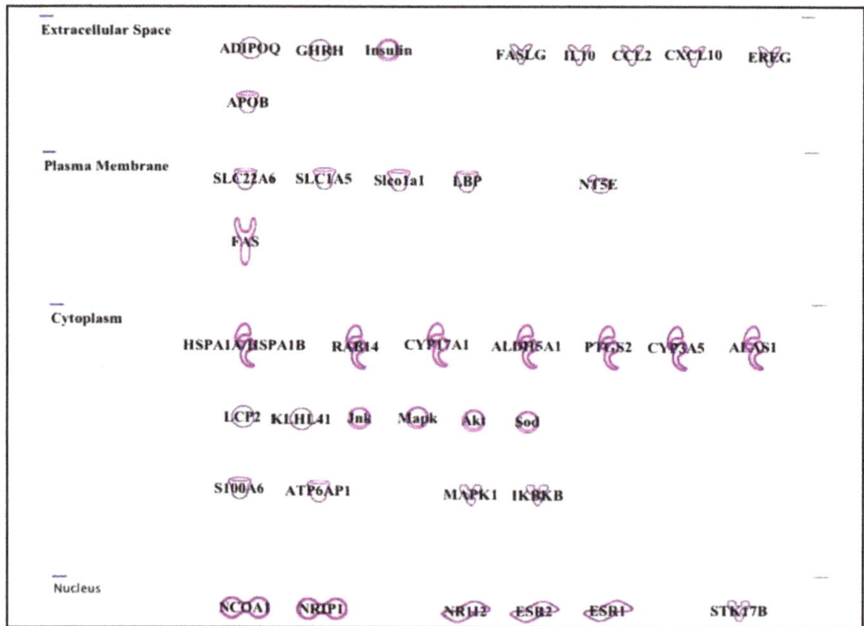

B: Organismal Injury

Figure 4. *Cont.*

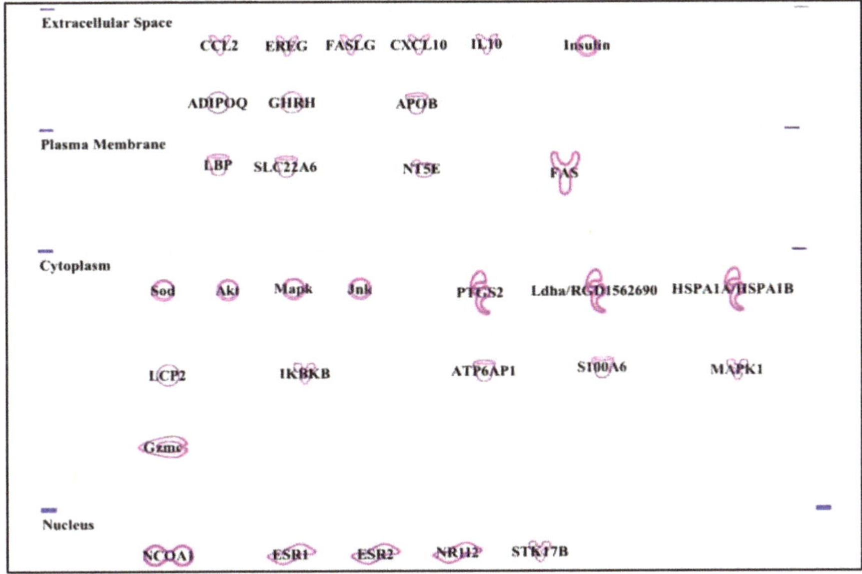

C: Cell Death

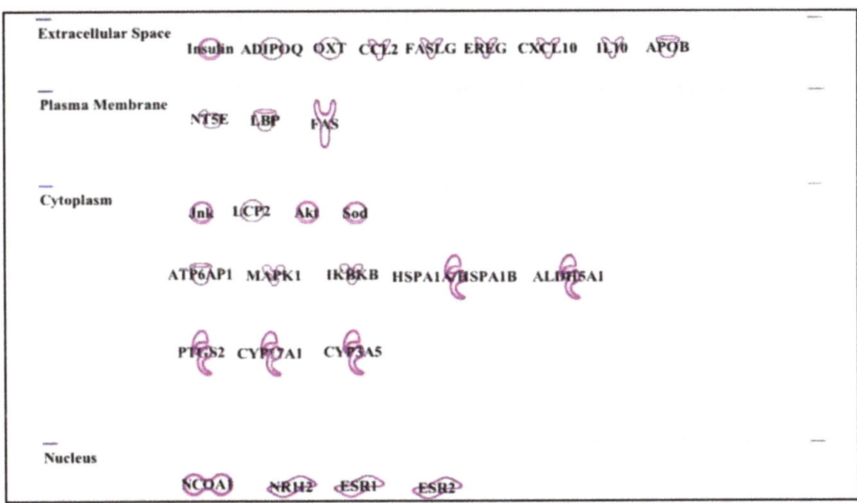

D: Organismal Injury

Figure 4. *Cont.*

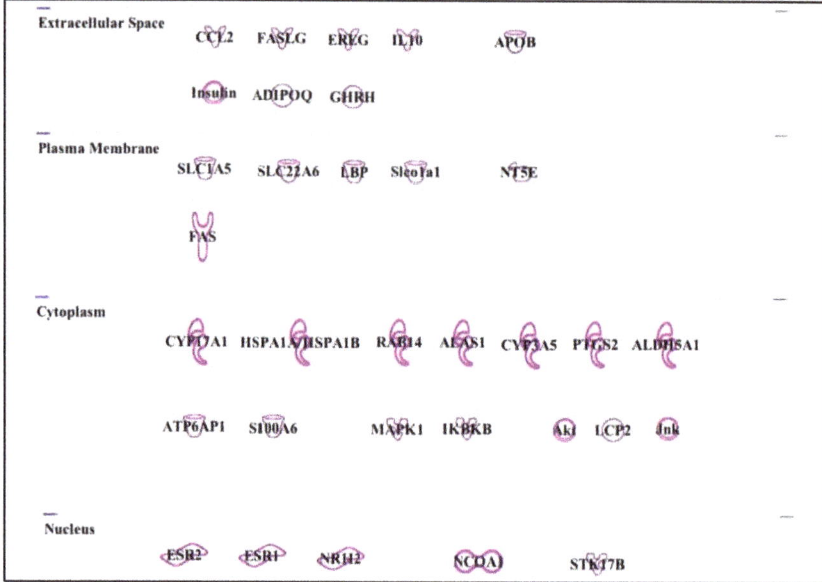

E: Cancer

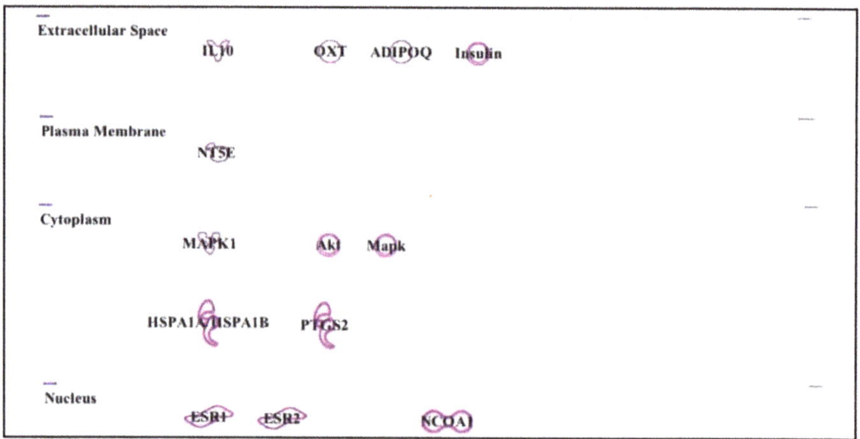

F: Behavior

Figure 4. Nonyl phenol. Panel **A**: Structure of nonyl phenol and legend of symbols used for panels **B** to **F** (as provided by IPA modeling "Bioprofiler" function). Cellular location and names of genes affected by nonyl phenol exposure linked to, respectively, Panel **B**: Cellular response to organismal injury, Panel **C**: Cell death, Panel **D**: Organismal survival, Panel **E**: Cancer, and Panel **F**: Behavior. Note that all symbols and genes names and functions are described in Supplementary Materials Tables S10–S13. The data were generated using human and animal model data. They include information from in vivo and in vitro experiments.

3.5. *Bisphenol A: Endocrine Disruptor*

The rise of safety concerns related to chemicals released from consumer polycarbonate plastics has led to a concomitant increase in the study of their potential toxic effects. Bisphenol A (BPA) is perhaps the most widely recognized plastic-derived toxin (for structure, see Figure 5A). Although most of the major studies regarding the potential toxic effects of BPA used rats as an animal model, it has since been found that primates metabolize BPA into an inactive form more efficiently than rodents, suggesting that humans may be less susceptible to its toxic effects than initially believed [110]. Fortunately, despite the uncertainty on the extent of BPA exposure to human toxicity, industries have ceased using BPA in the manufacture of consumer plastics based on an amendment of FDA restrictions ([111] Fed. Reg. 41,899; 78 Fed. Reg. 41,840).

Although it appears that current and future plastics may be largely free of BPA, due to the non-biodegradable nature of plastics, there is still concern regarding less direct routes of exposure. In 2011, 19 landfills had their leachate tested for potential contaminants and 95% of these samples tested positive for BPA [112]. It was also detected at levels known to negatively affect some invertebrate species in untreated and treated wastewater, as well as in river water downstream from a paper factory [113]. In a recent study to determine the possibility of food-borne exposure toxicity, researchers grew vegetable plants using BPA-contaminated water and found that ingestion of contaminated agriculture products would result in exposure to physiologically significant doses of BPA known to cause observable developmental changes experimentally [114]. Thus, despite the absence of BPA in modern plastics, former industrial practices may continue to have a negative impact on human health through soil, water, and agriculture contamination.

The health risks associated with BPA exposure are primarily related to endocrine disruption, as its main mechanism of action involves binding to estrogen receptors (ER), thus acting as a weak estrogen [115,116]. A few of the major health research foci regarding the effect of BPA's endocrine disrupting activities are obesity, reproductive system dysfunction, immune dysfunction, and cancers of the breast, prostate, and uterus [117]. The ability of BPA to disrupt the endocrine system may also interfere with progesterone receptor expression in human endometrial tissue [116], leading to an increased body mass and decreased glucose tolerance in males [118], negatively impacting male fertility [119], and accelerated growth during childhood [120]. In addition to these effects, as an EDC, BPA also targets genes involved in organismal injury, cell death, cancer, and behavior Figure 5B–F.

(A) Organismal injury: BPA can affect an organism's ability to heal and respond to injury by the disruption and alteration of the endocrine system (Figure 5B and Supplementary Materials Table S14). Chronic BPA exposure has also been shown to reduce cardiac remodeling in mice [120], and is toxic to bone mesenchymal stem cells [116]. In addition to affecting these systems, exposure to BPA also has an impact on the Sertoli cells of the testes, [121] as well as on ovarian cells [81].

(B) Cell/organismal death: In addition to overt organismal injury, BPA exposure can lead to apoptosis of spermatogenic cells [122]. In fact, many of the most toxic effects of BPA are observed in cells of the reproductive system (Figure 5C,D and Supplementary Materials Tables S15 and S16). To this end, BPA has been shown to negatively impact embryo and oocyte quality in mice by increasing apoptosis [81,116]. Additionally, BPA exposure can result in the induction of apoptosis in mouse pancreatic islet cells [116].

(C) Cancer: BPA exposure is specifically associated with breast cancers (Figure 5E and Supplementary Materials Table S17), through the increased expression of *HOXB9*, a gene important in mammary gland development known for its potential oncogenic properties [121]. This effect seems to be related to its ability to mimic or activate similar pathways as those mediated by endogenous estrogens. In addition, BPA exposure can also increase the expression of two matrix metalloproteinases, MMP2 and MMP9 [111], the functions of which are linked with enhanced breast cancer migration and invasion.

(D) Behavior: Sharing similar mechanisms with endogenous estrogens, it is not surprising that BPA has behavioral effects (Figure 5F and Supplementary Materials Table S18). A systematic review on BPA exposure in Canadian school-age children revealed that early-life and prenatal exposure was associated with increased behavioral disorders, such as anxiety, depression, hyperactivity, inattention, and conduct problems [115]. Additionally, children taking psychotropic medications were found to be more likely to be susceptible to BPA exposure than those not taking such medications [120].

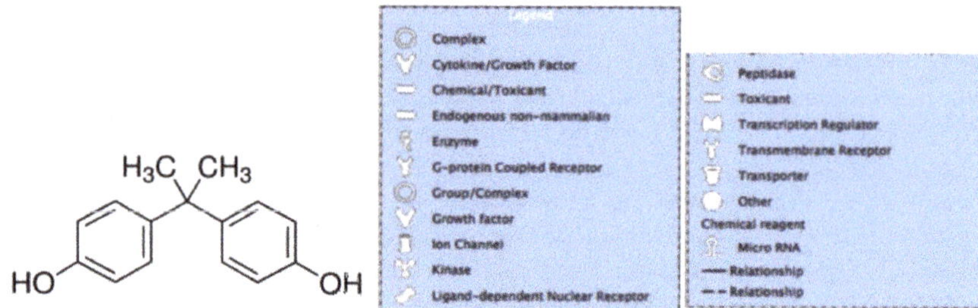

A: Bisphenol A

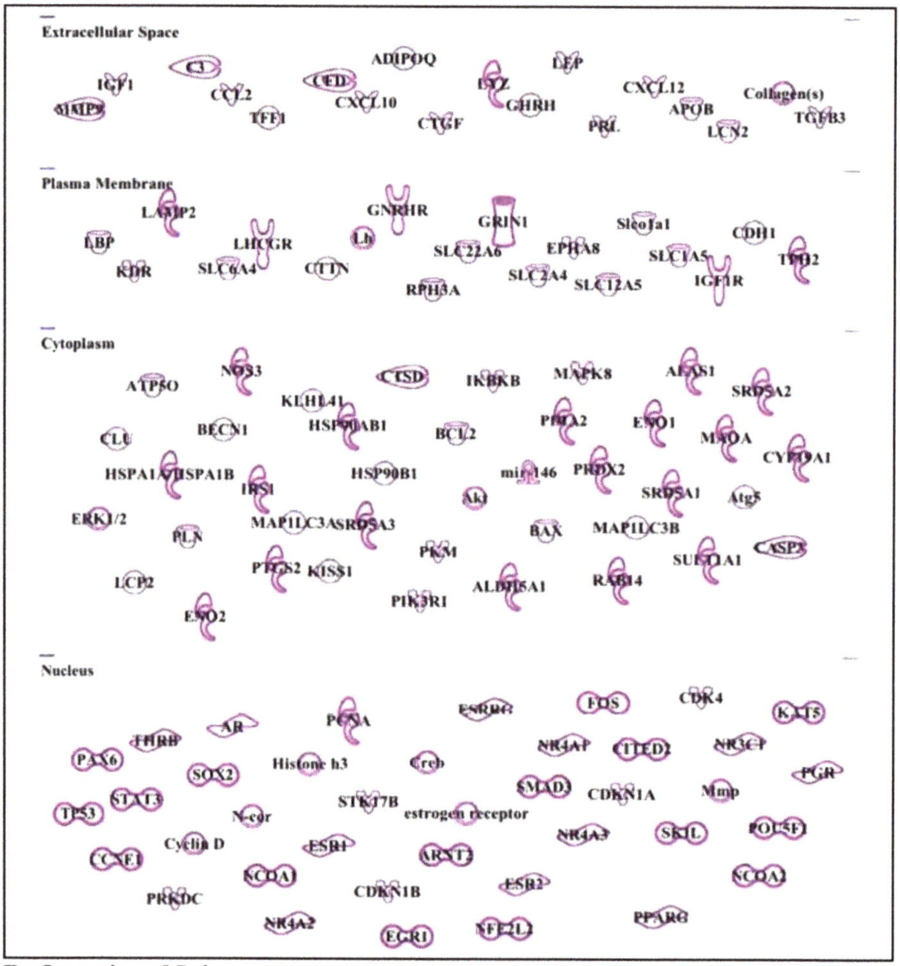

Extracellular Space

ADIPOQ LEP C3 IGF1 CCL2 CFD LYZ CXCL10 CXCL12 Collagen(s) MMP9 TFF1 CTGF GHRH PRL APOB LCN2 TGFB3

Plasma Membrane

LAMP2 GNRHR GRIN1 Slco1a1 CDH1 LBP LHCGR Lh SLC22A6 EPHA8 SLC1A5 TPO2 KDR SLC6A4 CTTN SLC2A4 SLC12A5 IGF1R RPH3A

Cytoplasm

ATP5O NOS3 CTSD IKBKB MAPK8 ALAS1 SRD5A2 KLHL41 CLU BECN1 HSP90AB1 BCL2 PTMA2 ENO1 MAOA CYP19A1 HSPA1A/HSPA1B HSP90B1 mir-146 PRDX2 IRS1 Akt SRD5A1 Atg5 ERK1/2 PLN MAP1LC3A SRD5A3 BAX MAP1LC3B PKM SULT1A1 CASP3 PTGS2 KISS1 LCP2 ALDH5A1 RAB14 PIK3R1 ENO2

Nucleus

ESRRG FOS CDK4 KAT5 PCNA AR THRB NR4A1 CITED2 NR3C1 PAX6 SOX2 Histone h3 Creb PGR SMAD3 CDKN1A Mmp STAT3 TP53 N-cor STK17B estrogen receptor NR4A3 SKIL POU5F1 CCNE1 Cyclin D ESR1 ARNT2 NCOA1 NCOA2 ESR2 PRKDC CDKN1B PPARG NR4A2 EGR1 NFE2L2

B: Organismal Injury

Figure 5. *Cont.*

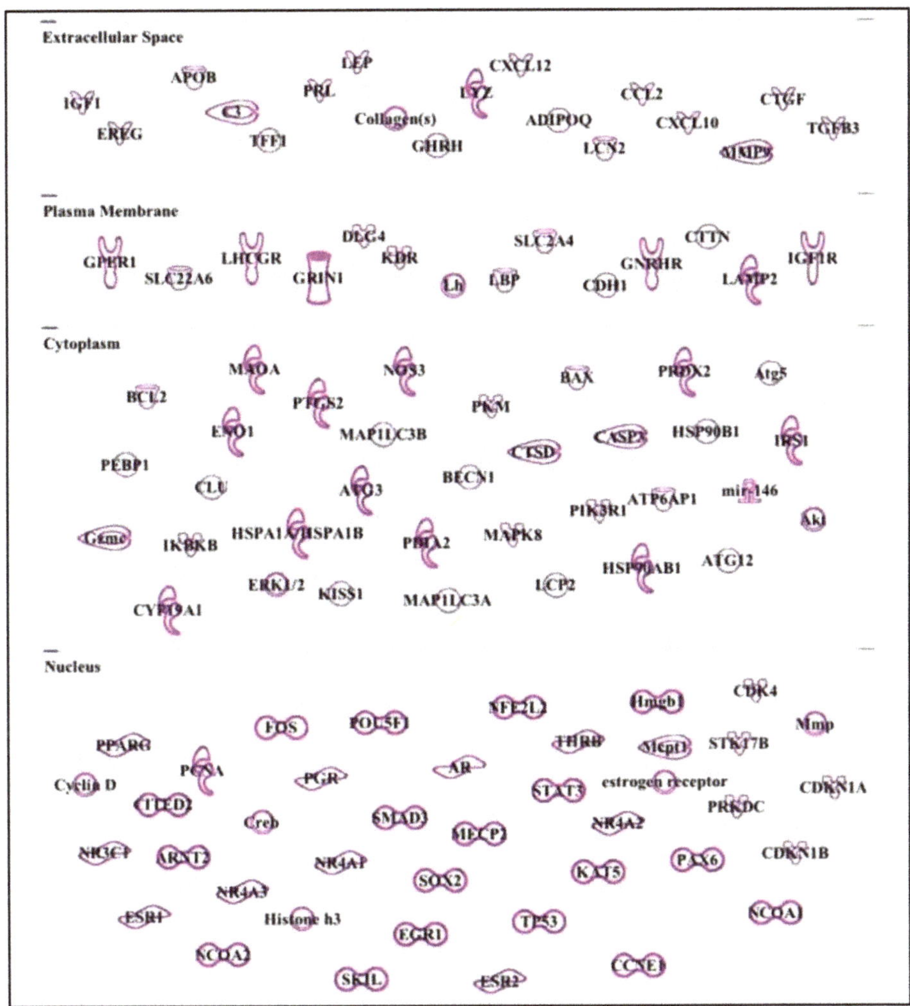

Figure 5. *Cont.*

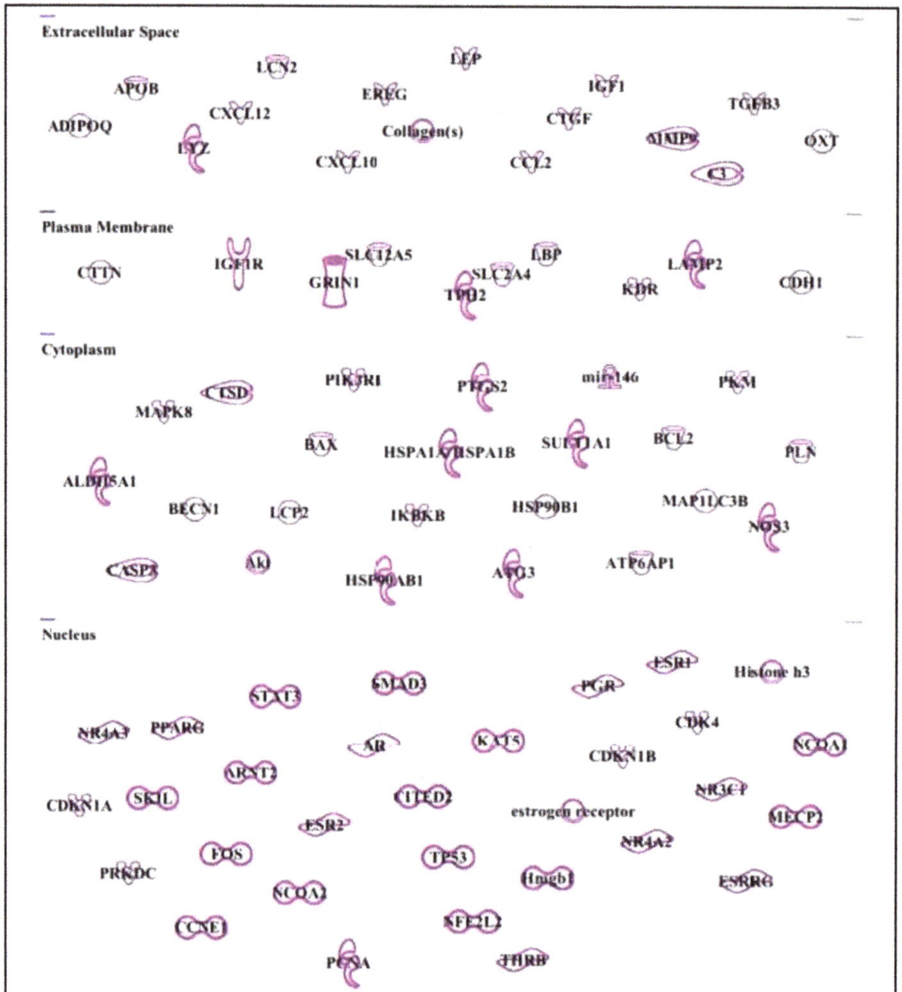

D: Organismal Death

Figure 5. *Cont.*

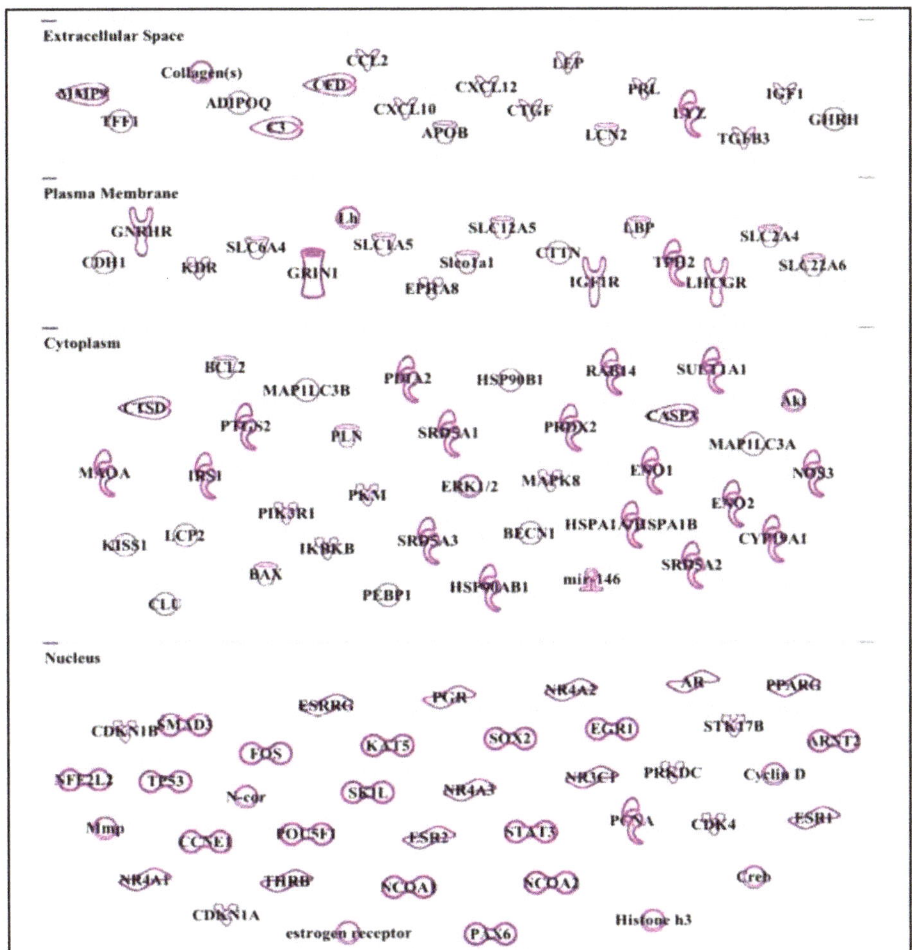

E: Cancer

Figure 5. *Cont.*

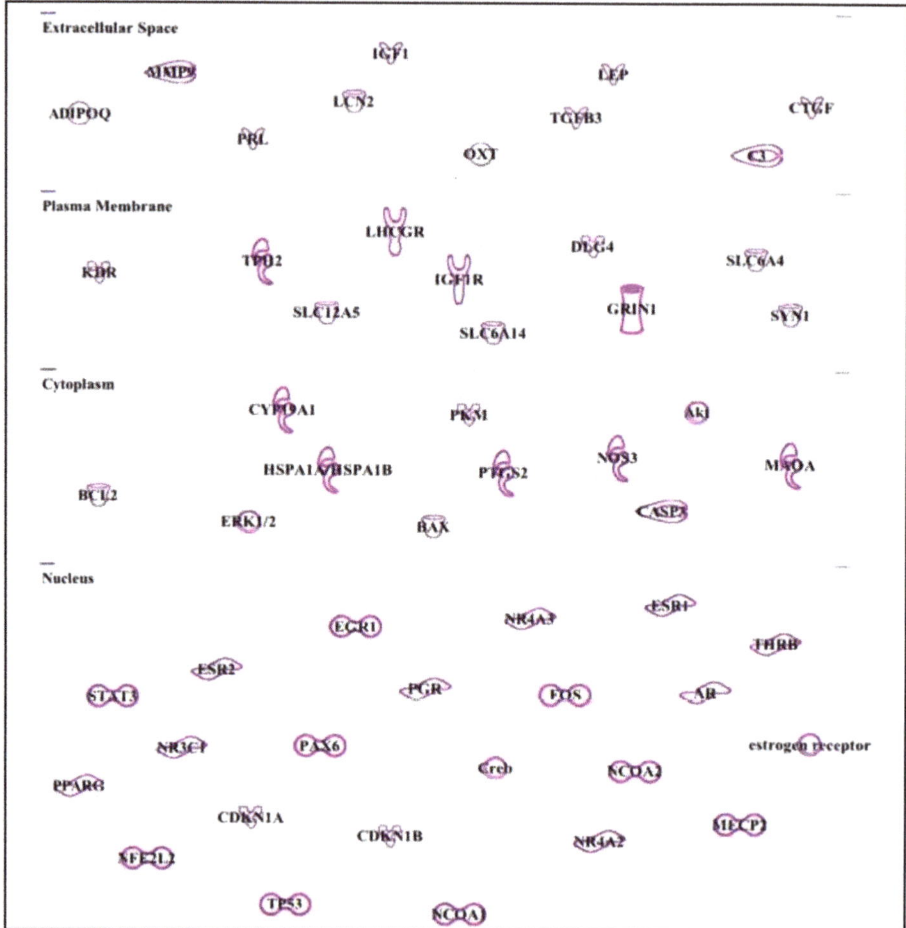

Figure 5. Bisphenol A. Panel **A**: Structure of bisphenol A (BPA) and legend of symbols used for panels **B** to **F** (as provided by IPA modeling "Bioprofiler" function). Cellular location and names of genes affected by bisphenol A exposure linked to, respectively, Panel **B**: Cellular response to organismal injury, Panel **C**: Cell death, Panel **D**: Organismal death, Panel **E**: Cancer, and Panel **F**: Behavior. Note that all symbols and genes names and functions are described in Supplementary Materials Tables S14–S18. The data were generated using human and animal model data. They include information from in vivo and in vitro experiments.

BPA is probably the most notorious EDC and has been known for several years to be a cause of cellular toxicity. As mentioned earlier, the recent banning of its use will hopefully lead to a decrease in its abundance in major river basins, as well as in tap/drinking water. Studies involving BPA degradation during chemical remediation using Persulfate have indicated that, despite the efficient degradation of BPA itself, its degradation products have genotoxic effects of their own [123,124]. This highlights the fact that the risks associated with prior BPA use are still a serious concern for human health. Appearing to be a common theme for EDC exposure, BPA's mode of action is also likely correlated with increased oxidative stress, which would contribute to a BPA-associated increase in DNA damage in mammals [125].

3.6. Buprenorphine: Opioid

Buprenorphine is now considered to be a substance of interest as a potential emerging water contaminant (for structure, see Figure 6A), based on its increased usage and prevalence as a treatment option for opioid addiction. Pregnant women abusing opioids present an additional problem compared to other drug abusers, as supporting these women in overcoming their addiction must be balanced with the serious concerns in protecting the fetus from the dangers of withdrawal. To palliate the negative effects of opioid abuse, one of the preferred agents of maintenance therapy in these pregnant patients is the opioid derivative buprenorphine, as it has a higher potential to reduce neonate abstinence syndrome (NAS, [126]) compared to methadone. Buprenorphine is a mixed agonist-antagonist of opioid receptors—a partial agonist of μ opioid receptors (MOR), and an antagonist of κ opioid receptors (KOR) and δ opioid receptors (DOR). In addition to its primary effects, the metabolic products of buprenorphine degradation, though differing in opioid receptor selectivity, half-life, and potency, may also be considered of concern for water quality.

Despite its status as a controlled substance, there is potential that physiologically relevant amounts of buprenorphine may be contaminating drinking water. In France, for example, buprenorphine was initially detected at levels of 40 ng/L in sewage water effluent at one study site, and at levels varying from 42 ng/L to 195 ng/L in sewage water influent samples at three additional study sites [26]. An earlier study, focusing on the Paris area, also detected low levels of buprenorphine from samples of wastewater entering four treatment plants [85]. While these levels correspond to small therapeutic doses compared to those prescribed to human patients, there is the possibility of a cumulative long-term effect through chronic exposure to the contaminated drinking water. The recent increase in buprenorphine use may eventually enhance its presence in water supplies. Therefore, it is important to identify the genes which may mediate any chronic effects of low-dose and long-term buprenorphine exposure. Consequently, as with our selected EDCs, we used our IPA® analysis to identify the buprenorphine-targeted genes involved in organismal injury, cell death, cancer, and behavior (Figure 6).

(A) Organismal injury/survival and cell death: Aside from the analgesic properties of opioids, an important secondary effect is increasing cell death, particularly of cells in the nervous and immune systems (Figure 6B,C and Supplementary Materials Tables S19–S21). Consequently, opioid addicts have been found to have decreased number of circulating progenitor stem cells [116]. Intriguingly, buprenorphine seems to exert a dose-dependent effect, either promoting cell survival and differentiation or increasing cell death and apoptosis [6,119,127]. Buprenorphine exposure in neuronal cells has also been shown to lead to caspase-3 activation and efflux of cytochrome c from the mitochondria, indicative of activation of the mitochondrial pathway of apoptosis [118].

(B) Cancer: Because of the associated addiction risks, medical practice for chronic pain in the United States is moving away from the use of opioids as treatment, except in cases where the benefits outweigh the risk [128] (Figure 6D and Supplementary Materials Table S22). However, due to the decreased abuse and addiction potential associated with buprenorphine compared to other pain management drugs, it is being investigated as an attractive substitute for managing chronic pain associated with cancer. However, given the previously discussed effects of opioids on cell populations, it is important to consider all the potential effects that buprenorphine might have on cancer cell populations.

For some time, it has been known that buprenorphine has a dose-dependent effect on serum prolactin levels in human, as low doses increase serum prolactin production levels while high doses of buprenorphine decrease them. Importantly, both the positive and negative effects on serum prolactin production are blocked by the opioid receptor antagonist naloxone [111]. Aside from the behavioral implications associated with this effect, prolactin has also been implicated in breast cancer [98] as high levels of prolactin receptor (PRLR) are expressed in some breast cancers, leading to increased cancer cell invasiveness [129].

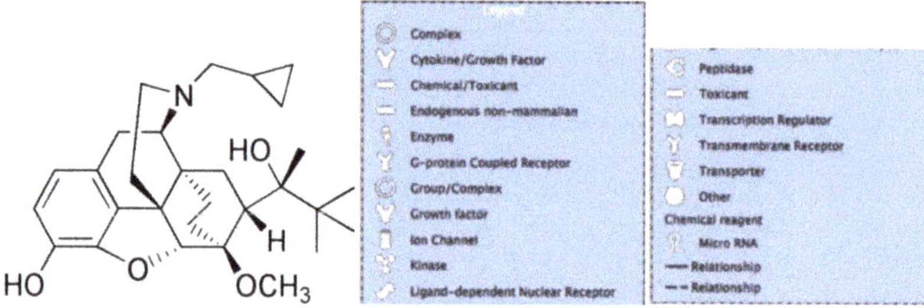

A: Buprenorphine

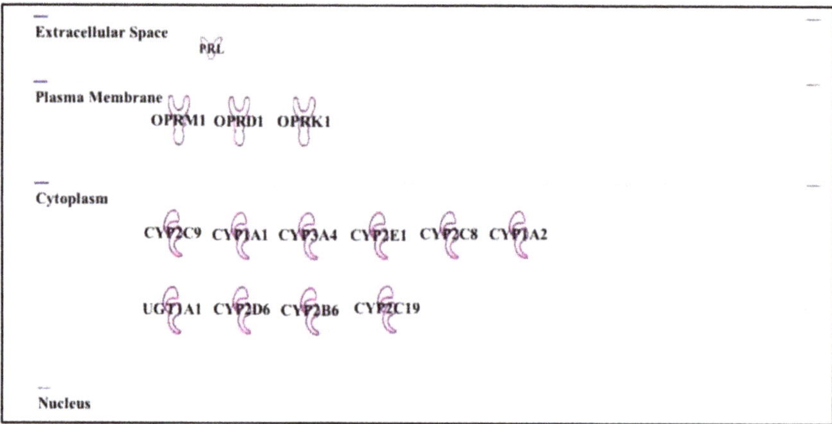

B: Organismal Injury/Survival

C: Cell Death

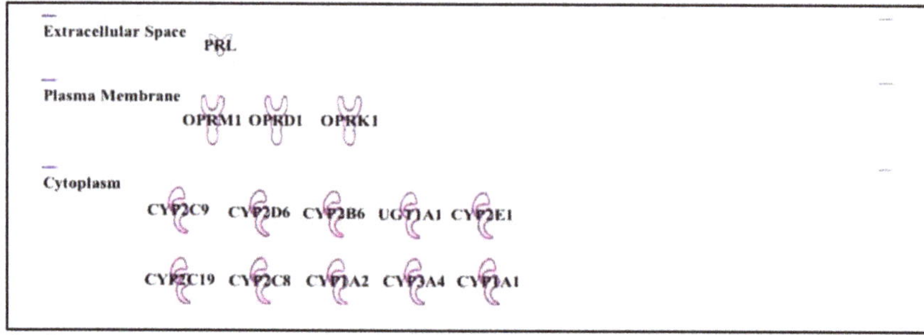

D: Cancer

Figure 6. *Cont.*

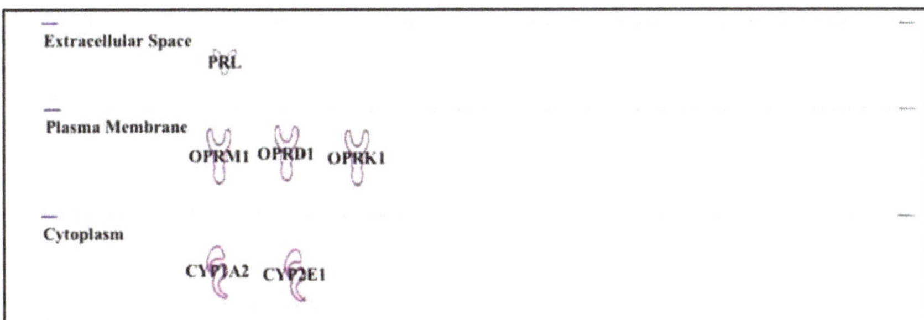

E: Behavior

Figure 6. Buprenorphine. Panel **A**: Structure of buprenorphine and legend of symbols used for panels **B** to **E** (as provided by IPA modeling "Bioprofiler" function). Cellular location and names of genes affected by buprenorphine exposure linked to, respectively, Panel **B**: Cellular response to organismal injury and survival, Panel **C**: Cell death, Panel **D**: Cancer, and Panel **E**: Behavior. Note that all symbols and genes names and functions are described in Supplementary Materials Tables S19–S23. The data were generated using human and animal model data. They include information from in vivo and in vitro experiments.

(C) Behavior: Although prolactin is most directly associated with lactation and some aspects of maternal behavior [128], the other pathways involved in prolactin release should also be considered when analyzing the effects of buprenorphine (Figure 6E and Supplementary Materials Table S23). One of the key feedback pathways involved in prolactin release is dopamine release from the hypothalamus [130], as high dopamine levels exert an inhibitory effect on the release of prolactin [131]. The fact that buprenorphine can both increase or decrease prolactin levels supports the hypothesis that it interacts with the dopaminergic systems of the brain, and, also supporting this conclusion, it has been demonstrated that buprenorphine exposure affects the dopamine receptor density in the striatum [123]. In turn, dopamine is widely implicated in the pathology of psychiatric disorders.

Though buprenorphine is not a bona fide EDC, its potential to become a major water contaminant justifies its inclusion in our study. Its presence in the tap/drinking water in large urban areas (see Table 1) suggests that buprenorphine contamination is becoming a serious issue in regions currently dealing with any type of "opioid crisis". As opposed to the EDCs that we investigated, which seem to share a mode of action strongly involving oxidative stress and DNA damage, buprenorphine does not appear to work in a similar manner, despite a few reports which link its toxicity with increased apoptosis [132].

4. Conclusions

As has become obvious over the last few years, water quality and emerging contaminants have become a serious concern for public health. Our ontology analysis clearly demonstrates that our selected examples have a plethora of negative effects on cellular systems, spanning from micro-organisms to human. The IPA® analytical settings used for our study outlined the most commonly observed effects, based on the current available literature, and we decided to focus on the to five "families" of function most often outlined by our search. The deleterious effects related to endocrine mimicking properties are a major concern directly related to exposure to these emerging contaminants. Even very low concentrations, such as those detected in the Ohio River (see Table 1), can lead to serious behavioral changes, as exemplified by the effects of buprenorphine anxiety, depression, hyperactivity, inattention, and conduct problems, as shown by Ejaredar et al. [133]. As expected, exposure to EDCs affects endocrine functions, mostly through standard interactions with hormone receptors, acting either as activators or inhibitors. These hormonal disruptions cause increased risks of developing various types of cancer (for example, but not limited to: prostate, breast, and ovarian cancers). In addition to these oncogenic properties, exposure to EDCs in our water supply clearly affects basic cellular functions as well, such as the ability to control cell death and survival (commonly mediated through modifications of genes involved in the apoptotic pathway).

From our analysis and based on recent reports, a common theme emerges connecting EDC exposure and increased oxidative stress which eventually leads to DNA damage. All the investigated EDCs in this study share a common link with this mode of action, as evidenced by studies on various organisms, from invertebrates to humans. This increased inability to compensate for EDC-induced DNA damage may play an important role in the toxicity of these products. Therefore, the cytotoxicity of these emerging contaminants might be provoked by, or be the consequence of, an underlying genotoxicity manifesting itself in a variety of ways, affecting a litany of genes involved in nearly all aspects of cell development and viability.

A potential explanation for why so many different genes can be affected by EDC or opioid exposure is that their mode of action could possibly be associated with epigenetic changes, although this possibility currently remains understudied. DNA methylation, as well as histone post-translational modifications (PTM), are critical epigenetic elements involved in regulating gene expression (for recent review, see Corella and Ordova, 2017 [134]). and an increasing number of publications have now demonstrated a connection between EDC exposure and the deregulation of DNA methylation at specific chromosomal locations.

Such connections between epigenetic regulations and EDC exposure have been reported for atrazine, the mode of action of which appears to affect the global levels of DNA methylation [135,136] as well as that of specific histone PTMs [137]. Highlighting the negative impact of EDC exposure, these epigenetic events have even been shown to affect rat sperm cells in a transgenerational manner, changing the DNA methylation pattern for up to three generations [138].

Endosulfan has also been shown to affect the expression of specific genes in an epigenetic manner [139,140], potentially modulating the expression profile of the histone modifiers histone deacetylase (HDAC) HDAC1 and 3, as well as histone methyltransferase PRMT5 and EZH2, in breast cancer cells [140]. This results in an increase in the histone PTMs tri-methylated histone H3 Lysine 27 (H3K27me3) and di-methylation of histone H4 arginine 3 (H4R3me2), two epigenetic modifications that are known to be involved in the regulation of gene expression by promoting the recruitment of activator or repressor proteins. EDC-mediated epigenetic changes in levels of DNA methylation targeting ERα have also been linked to impaired fertility in female rats mediated by endosulfan activity [139].

Nonyl phenol exposure can have an effect both on DNA methylation and on histone PTM patterns. The changes in DNA methylation were the first reported by [107], but NP exposure was more recently found to result in the deregulation of histone PTMs such as H3 and H4 acetylation, as well as the trimethylation of histone H3 Lysine 4 (H3K4me3) [141] in dendritic [141,142] and testicular cells [143].

BPA exposure can also lead to disruptions in DNA methylation and histone acetylation patterns [144]. Also, as was shown with atrazine exposure, a potential epigenetic transgenerational effect (on insulin production in this case) has been reported in F1 and F2 mouse offspring after maternal BPA exposure [145]. Additionally, this effect may be linked with changes in germ line cells affecting the animal's fertility (see Chianese et al., 2017 for review [146]).

Buprenorphine has also been suspected to act at an epigenetic level [147,148], as it has been linked with changes in gene expression of the methyl-DNA binding protein MeCP2, an important regulator of brain development, as well as several histone PTMs ([131,149], Georgel's laboratory, unpublished data).

Although these changes have not been fully characterized, a growing number of studies point towards complex regulatory mechanisms on an epigenetic level being strongly affected by novel emerging water contaminants. A systematic approach will be required to better understand all the implications which have been revealed to link EDCs and opioid water contamination with epigenetic events. Such an approach would potentially lead to a better understanding of the various, and often unrelated, cytotoxic and genotoxic effects mediated by emerging contaminants in our water supply.

The evidence collected through multiple studies, using a variety of model systems, including human cell lines, indicate that exposure to EDCs and buprenorphine, as expected, influences the expression of many genes involved in a variety of cellular steroid response events. The effects of EDCs have also been commonly linked with inflammation response, as well as changes in the expression of genes involved in the regulation of the cellular Red-ox potential (possibly linked to an increase in the generation of DNA-damaging radicals). The apparent link between EDC exposure and DNA damage was not as easily predictable when considering the chemical nature and makeup of EDCs. Such an increase in cellular DNA damage, which is non-sequence-specific in nature, may partially explain why such a variety of genes in various and seemingly unrelated cellular functions can be affected. The implied correlations from evidence which suggests the involvement of epigenetic re-mapping may also contribute to explaining the mechanisms of action for EDCs and opioids. The epigenetic changes over potentially large sections of the genome would also contribute to explaining the pleiotropic nature of the cells' and organisms' responses to EDC exposure.

If each emerging contaminant included in our analysis can individually affect at the level of gene expression of specific genes and contributes to DNA damage and epigenomic changes, the combinatorial effect of EDCs might prove to be additive or even synergistic in nature. This consideration would require a new and more global research strategy to be developed to investigate the holistic effects of exposure to multiple EDCs, as the combined effects of exposure to these chemicals may be more detrimental and widespread than initially anticipated. Long-term exposure studies involving the monitoring of genome stability and global epigenetic events would likely provide us with an improved view of the global scale of the cellular and organismal responses to these emerging contaminants.

Supplementary Materials: The following are available online at http://www.mdpi.com/2073-4441/11/12/2615/s1, List of tables (results from ontologic/metabolomic analysis using Ingenuity Pathways Analysis): Table S1: Atrazine and Cell Death; Table S2: Atrazine and Organismal Injury; Table S3: Atrazine and Survival; Table S4: Atrazine and Cancer; Table S5: Chloropyrifos and Organismal Injury; Table S6: Chloropyrifos and Cell Death and Survival; Table S7: Chloropyrifos and Cancer; Table S8: Chloropyrifos and Behavior; Table S9: Endosulfan; Table S10: Nonyl Phenol and Cell Death; Table S11: Nonyl Phenol and Organismal Injury; Table S12: Nonyl Phenol and Organismal Survival; Table S13: Nonyl Phenol and Cancer; Table S14: BPA and Organismal Injury; Table S15: BPA and Cell Death; Table S16: BPA and Organismal Death; Table S17: BPA and Cancer; Table S18: BPA and Behavior; Table S19: Buprenorphine and Organismal Injury; Table S20: Buprenorphine and Organismal Survival; Table S21: Buprenorphine and Cell Death; Table S22: Buprenorphine and Cancer; Table S23: Buprenorphine and Behavior.

Author Contributions: J.K. ran the IPA analysis and prepared the initial set of figures. D.D. wrote the endosulfan and atrazine section. D.C. contributed to the preparation of the final version of the Figures and Tables. R.G. contributed to the preparation of the final version of the Figures and Tables. P.T.G. wrote the Introduction and all result sections and the discussion other than "endosulfan and atrazine" (see Diane Dawley's contribution), as well as the conclusion.

Funding: This work was supported by the National Science Foundation, Award Number: 1458952.Proposal Title: RII Track-1: Gravitational Wave Astronomy and the Appalachian Freshwater Initiative, the MU Genomics Core, Bioinformatics Core and the WV-INBRE grant (P20GM103434). The license for IPA was provided by the WV-INBRE bioinformatics core, supported by the WV-INBRE grant (P20GM103434) NIH/NIGMS.Thanks to Emily Gillepsie and J. Adam Hall for critical reading and editing.

Conflicts of Interest: The authors declare no conflict of interest.

References

1. Cicmanec, J.L.S.D.; Gebbard, P.; Li, Y.; George, J.E.; Music, C.; Luna, G.; Tieman, G.; Hope, J.B. Measurement of Endocrine Disrupting Chemicals in West Virginia's Waterways: Seasonal Comparisons for Agricultural, Industrial and Residential Areas. 2002. Available online: http://info.ngwa.org/GWOL/pdf/030777111.pdf (accessed on 11 December 2019).
2. Ankley, G.E.; Francis, E.; Gray, R.; Kavlock, S.; McMaster, D.; Reese, G.; Sayles, A.; Sergeant, A.; Vallero, D. *Research Plan for Endocrine Disruptors*; U.S. Epa: Washington, DC, USA, 1998.
3. Lin, A.Y.; Wang, X.H.; Lin, C.F. Impact of wastewaters and hospital effluents on the occurrence of controlled substances in surface waters. *Chemosphere* **2010**, *81*, 562–570. [CrossRef] [PubMed]

4. Habauzit, D.; Flouriot, G.; Pakdel, F.; Saligaut, C. Effects of estrogens and endocrine-disrupting chemicals on cell differentiation-survival-proliferation in brain: contributions of neuronal cell lines. *J. Toxicol. Environ. Health. Part B* **2011**, *14*, 300–327. [CrossRef] [PubMed]
5. Vandenberg, L.N.; Colborn, T.; Hayes, T.B.; Heindel, J.J.; Jacobs, D.R.; Lee, D.H.; Shioda, T.; Soto, A.M.; vom Saal, F.S.; Welshons, W.V.; et al. Hormones and endocrine-disrupting chemicals: low-dose effects and nonmonotonic dose responses. *Endocr. Rev.* **2012**, *33*, 378–455. [CrossRef] [PubMed]
6. Amaro, A.A.; Esposito, A.I.; Mirisola, V.; Mehilli, A.; Rosano, C.; Noonan, D.M.; Albini, A.; Pfeffer, U.; Angelini, G. Endocrine disruptor agent nonyl phenol exerts an estrogen-like transcriptional activity on estrogen receptor positive breast cancer cells. *Curr. Med. Chem.* **2014**, *21*, 630–640. [CrossRef] [PubMed]
7. Rogers, J.A.; Metz, L.; Yong, V.W. Review: Endocrine disrupting chemicals and immune responses: a focus on bisphenol-A and its potential mechanisms. *Mol. Immunol.* **2013**, *53*, 421–430. [CrossRef]
8. Vom Saal, F.S.; Nagel, S.C.; Coe, B.L.; Angle, B.M.; Taylor, J.A. The estrogenic endocrine disrupting chemical bisphenol A (BPA) and obesity. *Mol. Cell. Endocrinol.* **2012**, *354*, 74–84. [CrossRef]
9. Ford, A.T. Intersexuality in Crustacea: an environmental issue? *Aquat. Toxicol.* **2012**, *108*, 125–129. [CrossRef]
10. Waye, A.; Trudeau, V.L. Neuroendocrine disruption: more than hormones are upset. *J. Toxicol. Environ. Health. Part B* **2011**, *14*, 270–291. [CrossRef]
11. Leon-Olea, M.; Martyniuk, C.J.; Orlando, E.F.; Ottinger, M.A.; Rosenfeld, C.; Wolstenholme, J.; Trudeau, V.L. Current concepts in neuroendocrine disruption. *Gen. Comp. Endocrinol.* **2014**, *203*, 158–173. [CrossRef]
12. Garcia-Closas, M.; Thompson, W.D.; Robins, J.M. Differential misclassification and the assessment of gene-environment interactions in case-control studies. *Am. J. Epidemiol.* **1998**, *147*, 426–433. [CrossRef]
13. Balik-Meisner, M.; Truong, L.; Scholl, E.H.; La Du, J.K.; Tanguay, R.L.; Reif, D.M. Elucidating Gene-by-Environment Interactions Associated with Differential Susceptibility to Chemical Exposure. *Environ. Health Perspect.* **2018**, *126*, 067010. [CrossRef] [PubMed]
14. Kundakovic, M.; Champagne, F.A. Epigenetic perspective on the developmental effects of bisphenol A. *Brain Behav. Immun.* **2011**, *25*, 1084–1093. [CrossRef] [PubMed]
15. Stone, W.W.; Gilliom, R.J.; Ryberg, K.R. Pesticides in U.S. streams and rivers: Occurrence and trends during 1992–2011. *Environ. Sci. Technol.* **2014**, *48*, 11025–11030. [CrossRef] [PubMed]
16. Ryberg, K.R.; Vecchia, A.V.; Gilliom, R.J.; Martin, J.D. *Pesticide Trends in Major Rivers of the United States, 1992–2010*; US Geological Survey Scientific Investigations Report; USGS: Reston, VA, USA, 2014.
17. Fawell, J.K.; Ohanian, E.; Giddings, M.; Toft, P.; Magara, Y.; Jackson, P. *Endosulfan in Drinking Water*; WHO/SDE/WSH/03.04/92; World Health Organization: Geneva, Switzerland, 2004.
18. Shivaramaiah, H.M.; Sanchez-Bayo, F.; Al-Rifai, J.; Kennedy, I.R. The fate of endosulfan in water. *J. Environ. Sci. Health B* **2005**, *40*, 711–720. [CrossRef] [PubMed]
19. Kolpin, D.W.; Furlong, E.T.; Meyer, M.T.; Thurman, E.M.; Zaugg, S.D.; Barber, L.B.; Buxton, H.T. Pharmaceuticals, hormones, and other organic wastewater contaminants in U.S. streams, 1999–2000: A national reconnaissance. *Environ. Sci. Technol.* **2002**, *36*, 1202–1211. [CrossRef] [PubMed]
20. Santhi, V.A.; Sakai, N.; Ahmad, E.D.; Mustafa, A.M. Occurrence of bisphenol A in surface water, drinking water and plasma from Malaysia with exposure assessment from consumption of drinking water. *Sci. Total Environ.* **2012**, *427–428*, 332–338. [CrossRef]
21. Arnold, S.M.; Clark, K.E.; Staples, C.A.; Klecka, G.M.; Dimond, S.S.; Caspers, N.; Hentges, S.G. Relevance of drinking water as a source of human exposure to bisphenol A. *J. Expo. Sci. Environ. Epidemiol.* **2013**, *23*, 137–144. [CrossRef]
22. Rice, C.P.; Schmitz-Afonso, I.; Loyo-Rosales, J.E.; Link, E.; Thoma, R.; Fay, L.; Altfater, D.; Camp, M.J. Alkylphenol and alkylphenol-ethoxylates in carp, water, and sediment from the Cuyahoga River, Ohio. *Environ. Sci. Technol.* **2003**, *37*, 3747–3754. [CrossRef]
23. Mao, Z.; Zheng, X.F.; Zhang, Y.Q.; Tao, X.X.; Li, Y.; Wang, W. Occurrence and biodegradation of nonylphenol in the environment. *Int. J. Mol. Sci.* **2012**, *13*, 491–505. [CrossRef]
24. Terzic, S.; Senta, I.; Ahel, M. Illicit drugs in wastewater of the city of Zagreb (Croatia)—Estimation of drug abuse in a transition country. *Environ. Pollut.* **2010**, *158*, 2686–2693. [CrossRef]
25. Tenore, P.L. Psychotherapeutic benefits of opioid agonist therapy. *J. Addict. Dis.* **2008**, *27*, 49–65. [CrossRef]
26. Nefau, T.; Karolak, S.; Castillo, L.; Boireau, V.; Levi, Y. Presence of illicit drugs and metabolites in influents and effluents of 25 sewage water treatment plants and map of drug consumption in France. *Sci. Total. Environ.* **2013**, *461–462*, 712–722. [CrossRef]

27. Konijnenberg, C.; Melinder, A. Prenatal exposure to methadone and buprenorphine: A review of the potential effects on cognitive development. *Child Neuropsychol.* **2011**, *17*, 495–519. [CrossRef] [PubMed]
28. European Commission. Commission Decision of 10 March 2004 concerning the non-inclusion of atrazine in Annex I to Council Directive 91/414/EEC and the withdrawal of authorisations for plant protection products containing this active substance. In *C(2004) 731. 2004/248/EC*; European Commission: Brussels, Belgium, 2004.
29. Boffetta, P.; Adami, H.O.; Berry, S.C.; Mandel, J.S. Atrazine and cancer: A review of the epidemiologic evidence. *Eur. J. Cancer Prev.* **2013**, *22*, 169–180. [CrossRef]
30. Guengerich, F.P. Cytochrome p450 and chemical toxicology. *Chem. Res. Toxicol.* **2008**, *21*, 70–83. [CrossRef]
31. Reichert, K.; Menzel, R. Expression profiling of five different xenobiotics using a Caenorhabditis elegans whole genome microarray. *Chemosphere* **2005**, *61*, 229–237. [CrossRef] [PubMed]
32. Andriani, G.A.; Almeida, V.P.; Faggioli, F.; Mauro, M.; Tsai, W.L.; Santambrogio, L.; Maslov, A.; Gadina, M.; Campisi, J.; Vijg, J.; et al. Whole Chromosome Instability induces senescence and promotes SASP. *Sci. Rep.* **2016**, *6*, 35218. [CrossRef]
33. Liu, X.M.; Shao, J.Z.; Xiang, L.X.; Chen, X.Y. Cytotoxic effects and apoptosis induction of atrazine in a grass carp (*Ctenopharyngodon idellus*) cell line. *Environ. Toxicol.* **2006**, *21*, 80–89. [CrossRef]
34. Jia, X.; Wang, D.; Gao, N.; Cao, H.; Zhang, H. Atrazine Triggers the Extrinsic Apoptosis Pathway in Lymphocytes of the Frog Pelophylax nigromaculata in Vivo. *Chem. Res. Toxicol.* **2015**, *28*, 2010–2018. [CrossRef]
35. Zhang, X.; Wang, M.; Gao, S.; Ren, R.; Zheng, J.; Zhang, Y. Atrazine-induced apoptosis of splenocytes in BALB/C mice. *BMC Med.* **2011**, *9*, 117. [CrossRef]
36. Lee, E.J.; Jang, Y.; Kang, K.; Song, D.H.; Kim, R.; Chang, H.W.; Lee, D.E.; Song, C.K.; Choi, B.; Kang, M.J.; et al. Atrazine induces endoplasmic reticulum stress-mediated apoptosis of T lymphocytes via the caspase-8-dependent pathway. *Environ. Toxicol.* **2016**, *31*, 998–1008. [CrossRef]
37. Smith, L.K.; Shah, R.R.; Cidlowski, J.A. Glucocorticoids modulate microRNA expression and processing during lymphocyte apoptosis. *J. Biol. Chem.* **2010**, *285*, 36698–36708. [CrossRef] [PubMed]
38. Kay, P.; Schlossmacher, G.; Matthews, L.; Sommer, P.; Singh, D.; White, A.; Ray, D. Loss of glucocorticoid receptor expression by DNA methylation prevents glucocorticoid induced apoptosis in human small cell lung cancer cells. *PLoS ONE* **2011**, *6*, e24839. [CrossRef] [PubMed]
39. Foradori, C.D.; Hinds, L.R.; Quihuis, A.M.; Lacagnina, A.F.; Breckenridge, C.B.; Handa, R.J. The differential effect of atrazine on luteinizing hormone release in adrenalectomized adult female Wistar rats. *Biol. Reprod.* **2011**, *85*, 684–689. [CrossRef] [PubMed]
40. Chen, Q.; Cederbaum, A.I. Cytotoxicity and apoptosis produced by cytochrome P450 2E1 in Hep G2 cells. *Mol. Pharm.* **1998**, *53*, 638–648. [CrossRef]
41. Alavanja, M.C.; Ross, M.K.; Bonner, M.R. Increased cancer burden among pesticide applicators and others due to pesticide exposure. *CA Cancer J. Clin.* **2013**, *63*, 120–142. [CrossRef]
42. Fan, W.; Yanase, T.; Morinaga, H.; Gondo, S.; Okabe, T.; Nomura, M.; Komatsu, T.; Morohashi, K.; Hayes, T.B.; Takayanagi, R.; et al. Atrazine-induced aromatase expression is SF-1 dependent: implications for endocrine disruption in wildlife and reproductive cancers in humans. *Environ. Health Perspect.* **2007**, *115*, 720–727. [CrossRef]
43. Simpkins, J.W.; Swenberg, J.A.; Weiss, N.; Brusick, D.; Eldridge, J.C.; Stevens, J.T.; Handa, R.J.; Hovey, R.C.; Plant, T.M.; Pastoor, T.P.; et al. Atrazine and breast cancer: A framework assessment of the toxicological and epidemiological evidence. *Toxicol. Sci.* **2011**, *123*, 441–459. [CrossRef]
44. Huang, P.; Yang, J.; Ning, J.; Wang, M.; Song, Q. Atrazine Triggers DNA Damage Response and Induces DNA Double-Strand Breaks in MCF-10A Cells. *Int. J. Mol. Sci.* **2015**, *16*, 14353–14368. [CrossRef]
45. Midic, U.; Vincent, K.A.; VandeVoort, C.A.; Latham, K.E. Effects of long-term endocrine disrupting compound exposure on Macaca mulatta embryonic stem cells. *Reprod. Toxicol.* **2016**, *65*, 382–393. [CrossRef] [PubMed]
46. Kang, C.; Xu, Q.; Martin, T.D.; Li, M.Z.; Demaria, M.; Aron, L.; Lu, T.; Yankner, B.A.; Campisi, J.; Elledge, S.J. The DNA damage response induces inflammation and senescence by inhibiting autophagy of GATA4. *Science* **2015**, *349*, aaa5612. [CrossRef] [PubMed]
47. Giusi, G.; Facciolo, R.M.; Canonaco, M.; Alleva, E.; Belloni, V.; Dessi'-Fulgheri, F.; Santucci, D. The endocrine disruptor atrazine accounts for a dimorphic somatostatinergic neuronal expression pattern in mice. *Toxicol. Sci.* **2006**, *89*, 257–264. [CrossRef] [PubMed]

48. Belloni, V.; Dessi-Fulgheri, F.; Zaccaroni, M.; Di Consiglio, E.; De Angelis, G.; Testai, E.; Santochirico, M.; Alleva, E.; Santucci, D. Early exposure to low doses of atrazine affects behavior in juvenile and adult CD1 mice. *Toxicology* **2011**, *279*, 19–26. [CrossRef] [PubMed]
49. Vogel, A.; Jocque, H.; Sirot, L.K.; Fiumera, A.C. Effects of atrazine exposure on male reproductive performance in Drosophila melanogaster. *J. Insect. Physiol.* **2015**, *72*, 14–21. [CrossRef]
50. Schmidel, A.J.; Assmann, K.L.; Werlang, C.C.; Bertoncello, K.T.; Francescon, F.; Rambo, C.L.; Beltrame, G.M.; Calegari, D.; Batista, C.B.; Blaser, R.E.; et al. Subchronic atrazine exposure changes defensive behaviour profile and disrupts brain acetylcholinesterase activity of zebrafish. *Neurotoxicol. Teratol.* **2014**, *44*, 62–69. [CrossRef]
51. Jowa, L.; Howd, R. Should atrazine and related chlorotriazines be considered carcinogenic for human health risk assessment? *J. Environ. Sci. Health Part C* **2011**, *29*, 91–144. [CrossRef]
52. Mitra, A.K.; Faruque, F.S. Breast cancer incidence and exposure to environmental chemicals in 82 counties in Mississippi. *South Med. J.* **2004**, *97*, 259–263. [CrossRef]
53. Mitra, A.K.; Faruque, F.S.; Avis, A.L. Breast cancer and environmental risks: Where is the link? *J. Environ. Health* **2004**, *66*, 24.
54. Van Der Kraak, G.J.; Hosmer, A.J.; Hanson, M.L.; Kloas, W.; Solomon, K.R. Effects of atrazine in fish, amphibians, and reptiles: an analysis based on quantitative weight of evidence. *Crit. Rev. Toxicol.* **2014**, *44* Suppl. S5, 1–66. [CrossRef]
55. Goodman, M.; Mandel, J.S.; DeSesso, J.M.; Scialli, A.R. Atrazine and pregnancy outcomes: A systematic review of epidemiologic evidence. *Birth Defects Res. Part B* **2014**, *101*, 215–236. [CrossRef] [PubMed]
56. Kuckelkorn, J.; Redelstein, R.; Heide, T.; Kunze, J.; Maletz, S.; Waldmann, P.; Grummt, T.; Seiler, T.B.; Hollert, H. A hierarchical testing strategy for micropollutants in drinking water regarding their potential endocrine-disrupting effects-towards health-related indicator values. *Environ. Sci. Pollut. Res. Int.* **2017**. [CrossRef] [PubMed]
57. Pinchuk, L.M.; Lee, S.R.; Filipov, N.M. In vitro atrazine exposure affects the phenotypic and functional maturation of dendritic cells. *Toxicol. Appl. Pharmacol.* **2007**, *223*, 206–217. [CrossRef] [PubMed]
58. Kravchenko, J.; Corsini, E.; Williams, M.A.; Decker, W.; Manjili, M.H.; Otsuki, T.; Singh, N.; Al-Mulla, F.; Al-Temaimi, R.; Amedei, A.; et al. Chemical compounds from anthropogenic environment and immune evasion mechanisms: potential interactions. *Carcinogenesis* **2015**, *36* Suppl. S1, S111–S127. [CrossRef] [PubMed]
59. Toughan, H.; Khalil, S.R.; El-Ghoneimy, A.A.; Awad, A.; Seddek, A.S. Effect of dietary supplementation with Spirulina platensis on Atrazine-induced oxidative stress- mediated hepatic damage and inflammation in the common carp (Cyprinus carpio L.). *Ecotoxicol. Environ. Saf.* **2017**, *149*, 135–142. [CrossRef]
60. Yuan, B.; Liang, S.; Jin, Y.X.; Zhang, M.J.; Zhang, J.B.; Kim, N.H. Toxic effects of atrazine on porcine oocytes and possible mechanisms of action. *PLoS ONE* **2017**, *12*, e0179861. [CrossRef]
61. Ruiz-Guzman, J.A.; Gomez-Corrales, P.; Cruz-Esquivel, A.; Marrugo-Negrete, J.L. Cytogenetic damage in peripheral blood lymphocytes of children exposed to pesticides in agricultural areas of the department of Cordoba, Colombia. *Mutat. Res.* **2017**, *824*, 25–31. [CrossRef]
62. Eyer, P. The role of oximes in the management of organophosphorus pesticide poisoning. *Toxicol. Rev.* **2003**, *22*, 165–190. [CrossRef]
63. Casida, J.E. Esterase inhibitors as pesticides. *Science* **1964**, *146*, 1011–1017. [CrossRef]
64. Eddleston, M.; Phillips, M.R. Self poisoning with pesticides. *Bmj* **2004**, *328*, 42–44. [CrossRef]
65. Namba, T. Cholinesterase inhibition by organophosphorus compounds and its clinical effects. *Bull. World Health Organ.* **1971**, *44*, 289–307. [PubMed]
66. Jamal, G.A.; Hansen, S.; Julu, P.O. Low level exposures to organophosphorus esters may cause neurotoxicity. *Toxicology* **2002**, *181–182*, 23–33. [CrossRef]
67. Lotti, M.; Moretto, A. Organophosphate-induced delayed polyneuropathy. *Toxicol. Rev.* **2005**, *24*, 37–49. [CrossRef] [PubMed]
68. Traverso, N.; Ricciarelli, R.; Nitti, M.; Marengo, B.; Furfaro, A.L.; Pronzato, M.A.; Marinari, U.M.; Domenicotti, C. Role of glutathione in cancer progression and chemoresistance. *Oxid. Med. Cell. Longev.* **2013**, *2013*, 972913. [CrossRef] [PubMed]
69. Poet, T.S.; Timchalk, C.; Hotchkiss, J.A.; Bartels, M.J. Chlorpyrifos PBPK/PD model for multiple routes of exposure. *Xenobiotica* **2014**, *44*, 868–881. [CrossRef]

70. Foxenberg, R.J.; Ellison, C.A.; Knaak, J.B.; Ma, C.; Olson, J.R. Cytochrome P450-specific human PBPK/PD models for the organophosphorus pesticides: Chlorpyrifos and parathion. *Toxicology* **2011**, *285*, 57–66. [CrossRef]
71. Gao, J.; Naughton, S.X.; Beck, W.D.; Hernandez, C.M.; Wu, G.; Wei, Z.; Yang, X.; Bartlett, M.G.; Terry, A.V., Jr. Chlorpyrifos and chlorpyrifos oxon impair the transport of membrane bound organelles in rat cortical axons. *Neurotoxicology* **2017**, *62*, 111–123. [CrossRef]
72. Slotkin, T.A.; Skavicus, S.; Card, J.; Levin, E.D.; Seidler, F.J. Diverse neurotoxicants target the differentiation of embryonic neural stem cells into neuronal and glial phenotypes. *Toxicology* **2016**, *372*, 42–51. [CrossRef]
73. Sultana Shaik, A.; Shaik, A.P.; Jamil, K.; Alsaeed, A.H. Evaluation of cytotoxicity and genotoxicity of pesticide mixtures on lymphocytes. *Toxicol. Mech. Methods* **2016**, *26*, 588–594. [CrossRef]
74. Ernst, W.R.; Jonah, P.; Doe, K.; Julien, G.; Hennigar, P. Toxicity to aquatic organisms of off-target deposition of endosulfan applied by aircraft. *Environ. Toxicol. Chem.* **1991**, *10*, 103–114. [CrossRef]
75. Naqvi, S.M.; Vaishnavi, C. Bioaccumulative potential and toxicity of endosulfan insecticide to non-target animals. *Comp. Biochem. Physiol. Part C Comp. Pharmacol. Toxicol.* **1993**, *105*, 347–361. [CrossRef]
76. Moses, V.; Peter, J.V. Acute intentional toxicity: endosulfan and other organochlorines. *Clin. Toxicol.* **2010**, *48*, 539–544. [CrossRef] [PubMed]
77. Chakravorty, S.L.B.; Singh, T.P. Effect of endosulfan (thiodan) on vitellogenesis and its modulation by different hormones in the vitellogenic catfish Claria batrachus. *Toxicology* **1992**, *75*, 191–198. [CrossRef]
78. Colborn, T.; vom Saal, F.S.; Soto, A.M. Developmental effects of endocrine-disrupting chemicals in wildlife and humans. *Environ. Health Perspect.* **1993**, *101*, 378–384. [CrossRef]
79. Rousseau, J.; Cossette, L.; Grenier, S.; Martinoli, M.G. Modulation of prolactin expression by xenoestrogens. *Gen. Comp. Endocrinol.* **2002**, *126*, 175–182. [CrossRef]
80. Caride, A.; Lafuente, A.; Cabaleiro, T. Endosulfan effects on pituitary hormone and both nitrosative and oxidative stress in pubertal male rats. *Toxicol. Lett.* **2010**, *197*, 106–112. [CrossRef]
81. Bradlow, H.L.; Davis, D.L.; Lin, G.; Sepkovic, D.; Tiwari, R. Effects of pesticides on the ratio of 16 alpha/2-hydroxyestrone: a biologic marker of breast cancer risk. *Environ. Health Perspect.* **1995**, *103* Suppl. S7, 147–150. [CrossRef]
82. Suzuki, T.; Ide, K.; Ishida, M. Response of MCF-7 human breast cancer cells to some binary mixtures of oestrogenic compounds in-vitro. *J. Pharm. Pharmacol.* **2001**, *53*, 1549–1554. [CrossRef]
83. Hu, L.; Xia, L.; Zhou, H.; Wu, B.; Mu, Y.; Wu, Y.; Yan, J. TF/FVIIa/PAR2 promotes cell proliferation and migration via PKCalpha and ERK-dependent c-Jun/AP-1 pathway in colon cancer cell line SW620. *Tumour Biol.* **2013**, *34*, 2573–25781. [CrossRef]
84. Kannan, K.; Holcombe, R.F.; Jain, S.K.; Alvarez-Hernandez, X.; Chervenak, R.; Wolf, R.E.; Glass, J. Evidence for the induction of apoptosis by endosulfan in a human T-cell leukemic line. *Mol. Cell. Biochem.* **2000**, *205*, 53–66. [CrossRef]
85. Ahmed, T.; Tripathi, A.K.; Ahmed, R.S.; Das, S.; Suke, S.G.; Pathak, R.; Chakraboti, A.; Banerjee, B.D. Endosulfan-induced apoptosis and glutathione depletion in human peripheral blood mononuclear cells: Attenuation by N-acetylcysteine. *J. Biochem. Mol. Toxicol.* **2008**, *22*, 299–304. [CrossRef] [PubMed]
86. Benachour, N.; Séralini, G.E. Glyphosate formulations induce apoptosis and necrosis in human umbilical, embryonic, and placental cells. *Chem. Res. Toxicol.* **2008**, *22*, 97–105. [CrossRef] [PubMed]
87. Altamirano, G.A.; Delconte, M.B.; Gomez, A.L.; Alarcon, R.; Bosquiazzo, V.L.; Luque, E.H.; Munoz-de-Toro, M.; Kass, L. Early postnatal exposure to endosulfan interferes with the normal development of the male rat mammary gland. *Toxicol. Lett.* **2017**, *281*, 102–109. [CrossRef] [PubMed]
88. Gomez-Gimenez, B.; Llansola, M.; Cabrera-Pastor, A.; Hernandez-Rabaza, V.; Agusti, A.; Felipo, V. Endosulfan and Cypermethrin Pesticide Mixture Induces Synergistic or Antagonistic Effects on Developmental Exposed Rats Depending on the Analyzed Behavioral or Neurochemical End Points. *ACS Chem. Neurosci.* **2017**. [CrossRef] [PubMed]
89. Sebastian, R.; Raghavan, S.C. Molecular mechanism of Endosulfan action in mammals. *J. Biosci.* **2017**, *42*, 149–153. [CrossRef] [PubMed]
90. Soares, A.; Guieysse, B.; Jefferson, B.; Cartmell, E.; Lester, J.N. Nonylphenol in the environment: A critical review on occurrence, fate, toxicity and treatment in wastewaters. *Environ. Int.* **2008**, *34*, 1033–1049. [CrossRef] [PubMed]
91. OSPAR Convention. *Recommendation on Nonylphenol-Ethoxylates*; OSPAR Convention: London, UK, 2000.

92. Directive, EU Cement. Amending for the 26th time the Council directive 76/769/EEC relating to restrictions on the marketing and use of certain dangerous substances and preparations (nonylphenol, nonylphenol ethoxylate and cement). In *Directive 2003/53/EC*; Off. J. Eur. Commun. L; European Commission: Luxembourg, 2003.
93. Li, M.H. Effects of nonylphenol on cholinesterase and carboxylesterase activities in male guppies (Poecilia reticulata). *Ecotoxicol. Environ. Saf.* **2008**, *71*, 781–786. [CrossRef]
94. Gejlsbjerg, B.; Klinge, C.; Samsoe-Petersen, L.; Madsen, T. Toxicity of linear alkylbenzene sulfonates and nonylphenol in sludge-amended soil. *Environ. Toxicol. Chem.* **2001**, *20*, 2709–2716. [CrossRef]
95. Park, S.Y.; Choi, J. Genotoxic effects of nonylphenol and bisphenol A exposure in aquatic biomonitoring species: Freshwater crustacean, Daphnia magna, and aquatic midge, Chironomus riparius. *Bull. Environ. Contam. Toxicol.* **2009**, *83*, 463–468. [CrossRef]
96. Shen, W.Y.; Zhou, Z.L.; Li, X.J. Effects of long term exposure to bisphenol A and nonylphenol on the reproduction of zebrafish (Danio rerio). *J. Fish. China* **2007**, *31*, 59–64.
97. Vazquez-Duhalt, R.; Marquez-Rocha, F.; Ponce, E.; Licea, A.F.; Viana, M.T. Nonylphenol, an integrated vision of a pollutant. *Appl. Ecol. Environ. Res.* **2005**, *4*, 1–25. [CrossRef]
98. Aoki, M.; Kurasaki, M.; Saito, T.; Seki, S.; Hosokawa, T.; Takahashi, Y.; Fujita, H.; Iwakuma, T. Nonylphenol enhances apoptosis induced by serum deprivation in PC12 cells. *Life Sci.* **2004**, *74*, 2301–2312. [CrossRef] [PubMed]
99. Wong, C.T.; Wais, J.; Crawford, D.A. Prenatal exposure to common environmental factors affects brain lipids and increases risk of developing autism spectrum disorders. *Eur. J. Neurosci.* **2015**, *42*, 2742–2760. [CrossRef] [PubMed]
100. Soto, A.M.; Justicia, H.; Wray, J.W.; Sonnenschein, C. p-Nonyl-phenol: an estrogenic xenobiotic released from "modified" polystyrene. *Environ. Health Perspect.* **1991**, *92*, 167–173. [CrossRef] [PubMed]
101. White, R.; Jobling, S.; Hoare, S.A.; Sumpter, J.P.; Parker, M.G. Environmentally persistent alkylphenolic compounds are estrogenic. *Endocrinology* **1994**, *135*, 175–182. [CrossRef]
102. Mendes, J.A. The endocrine disrupters: a major medical challenge. *Food Chem. Toxicol.* **2002**, *40*, 781–788. [CrossRef]
103. Ferguson, S.A.; Flynn, K.M.; Delclos, K.B.; Newbold, R.R. Maternal and offspring toxicity but few sexually dimorphic behavioral alterations result from nonylphenol exposure. *Neurotoxicol. Teratol.* **2000**, *22*, 583–591. [CrossRef]
104. Ponzo, O.J.; Silvia, C. Evidence of reproductive disruption associated with neuroendocrine changes induced by UV-B filters, phthalates and nonylphenol during sexual maturation in rats of both gender. *Toxicology* **2013**, *311*, 41–51. [CrossRef]
105. Negishi, T.; Kawasaki, K.; Suzaki, S.; Maeda, H.; Ishii, Y.; Kyuwa, S.; Kuroda, Y.; Yoshikawa, Y. Behavioral alterations in response to fear-provoking stimuli and tranylcypromine induced by perinatal exposure to bisphenol A and nonylphenol in male rats. *Environ. Health Perspect.* **2004**, *112*, 1159–1164. [CrossRef]
106. Ferguson, S.A.; Flynn, K.M.; Delclos, K.B.; Newbold, R.R.; Gough, B.J. Effects of lifelong dietary exposure to genistein or nonylphenol on amphetamine-stimulated striatal dopamine release in male and female rats. *Neurotoxicol. Teratol.* **2002**, *24*, 37–45. [CrossRef]
107. Nagao, T.; Wada, K.; Marumo, H.; Yoshimura, S.; Ono, H. Reproductive effects of nonylphenol in rats after gavage administration: a two-generation study. *Reprod. Toxicol.* **2001**, *15*, 293–315. [CrossRef]
108. Kim, H.; Oh, S.; Gye, M.C.; Shin, I. Comparative toxicological evaluation of nonylphenol and nonylphenol polyethoxylates using human keratinocytes. *Drug Chem. Toxicol.* **2017**. [CrossRef] [PubMed]
109. Derakhshesh, N.; Movahedinia, A.; Salamat, N.; Hashemitabar, M.; Bayati, V. Using a liver cell culture from Epinephelus coioides as a model to evaluate the nonylphenol-induced oxidative stress. *Mar. Pollut. Bull.* **2017**, *122*, 243–252. [CrossRef] [PubMed]
110. Doerge, D.R.; Twaddle, N.C.; Vanlandingham, M.; Fisher, J.W. Pharmacokinetics of bisphenol A in neonatal and adult Sprague-Dawley rats. *Toxicol. Appl. Pharmacol.* **2010**, *247*, 158–165. [CrossRef]
111. Antherieu, S.; Ledirac, N.; Luzy, A.P.; Lenormand, P.; Caron, J.C.; Rahmani, R. Endosulfan decreases cell growth and apoptosis in human HaCaT keratinocytes: partial ROS-dependent ERK1/2 mechanism. *J. Cell. Physiol.* **2007**, *213*, 177–186. [CrossRef]

112. Masoner, J.R.; Kolpin, D.W.; Furlong, E.T.; Cozzarelli, I.M.; Gray, J.L.; Schwab, E.A. Contaminants of emerging concern in fresh leachate from landfills in the conterminous United States. *Environ. Sci. Process. Impacts* **2014**, *16*, 2335–2354. [CrossRef]
113. Dsikowitzky, L.; Botalova, O.; Illgut, S.; Bosowski, S.; Schwarzbauer, J. Identification of characteristic organic contaminants in wastewaters from modern paper production sites and subsequent tracing in a river. *J. Hazard. Mater.* **2015**, *300*, 254–262. [CrossRef]
114. Lu, J.; Wu, J.; Stoffella, P.J.; Wilson, P.C. Uptake and distribution of bisphenol A and nonylphenol in vegetable crops irrigated with reclaimed water. *J. Hazard. Mater.* **2015**, *283*, 865–870. [CrossRef]
115. Adewale, H.B.; Jefferson, W.N.; Newbold, R.R.; Patisaul, H.B. Neonatal bisphenol-a exposure alters rat reproductive development and ovarian morphology without impairing activation of gonadotropin-releasing hormone neurons. *Biol. Reprod.* **2009**, *81*, 690–699. [CrossRef]
116. Aldad, T.S.; Rahmani, N.; Leranth, C.; Taylor, H.S. Bisphenol-A exposure alters endometrial progesterone receptor expression in the nonhuman primate. *Fertil. Steril.* **2011**, *96*, 175–179. [CrossRef] [PubMed]
117. Heindel, J.J.; Newbold, R.R.; Bucher, J.R.; Camacho, L.; Delclos, K.B.; Lewis, S.M.; Vanlandingham, M.; Churchwell, M.I.; Twaddle, N.C.; McLellen, M.; et al. NIEHS/FDA CLARITY-BPA research program update. *Reprod. Toxicol.* **2015**, *58*, 33–44. [CrossRef] [PubMed]
118. Angle, B.M.; Do, R.P.; Ponzi, D.; Stahlhut, R.W.; Drury, B.E.; Nagel, S.C.; Welshons, W.V.; Besch-Williford, C.L.; Palanza, P.; Parmigiani, S.; et al. Metabolic disruption in male mice due to fetal exposure to low but not high doses of bisphenol A (BPA): evidence for effects on body weight, food intake, adipocytes, leptin, adiponectin, insulin and glucose regulation. *Reprod. Toxicol.* **2013**, *42*, 256–268. [CrossRef] [PubMed]
119. Amoroso, S.; Di Renzo, G.; Cuocolo, R.; Amantea, B.; Leo, A.; Taglialatela, M.; Annunziato, L. Evidence for a differential interaction of buprenorphine with opiate receptor subtypes controlling prolactin secretion. *Eur. J. Pharmacol.* **1988**, *145*, 257–260. [CrossRef]
120. Ahangarpour, A.; Afshari, G.; Mard, S.A.; Khodadadi, A.; Hashemitabar, M. Preventive effects of procyanidin A2 on glucose homeostasis, pancreatic and duodenal homebox 1, and glucose transporter 2 gene expression disturbance induced by bisphenol A in male mice. *J. Physiol. Pharmacol.* **2016**, *67*, 243–252.
121. Baker, D.R.; Barron, L.; Kasprzyk-Hordern, B. Illicit and pharmaceutical drug consumption estimated via wastewater analysis. Part A: Chemical analysis and drug use estimates. *Sci. Total Environ.* **2014**, *487*, 629–641. [CrossRef]
122. Xie, M.; Bu, P.; Li, F.; Lan, S.; Wu, H.; Yuan, L.; Wang, Y. Neonatal bisphenol A exposure induces meiotic arrest and apoptosis of spermatogenic cells. *Oncotarget* **2016**, *7*, 10606–10615. [CrossRef]
123. Arslan-Alaton, I.; Olmez-Hanci, T.; Dogan, M.; Ozturk, T. Zero-valent aluminum-mediated degradation of Bisphenol A in the presence of common oxidants. *Water Sci. Technol.* **2017**, *76*, 2455–2464. [CrossRef]
124. Olmez-Hanci, T.; Arslan-Alaton, I.; Dogan, M.; Khoei, S.; Fakhri, H.; Korkmaz, G. Enhanced degradation of micropollutants by zero-valent aluminum activated persulfate: assessment of toxicity and genotoxic activity. *Water Sci. Technol.* **2017**, *76*, 3195–3204. [CrossRef]
125. Ding, Z.M.; Jiao, X.F.; Wu, D.; Zhang, J.Y.; Chen, F.; Wang, Y.S.; Huang, C.J.; Zhang, S.X.; Li, X.; Huo, L.J. Bisphenol AF negatively affects oocyte maturation of mouse in vitro through increasing oxidative stress and DNA damage. *Chem. Biol. Interact* **2017**, *278*, 222–229. [CrossRef]
126. Jones, H.E.; Johnson, R.E.; Jasinski, D.R.; O'Grady, K.E.; Chisholm, C.A.; Choo, R.E.; Crocetti, M.; Dudas, R.; Harrow, C.; Huestis, M.A.; et al. Buprenorphine versus methadone in the treatment of pregnant opioid-dependent patients: effects on the neonatal abstinence syndrome. *Drug Alcohol Depend.* **2005**, *79*, 1–10. [CrossRef] [PubMed]
127. Allouche, S.; Le Marec, T.; Coquerel, A.; Noble, F.; Marie, N. Striatal dopamine D1 and D2 receptors are differentially regulated following buprenorphine or methadone treatment. *Psychopharmacology* **2015**, *232*, 1527–1533. [CrossRef] [PubMed]
128. Annamalai, J.; Namasivayam, V. Endocrine disrupting chemicals in the atmosphere: Their effects on humans and wildlife. *Environ. Int.* **2015**, *76*, 78–97. [CrossRef] [PubMed]
129. Pedraz-Cuesta, E.; Fredsted, J.; Jensen, H.H.; Bornebusch, A.; Nejsum, L.N.; Kragelund, B.B.; Pedersen, S.F. Prolactin Signaling Stimulates Invasion via Na(+)/H(+) Exchanger NHE1 in T47D Human Breast Cancer Cells. *Mol. Endocrinol.* **2016**, *30*, 693–708. [CrossRef]
130. Fitzgerald, P.; Dinan, T.G. Prolactin and dopamine: what is the connection? A review article. *J. Psychopharmacol.* **2008**, *22*, 12–19. [CrossRef]

131. Ausio, J. MeCP2 and the enigmatic organization of brain chromatin. Implications for depression and cocaine addiction. *Clin. Epigenet.* **2016**, *8*, 58. [CrossRef]
132. Kugawa, F.; Arae, K.; Ueno, A.; Aoki, M. Buprenorphine hydrochloride induces apoptosis in NG108-15 nerve cells. *Eur. J. Pharmacol.* **1998**, *347*, 105–112. [CrossRef]
133. Ejaredar, M.; Lee, Y.; Roberts, D.J.; Sauve, R.; Dewey, D. Bisphenol A exposure and children's behavior: A systematic review. *J. Expo. Sci. Environ. Epidemiol.* **2016**. [CrossRef]
134. Corella, D.; Ordovas, J.M. Basic Concepts in Molecular Biology Related to Genetics and Epigenetics. *Rev. Esp. Cardiol.* **2017**, *70*, 744–753. [CrossRef]
135. Gely-Pernot, A.; Hao, C.; Becker, E.; Stuparevic, I.; Kervarrec, C.; Chalmel, F.; Primig, M.; Jegou, B.; Smagulova, F. The epigenetic processes of meiosis in male mice are broadly affected by the widely used herbicide atrazine. *BMC Genom.* **2015**, *16*, 885. [CrossRef]
136. Wirbisky-Hershberger, S.E.; Sanchez, O.F.; Horzmann, K.A.; Thanki, D.; Yuan, C.; Freeman, J.L. Atrazine exposure decreases the activity of DNMTs, global DNA methylation levels, and dnmt expression. *Food Chem. Toxicol.* **2017**, *109*, 727–734. [CrossRef] [PubMed]
137. Hao, C.; Gely-Pernot, A.; Kervarrec, C.; Boudjema, M.; Becker, E.; Khil, P.; Tevosian, S.; Jegou, B.; Smagulova, F. Exposure to the widely used herbicide atrazine results in deregulation of global tissue-specific RNA transcription in the third generation and is associated with a global decrease of histone trimethylation in mice. *Nucl. Acids Res.* **2016**, *44*, 9784–9802. [CrossRef] [PubMed]
138. McBirney, M.; King, S.E.; Pappalardo, M.; Houser, E.; Unkefer, M.; Nilsson, E.; Sadler-Riggleman, I.; Beck, D.; Winchester, P.; Skinner, M.K. Atrazine induced epigenetic transgenerational inheritance of disease, lean phenotype and sperm epimutation pathology biomarkers. *PLoS ONE* **2017**, *12*, e0184306. [CrossRef] [PubMed]
139. Milesi, M.M.; Varayoud, J.; Ramos, J.G.; Luque, E.H. Uterine ERalpha epigenetic modifications are induced by the endocrine disruptor endosulfan in female rats with impaired fertility. *Mol. Cell. Endocrinol.* **2017**, *454*, 1–11. [CrossRef]
140. Ghosh, K.; Chatterjee, B.; Jayaprasad, A.G.; Kanade, S.R. The persistent organochlorine pesticide endosulfan modulates multiple epigenetic regulators with oncogenic potential in MCF-7 cells. *Sci. Total Environ.* **2017**. [CrossRef]
141. Hung, C.H.; Yang, S.N.; Kuo, P.L.; Chu, Y.T.; Chang, H.W.; Wei, W.J.; Huang, S.K.; Jong, Y.J. Modulation of cytokine expression in human myeloid dendritic cells by environmental endocrine-disrupting chemicals involves epigenetic regulation. *Environ. Health Perspect.* **2010**, *118*, 67–72. [CrossRef]
142. Hung, C.H.; Yang, S.N.; Wang, Y.F.; Liao, W.T.; Kuo, P.L.; Tsai, E.M.; Lee, C.L.; Chao, Y.S.; Yu, H.S.; Huang, S.K.; et al. Environmental alkylphenols modulate cytokine expression in plasmacytoid dendritic cells. *PLoS ONE* **2013**, *8*, e73534. [CrossRef]
143. Ajj, H.; Chesnel, A.; Pinel, S.; Plenat, F.; Flament, S.; Dumond, H. An alkylphenol mix promotes seminoma derived cell proliferation through an ERalpha36-mediated mechanism. *PLoS ONE* **2013**, *8*, e61758. [CrossRef]
144. Zheng, H.; Zhou, X.; Li, D.K.; Yang, F.; Pan, H.; Li, T.; Miao, M.; Li, R.; Yuan, W. Genome-wide alteration in DNA hydroxymethylation in the sperm from bisphenol A-exposed men. *PLoS ONE* **2017**, *12*, e0178535. [CrossRef]
145. Bansal, A.; Rashid, C.; Xin, F.; Li, C.; Polyak, E.; Duemler, A.; van der Meer, T.; Stefaniak, M.; Wajid, S.; Doliba, N.; et al. Sex- and Dose-Specific Effects of Maternal Bisphenol A Exposure on Pancreatic Islets of First- and Second-Generation Adult Mice Offspring. *Environ. Health Perspect.* **2017**, *125*, 097022. [CrossRef]
146. Chianese, R.; Troisi, J.; Richards, S.; Scafuro, M.; Fasano, S.; Guida, M.; Pierantoni, R.; Meccariello, R. Bisphenol A in reproduction: Epigenetic effects. *Curr. Med. Chem.* **2017**. [CrossRef] [PubMed]
147. McLemore, G.L.; Lewis, T.; Jones, C.H.; Gauda, E.B. Novel pharmacotherapeutic strategies for treatment of opioid-induced neonatal abstinence syndrome. *Semin. Fetal Neonatal Med.* **2013**, *18*, 35–41. [CrossRef] [PubMed]
148. McCarthy, J.J.; Leamon, M.H.; Finnegan, L.P.; Fassbender, C. Opioid dependence and pregnancy: Minimizing stress on the fetal brain. *Am. J. Obs. Gynecol.* **2017**, *216*, 226–231. [CrossRef] [PubMed]
149. Ausio, J.; Georgel, P.T. MeCP2 and CTCF: Enhancing the cross-talk of silencers. *Biochem. Cell Biol.* **2017**, *95*, 593–608. [CrossRef] [PubMed]

Article

Flow-Mediated Vulnerability of Source Waters to Elevated TDS in an Appalachian River Basin

Eric R. Merriam [1,*], J. Todd Petty [2], Melissa O'Neal [3] and Paul F. Ziemkiewicz [3]

1 Pittsburgh District, U.S. Army Corps of Engineers, Pittsburgh, PA 15222, USA
2 School of Natural Resources, West Virginia University, Morgantown, WV 26506-6125, USA; Todd.Petty@mail.wvu.edu
3 West Virginia Water Research Institute, West Virginia University, Morgantown, WV 26506, USA; Melissa.O'Neal@mail.wvu.edu (M.O.); paul.ziemkiewicz@mail.wvu.edu (P.F.Z.)
* Correspondence: eric.r.merriam@usace.army.mil

Received: 31 December 2019; Accepted: 30 January 2020; Published: 31 January 2020

Abstract: Widespread salinization—and, in a broader sense, an increase in all total dissolved solids (TDS)—is threatening freshwater ecosystems and the services they provide (e.g., drinking water provision). We used a mixed modeling approach to characterize long-term (2010–2018) spatio-temporal variability in TDS within the Monongahela River basin and used this information to assess the extent and drivers of vulnerability. The West Fork River was predicted to exceed 500 mg/L a total of 133 days. Occurrence and duration (maximum = 28 days) of—and thus vulnerability to—exceedances within the West Fork River were driven by low flows. Projected decreases in mean daily discharge by ≤10 cfs resulted in an additional 34 days exceeding 500 mg/L. Consistently low TDS within the Tygart Valley and Cheat Rivers reduced vulnerability of the receiving Monongahela River to elevated TDS which was neither observed (maximum = 419 mg/L) nor predicted (341 mg/L) to exceed the secondary drinking water standard of 500 mg/L. Potential changes in future land use and/or severity of low-flow conditions could increase vulnerability of the Monongahela River to elevated TDS. Management should include efforts to increase assimilative capacity by identifying and decreasing sources of TDS. Upstream reservoirs could be managed toward low-flow thresholds; however, further study is needed to ensure all authorized reservoir purposes could be maintained.

Keywords: water quality; vulnerability; total dissolved solids; drinking water

1. Introduction

Anthropogenic activities are contributing to widespread salinization of streams and rivers [1]. Salinization—and, in a broader sense, an increase in total dissolved solids (TDS)—has important implications for biodiversity [2–4] and ecosystem processes [5]. Elevated TDS also impacts human health and well-being by degrading drinking water quality and/or increasing the cost of water treatment [6], degrading infrastructure, and altering ecosystem goods and services [7].

Elevated TDS concentrations within riverine systems are primarily driven by the spatial pattern and extent of land use change, such as resource extraction, urban development, and agriculture and associated pollutants (acid mine drainage, agricultural and storm water runoff) throughout the watershed [1,8]. Numerous studies have also documented the importance of temporal variability in flow—both natural [9] and anthropogenic [10]—in controlling water quality and its impacts on freshwater ecosystems. Understanding the extent to which temporal variability in flow modulates vulnerability of freshwater ecosystems to elevated TDS across space will be critical to ensuring the resiliency and well-being of individuals and communities that rely on the services they provide [11,12]. This is particularly true given uncertainty regarding the effects of climate change on flow variation, which will likely further impact water quality and alter system vulnerability [13].

Herein, we present results of an analysis of temporal variation in TDS concentrations and its relationship with flow variability in the upper Monongahela River basin in West Virginia, USA. Elevated TDS within the Monongahela River and its major tributaries is primarily the result of historic coal mining and contemporary oil and gas development and has been shown to impact drinking water during low-flow events [14,15]. In 2009, one of the authors (Paul Ziemkiewicz), Director of West Virginia University's Water Research Institute, initiated the Three Rivers Quest (3RQ) program in response to rising TDS levels in the Monongahela River. The result was the identification of treated mine drainage, rich in Ca, Na and SO_4 as the controlling factor in the River's TDS load. A discharge management model and agreement among operators of mine discharge treatment facilities was developed and implemented in January 2010. The model allowed the treatment unit operators to modulate discharge load based on flow in the river with allocation of TDS load to ensure that the Monongahela River mainstem would not exceed the secondary drinking water standards for sulfate or TDS (250 and 500 mg/L respectively) [16]. Since then, flows within the Monongahela River and its tributaries inform management decisions, that include timing discharge of treated mine drainage based on assimilative capacity and flow augmentation from upstream reservoirs [17] designed to improve downstream water quality, resulting in additional spatio-temporal variability and complexity in TDS. Consequently, the upper Monongahela River drainage provides a unique and relevant opportunity to characterize how complex spatial and temporal controls over TDS contribute to vulnerability. We use a mixed modeling approach to characterize long-term (2010–2019) spatial and temporal variability within the Monongahela River and its tributaries and use this information to characterize system vulnerability.

2. Materials and Methods

2.1. Study Area

The study area was defined as the upper Monongahela River drainage upstream of Masontown, Pennsylvania, and drains approximately 11,700 km^2 within Pennsylvania and West Virginia. The study area includes the Cheat River, Tygart Valley River, West Fork River, and Upper Monongahela 8 digit hydrologic unit code (HUC) watersheds (Figure 1). The study area is predominately forested (78%). The drainage network is influenced by pre- and post-Surface Mine Control and Reclamation Act (SMCRA) surface mining (3%) and residential and urban development (2%). The stream network also drains 1020 deep mine national pollution discharge elimination system permits, as well as 25,675 conventional and 954 unconventional oil and gas wells [18,19]. Flows within the upper Monongahela River drainage are regulated by two U.S. Army Corps of Engineers reservoirs—Stonewall Jackson Lake on the West Fork River and Tygart Lake on the Tygart River (Figure 1). Both reservoirs were authorized for flood protection and water supply purposes. Water releases from both reservoirs during low-flow periods help maintain commercial navigation and improve water quality for domestic (e.g., drinking water) and ecological (e.g., fish and wildlife habitat) purposes. Flows within the Cheat River are regulated by a private hydroelectric dam (Figure 1).

2.2. Data Collection

2.2.1. Water Quality Data

We compiled 969 unique water quality sampling records collected bi-monthly (2010–2014) and monthly (2015–2018) from six sites [three along the Monongahela River (M1, M2, M3), and near the mouths of the Cheat River (CR), Tygart Valley River (TV), and West Fork (WF); Table 1, Figure 1] within the Monongahela River system as part of the Three Rivers Quest water quality monitoring program (https://3riversquest.wvu.edu/).

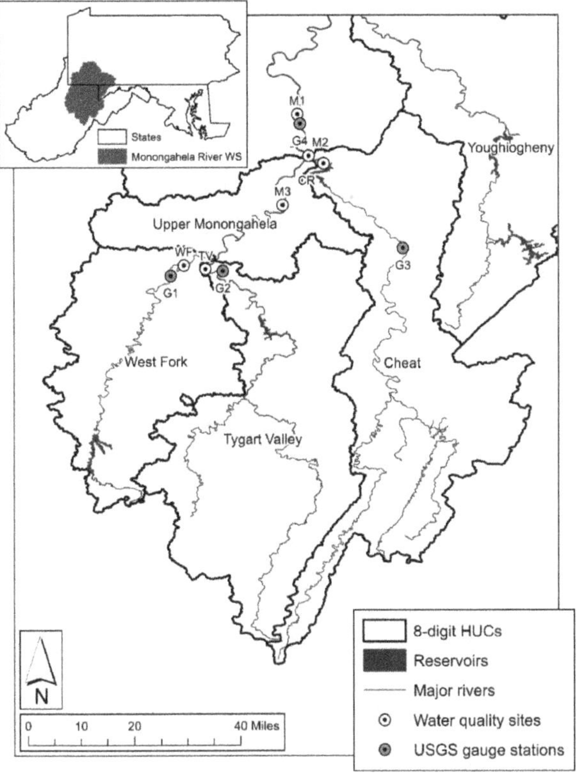

Figure 1. Location of water quality sampling sites (n = 6) and United States Geological Survey (USGS) gauging stations (n = 4) used to characterize and predict spatio-temporal variability in total dissolved solids (TDS) within 8 digit hydrologic unit code (HUC) watersheds comprising the upper Monongahela River basin.

Table 1. Basin areas, collection dates (month/year), and number of samples (n; days of data) for the six sample sites. Site codes match those presented in Figure 1.

Site Name	Site Code	BA (km^2)	Collection Dates	*n*
Monongahela River 1	M1	11,699	January 2010–December 2018	175
Monongahela River 2	M2	7049	October 2010–December 2018	154
Monongahela River 3	M3	6670	October 2010–December 2018	155
Cheat River	CR	3645	October 2010–December 2018	155
Tygart Valley River	TV	3577	October 2010–December 2018	155
West Fork River	WF	2135	January 2010–December 2018	175

Samples were collected and analyzed for dissolved alkalinity (mg/L $CaCO_3$ equivalents; EPA method SM-2320B), dissolved Al, Ca, Fe, Mn, Mg, and Na (mg/L; EPA method 6010B), and dissolved Br, Cl and SO_4 (mg/L, EPA method 300.0). Total dissolved solids (TDS) was calculated as the sum of the concentrations of all measured dissolved constituents. In situ measures of temperature, electrical conductivity, and pH were obtained using a YSI 556 multiprobe (Yellow Springs Instruments, Yellow Springs, Ohio).

2.2.2. Hydrologic Data

We associated mean daily discharge data from four United States Geological Survey (USGS) gauging stations with the six temporal water quality sites. Four water quality sampling locations [WF (USGS gauge #03061000), TV (03057000), CR (03070260), M1 (03072655)] were associated with individual USGS gauges located on the same stream (Figure 1). We estimated mean daily discharge for the two most upstream sites on the Monongahela River (M2, M3) as the sum of discharges at WF and TV.

2.3. Statistical Analyses

2.3.1. Water Quality Characteristics

We calculated summary statistics for observed TDS and its individual constituents to characterize spatial variability among sites and temporal variability within sites.

2.3.2. Water Quality Modeling

We used a hierarchical linear-mixed effect model to predict TDS from mean daily discharge. We log[x] transformed TDS and discharge to approximate normality. We fit a 'beyond optimal model' (BOM) to the training set that allowed slopes and intercepts to vary among sites, years, and years within sites [20]. This random effects structure enabled us to account for site-specific (e.g., upstream land use) and temporal (e.g., annual precipitation) characteristics affecting TDS within the Monongahela River watershed, as well as differences within sites among years (e.g., implementation of regulation or management strategies). We identified the optimal random effects structure by first iteratively dropping random slope terms and comparing the less parameterized model to the BOM [21]. We then iteratively dropped the random intercept terms and compared the less parameterized model back to the optimal random slopes model. We retained random effects with ΔAIC > 2. We assessed model performance by calculating root mean square error (RMSE) and marginal (variance explained by fixed effects) and conditional (variance explained by fixed and random effects) coefficients of determination (R^2). We compared observed and expected TDS within the test set to assess predictive accuracy [22]. We used functions in package 'lme4' [23] and 'lmerTest' [24] for model construction and selection. We used package 'MuMIN' to calculate R^2 values [25]. We performed all analyses in Program R [26].

2.3.3. Current Conditions and Vulnerability

We applied the optimal model to predict TDS at each sampling location as a function of observed mean daily discharge. We calculated the number of days and number of consecutive days each year with predicted TDS exceeding the EPA secondary drinking water standard of ≥500 mg/L [16]. We calculated the threshold discharge predicted to result in exceedance of the 500 mg/L drinking water standard. We calculated the difference between the threshold and observed mean daily discharge values (Δ discharge). Positive Δ discharge values indicate the additional mean daily discharge needed to decrease TDS concentrations below 500 mg/L. Negative Δ discharge values represent the decrease in mean daily discharge required to result in additional exceedances. We calculated the number of additional days predicted to exceed 500 mg/L when Δ discharge was ≤100% of the observed value.

3. Results

3.1. Water Quality Characteristics

Measured water quality was highly variable both within and among study sites (Table 2). Mean observed TDS was highest at WF and exceeded 500 mg/L (maximum = 748 mg/L). TDS was most variable within M1; however, TDS was not observed in excess of 500 mg/L in the Monongahela River (maximum at M1 = 391 mg/L, maximum at M2 = 419 mg/L, and maximum at M3 = 388 mg/L). TDS was consistently low in in both CR and TV. SO_4 was the dominant ion contributing to TDS at all sites. Concentrations of dissolved Al, Fe, and Mn were highest and most variable in WF.

Table 2. Means (and standard deviations) of total dissolved solids (TDS) and its contributing constituents across all water samples taken within the Cheat (CR), Tygart Valley (TV), West Fork (WF), and Monongahela (M) Rivers. Refer to Figure 1 and Table 1 for site location and sample information. Means are reported in mg/L. Alkalinity (Alk) is reported in mg/L $CaCO_3$ equivalents.

	M1	M2	M3	CR	TV	WF
TDS	172 (71)	185 (74)	176 (69)	53 (13)	70 (25)	363 (127)
Alk	37.2 (15.9)	44.6 (13.9)	46.6 (13.6)	15.4 (6.1)	25.4 (12.4)	87.8 (22.1)
Br	0.04 (0.08)	0.04 (0.09)	0.02 (0.03)	0.00 (0.01)	0.01 (0.02)	0.05 (0.12)
Al	0.09 (0.11)	0.10 (0.31)	0.09 (0.27)	0.08 (0.13)	0.05 (0.06)	0.13 (0.70)
Fe	0.11 (0.15)	0.16 (0.54)	0.12 (0.33)	0.09 (0.16)	0.08 (0.08)	0.21 (1.26)
Mn	0.10 (0.07)	0.09 (0.07)	0.07 (0.06)	0.08 (0.06)	0.05 (0.04)	0.08 (0.05)
Cl	11.2 (7.1)	11.2 (4.9)	9.8 (5.0)	3.5 (1.4)	5.5 (2.6)	15.4 (8.7)
Ca	27.3 (9.1)	31.2 (9.6)	30.7 (11.0)	12.1 (3.7)	14.6 (3.6)	65.5 (21.7)
Na	21.1 (13.2)	20.2 (11.4)	21.2 (16.2)	3.3 (2.4)	7.2 (3.1)	30.6 (16.8)
Mg	7.0 (2.8)	8.2 (3.2)	7.7 (3.1)	2.6 (1.0)	3.1 (0.8)	17.6 (6.5)
SO_4	86 (42)	91 (45)	83 (39)	23 (7)	27 (18)	189 (82)

3.2. Water Quality Modeling

The optimal random effects structure for predicting TDS included random intercepts among years and sites, as well as random slopes among sites (Table 3).

Table 3. Step-down model selection results for linear-mixed models predicting log[x]-transformed total dissolved solids within the Monongahela River. Asterisks denote random effects retained in the final model (i.e., ΔAIC >2).

Model Structure	AIC	ΔAIC
Beyond optimal model [a]	239	–
Dropped random slopes		
Log(Q) | Year	239	0
* log(Q) | Site	257	18
Log(Q) | Site:Year	249	−2
Optimal random slope model [b]	237	–
Dropped random intercepts		
* 1 | Year	243	6
* 1 | Site	400	163
1 | Site:Year	235	−2

[a] Beyond optimal model = log(Q) + 1|Site + 1|Year + 1|Site: Year + log(Q)|Site + log(Q)|Year + (Q)|Site: Year. [b] Optimal random slope model = log(Q) + 1|Site + 1|Year + 1|Site: Year + log(Q)|Site.

Mean daily discharge had a significant negative effect on TDS concentration (Table 4). Together, fixed and random effects explained 95% of the total variation in TDS (i.e., marginal $R^2 = 0.95$). Mean daily discharge explained 5% of the overall variation (i.e., conditional $R^2 = 0.05$). The final model had a root mean square error (RMSE) of 0.25 when predicting log[x]-transformed TDS within the test set. We saw a strong relationship ($y = 0.00 + 1.01x$; $R^2 = 0.88$) between observed and predicted TDS within the test set. Predictive accuracy was similar within tributary (RMSE = 0.23) and Monongahela River (RMSE = 0.26) sites.

Table 4. Parameter estimates for the best supported model predicting log[x]-transformed total dissolved solids in the Monongahela River.

Parameter	Estimate	SE	*t*-Value	*p*-Value
Fixed effects				
Intercept	6.64	0.44	15.1	<0.001
Log(Q)	−0.23	0.03	−9.16	<0.001
Random effects				
$\sigma^{1\|Year}$	0.047	–	–	–
$\sigma^{1\|Site}$	1.06	–	–	–
$\sigma^{Q\|Site}$	0.057	–	–	–
$\sigma^{Residual}$	0.267	–	–	–

3.3. *Current Conditions and Vulnerability*

Predicted TDS was highly variable among sites (Figure 2).

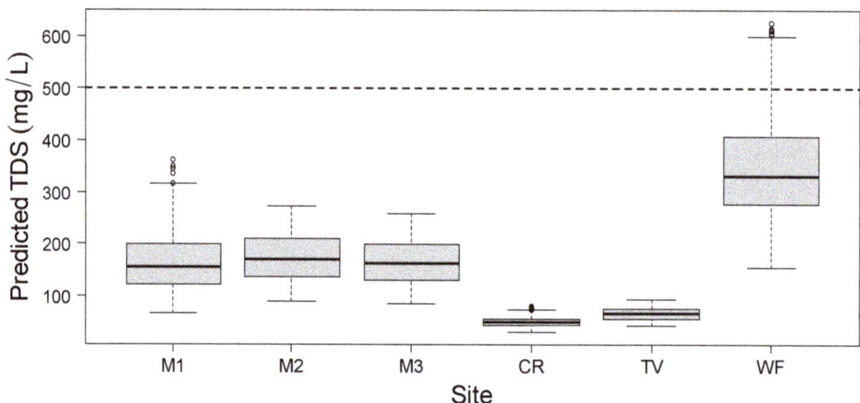

Figure 2. Variability in predicted total dissolved solids (TDS) concentrations across all days and years within and among the six sites on the Cheat (CR), Tygart Valley (TV), West Fork (WF), and Monongahela (M) Rivers. Refer to Figure 1 for site locations.

Median TDS was similar across M1 (155 mg/L), M2 (171 mg/L), and M3 (163 mg/L); however, TDS was slightly more variable at M1 (range = 68–362 mg/L) than M2 (90–271 mg/L) and M3 (86–259 mg/L) (Figure 2). Predicted TDS within the Monongahela River never exceeded the EPA secondary drinking water standard of 500 mg/L (Figure 2). Predicted TDS was consistently low within CR (range = 31–80 mg/L, median = 51 mg/L) and TV (range = 43–94 mg/L, median = 67 mg/L) and did not exceed the EPA secondary drinking water standard of 500 mg/L. Predicted TDS was greater (median = 331 mg/L) and more variable (range = 154–626 mg/L) within WF (Figure 2). TDS within WF exceeded 500 mg/L a total of 133 days over 24 events (Figure 3).

Exceedances occurred during six of the nine years, with the number of days exceeding 500 mg/L each year ranging from 2 (2017) to 65 (2010) (Figure 3). The number of consecutive days exceeding 500 mg/L within WF also varied among years and ranged from 1 to 28 (2010) (Figure 3). Inter- and intra-annual variability in exceedances were driven by the frequency and duration of low-flow conditions (Figure 4).

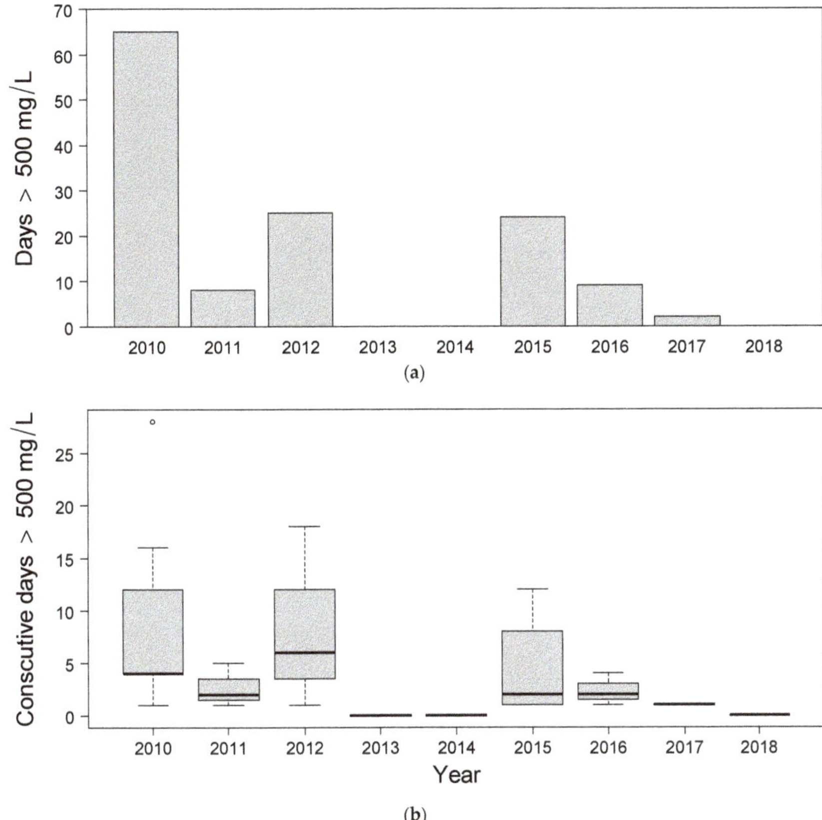

Figure 3. (**a**) Histogram showing number of days each year with total dissolved solids (TDS) exceeding 500 mg/L within the West Fork (WF) River; (**b**) Box plots showing the number of consecutive days TDS exceeded 500 mg/L during exceedance events occurring each year.

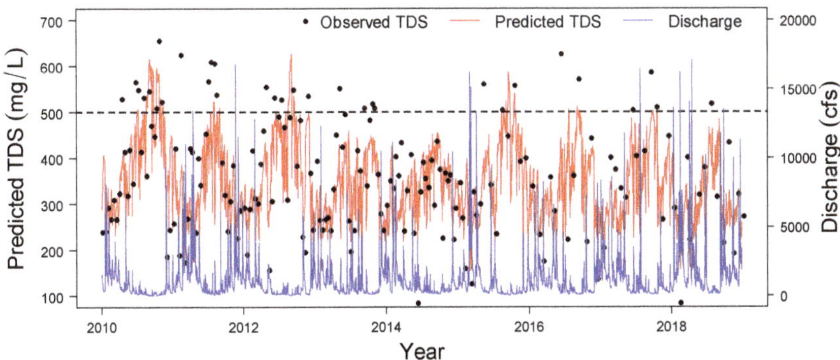

Figure 4. Time series of predicted TDS and observed discharge at the West Fork water quality and United States Geological Survey (USGS) gauge station (refer to Figure 1 for locations). Observed TDS is shown for reference.

Threshold discharges predicted to result in exceedance within WF varied among years and ranged from 84 cfs (2013) to 141 cfs (2010). Additional mean daily discharge required to decrease TDS below 500 mg/L (i.e., Δ discharge) during the 24 exceedance events ranged from 1 to 83 cfs (Figure 5).

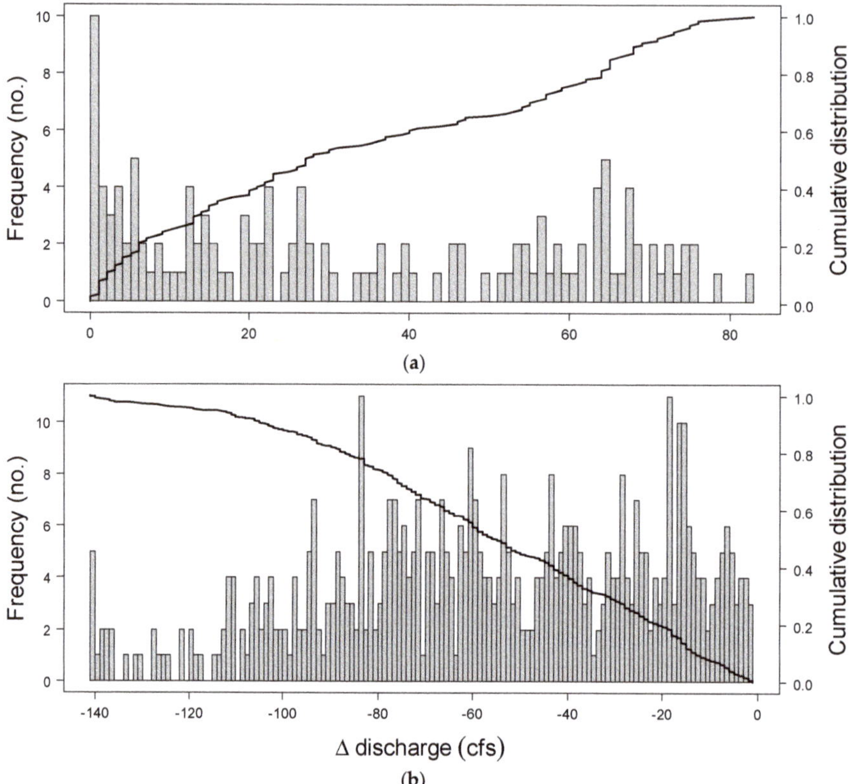

Figure 5. (**a**) Frequency and cumulative distribution of increased mean daily discharge (positive Δ discharge) required to decrease predicted TDS below 500 mg/L across the 133 days where the secondary drinking water standard was exceeded; (**b**) Frequency and cumulative distribution of decreased mean daily discharge (negative Δ discharge) required to result in exceedance of the 500 mg/L across the 499 days with Δ discharge < 100% of the observed value.

Mean daily discharge within WF fell short of the exceedance threshold by ≤1 cfs 10 times (8% of total exceedances). Twenty-four percent (26%; 34 days) of exceedances were characterized by Δ discharge ≤10 cfs (Figure 5). A total of 499 additional days were predicted to exceed 500 mg/L with simulated decreases in mean daily discharge (Δ discharge) ≤100% of the observed value (Figure 5). Of those, 39 days (8%) were predicted to exceed 500 mg/L with a decrease in observed discharge of ≤10 cfs (i.e., Δ discharge ≥−10) (Figure 5).

4. Discussion

We characterized a high degree of spatial and temporal variability in TDS that provided critical insight into vulnerability of the upper Monongahela River basin to elevated TDS. TDS within the West Fork River (WF) was predicted to exceed the secondary drinking water standard of 500 mg/L a total of 133 days from 2010 to 2018, with the frequency and duration of exceedances being closely tied to low-flow conditions. Consistently low TDS within the Tygart Valley River (TV) and Cheat River (CR) reduced vulnerability of the Monongahela River (M1, M2, M3) to elevated TDS, which was neither observed (maximum = 419 mg/L) nor predicted (341 mg/L) to exceed 500 mg/L.

Dominance of SO_4 in the ionic signature and variability in elevated Al, Fe, and Mn suggest and support previous studies identifying mining as the main contributor to elevated TDS in the Monongahela River basin [27,28]. Elevated concentrations of and variability in other major ions (Ca, Mg, Na, Cl) could be attributed to other land use activities such as oil and gas development [14,15] and urban development [27]. The extent to which these landscape features contribute to elevated TDS ultimately controls vulnerability of the system to variability in flow. TDS concentrations within CR and TV remained below 500 mg/L regardless of flow, suggesting limited vulnerability to variability in flow. In contrast, TDS concentrations within WF were elevated to the point that temporal variability in TDS associated with changes in flow were enough to result in exceedance of the 500 mg/L drinking water criterion. These results corroborate previous work documenting the importance of flow variability—both natural (e.g., precipitation patterns [9]) and anthropogenic (i.e., discharge management and reservoir releases [10])—in modulating vulnerability of this and other freshwater systems to degraded water quality [12,29].

TDS was predicted to exceed the 500 mg/L drinking water criterion a total of 133 times within the West Fork River. The number of consecutive days with predicted TDS exceeding 500 mg/L ranged from 1 to 28. TDS concentrations exceeding 500 mg/L can lead to pipe corrosion, increased scaling and sedimentation, and taste and odor problems [15], potentially increasing the cost of water treatment and maintenance of degraded infrastructure. Thus, TDS concentrations observed and predicted in the current study have important implications for the individuals and communities obtaining their drinking water from the West Fork River (e.g., 60,000 individuals serviced by a single water supplier). Despite these concerns, TDS is not specifically targeted for removal during conventional drinking water treatment, nor is the 500 mg/L secondary drinking water standard mandatory or enforceable [15,16]. Consequently, it is incumbent upon water resource agencies to manage TDS in an effort to help ensure continuity of vital ecosystem services within the West Fork and Monongahela Rivers both now and into the future.

Programs like 3RQ will be most successful if done within the context of flow variability and assimilative capacity. The threshold discharge (i.e., discharge resulting in exceedance of the 500 mg/L) identified for the West Fork River in the current study was higher in 2010 (141 cfs) than subsequent years (84–140 cfs), suggesting increased assimilative capacity in the West Fork and Monongahela Rivers following implementation of the discharge management model. This assertion is further supported by a predicted decrease in the frequency and duration of drinking water standard (i.e., 500 mg/L) exceedances within the West Fork River following 2010. Defining a pollutant's chemical footprint—the volume of water available to dilute additional pollution loading or required to dilute existing loads to allowable levels (e.g., TDS of 500 mg/L)—represents another tool for managing water quality within the context of assimilative capacity [30,31]. This information can then be used to strategically target management efforts (i.e., reducing current pollution loads and/or protecting areas for maintenance of downstream assimilative capacity) that maximize system resiliency and decrease vulnerability.

Our results could also be used to inform updated management of the Stonewall Jackson and Tygart River Reservoirs. Augmenting flows through additional reservoir releases during critical low-flow periods could decrease current and future vulnerability of the West Fork and Monongahela Rivers to elevated TDS. The minimum flow thresholds required to maintain TDS below 500 mg/L identified in this study could inform such an effort and be incorporated into an adaptive management plan [32]. Both reservoirs currently augment low flows to maintain downstream water quality and are vital to the maintenance of drinking water standards in both rivers. However, any effort to alter reservoir releases must not affect the reservoirs' capacity to maintain all authorized purposes (i.e., flood protection, water supply for maintenance of water quality and navigation, recreation, and fish and wildlife enhancement) over both the short- and long-term. Additional research is needed to characterize how altering reservoir releases would affect all ecological (e.g., reservoir and downstream water quality and habitat) and socioeconomic (e.g., recreation, water supply) systems under both realized and potential future low-flow conditions.

Our study represents one of only a few to assess how spatio-temporal variability in TDS concentrations contribute to vulnerability of freshwater ecosystems and how they should be managed (e.g., drinking water) (but see [11,12]). Our results corroborate previous studies documenting increased vulnerability of surface waters to elevated ionic concentrations during periods of low flow [12]. Salts are generally transported to streams via subsurface flow through natural or altered (i.e., mined) landscapes. Ionic concentrations are generally greatest during dry periods when stream flow is dominated by subsurface flow and becomes diluted with increasing surface runoff [33]. In contrast, streams and rivers are often most vulnerable to pollutants transported to streams via surface runoff (e.g., nutrients) during high flow events [12]. Spatial and temporal complexities make managing toward individual pollutant criteria difficult [11]. Spatial and temporal complexities among multiple pollutants make maintaining functional ecosystems and ecosystems services difficult. Future research should focus on characterizing spatio-temporal variability of key pollutants and how that variability controls vulnerability of critical source waters.

The methodology described herein can be used to assess current and future vulnerability of, and create management plans for, systems impacted by any number of pollutants. The mixed modeling framework used in this study enabled us to quantify and predict the effects of discharge (i.e., fixed effects), while accounting for site- (e.g., upstream land use) and year-specific (e.g., temporal changes in water quality management) factors that affect spatio-temporal patterns in TDS and response to flow variability. This approach enabled us to predict TDS concentrations with a high degree of certainty and accuracy (R^2 = 0.95; test set RMSE = 0.25). Our study also highlights the value and utility of long-term monitoring data for providing insight into continuous water quality conditions in systems where such data are unavailable.

Our results suggest the upper Monongahela River basin may be vulnerable to even minor changes in TDS and/or discharge. Decreases in mean daily discharge by ≤10 cfs resulted in an additional 34 days exceeding 500 mg/L within the West Fork River. Although the Monongahela River was never observed or predicted to exceed 500 mg/L, potential changes in future land use impacts (e.g., continued expansion of unconventional oil and gas, additional mine drainage contributions or reduced treatment of current mine discharge) and/or discharge could elevate TDS levels to the point that this system becomes vulnerable to variability in flow. Elevated temperatures (i.e., increased evapotranspiration [33]) and increased variability in precipitation under climate change have the potential to exacerbate drought and low-flow conditions, further reducing assimilative capacity within the upper Monongahela River basin during critical low-flow periods [34–36]. In that event, the discharge management model would require adjustment to account for changes in either TDS load or assimilative capacity. Climate change is also expected to interact with and amplify hydrologic modifications in the built environment (e.g., increased stormwater runoff, reservoir systems), as well as other anthropogenic activities (e.g., mine drainage and increased water extraction) to further impact water quality and associated services [13,37–39]. Given the role flow variability plays in modulating vulnerability of receiving waters to TDS, uncertainty regarding the effects of climate change on flow within this region makes this an area of concern. This is particularly true given the role Appalachian headwater streams will play in contributing to and securing regional water supply throughout the 21st century [40].

5. Conclusions

Results of the present study demonstrate the myriad of spatial (e.g., land use) and temporal (e.g., variability in flow) factors that control vulnerability of source waters to elevated TDS. Management decisions that do not incorporate these complexities risk ineffective or inappropriate actions. Management of TDS within this and other systems should first seek to identify and prioritize areas for reducing current sources of TDS throughout the watershed and for protection of minimally impacted streams to maintain current and future assimilative capacity. It could also be possible to leverage existing water management and reservoir systems to maintain assimilative capacity during critical low-flow periods; however, additional study would be needed to verify that these reservoirs have the capacity to maintain all authorized purposes (e.g., navigation, fish and wildlife habitat enhancement) while providing additional low-flow augmentation for maintenance of water quality. Given widespread salinization of streams and rivers [1], an important avenue of continued research will be to characterize the spatio-temporal vulnerability of this and other critical source waters to elevated TDS. It will be particularly important to characterize vulnerability to TDS within the context of other pollutants and under a range of future land use and climate change (i.e., flow variability) scenarios. Such efforts will be critical toward effectively ensuring sustainability of aquatic ecosystems and the vital services they provide (e.g., drinking water provision).

Author Contributions: Conceptualization, E.R.M. and J.T.P.; Formal Analysis, E.R.M.; Investigation, M.O.; Resources, P.F.Z.; Data Curation, M.O.; Writing—Original Draft Preparation, E.R.M.; Writing—Review and Editing, J.T.P., P.F.Z., and M.O.; Project Administration, E.R.M., J.T.P, and M.O.; Funding Acquisition, J.T.P., P.F.Z., M.O., and E.R.M. All authors have read and agreed to the published version of the manuscript.

Funding: This research was funded by the National Science Foundation, award number OIA-1458952. This work was partially supported by the U.S. Army Corps of Engineers Planning Assistance to States Program. Any opinions, findings, and conclusions or recommendations expressed in this material are those of the authors and do not necessarily reflect the views of the National Science Foundation or U.S. Army Corps of Engineers. Funding in support of the 3RQ program was provided by the Colcom Foundation, Pittsburgh PA and through the U.S. Geological Survey 104b program.

Acknowledgments: Field sampling by West Virginia University staff: Jason Fillhart, Benjamin Mack, Benjamin Pursglove; undergraduate students: Kaylynn Kotlar, Reva Dickson, Joshua Ash, Anthony Diamario, Jason Eulberg, Tyler Richards, Jude Platz, Cullen Platz, and Alex Pall; graduate students: Eric Baker, Zac Zacavish, Chance Chapman, Madison Cogar, Joseph Kingsburg, and Levi Cyphers. The authors would also like to thank Rosemary Reilly (Pittsburgh District, U.S. Army Corps of Engineers) and two anonymous reviewers for providing comments that greatly improved the quality of the manuscript.

Conflicts of Interest: The authors declare no conflict of interest.

References

1. Kaushal, S.S.; Likens, G.E.; Pace, M.L.; Utz, R.M.; Haq, S.; Gorman, J.; Grese, M. Freshwater salinization syndrome on a continental scale. *Procl. Natl. Acad. Sci.* **2018**, *115*, E574–E583. [CrossRef]
2. Szocs, E.; Coring, E.; Bathe, J.; Schafer, R.B. Effects of anthropogenic salinization on biological traits and community composition of stream macroinvertebrates. *Sci. Tot. Environ.* **2014**, *468–469*, 943–949. [CrossRef]
3. Castillo, A.M.; Sharpe, D.M.T.; Chalambor, C.K.; De Leon, L.F. Exploring the effects of salinization on trophic diversity in freshwater ecosystems: a quantitative review. *Hydrobiologia* **2018**, *807*, 1–17. [CrossRef]
4. Timpano, A.J.; Schoenholtz, S.H.; Soucek, D.J.; Zipper, C.E. Benthic macroinvertebrate community response to salinization in headwater streams in Appalachia USA over multiple years. *Ecol. Indic.* **2018**, *91*, 645–656. [CrossRef]
5. Berger, E.; Fror, O.; Schafer, R.B. Salinity impacts on river ecosystem processes: a critical mini-review. *Phil. Trans. R. Soc. B* **2018**, *374*. [CrossRef]
6. Kaushal, S.S. Increased salinization decreases safe drinking water. *Environ. Sci. Technol.* **2016**, *50*, 2765–2766. [CrossRef] [PubMed]
7. Canedo-Arguelles, M.; Hawkins, C.P.; Kefford, B.J.; Schafer, R.B.; Dyack, B.J.; Brueet, S.; Buchwalter, D.; Dunlop, J.; Fror, O.; Lazorchak, J.; et al. Saving freshwater from salts. *Science* **2016**, *351*, 914–916. [CrossRef]
8. Kaushal, S.S.; Groffman, P.M.; Likens, G.E.; Belt, K.T.; Stack, W.P.; Kelly, V.R.; Band, L.E.; Fisher, G.T. Increased salinization of fresh water in the northeastern United States. *Procl. Natl. Acad. Sci.* **2005**, *102*, 13517–13520. [CrossRef] [PubMed]
9. Petty, J.T.; Barker, J. Water quality variability in tributaries of the Cheat River, a mined Appalachian watershed. In Proceedings of the Joint Conference of the 21st Annual Meeting of the American Society of Mining and Reclamation and 25th West Virginia Surface Mine Drainage Task Force Symposium, Morgantown, WV, USA, 18–22 April 2004.
10. Nilsson, C.; Malm-Renofalt, B. Linking flow regime and water quality in rivers: a challenge to adaptive catchment management. *Ecol. Soc.* **2008**, *13*, 18. [CrossRef]

11. Sheldon, F.; Fellows, C.S. Water quality in two Australian dry land rivers: spatial and temporal variability and the role of flow. *Mar. Freshwater Res.* **2010**, *61*, 864–874. [CrossRef]
12. Guo, D.; Lintern, A.; Webb, J.A.; Ryu, D.; Liu, S.; Bende-Michl, U.; Leahy, P.; Wilson, P.; Western, A.W. Key factors affecting temporal variability in stream water quality. *Water Resour. Res.* **2019**, *55*, 112–129. [CrossRef]
13. Whitehead, P.G.; Wilby, R.L.; Battarbee, R.W.; Kernan, M.; Wade, A.J. A review of the potential impacts of climate change on surface water quality. *Hydrolog. Sci. J.* **2009**, *54*, 101–123. [CrossRef]
14. Wilson, J.M.; VanBriesen, J.M. Oil and gas produced water management and surface drinking water sources in Pennsylvania. *Environ. Pract.* **2012**, *14*, 288–300. [CrossRef]
15. Wilson, J.M.; Wang, Y.; VanBriesen, J.M. Sources of high total dissolved solids to drinking water supply in southwestern Pennsylvania. *J. Environ. Eng.* **2014**, *140*. [CrossRef]
16. Environmental Protection Agency (EPA). Secondary drinking water regulations: guidance for nuisance chemicals. Available online: https://www.epa.gov/dwstandardsregulations/secondary-drinking-water-standards-guidance-nuisance-chemicals (accessed on 30 December 2019).
17. U.S. Army Corps of Engineers. Stonewall Jackson Lake, West Fork River, West Virginia, Design Memorandum No. 1, General Design Memorandum, Pittsburgh District, Pittsburgh Pennsylvania. 1971. Available online: https://www.lrp.usace.army.mil/Portals/72/docs/HotProjects/SWJ%20Public%20release%20info1.pdf (accessed on 30 December 2019).
18. West Virginia Department of Environmental Protection GIS Server, Data Download. Available online: https://tagis.dep.wv.gov/home/Downloads (accessed on 31 December 2019).
19. Pennsylvania Department of Environmental Protection, Open Data Portal. Available online: https://newdata-padep-1.opendata.arcgis.com (accessed on 31 December 2019).
20. Zuur, A.F.; Ieno, E.N.; Walker, N.J.; Saveliev, A.A.; Smith, G.M. *Mixed Effects Models and Extensions in Ecology with R*; Springer: New York, NY, USA, 2009.
21. Arnold, T.W. Uninformative parameters and model selection using Akaike's information criterion. *J. Wildl. Manag.* **2010**, *74*, 1175–1178. [CrossRef]
22. Pineiro, G.; Perelman, S.; Guerschman, J.P.; Paruelo, J.M. How to evaluate models: observed vs. predicted or predicted vs. observed? *Ecol. Model.* **2008**, *216*, 316–322. [CrossRef]
23. Bates, D.; Maechler, M.; Bolker, B.; Walker, S. Fitting linear mixed-effects models using lme4. *J. Stat. Softw.* **2015**, *67*, 1–48. [CrossRef]
24. Kuznetsova, A.; Brockhoff, P.B.; Christensen, R.H.B. lmerTest Package: tests in linear mixed effects models. *J. Stat. Softw.* **2017**, *82*, 1–26. [CrossRef]
25. Barton, K. MuMIn: Multi-model inference. R package version 1.42.1. 2018. Available online: http://CRAN.R-project.org/package=MuMIn (accessed on 30 January 2020).
26. R Core Team. Available online: https://www.R-project.org/ (accessed on 31 December 2019).
27. Petty, J.T.; Fulton, J.B.; Strager, M.P.; Merovich, G.T.; Stiles, J.M.; Ziemkiewicz, P.F. Landscape indicators and thresholds of stream ecological impairment in an intensively mined Appalachian watershed. *J. N. Am. Benthol. Soc.* **2010**, *29*, 1292–1309. [CrossRef]
28. Merovich, G.T.; Petty, J.T.; Strager, M.P.; Fulton, J.B. Hierarchical classification of stream condition: a house-neighborhood framework for establishing conservation priorities in complex riverscapes. *Freshw. Sci.* **2013**, *32*, 874–891. [CrossRef]
29. Merriam, E.R.; Fernandez, R.; Petty, J.T.; Zegre, N. Can brook trout survive climate change in large rivers? If it rains. *Sci. Tot. Environ.* **2017**, *607–608*, 1225–1236. [CrossRef] [PubMed]
30. Bjorn, A.; Diamond, M.; Birkved, M.; Hauschild, M.Z. Chemical footprint method for improved communication of freshwater ecotoxicity impacts in the context of ecological limits. *Environ. Sci. Technol.* **2014**, *48*, 13253–13262. [CrossRef] [PubMed]
31. Zijp, M.C.; Posthuma, L.; van de Meent, D. Definition and applications of a versatile chemical pollution footprint methodology. *Environ. Sci. Technol.* **2014**, *48*, 10588–10597. [CrossRef] [PubMed]
32. Warner, A.T.; Bach, L.B.; Hickey, J.T. Restoring environmental flows through adaptive reservoir management: planning, science, and implementation through the Sustainable Rivers Project. *Hydrolog. Sci. J.* **2014**, *59*, 770–785. [CrossRef]
33. Gaertner, B.A.; Zegre, N.; Warner, T.; Fernandez, R.; He, Y.; Merriam, E.R. Climate, forest growing season, and evapotranspiration changes in the central Appalachian Mountains, USA. *Sci. Tot. Environ.* **2019**, *650*, 1371–1381. [CrossRef]

34. Sereda, J.; Bogard, M.; Hudson, J.; Helps, D.; Dessouki, T. Climate warming and the onset of salinization: rapid changes in the limnology of two northern plains lakes. *Limnologica* **2011**, *41*, 1–9. [CrossRef]
35. Canedo-Arguelles, M.; Kefford, B.J.; Piscart, C.; Prat, N.; Schafer, R.B.; Schulz, C.-J. Salinisation of rivers: An urgent ecological issue. *Environ. Pollut.* **2013**, *173*, 157–167. [CrossRef]
36. Herbert, E.R.; Boon, P.; Burgin, A.J.; Neubauer, S.C.; Franklin, R.B.; Ardon, M.; Hopfensperger, K.N.; Lamers, L.P.M.; Gell, P. A global perspective on wetland salinization: ecological consequences of a growing threat to freshwater wetlands. *Ecosphere* **2015**, *6*, 1–43. [CrossRef]
37. Delpla, I.; Jung, A.-V.; Baures, E.; Clement, M.; Thomas, O. Impacts of climate change on surface water quality in relation to drinking water production. *Environ. Int.* **2009**, *35*, 1225–1233. [CrossRef]
38. Kaushal, S.S.; Gold, A.J.; Mayer, P.M. Land use, climate, and water resources—global stages of interaction. *Water* **2017**, *9*, 815. [CrossRef]
39. Merriam, E.R.; Petty, J.T.; Clingerman, J. Conservation planning at the intersection of landscape and climate change: Brook trout in the Chesapeake Bay watershed. *Ecosphere* **2019**, *10*. [CrossRef]
40. Fernandez, R.; Zegre, N. Seasonal changes in water and energy balances over the Appalachian Region and beyond throughout the twenty-first century. *J. Appl. Meteorol. Clim.* **2019**, *58*, 1079–1102. [CrossRef]

Article

A Spatially Distributed Investigation of Stream Water Temperature in a Contemporary Mixed-Land-Use Watershed

Jason P. Horne [1] and Jason A. Hubbart [1,2,*]

1 Davis College of Agriculture, Natural Resources and Design, Schools of Agriculture and Food, and Natural Resources, West Virginia University, 3109 Agricultural Sciences Building, Morgantown, WV 26506, USA; jph0021@mix.wvu.edu

2 Institute of Water Security and Science, West Virginia University, 3109 Agricultural Sciences Building, Morgantown, WV 26506, USA

* Correspondence: Jason.Hubbart@mail.wvu.edu; Tel.: +1-304-293-2472

Received: 18 May 2020; Accepted: 16 June 2020; Published: 20 June 2020

Abstract: Stream water temperature (°C) is an important physical variable that influences many biological and abiotic water quality processes. The intermingled mosaic of land-use/land-cover (LULC) types and corresponding variability in stream water temperature (Tw) processes in contemporary mixed-land-use watersheds necessitate research to advance management and policy decisions. Water temperature was analyzed from 21 gauging sites using a nested-scale experimental watershed study design. Results showed that forested land use was negatively correlated ($\alpha = 0.05$) with mean and maximum Tw. Agricultural land use was significantly positively correlated ($\alpha = 0.05$) with maximum Tw except during the spring season. Mixed development and Tw were significantly correlated ($\alpha = 0.05$) at quarterly and monthly timescales. Correlation trends in some reaches were reversed between the winter and summer seasons, contradicting previous research. During the winter season, mixed development showed a negative relationship with minimum Tw and mean Tw. During the summer season, higher minimum, maximum, and mean Tw correlations were observed. Advanced understanding generated through this high-resolution investigation improves land managers' ability to improve conservation strategies in freshwater aquatic ecosystems of contemporary watersheds.

Keywords: stream water temperature; land use; hydrology; experimental watershed; water resources; water quality; watershed management

1. Introduction

Stream water temperature (Tw) affects abiotic and biotic processes in aquatic ecosystems [1,2]. Abiotic variables influenced by Tw include dissolved oxygen concentration, chemical reaction rates, viscosity, density, and surface tension [3]. Biological processes influenced by Tw include the growth rate of fish [4] and rates of primary production in some autotrophic species [5]. Water temperature therefore impacts multiple trophic levels of the aquatic food web, including periphyton, benthic macroinvertebrates, and fishes [6]. This is of further relevance given that many aquatic organisms (e.g., mottled sculpin (*Cottus bairdii*), *Escherichia coli*, and flathead mayfly (*Heptageniidae* sp.)) have specific tolerance ranges for Tw [4]. Water temperature tolerances/preferences also directly affect species distributions and population densities, thus affecting many important aquatic industries (i.e., fishing) [7–9]. Globally, water temperature has traditionally received less attention relative to other water quality parameters, such as suspended sediment and water chemistry [1]. However, water temperature's ecohydrological importance and susceptibility to anthropogenic disturbance make it a critical variable of concern for resource managers [4,10].

Most contemporary watersheds include many different land-use/land-cover types (LULC) that influence rainfall–runoff temperatures and therefore receiving water temperature [11–13]. For example, vegetation intercepts incoming shortwave radiation, reducing the amount of radiation reaching the surface and thereby reducing surface temperatures [6,14,15]. This is important because nearly every published study on the subject showed that incident shortwave radiation is the variable of greatest influence on stream Tw [6,14,16]. For example, Webb and Zhang [17,18] showed that incident shortwave radiation accounted for 70% of a stream's thermal inputs, whereas other significant sources of energy come from longwave radiation emitted by atmospheric water vapor, and advected energy from the stream bed/bank. Forested land-use practices are commonly associated with lower Tw fluctuations and lower overall Tw, particularly in the warmer months. In almost every study, an increase in stream temperature was observed subsequent to the removal of riparian buffers [14,17,19–22]. While not the only influencing variable, it is well accepted that forest canopies attenuate the diel fluctuation of ambient air temperatures (Ta) and therefore Tw [6,14,15,22,23].

Agricultural land-use types are typically associated with increased sedimentation, increased nutrient loading, pesticide contamination, increased suspended and dissolved solids, and pathogens such as *Escherichia coli* [24–26]. Turbid water associated with agricultural land-use practices has been shown to have higher concentrations of total suspended solids (TSS) and thus higher Tw relative to clear water [24] due to particulate matter heat adsorption [15,24]. Younus et al. [27] created a numerical model to compute free surface flow hydrodynamics and coupled stream temperature dynamics in an agricultural watershed. Model results showed that an increase in subsurface lateral flow and shortwave radiation accounted for the most significant contributors to the stream heat budget [27]. Interestingly, the influence of subsurface flow on the stream heat budget was comparable to shortwave radiation [27]. This strong influence by subsurface flow is attributable to water being heated while infiltrating and percolating through the surface soil during periods of bare fields and high quarterly radiation. Oke [15] noted that crops can provide some shade through increased shortwave interception and reduce runoff volumes through transpiration [28]. This is as opposed to following harvest, when runoff volumes are higher and the soil is exposed to greater amounts of shortwave radiation, reaching higher temperatures [15,27,29]. Water then contacting and/or infiltrating into the soil within the shallow vadose zone is heated and transported to surrounding streams via runoff or subsurface lateral flows [27]. A similar runoff heating process also occurs in mixed development land-use types [30–32].

Previous research showed that mixed development land-use types increase the volume of heated runoff entering adjacent streams during summer precipitation events [32–34]. These findings have been attributed to stormwater runoff contacting heated impervious surfaces [17,32,33,35]. As heated runoff enters adjacent water bodies, it can result in a thermal surge [32–34]. Thermal surges were characterized as a 1 °C increase in Tw within fifteen minutes following a rain event [32,34]. Rice et al. [34] supported this finding, showing a linear relationship between percent impervious surface and mean thermal surge amplitude, where surges of 3.3 °C were observed in areas with 75% impervious surface cover [34]. Zeiger and Hubbart [22] observed surge increases of 4 °C lasting for up to five hours in work conducted in the central United States. Other authors showed thermal point source inputs in mixed development land-use types, particularly those from stormwater or wastewater outlets [30]. Kinouchi et al. [30] showed a positive correlation between increasing annual temperature and an increase in heat input from wastewater treatment plants. This increase was 0.1–0.2 °C/year resulting in an overall increase of 4.2 °C over twenty years [30]. Other research showed that mixed-development (urban) areas can increase subsurface temperatures through above ground (urban heat island effect) to below ground advection [15,32,36].

Ambient air temperature (Ta) has been shown to have a substantial influence on Tw [32,34,37,38]. This is important because the Ta and Tw relationship could mask impacts of land use on Tw. For example, research has shown conclusively that as water moves downstream, it is impacted less by the various inflows from runoff/tributaries, and more so by ambient air temperature (Ta) [37,38]. Intercatchment variations in riparian vegetation, geology, urbanization, aspect, elevation, and catchment size have also been shown to differentially affect Tw [39]. Rice et al. [34] showed that increasing watershed urbanization causes the Tw and Ta relationship to break down. This is offset, however, given that higher channel volumes downstream are often sufficient enough to attenuate thermal inputs [2,40]. Thus, stream inputs (e.g., groundwater, surface runoff, and confluence) from surrounding land-use types may have a greater influence upstream near headwaters relative to downstream and during periods of low(er) flow relative to periods of flooding [6,9]. Ultimately, water volume and Ta are not the only influencing factors affecting Tw, and the impacts of complex LULC must be considered [6,14,27,32,34].

The Appalachian region of the United States is an example of a physiographically complex region with high relief relative to surrounding areas [41]. Factors such as higher relief more greatly alter the rate of runoff from surrounding land-use types relative to flatter terrain [42]. Thus, a physiographically complex region, such as Appalachia, is well suited to advance existing knowledge gaps in the relationship between land-use types and Tw for many regions. In recent years, Appalachia, in particular the state of West Virginia, has been impacted by increased flood frequency, largely attributed to an increase in percent mixed development land-use types [43]. As mixed development land-use types expand, they often replace forested land-use types [44]. This is important because previous research showed these two land-use classifications have distinct effects on Tw. Furthermore, the interactions between competing mixed land-use types are largely unknown. Thus, studies in a contemporary mixed-land-use watershed designed to assess land-use impacts on hydrologic variables, including Tw, are needed in general (globally) and in the region (specifically) [45–48].

There is an ongoing need for high-resolution studies in contemporary watersheds that include multiple land-use practices [45–48]. Such studies will advance spatial and temporal understanding of Tw regimes, and therefore improve management decision making in contemporary (municipal) watersheds [13,45–48]. An effective method used to assess hydrologic processes in contemporary mixed-land-use watersheds is the experimental watershed study design [13,25,32,45,47–51]. The experimental watershed study design is effective at addressing both site-specific management questions and assisting in predictive model development, calibration, and validation [48–50]. Using this study design, researchers can partition a larger catchment into individual sub-catchments, enabling quantification of specific land-use impacts on variables of interest [13,48–51]. There is a lack of municipal watershed-scale studies that utilize this broadly accepted method (experimental watershed study design) in Appalachia, or elsewhere. The lack of such studies may be in part attributable to preconceptions of historic high cost of instrumentation, time-consuming data collection, and challenges with the transferability of results [48–51]. However, advantages of successful outcomes of the experimental watershed study design (e.g., high-temporal and spatial-resolution data) have been repeatedly shown to far exceed disadvantages [48–51].

The overall objective of this study was to use a highly instrumented (n = 22 gauging sites) experimental watershed study design to investigate land-use practice impacts on Tw, particularly maximum Tw, spatially and temporally in a representative contemporary mixed-land-use watershed. Sub-objectives included quantifying (a) annual, (b) quarterly, and (c) monthly relationships between LULC and Tw. Increased understanding generated through work such as this, better informs land managers wishing to improve the conservation and preservation of aquatic ecosystems [1,2,7–9,47–51].

2. Materials and Methods

2.1. Study Site

West Run Watershed (WRW) is located in the Monongahela Watershed and is categorized as a Hydrologic Group D watershed (HUC #05020003) located near Morgantown, West Virginia, USA [12]. The watershed spans 23 km^2, and the main drainage, West Run Creek, drains directly into the Monongahela River [27,35]. Anecdotally observed by the project managers of the watershed-based plan for West Run of the Monongahela River [52], the increase in urban sprawl stemming from the surrounding city of Morgantown continues to increase the severity and frequency of flooding.

The channel of West Run, according to habitat surveys, lacks sinuosity in many reaches and possesses a low channel slope of 1.1%, with back watering (flooding) near the terminus [12,52]. For this reason, site 22 (Figure 1) was excluded from the current study due to potential backwatering with the Monongahela River, and subsequent confounding analyses at that location. The average net radiation in Morgantown, measured during a study by Arguez et al. [53], from 1981 to 2010 was 130.68 W/m^2. The average recorded precipitation depth in Morgantown is approximately 1096 mm/year (1981–2010). Precipitation falls throughout the year but increases in quantity during the spring and summer months [54]. Previous studies showed that precipitation has significantly ($p = 0.01$) increased in the Appalachian region by 2.2% over the past 111 years [27,54,55]. Precipitation is frequently generated via frontal storm convergence systems, but during the summer months in particular, precipitation also occurs via orographic and/or convective processes [12,54–56]. Dynamic water parameters (i.e., base flow, Tw, and storm flow) were first monitored in the WRW in 2016 when stilling wells were installed [12], using a paired and nested-scale experimental watershed study design. Each site was equipped with a Solinst Levelogger Gold pressure transducer that logged and stored Tw (°C) data, with an accuracy of ±0.05 °C, and stage (water depth, cm), with an accuracy of ±0.3 cm, at five-minute intervals. During the study period, climate data were recorded using research-grade climate instrumentation located within approximately 100 m of site 13 (Figure 1). Climate variables (recorded at a height of 3 m) included precipitation (TE525 Tipping Bucket Rain Gauge), average air temperature and relative humidity (Campbell Scientific HC2S3 Temperature and Relative Humidity Probe), average wind speed (Met One 034B Wind Set instrument), and net radiation (Campbell Scientific NR01 Four-Component Net Radiation Sensor).

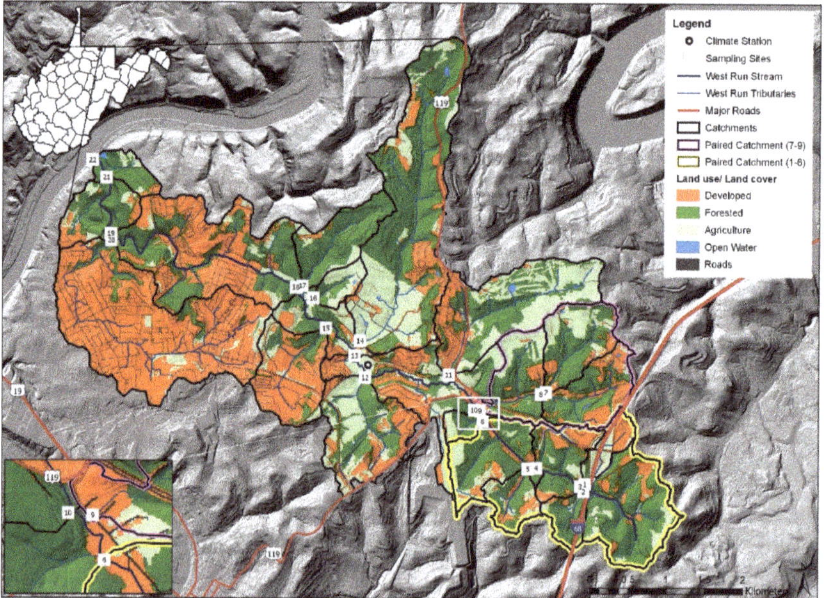

Figure 1. Land use and land cover of West Run Watershed (WRW), West Virginia, USA, including 22 nested gauging sites and corresponding sub-basins.

2.2. LULC Data

LULC data were derived from National Agricultural Imaging Program (NAIP) 2016 data. Initially, LULC types included 16 different assignations. For the current work, each of the original 16 LULC types was further grouped (lumped) into one of four LULC categories mixed-developed, agriculture, forested, and open water (Table 1). Using Arc GIS, watershed and sub-catchment boundaries were delineated, and LULC data incorporated. Each pixel representing 5 m^2 was counted and converted to km^2 and used to estimate percent LULC for each sub-catchment draining to each monitoring site; see Table 2 and Figure 2. Overall land use in West Run Watershed assessed using Arc GIS projected a composition of 19.4% agriculture, 42.7% forest, and 37.7% urban/suburban, with the portion of agricultural land use mainly comprised of animal husbandry (i.e., cattle) and crop fields (i.e., corn, soybeans, and cover crops) [12,27]. The presence of different land-use types coupled with regular land-use manipulation via human development justifies WRW categorization as a representative contemporary mixed-land-use watershed [12]. A combination of agricultural, mining, and industrial land use in and surrounding WRW has contributed to ongoing land and water resource degradation [12]. As per The United States Census Bureau report from 2019, the current population of Morgantown is 30,539 [57].

Table 1. Land-use/land-cover (LULC) (km^2) of West Run Watershed, West Virginia, USA, from National Agricultural Imaging Program (NAIP) 2016 data with percent LULC types in parenthesis for each sub-basin and respective site in West Run Watershed. Sub-basin percent in parenthesis is the proportional of the total watershed area. Cumulative open water (0.295 km^2) excluded and all values rounded to the tenths. Bold values represent dominant LULC type.

Site	Mixed Development (km^2 (%))	Agriculture (km^2 (%))	Forested (km^2 (%))	Sub-Basin (km^2 (%)) Total
1	**0.2 (53.2)**	0.1 (38.7)	0.0 (8.1)	0.3 (1.3)
2	0.0 (13.6)	0.0 (12.2)	**0.2 (74.2)**	0.3 (1.3)
3	0.4 (22.4)	0.3 (16.7)	**1.1 (61.3)**	1.9 (8.0)
4	0.6 (25.9)	0.4 (14.9)	**1.5 (59)**	2.5 (10.7)
5	0.1 (23.4)	0.1 (25.5)	**0.2 (51.1)**	0.4 (1.6)
6	0.9 (23.9)	0.6 (17.2)	**2.2 (58.7)**	3.7 (16.0)
7	0.1 (16.3)	0.2 (28.6)	**0.4 (54.9)**	0.8 (3.4)
8	0.5 (30.8)	0.3 (16.5)	**0.8 (52.4)**	1.6 (6.7)
9	0.6 (23.9)	0.4 (19.3)	**1.2 (52.8)**	2.3 (9.8)
10	1.5 (24.9)	1.1 (18.4)	**3.5 (56.5)**	6.2 (26.6)
11	0.3 (18.2)	**0.7 (41.9)**	0.7 (39.2)	1.8 (7.5)
12	0.4 (31.8)	0.4 (33.7)	**0.4 (34.5)**	1.2 (7.5)
13	2.8 (26.8)	2.7 (25.8)	**5.0 (47.1)**	10.5 (45.3)
14	0.5 (16.2)	0.9 (26.4)	**1.9 (56.9)**	3.4 (14.4)
15	**0.7 (70.3)**	0.1 (10.3)	0.2 (19.4)	1.0 (4.2)
16	0.0 (5.4)	**0.1 (58.7)**	0.1 (35.2)	0.2 (1.1)
17	0.0 (4.8)	0.1 (9.4)	**0.6 (85.8)**	0.7 (3.2)
18	4.3 (26.0)	4.1 (24.9)	**8.0 (48.9)**	16.4 (70.6)
19	5.6 (29.4)	4.2 (22.5)	**9.0 (47.9)**	18.9 (81.2)
20	**3.0 (89.2)**	0.1 (4.2)	0.2 (6.6)	3.4 (14.7)
21	8.7 (38.1)	4.5 (19.5)	**9.7 (42.2)**	22.9 (98.7)
22	8.8 (37.7)	4.5 (19.4)	**9.9 (42.7)**	23.2 (100.0)

Table 2. Original LULC classifications from NAIP 2016 data and the reclassification developed for use when analyzing Tw and LULC relationships in West Run Watershed, West Virginia, USA.

Reclassified LULC	Original LULC Classification
Mixed Development	Roads, impervious, mixed development, barren
Agriculture	Low vegetation, hay pasture, cultivated crops
Forested	Mine grass, forest, mixed mesophytic forest, dry mesic oak forest, dry oak forest, small stream riparian habitats
Open water	Water, river floodplains, and wetlands PEM

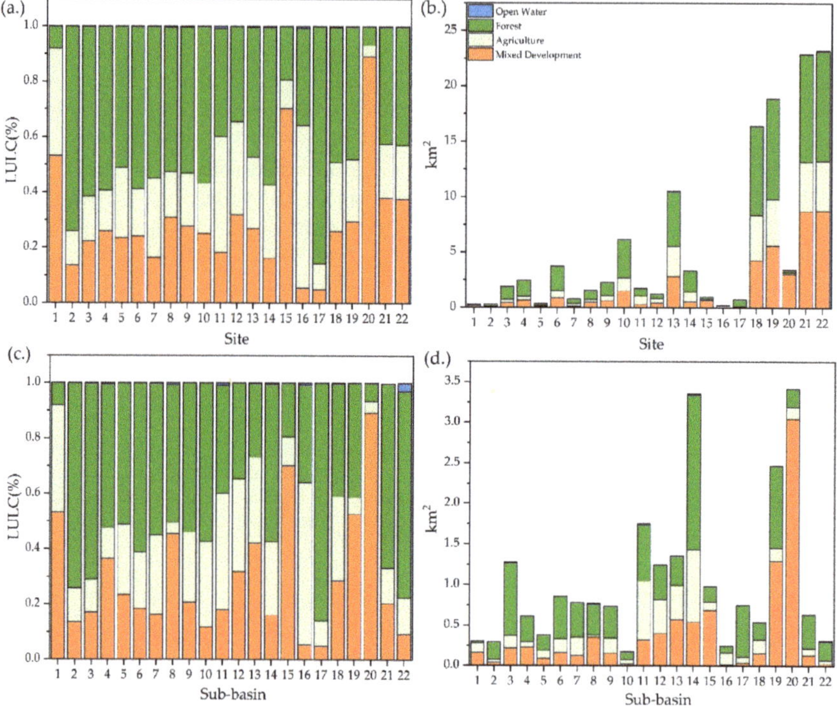

Figure 2. (**a**) Percent LULC of the sub-catchment and associated sites (cumulative), (**b**) the LULC area of the sub-catchment and each associated site (cumulative), (**c**) percent LULC of the sub-basin and associated sites, and (**d**) the LULC area of the sub-basin and each associated site in West Run Watershed, West Virginia, USA.

Forested land-use practices are the most abundant land-use classification in the WRW. For the current work, land-use/land-cover types grouped in the forested category included mature vegetation but were separated from mine grass classification, which is assumed to have succeeded to an intermediate successional stage [58,59]. Agricultural land-use practices were those classifications associated with early successional stages (i.e., low vegetation) or agriculturally maintained pastures and fields. The original classification grouped as mixed development land-use practices included mixed-development and all other LULC classifications associated with urban areas and/or impervious surfaces (e.g., barren, roads, and impervious) (Table 1) [34,35,60,61]. Figure 2 shows LULC proportions separated into percent area, and the LULC area of both the individual sites and each individual sub-basin.

2.3. Data Analysis

Data analyses included descriptive statistics (5 min data) of annual, quarterly, and monthly data in 2018 [47]. Quarterly time steps were delineated as 1 January–31 March (quarter 1), 1 April–30 June (quarter 2), 1 July–30 September (quarter 3), and 1 October–31 December (quarter 4). Data postprocessing included estimation of erroneous data or missing points (<0.2% of total data) by averaging between data points on either side of a gap, or by linear interpolation [62]. The Tw data were shown to be non-normally distributed using the Anderson–Darling test [62]. Therefore, land-use areas corresponding to each site were compared to maximum Tw using the Spearman rank correlation coefficient test ($\alpha = 0.05$). The open water land-use type was excluded from analyses due to its negligible areal coverage relative to other land-use types in the watershed. Three separate analyses were run including (a) all sites (b) tributaries only (i.e., sites 1, 2, 5, 7, 8, 9, 11, 12, 14, 15, 16, 17, and 20), and (c) mainstem West Run Creek only (i.e., sites 3, 4, 6, 10, 13, 18, and 21) to assess the varying effects of surrounding land-use practices on tributaries vs. mainstem sites. Due to the detrimental influence of maximum Tw on aquatic biological/geochemical processes, a pairwise comparison of daily maximum Tw was conducted on a site by site basis using a Kruskal–Wallis ANOVA [62]. Multiple principle component analyses (PCAs) were developed to illustrate the relationship between LULC types at an annual, quarterly, and monthly timescale, again excluding open water [63]. Using OriginPro Pro 9b Academic (OriginLab Corporation, Northampton, MA, USA), correlation biplots were generated with standardized data [63]. Due to the use of observed Tw data, no autoscaling preprocessing was needed to compare Tw and LULC data in the PCA analysis. However, due to the use of differing units (i.e., proportions (%) and temperature (°C)), a correlation matrix was used rather than a covariate matrix [63]. Following the Kaiser–Guttman criterion, eigenvalues greater than one were accepted as principle components above the threshold of importance [63].

3. Results

3.1. Climate during Study

Total precipitation recorded in West Run Watershed in 2018 was 1378 mm, which was 282 mm higher than average annual precipitation (1096 mm) over the previous 111 years [54]. The largest precipitation event (22.9 mm) occurred in a 5 min window on 9 September. The largest continuous precipitation event (83.2 mm) began on 8 September at 18:00 and lasted until 9 September at 16:30; see Figure 3b. Mean air temperature (Ta) in 2018 was 11.6 °C, which was 0.2 °C lower than the average annual temperature (11.4 °C) in West Virginia between 1990 and 2016 [54]. The coldest (−24.8 °C) and warmest (34.6 °C) recorded temperature occurred on January 1st at 8:00 and on 1 July at 17:00, respectively. The maximum net radiation was 1100 W/m^2, recorded on 7 May at 12:00. The mean near surface (1.5 m) net radiation was 139.5 W/m^2 (Figure 3), which was 8.82 W/m^2 higher than the mean annual net radiation in Morgantown between 1981 and 2010 (130.68 W/m^2) [53].

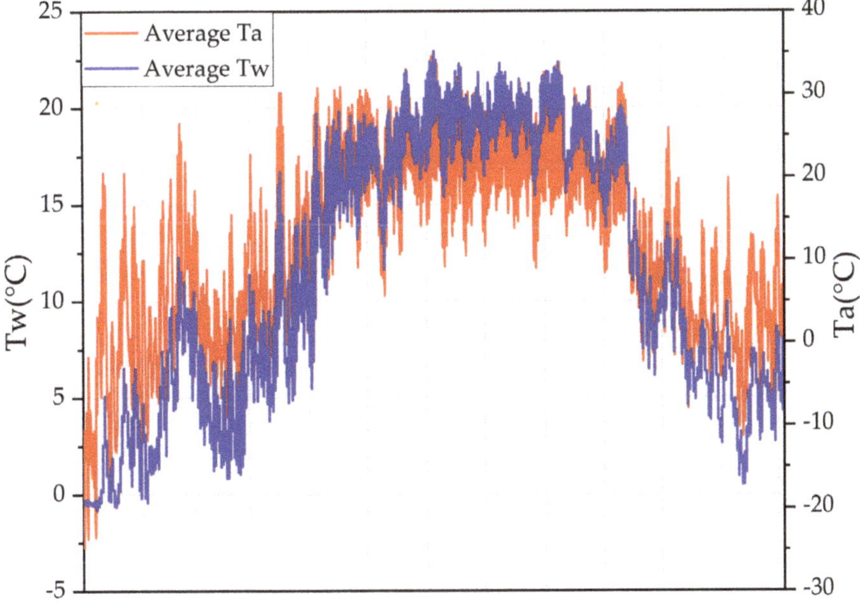

Figure 3. *Cont.*

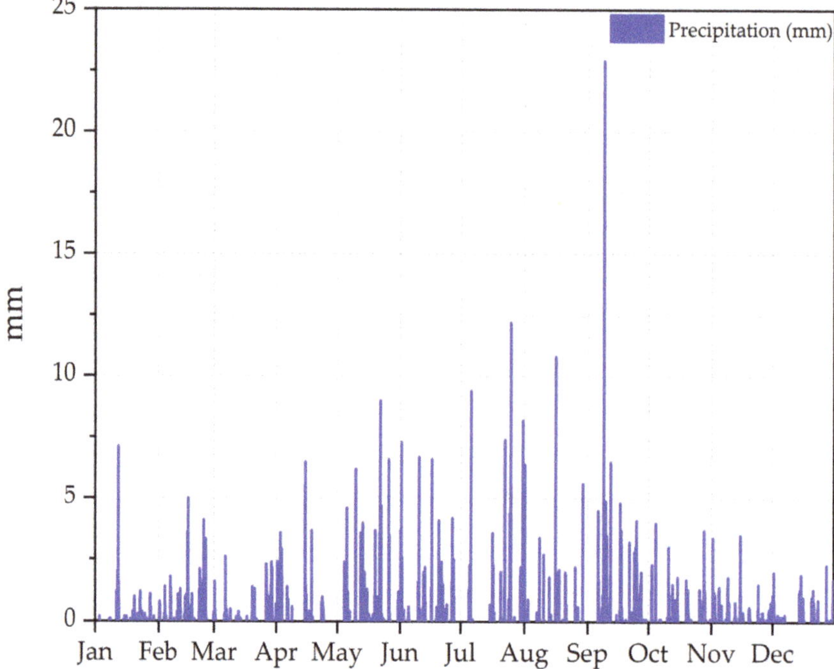

Figure 3. Thirty-minute timeseries of air temperature (Ta) and Tw in 2018 (**top graph**) and a thirty-minute timeseries of precipitation (**bottom graph**) in 2018 collected from a climate station located near site 13 (Figure 1) in West Run Watershed, West Virginia, USA.

3.2. Annual Stream Water Temperature

Annual average water temperature (Tw) across all sites varied by 2.4 °C, with a maximum of 12.5 °C at site 11 [dominant LULC agriculture; (41.9%)] and a minimum of 10.1 °C at site 17 [dominant LULC forested; (85.8%)] (Table 3). The lowest recorded Tw (−3.0 °C) occurred at site 16 [dominant LULC agriculture; (58.7%)] at 6:45 on 7 January, and the highest recorded Tw (27.4 °C) occurred at site 5 [dominant LULC forested; (51.1%)] at 16:05 on 3 July. Both were recorded less than a week after the lowest (−24.8 °C) and highest recorded Ta (34.6 °C).

Pairwise comparisons on a site by site basis comparing 21 of the 22 sites indicated that maximum Tw at site 1 was significantly different from site 5 (p = 0.02) [dominant LULC forested; (51.1%)], 11 (p = 0.00) [dominant LULC agriculture; (41.9%)], 12 (p = 0.00) [dominant LULC forested; (34.5%)], and 15 (p = 0.01) [dominant LULC mixed development; (70.3%)] (Table 4). Maximum Tw at site 2 [dominate LULC forested (74.2%)] was significantly different from 5 (p = 0.00) [dominate LULC forested; (51.1%)], 14 (p = 0.00) [dominant LULC forested; (56.9%)], 15 (p = 0.00), 18 (p = 0.00) [dominant LULC forested; (48.9%)], 19 (p = 0.00) [dominant LULC forested; (47.9%)], and 21 (p = 0.00) [dominant LULC forested; (42.2%)]. Maximum Tw at site 3 [dominant LULC forested; (61.3%)] was significantly different from site 5 (p = 0.01), 11 (p = 0.00), 12 (p = 0.00), and 15 (p = 0.01). Maximum Tw at site 4 [dominant LULC forested; (59.0%)] was significantly different from site 5 (p = 0.01), 11 (p = 0.00), 12 (p = 0.00), and 15 (p = 0.01). Maximum Tw at site 5 was significantly different from site 17 (p = 0.00). Maximum Tw at site 17 was significantly different from site 5 (p = 0.00), 6 (p = 0.01), 11 (p = 0.00), 12 (p = 0.00), 14 (p = 0.01), 15 (p = 0.00), 18 (p = 0.01), 19 (p = 0.02), 21 (p = 0.02). Spearman correlation coefficients of all twenty-one sites across the entire annual year indicated a significant (p = 0.01) positive correlation (rs = 0.6) between maximum Tw and agriculture LULC area, and a negative correlation (rs = −0.5) between maximum Tw and Forest LULCs ($p = 0.03$).

Table 3. Descriptive statistics of stream water temperature (°C), annual by site, quarterly (all sites), and monthly (all sites) collected in West Run Watershed, West Virginia, USA, in 2018.

Sites/Season/Month	Mean (°C)	Standard Deviation (°C)	Minimum (°C)	Maximum (°C)
Site 1	10.9	4.4	−0.3	22.2
Site 2	10.5	5.4	−0.2	20.6
Site 3	10.8	6.1	−0.5	21.6
Site 4	10.5	6.7	−1.9	22.8
Site 5	11.9	7.6	−0.6	27.4
Site 6	11.7	7.1	−0.7	25.9
Site 7	11.2	6.7	−2.9	23.2
Site 8	11.5	6.0	−1.1	22.3
Site 9	11.8	6.5	−1.2	23.4
Site 10	11.3	7.1	−1.1	23.5
Site 11	12.5	6.2	1.2	27.3
Site 12	11.3	7.9	−1.3	27.4
Site 13	11.4	7.0	−1.3	23.7
Site 14	11.4	7.5	−1.1	26.4
Site 15	12.4	6.7	−1.1	24.1
Site 16	10.8	7.0	−3.0	26.7
Site 17	10.1	6.9	−1.4	21.5
Site 18	12.0	7.3	−0.9	24.8
Site 19	11.8	7.6	−1.1	26.3
Site 20	11.3	7.6	−2.0	23.8
Site 21	11.7	7.7	−1.2	25.8
Site 22	11.7	7.2	−0.6	22.3
Quarter 1	3.2	3.8	−3.0	14.0
Quarter 2	14.0	4.7	1.6	27.3
Quarter 3	19.2	2.2	12.5	27.4
Quarter 4	8.5	4.6	−1.4	22.8
January	1.4	2.3	−3.0	8.3
February	5.5	3.2	−2.6	14.0
March	4.6	2.5	−1.4	13.2
April	4.6	2.5	−1.4	13.2
May	9.2	3.3	1.6	23.4
June	16.1	2.3	9.0	24.2
July	18.0	2.5	10.3	27.4
August	19.3	2.1	12.8	26.3
September	19.6	1.9	13.2	27.1
October	18.2	2.1	12.5	26.7
November	7.4	2.7	1.0	15.3
December	4.9	2.2	−1.4	11.4
All Sites	11.4	6.9	−3.0	27.4

Table 4. Site by site comparison (i.e., pairwise comparison) of annual maximum daily Tw and dominant LULC percent (parenthetic) at twenty-one sites at West Run Watershed, West Virginia, USA, in 2018. Mix Dev represents the LULC type mixed development. All *p*-values displayed are significant at an $\alpha = 0.05$.

Site Comparison			*p*-Value (α = 0.05)
WRW 1 (22.2 °C) Mix Dev; (53.2%)	vs.	WRW 5 (27.4 °C) Forested; (51.1%)	0.02
WRW 1 (22.2 °C) Mix Dev; (53.2%)	vs.	WRW 11 (27.3 °C) Agriculture (41.9%)	0
WRW 1 (22.2 °C) Mix Dev; (53.2%)	vs.	WRW 12 (27.4) Forested; (34.5%)	0
WRW 1 (22.2 °C) Mix Dev; (53.2%)	vs.	WRW 15 (24.1 °C) Mix Dev (70.3%)	0.01
WRW 2 (22.2 °C) Forested; (74.2%)	vs.	WRW 5 (27.4 °C) Forested; (51.1%)	0
WRW 2 (20.6 °C) Forested; (74.2%)	vs.	WRW 6 (25.9 °C) Forested; (58.7%)	0
WRW 2 (20.6 °C) Forested; (74.2%)	vs.	WRW 9 (23.4 °C) Forested; (52.8%)	0.01
WRW 2 (20.6 °C) Forested; (74.2%)	vs.	WRW 11 (27.3 °C) Agriculture (41.9%)	0
WRW 2 (20.6 °C) Forested; (74.2%)	vs.	WRW 12 (27.4 °C) Forested; (34.5%)	0
WRW 2 (20.6 °C) Forested; (74.2%)	vs.	WRW 14 (26.4 °C) Forested (56.9%)	0
WRW 2 (20.6 °C) Forested; (74.2%)	vs.	WRW 15 (24.1 °C) Mix Dev (70.3%)	0
WRW 2 (20.6 °C) Forested; (74.2%)	vs.	WRW 18 (24.8 °C) Forested; (48.9%)	0
WRW 2 (20.6 °C) Forested; (74.2%)	vs.	WRW 19 (26.3 °C) Forested; (47.9%)	0
WRW 2 (20.6 °C) Forested; (74.2%)	vs.	WRW 21 (25.8 °C) Forested; (42.2%)	0.01
WRW 3 (21.6 °C) Forested; (61.3%)	vs.	WRW 5 (27.4 °C) Forested; (51.1%)	0.01
WRW 3 (21.6 °C) Forested; (61.3%)	vs.	WRW 11 (27.3 °C) Agriculture (41.9%)	0
WRW 3 (21.6 °C) Forested; (61.3%)	vs.	WRW 12 (27.4 °C) Forested; (34.5%)	0
WRW 3 (21.6 °C) Forested; (61.3%)	vs.	WRW 15 (24.1 °C) Mix Dev (70.3%)	0.01
WRW 4 (22.8 °C) Forested; (59%)	vs.	WRW 5 (27.4 °C) Forested; (51.1%)	0.01
WRW 4 (22.8 °C) Forested; (59%)	vs.	WRW 11 (27.3 °C) Agriculture (41.9%)	0
WRW 4 (22.8 °C) Forested; (59%)	vs.	WRW 12 (27.4 °C) Forested (34.5%)	0
WRW 4 (22.8 °C) Forested; (59%)	vs.	WRW 15 (24.1 °C) Forested (85.8%)	0.01
WRW 5 (27.4 °C) Forested; (51.1%)	vs.	WRW 17 (21.5 °C) Forested (85.8%)	0
WRW 6 (25.9 °C) Forested; (58.7%)	vs.	WRW 17 (21.5 °C) Forested (85.8%)	0.01
WRW 11 (27.3 °C) Agriculture (41.9%)	vs.	WRW 17 (21.5 °C) Forested (85.8%)	0
WRW 12 (27.4 °C) Forested; (34.5%)	vs.	WRW 17 (21.5 °C) Forested (85.8%)	0
WRW 14 (26.4 °C) Forested (56.9%)	vs.	WRW 17 (21.5 °C) Forested (85.8%)	0.01
WRW 15 (24.1 °C) Mix Dev (70.3%)	vs.	WRW 17 (21.5 °C) Forested (85.8%)	0
WRW 17 (21.5 °C) Forested (85.8%)	vs.	WRW 18 (24.8 °C) Forested; (48.9%)	0.01
WRW 17 (21.5 °C) Forested (85.8%)	vs.	WRW 19 (26.9 °C) Forested; (47.9%)	0.02
WRW 17 (21.5 °C) Forested (85.8%)	vs.	WRW 21 (25.8 °C) Forested; (42.2%)	0.02

3.3. *Quarterly Stream Water Temperature*

Seasonal Tw regimes showed that quarter 1 (1 January–31 March) had the lowest minimum Tw (−3.0 °C, site 16), with a mean of 3.2 °C, a maximum Tw of 14.0 °C (site 12), and a standard deviation of 3.8 °C (Table 3, Figure 4). Quarter 2 (1 April–30 June) included the highest standard deviation (4.7 °C), and a mean Tw of 14.6 °C, a maximum Tw of 27.3 °C (site 11), and a minimum Tw of 1.6 °C (site 17). Quarter 3 (1 July–30 September) had the highest mean Tw (19.2 °C) and maximum Tw (27.4 °C) (site 12), with a standard deviation of 2.2 °C, and a minimum of 12.5 °C (site 4). Results from the Kruskal–Wallis ANOVA showed that all 21 sites were significantly different from each other during one of the four quarters based on daily maximum Tw. The most significant differences were observed in quarter 3, with 130 significant differences (all, $p \leq 0.04$). Quarter 4 had three significant differences—the lowest number of returned significant differences of all four quarters. All three of the returned significant differences involved site 1 [site 4 (p = 0.01), 5 (p = 0.05), and 17 (p = 0.00)]. Significant Spearman correlation coefficients (α = 0.05) between LULC types and maximum Tw occurred during quarter 1, 3, and 4. Agriculture LULCs showed a significant positive correlation with maximum Tw during quarter 1 (rs = 0.5) (p = 0.03) and quarter 3 (rs = 0.5) (p = 0.01), specifically at sites 1, 11, 12, and 16. Forested LULCs showed a significant negative correlation with maximum Tw during quarter 3 (rs = −0.5) (p = 0.04), specifically at sites 2 and 17.

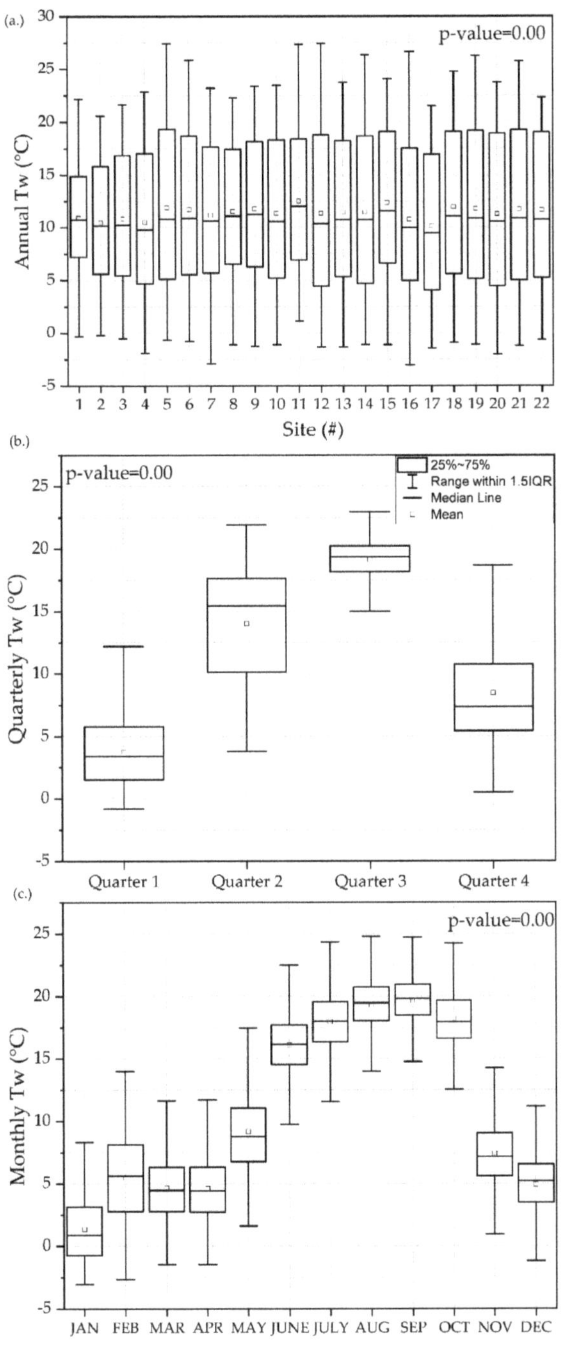

Figure 4. Five-minute water temperature data shown on a (**a**) site by site, (**b**) quarterly, and (**c**) monthly basis collected from 22 gauging sites in West Run Watershed, West Virginia, USA, in 2018.

3.4. Monthly Stream Water Temperature

January included the lowest minimum Tw (−3.0 °C) (site 16), with a standard deviation of 2.4 °C (all 21 sites), a mean Tw of 1.4 °C (All 21 sites), and a maximum of 8.3 °C (site 8) (Table 3). May included the highest standard deviation (3.3 °C), with a mean Tw of 21.8 °C, a maximum of 23.4 °C (site 12), and a minimum of 1.6 °C (site 17). July included the highest maximum Tw (27.4°C) (site 12), with a mean Tw of 16.1 °C, a minimum Tw of 10.3 °C (site 17), and a standard deviation of 2.3 °C. September included the highest mean Tw (19.6 °C), a minimum Tw of 13.2 °C (site 17), a maximum Tw of 27.1 °C (site 16), and a standard deviation of 1.9 °C (Table 3). LULC and Tw analysis (n = 21 sites) showed 12 significant ($\alpha = 0.05$) correlations with Tw variables. Results from the Kruskal–Wallis ANOVA showed that all 21 sites were significantly different from each other during at least one of the twelve months based on daily maximum Tw. The most significant differences were shown in July, with 108 significant differences ($p \leq 0.04$). Both February and October had no significant differences between sites. Significant Spearman correlation coefficients ($\alpha = 0.05$), between LULC types and maximum Tw, occurred during February, April, June, July, August, September, October, November, and December. Agriculture LULCs showed a significant positive correlation with maximum Tw during February (rs = 0.5) ($p = 0.03$), June (rs = 0.6) ($p = 0.03$), August, (rs = 0.7) ($p = 0.04$), September (rs = 0.7) ($p = 0.05$), November (rs = 0.8) ($p = 0.03$), and December (rs = 0.5) ($p = 0.05$), specifically at sites 1, 11, 12 and 16. Forested LULCs showed a significant negative correlation with maximum Tw during August (rs = −0.9) ($p = 0.01$), September (rs = −0.9) ($p = 0.00$), and December (rs = −0.5) ($p = 0.02$), specifically at sites 2 and 17 (all p values ≤ 0.04). Mixed-development LULCs showed a significant positive correlation with maximum Tw during August, (rs = 0.9) ($p = 0.01$), September (rs = 0.9) ($p = 0.00$), and November (rs = 0.6) ($p = 0.04$) at sites 1, 8, 12, 15, and 20.

3.5. PCAs

Annual timeseries comparisons between LULC areas and Tw variables at site 1–21 all showed two eigenvalues above the threshold of importance. If all variables (n = 4) exerted equal influence for the annual PCA, the values would be 0.5 (i.e., $\sqrt{(1/n)}$) [63]. For mean Tw annual PCA, PC1 had an eigenvalue of 2.0 and explained 50.4% of the variance and PC2 had an eigenvalue of 1.2 and explained 31.0% of the variance. PC1 was comprised of a positive relationship between mean Tw and mixed-development LULC and a negative relationship with forested LULC. PC2 was comprised of a positive relationship between agricultural LULC and mean Tw. For the minimum Tw annual PCA, PC1 had an eigenvalue of 1.8 and explained 45.1% of the variance and PC2 had an eigenvalue of 1.2 and explained 30.5% of the variance. PC1 was comprised of a negative correlation between minimum Tw and mixed-development LULC and a positive relationship with forested LULC. PC2 had a positive relationship between agricultural LULC and minimum Tw. For the maximum Tw annual PCA, PC1 had an eigenvalue of 1.8 and explained 45.8% of the variance and PC2 had an eigenvalue of 1.7 and explained 41.8% of the variance. PC1 was comprised of a positive relationship between maximum Tw and mixed-development LULC and a negative relationship with forested LULC. PC2 was comprised of a positive relationship between agricultural LULC and maximum Tw. Mean and maximum quarterly timeseries comparisons between LULC areas showed three eigenvalues above the threshold of importance. The mean Tw quarterly timeseries comparisons between LULC areas showed two eigenvalues above the threshold of importance. If all variables (n = 7) exerted equal influence, the eigenvalues would equal 0.38. For the mean Tw quarterly PCA, PC1 had an eigenvalue of 2.9 and explained 40.9% of the variance and PC2 had an eigenvalue of 2.0 and explained 28.4% of the variance. PC1 had a positive relationship between mixed-development LULC and quarter 1, and a negative relationship with forested LULC. PC2 was comprised of a positive relationship between mixed-development LULCs and quarters 2 and 3. For the minimum Tw quarterly PCA, PC1 had an eigenvalue of 2.5 and explained 35.3% of the variance and PC2 had an eigenvalue of 1.8 and explained 25.6% of the variance. PC1 showed a positive relationship between mixed-development LULC and quarter 1, 2 and 4, and a negative relationship with forested LULC. PC2 showed a positive relationship between mixed-development LULC and quarters 1 and 4, and a negative relationship with forested LULC. For the maximum Tw quarterly PCA, PC1 had an eigenvalue of 3.4 and explained 47.9% of the variance and PC2 had an eigenvalue of 1.8 and explained 26.3% of the variance. PC1 was comprised of a positive relationship between agricultural LULC and quarter 1, 2 and 4. PC2 was comprised of a positive relationship between mixed-development LULC and a negative relationship with forested LULC during quarter 4. Monthly timeseries comparison between LULC areas and Tw variables all showed four eigenvalues above the threshold of importance. If all variables (n = 15) exerted equal influence, the eigenvalues would equal 0.26. For the mean Tw monthly PCA, PC1 had an eigenvalue of 6.2 and explained 41.3% of the variance and PC2 had an eigenvalue of 5.5 and explained 36.8% of the variance. For the minimum Tw monthly PCA, PC1 had an eigenvalue of 6.4 and explained 42.9% of the variance and PC2 had an eigenvalue of 3.4 and explained 22.6% of the variance. For the maximum Tw monthly PCA, PC1 had an eigenvalue of 7.5 and explained 49.7% of the variance and PC2 had an eigenvalue of 2.7 and explained 17.8% of the variance. Extracted Eigenvector coefficients and plots for each monthly principle component analysis are included in the Appendix A (Tables A1–A3, Figures A1–A3).

4. Discussion

4.1. Climate during Study

Climate variables (e.g., Ta, precipitation, and net radiation) recorded in West Run Watershed in 2018 were average relative to historic climatic trends of West Virginia (i.e., 1900–2016) [54]. The average mean temperature (11.6 °C) differed only slightly (1.7%) from historic (1900–2016) averages observed in West Virginia from 1900 to 2016 (11.4 °C). During the study period, there was above average (20.5% higher) total precipitation relative to the historic (1900–2016) average (1096 mm) [54,64]. WRW did not include a dry season in 2018 and, a majority of the total precipitation fell during quarters 2 and 3 [46]. The overall climate in WRW during 2018 was predictably variable and consistent with historic climate trends (Figure 3) [46].

4.2. Stream Water Temperature

The highest maximum (27.4 °C) Tw was recorded during quarter 3, specifically for July. This is expected given the seasonal climate of WRW, which has the highest recorded Ta (34.6 °C) during July. While the highest Ta and Tw were recorded during quarter 3, the highest mean Ta was recorded during July, and the highest mean Tw was recorded during September (19.6 °C). In quarter 2, May had the highest Tw standard deviation (3.3 °C) (Figure 4) [64]. Further analysis into the Tw and Ta trends (Figure 3) showed that Tw followed but lagged behind Ta during May and across the entire study period [65–68]. This trend is constant with results of past studies that showed a strong relationship between Tw and Ta [39,66,67]. In May, Ta had a high standard deviation compared to other months (6.2 °C). The high Tw standard deviation in quarter 2 and May is likely attributable to the close relationship between Ta and Tw, as shown in many previous studies [38,39,66,67].

4.3. Stream Water Temperature LULC Relations

In general, results from both the PCA (Figure 5) and the Spearman rank correlation coefficient test (Table 5) showed that an increase in the proportion of forested LULC types is negatively correlated with all Tw variables, as confirmed in previous studies [14,22,65]. These results follow the same conclusions made by previous researchers that showed that forest harvest (e.g., clear cuts/canopy removal) increases Tw [6,14,16,22]. Additionally, although not surprising, during the winter season, a positive correlation was observed between the proportion of forest LULC and minimum Tw.

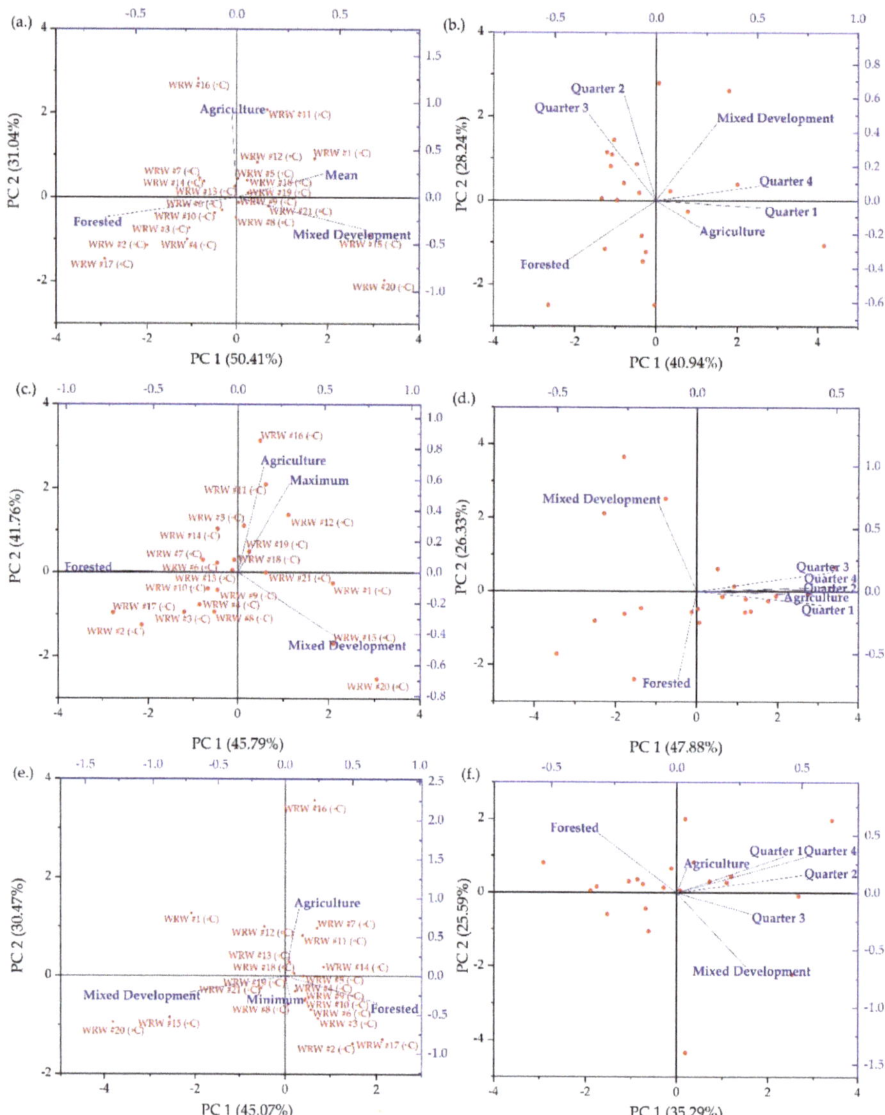

Figure 5. Results of the principle component analysis, showing biplots of the extracted principle components of annual water temperature data (mean (**a**), maximum (**c**), and minimum (**e**)), and biplots with extracted the principle components of quarterly water temperature data (mean (**b**), maximum (**d**), and minimum (**f**)) collected in 2018 at the 22 monitoring sites of West Run Watershed, WV, USA, and their corresponding LULC area (km^2).

Table 5. Spearman correlations between LULC types and the maximum Tw of all 21 sites, mainstem, and tributaries of West Run Watershed, West Virginia, USA, in 2018. Red (i.e., positive correlation) indicates an increase in Tw with increasing LULC and blue (i.e., negative correlation) indicates a decrease in Tw with increasing LULC. Bold values are significant correlations at $\alpha = 0.05$.

All Sites																
Max Correlation	**Quarter 1**	**Quarter 2**	**Quarter 3**	**Quarter 4**	**Jan**	**Feb**	**Mar**	**Apr**	**May**	**Jun**	**Jul**	**Aug**	**Sep**	**Oct**	**Nov**	**Dec**
Mixed Development	0.0	0.1	0.0	−0.1	0.2	0.1	0.0	0.0	0.2	0.1	0.1	0.2	0.2	−0.1	0.3	0.3
Agriculture	**0.5**	0.4	**0.5**	0.4	0.3	**0.5**	0.3	0.3	0.3	0.4	**0.5**	**0.5**	0.4	0.4	0.3	**0.5**
Forested	−0.1	−0.2	**−0.5**	−0.3	−0.3	−0.2	−0.3	−0.1	−0.3	−0.2	−0.3	−0.4	**−0.5**	−0.3	−0.4	**−0.5**
Mainstem																
Max Correlation	**Quarter 1**	**Quarter 2**	**Quarter 3**	**Quarter 4**	**Jan**	**Feb**	**Mar**	**Apr**	**May**	**Jun**	**Jul**	**Aug**	**Sep**	**Oct**	**Nov**	**Dec**
Mixed Development	0.0	0.4	0.5	0.4	−0.6	0.0	0.2	0.6	0.6	0.4	0.5	**0.9**	**0.9**	0.4	0.6	−0.1
Agriculture	0.5	0.3	0.5	**0.8**	−0.2	0.5	0.4	0.5	0.4	0.3	0.5	**0.7**	**0.7**	**0.8**	**0.8**	0.7
Forested	−0.3	−0.5	−0.6	−0.6	0.5	−0.3	−0.4	**−0.7**	−0.7	−0.5	−0.6	**−0.9**	**−0.9**	−0.6	−0.7	−0.2
Tributaries																
Max Correlation	**Quarter 1**	**Quarter 2**	**Quarter 3**	**Quarter 4**	**Jan**	**Feb**	**Mar**	**Apr**	**May**	**Jun**	**Jul**	**Aug**	**Sep**	**Oct**	**Nov**	**Dec**
Mixed Development	0.0	0.1	0.1	−0.2	0.2	0.1	−0.1	−0.3	0.3	0.3	0.4	0.3	0.3	0.1	**0.6**	0.4
Agriculture	0.5	0.3	0.4	0.3	−0.1	0.4	0.2	0.2	0.5	**0.6**	**0.6**	**0.6**	**0.6**	0.5	**0.6**	0.4
Forested	0.0	0.0	−0.4	−0.1	−0.4	0.4	0.2	0.4	0.3	0.5	0.3	0.2	0.1	0.3	0.4	0.2

Moore et al. [14] suggested that riparian vegetation insolates Tw by lowering convective heat loss to the above atmosphere, thereby cooling Tw in the summer and warming Tw in the winter via latent heat gain [15,23,68]. In the current work, positive correlations between maximum Tw and percent forest LULC were observed in specific tributaries (Figure 6), thus contradicting findings of previous research [6,65,68]. However, when sites 8 and 9 were removed from the data pool, correlations trends reverted to the expected negative correlation [6,65]. For sites 8 and 9 of the current work, the positive correlation may be attributed to the high proportion of directly adjacent mixed-development LULC types at both sites 8 and 9 prior to monitoring sites.

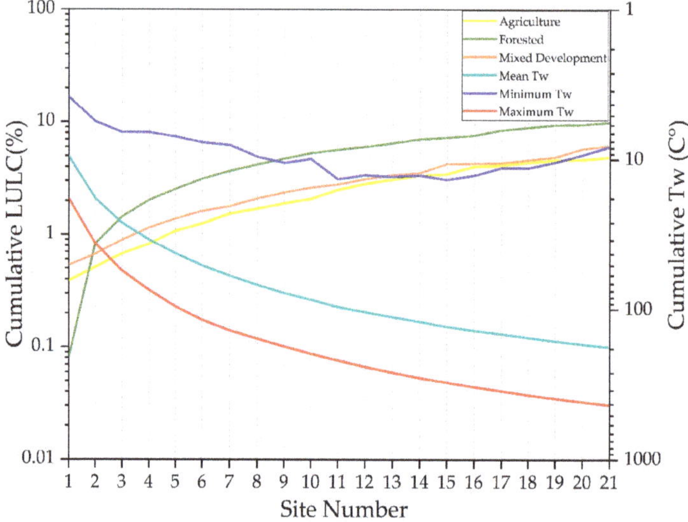

Figure 6. Average annual Tw variables (mean Tw, minimum Tw, and maximum Tw) vs. LULC types (forested, agriculture, and mixed development) moving from the headwaters to the terminus in West Run Watershed, West Virginia, USA, in 2018.

Results from both the PCA (Figure 5) analyses and the Spearman rank correlation coefficient test (Table 5) showed that mixed-development LULC types were significantly correlated with mean, minimum, and maximum Tw, with variable effects throughout the year [30–32,34,35,60,61]. While results of the mixed-development Tw analysis were similar to results of previous research, the current work provides important validation by means of the high number of sampling sites and high frequency sampling [32,34,47]. Interestingly, PCA results showed a negative correlation between minimum, maximum, and mean Tw during the winter and early spring months (i.e., January, February, March, April, November, and December), indicating overall lower Tw. These findings contradict results of other research analyzing the relationship between mixed-development LULC and Tw. Rice et al. [34] showed that mean Tw increased during the winter season in heavily urbanized catchments of Boone, North Carolina. Alternatively, to Rice et al. [34], lower Tw correlated with mixed development during cooler months (i.e., January, February, March, April, November, and December) could be explained by runoff from impervious surfaces, which during these months is often lower in temperature than the Tw of surrounding streams. Qun et al. [69] showed a negative correlation between urban impervious surface and land surface temperatures (i.e., lower surface temperatures) during winter daytimes. Conflicting results might be further explained by the complex physiographic mosaic of the study watershed in which mixed development LULC types are broken up by other forested and agricultural LULC types at varying relative positions on the landscape, thus leaving room for future investigations.

Agricultural LULC types had a positive Spearman correlation, with maximum Tw in every quarter and month, except January (Table 5). These findings may be, at least in part, due to the removal of riparian vegetation, increased subsurface lateral flow rates through drainage tiles, and/or increased soil shortwave radiation exposure during periods when fields are bare as per findings of previous literature [27,70]. Interestingly, a relationship appeared between maximum Tw and crop absence. Typically, in the study watershed, crops are planted in spring and harvested in middle to late July following the observed trend in Table 5. Therefore, the positive relationship between crop absence and maximum Tw could be explained, at least in part, by increased shortwave interception and reduce runoff volumes through evapotranspiration and interception of precipitation by crops [15,23]. Conversely, after harvest, the relationship becomes significant because runoff volumes increase and the soil is exposed to greater amounts of shortwave radiation thereby reaching higher temperatures [15,27,29]. Furthermore, during these periods, water contacting or infiltrating into the soil is heated and transported to surrounding streams via runoff or subsurface lateral flows [27]. Younus et al. [27] found that drainage tiles exacerbated transport of heated subsurface lateral flow into surrounding streams during irrigation or precipitation events. The agricultural fields in the current study watershed have drainage tiles installed. Thus, subsurface lateral flows to West Run Creek may be increased, further explaining the significant relationship between maximum Tw and agricultural LULC types.

4.4. LULC Tw Tipping Points

Forested LULC types influenced Tw most significantly in the current investigation (Figure 6). As the percent of forested LULC types decreased below 74.2%, associated maximum Tw began to increase (thus a potential tipping point). As the percent of forested LULC decreased below 61.1% mean Tw began to either increase or decrease depending on the time of year. Below 52.2% forested LULC, the minimum Tw of associated streams began to increase or decrease again depending on the time of year. Mixed-development LULC types had next greatest impacts on Tw.

As the percent of mixed-development LULC types increased above 14%, associated mean Tw increased (thus a potential mixed-development tipping point). Above 24.1% mixed development, maximum Tw began to either increase or decrease depending on the time of year, and above 26.8% mixed development, the minimum Tw of streams began to increase or decrease (depending on the time of year). Agricultural LULC types also influenced Tw significantly. As the percent of agriculture LULC types increased above 14.9%, associated maximum Tw began to increase. Above 16.0% agriculture LULC, mean Tw began to increase, and above 26.4% agriculture LULC, the minimum stream Tw began to increase or decrease (time of year dependent).

4.5. Study Implications and Future Directions

Implementing the nested experimental watershed study design coupled with the high-temporal and spatial-sampling regime used in this work allowed for a more comprehensive evaluation of surrounding LULC effects on associated Tw than is normally found in the primary literature. Forested LULC types were associated with overall lower Tw, whereas both mixed-development and agriculture LULC types had higher overall Tw. The finding of LULC tipping points (thresholds) for all three of the analyzed LULC classifications provides valuable information for both land managers and policy makers. For example, Tw tipping points emerged as forested LULC types dropped below 74.2%, due to the conversion to mixed development or agricultural LULC types. These tipping points can be used to guide management decisions in terms of development limits. Given the importance of maximum Tw for stream biota, a preliminary analysis of thermal surges was conducted. Select summer precipitation events were assessed as per Zeiger and Hubbart [32], Rice et al. [34], and Anderson et al. [71], where a Tw surge is defined as a greater than 1.0 °C increase within a 15 min time interval. Figure 7 is a stacked comparison of Tw on the y-axis and time on the x-axis showing time lag from the related precipitation event. Successive sites were added to Figure 7, showing the thermal plume moving through the watershed and eventually dissipating. Although this analysis was preliminary, it showed the existence of these events, thus conveying the need for future research into thermal surge dynamics. Future investigations may benefit from additional monitoring years to better understand the importance of climate and antecedent conditions with regard to Tw processes. In addition, future studies could focus on minimum Tw and mean Tw, and perhaps Tw variance, to understand relationships between land use, climate change and Tw regimes.

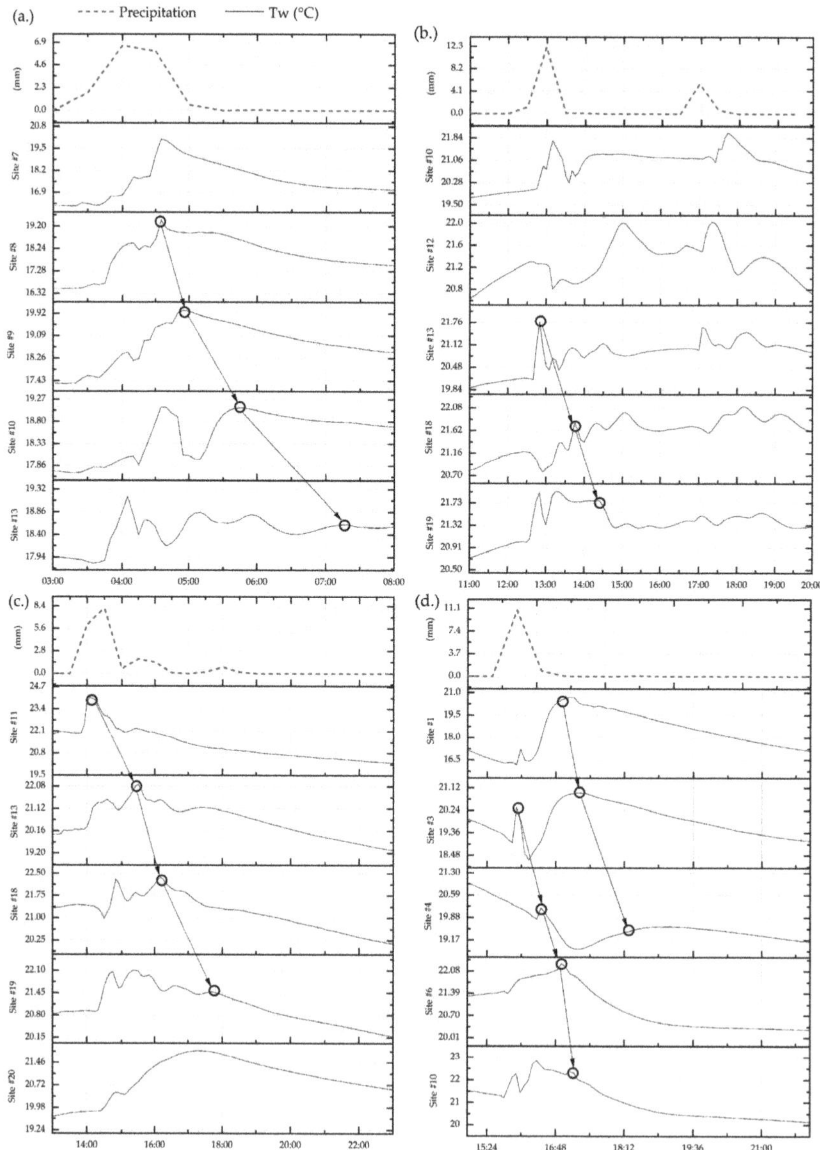

Figure 7. Stream water temperature surges sensed at sites in West Run Creek following a summer precipitation event during the summer of 2018. Black circles mark the peak Tw surges and arrows track the Tw surge as it moves downstream. (**a**) Stream water temperature surge one occurring on 06/17 at

04:15 (15.1 mm precipitation event) moving through West Run Creek. (**b**) Stream water temperature surge two occurring on 07/25 at 12:00 (13.5 mm precipitation event) moving through West Run Creek. (**c**) Stream water temperature surge three occurring on 07/16 at 13:00 (19.4 mm precipitation event) moving through West Run Creek. (**d**) Stream water temperature surge four (left) occurring on 08/16 at 15:55 (11.7 mm precipitation event) and stream water temperature surge five (right) occurring on 08/16 at 16:15 (11.7 mm precipitation).

5. Conclusions

No previous research investigating stream water temperature (Tw) in contemporary watersheds has used such a high-temporal and spatial-sampling regime as that included in the current investigation. Implementation of the experimental design used in the current research is necessary to provide both validation for previous results and the discovery of temporal variation in LULC characteristics influencing Tw. In the current work, the relationship between LULC types and Tw was investigated in the Appalachian region of the eastern United States. The analysis used five-minute Tw timeseries data collected at 22-site nested sites using an experimental watershed study design. Results indicated that LULC has varying effects on Tw both spatially and temporally. PCA results showed that forested LULC types typically lowered maximum and mean Tw, particularly in the late summer months, whereas Spearman correlation results showed significant ($p = 0.01$) negative correlations with maximum Tw (−0.9) during August and September. PCA results indicated that mixed-development LULC types typically increased maximum and mean Tw during the summer months, whereas Spearman correlation results showed significant ($p = 0.00$) positive correlation with maximum Tw (0.9) during August and September. PCA results showed that agriculture LULC types were correlated with maximum Tw in every month except January. Although results are specific to the study watershed, the finding of tipping points shows LULC thresholds that, when exceeded, may begin to impact associated Tw. These relationships likely exist in all watersheds, particularly contemporary (municipal) watersheds. Given the impact that maximum Tw can have on stream ecosystems, a preliminary investigation of thermal surge events was conducted. This investigation showed thermal surges in the study watershed, and therefore presents future research opportunities into the investigation of Tw surge dynamics. Both results and findings of this study will advance the decision-making success of land managers and policy makers concerned with the health of aquatic ecosystems. In particular, the high-resolution (n = 22) study design presented in this work facilitates identification of upland mitigation sites and corresponding greater certainty in fiscal investment outcomes.

Author Contributions: Conceptualization, J.A.H.; methodology, J.A.H.; formal analysis, J.P.H. and J.A.H.; investigation, J.A.H. and J.P.H.; resources, J.A.H.; data curation, J.A.H.; writing—original draft preparation, J.A.H. and J.P.H.; writing—review and editing, J.A.H. and J.P.H.; visualization, J.A.H. and J.P.H.; supervision, J.A.H.; project administration, J.A.H.; funding acquisition, J.A.H. The authors declare no conflict of interest for the current work. All authors have read and agreed to the published version of the manuscript.

Funding: This work was supported by the National Science Foundation under Award Number OIA−1458952, the USDA National Institute of Food and Agriculture, Hatch project accession number 1011536, and the West Virginia Agricultural and Forestry Experiment Station. Results presented may not reflect the views of the sponsors and no official endorsement should be inferred. The funders had no role in study design, data collection and analysis, decision to publish, or preparation of the manuscript.

Acknowledgments: Special thanks are due to the many scientists of the Interdisciplinary Hydrology Laboratory (https://www.researchgate.net/lab/The-Interdisciplinary-Hydrology-Laboratory-Jason-A-Hubbart), and the Institute of Water Security and Science (https://iwss.wvu.edu/). The authors also appreciate the feedback of anonymous reviewers whose constructive comments improved the article.

Conflicts of Interest: The authors declare no conflict of interest for the current work.

Appendix A

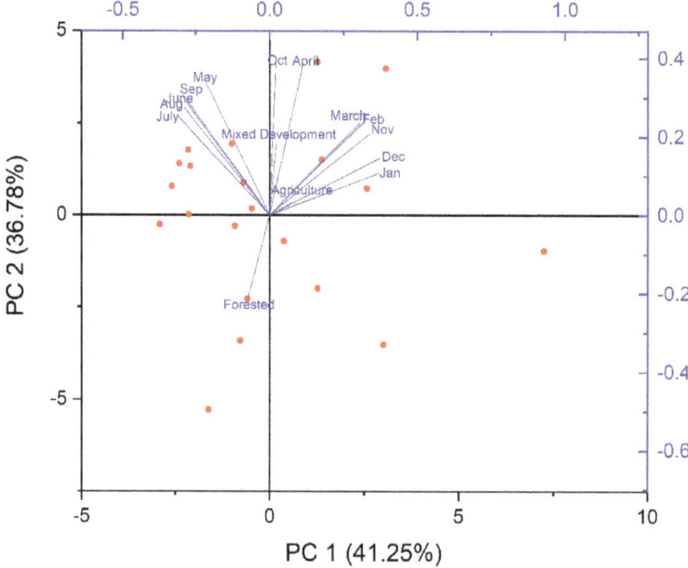

Figure A1. Results of the principle component analysis, showing biplots of the extracted principle components of mean monthly water temperature data collected in 2018 at the 22 monitoring sites of West Run Watershed, WV, USA, and their corresponding LULC area (km^2).

Table A1. Coefficients of principal components comprising 15 variables used to define 15 principal components of mean monthly water temperature data in 2018 in West Run Watershed, WV, USA.

Variables	Coefficients of PC1	Coefficients of PC2	Coefficients of PC3	Coefficients of PC4
Jan	0.37	0.11	0.04	0.07
Feb	0.31	0.24	0.14	0.13
March	0.32	0.24	0.06	0.04
April	0.12	0.39	0.12	0.08
May	−0.21	0.34	−0.03	0.08
June	−0.28	0.29	0.02	0.04
July	−0.31	0.25	0.15	−0.03
Aug	−0.29	0.27	0.15	−0.05
Sep	−0.28	0.29	0.03	−0.04
Oct	0.02	0.39	0.08	0.14
Nov	0.34	0.21	0.02	0.04
Dec	0.37	0.15	0.05	−0.01
Mixed Development	0.02	0.19	−0.72	−0.02
Agriculture	0.08	0.03	0.39	−0.80
Forested	−0.08	−0.22	0.49	0.55

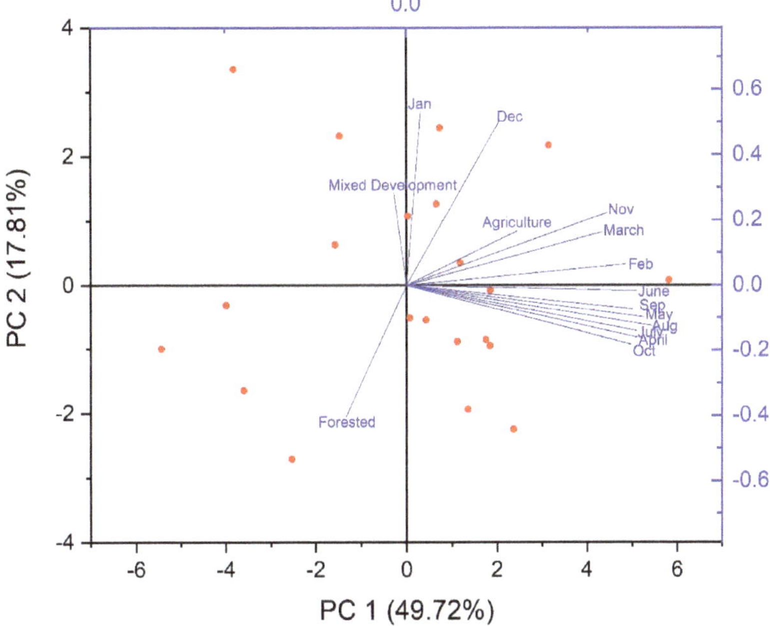

Figure A2. Results of the principle component analysis, showing biplots of the extracted principle components of maximum monthly water temperature data collected in 2018 at the 22 monitoring sites of West Run Watershed, WV, USA, and their corresponding LULC area (km^2).

Table A2. Coefficients of principal components comprising 15 variables used to define 15 principal components of maximum monthly water temperature data in 2018 in West Run Watershed, WV, USA.

Variables	Coefficients of PC1	Coefficients of PC2	Coefficients of PC3	Coefficients of PC4
Jan	0.02	0.53	0.23	0.10
Feb	0.30	0.06	0.29	0.13
March	0.27	0.16	0.15	0.02
April	0.32	−0.16	0.12	0.16
May	0.33	−0.10	−0.18	0.08
June	0.32	−0.02	0.02	0.24
July	0.32	−0.14	−0.13	0.00
Aug	0.34	−0.12	−0.17	−0.06
Sep	0.31	−0.07	−0.17	−0.29
Oct	0.31	−0.18	−0.06	−0.16
Nov	0.28	0.22	0.05	0.33
Dec	0.13	0.50	0.16	0.08
Mixed Development	−0.02	0.28	−0.63	0.17
Agriculture	0.15	0.17	0.29	−0.73
Forested	−0.08	−0.40	0.46	0.30

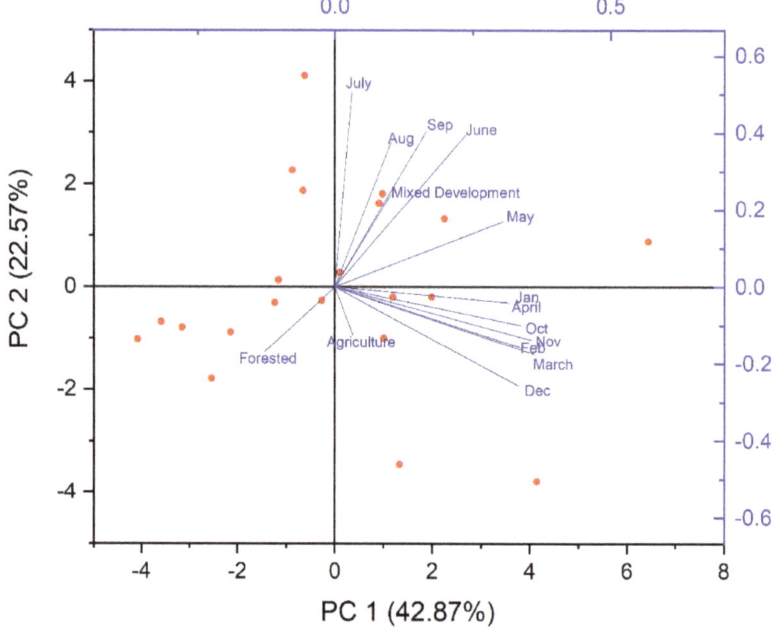

Figure A3. Results of the principle component analysis, showing biplots of the extracted principle components of minimum monthly water temperature data collected in 2018 at the 22 monitoring sites of West Run Watershed, WV, USA, and their corresponding LULC area (km^2).

Table A3. Coefficients of principal components comprising 15 variables used to define 15 principal components of minimum monthly water temperature data in 2018 in West Run Watershed, WV, USA.

Variables	Coefficients of PC1	Coefficients of PC2	Coefficients of PC3	Coefficients of PC4
Jan	0.30	−0.04	0.26	−0.22
Feb	0.34	−0.16	0.07	−0.04
March	0.36	−0.17	−0.08	−0.01
April	0.32	−0.04	0.00	−0.03
May	0.31	0.17	−0.03	−0.08
June	0.24	0.40	0.02	−0.10
July	0.03	0.51	0.14	0.10
Aug	0.10	0.38	0.31	0.27
Sep	0.17	0.40	0.16	0.01
Oct	0.34	−0.10	0.01	0.08
Nov	0.36	−0.14	0.09	0.00
Dec	0.33	−0.26	−0.09	0.09
Mixed Development	0.10	0.24	−0.62	−0.20
Agriculture	0.03	−0.12	0.06	0.82
Forested	−0.13	−0.17	0.61	−0.34

References

1. Webb, B.W.; Hannah, D.M.; Moore, R.D.; Brown, L.E.; Nobilis, F. Recent advances in stream and river temperature research. *Hydrol. Process.* **2008**, *918*, 902–918. [CrossRef]
2. Caissie, D. The thermal regime of rivers: A review. *Freshw. Biol.* **2006**, *51*, 1389–1406. [CrossRef]
3. Webb, B.W. Trends in stream and river temperature. *Hydrol. Process.* **1996**, *10*, 205–226. [CrossRef]

4. Coutant, C.C. *Perspectives on Temperature in the Pacific Northwest's Fresh Waters*; Oak Ridge National Laboratory: Oak Ridge, TN, USA, 1999.
5. Winder, M.; Sommer, U. Phytoplankton response to a changing climate. *Hydrobiologia* **2012**, *698*, 5–16. [CrossRef]
6. Pollock, M.M.; Beechie, T.J.; Liermann, M.; Bigley, E.R. Stream Temperature Relationships to Forest Harvest in the Olympic Peninsula. *Fish. Sci.* **2009**, *45*, 1–33.
7. Petty, J.T.; Thorne, D.; Huntsman, B.M.; Mazik, P.M. The temperature-productivity squeeze: Constraints on brook trout growth along an Appalachian river continuum. *Hydrobiologia* **2014**, *727*, 151–166. [CrossRef]
8. Martin, R.W.; Petty, J.T. Local stream temperature and drainage network topology interact to influence the distribution of smallmouth bass and brook trout in a Central appalachian watershed. *J. Freshw. Ecol.* **2009**, *24*, 497–508. [CrossRef]
9. Petty, J.T.; Hansbarger, J.L.; Huntsman, B.M.; Mazik, P.M. Brook trout movement in response to temperature, flow, and thermal refugia within a complex Appalachian riverscape. *Trans. Am. Fish. Soc.* **2012**, *141*, 1060–1073. [CrossRef]
10. Poole, G.C.; Berman, C.H. An ecological perspective on in-stream temperature: Natural heat dynamics and mechanisms of human-caused thermal degradation. *Environ. Manag.* **2001**, *27*, 787–802. [CrossRef]
11. Borman, M.; Larson, L. A case study of river temperature response to agricultural land use and environmental thermal patterns. *J. Soil Water Conserv.* **2003**, *58*, 8–12.
12. Kellner, E.; Hubbart, J.; Stephan, K.; Freedman, Z.; Kutta, E.; Kelly, C.; Morrissey, E. Characterization of sub-watershed-scale stream chemistry regimes in an Appalachian mixed-land-use watershed. *Environ. Monit. Assess.* **2018**, *190*. [CrossRef] [PubMed]
13. Mustafa, M.; Barnhart, B.; Babbar-Sebens, M.; Ficklin, D. Modeling landscape change effects on stream temperature using the Soil and Water Assessment Tool. *Water Switz.* **2018**, *10*, 1–18. [CrossRef]
14. Moore, R.D.; Spittlehouse, D.L.; Story, A. Riparian microclimate and stream temperature response to forest harvesting: A review. *J. Am. Water Resour. Assoc.* **2005**, *7*, 813–834. [CrossRef]
15. Oke, T.R. *Boundary Layer Climates*, 2nd ed.; Routledge: London, UK, 1987; ISBN 0-203-71545-4.
16. Gravelle, J.A.; Link, T.E. Influence of timber harvesting on headwater peak stream temperatures in a northern Idaho watershed. (Special issue: Science and management of forest headwater streams). *For. Sci.* **2007**, *53*, 189–205.
17. Webb, B.W.; Zhang, Y. Water temperatures and heat budgets in Dorset chalk water courses. *Hydrol. Process.* **1999**, *321*, 309–321. [CrossRef]
18. Webb, B.W.; Zhang, Y. Spatial and Seasonal Variablility in the Componeets of The River Heat Budget. *Hydrol. Process.* **1997**, *11*, 79–101. [CrossRef]
19. Bulliner, E.; Hubbart, J.A. An improved hemispherical photography model for stream surface shortwave radiation estimations in a central U.S. hardwood forest. *Hydrol. Process.* **2013**, *27*, 3885–3895. [CrossRef]
20. Imhoff, M.L.; Zhang, P.; Wolfe, R.E.; Bounoua, L. Remote sensing of the urban heat island effect across biomes in the continental USA. *Remote Sens. Environ.* **2010**, *114*, 504–513. [CrossRef]
21. Johnson, R.K.; Almlöf, K. Adapting boreal streams to climate change: Effects of riparian vegetation on water temperature and biological assemblages. *Freshw. Sci.* **2016**, *35*, 984–997. [CrossRef]
22. Johnson, S.L.; Jones, J.A. Stream temperature responses to forest harvest and debris flows in western Cascades, Oregon. *Can. J. Fish. Aquat. Sci.* **2011**, *57*, 30–39. [CrossRef]
23. Xuhui, L. *Fundamentals of Boundary-Layer Meteorology*; Springer: Gewerbestrasse, Switzerland, 2018; ISBN 978-3-319-60851-8.
24. Boyde, C. *Water Quality: An Introduction*; Springer: Gewerbestrasse, Switzerland, 2016; ISBN 978-3-319-70548-4.
25. Petersen, F.; Hubbart, J.A.; Kellner, E.; Kutta, E. Land-use-mediated Escherichia coli concentrations in a contemporary Appalachian watershed. *Environ. Earth Sci.* **2018**, *77*, 1–13. [CrossRef]
26. Wiebe, K.D.; Gollehon, N.R. *Agricultural Resources and Environmental Indicators*; Nova Publishers: Hauppauge, NY, USA, 2007; ISBN 978-1-60021-467-7.
27. Younus, M.; Hondzo, M.; Engel, B.A. Stream Temperature Dynamics in Upland Agricultural Watersheds. *J. Environ. Eng.* **2000**, *126*, 518–526. [CrossRef]
28. Campbell, G.S.; Norman, J.M. *Introduction to Environmental Biophysics*, 2nd ed.; Springer: New York, NY, USA, 1998; ISBN 978-0-387-94937-6.

29. Hatfield, J.L.; Sauer, T.J.; Prueger, J.H. Managing Soils to Achieve Greater Water Use Efficiency: A Review. *Agron. J.* **2001**, *93*, 10. [CrossRef]
30. Kinouchi, T.; Yagi, H.; Miyamoto, M. Increase in stream temperature related to anthropogenic heat input from urban wastewater. *J. Hydrol.* **2007**, *335*, 78–88. [CrossRef]
31. Palmer, M.A.; Nelson, K.C. Stream Temperature Surges under Urbanization. *J. Am. Water Resour. Assoc.* **2007**, *43*, 440–452. [CrossRef]
32. Zeiger, S.J.; Hubbart, J.A. Urban Stormwater Temperature Surges: A Central US Watershed Study. *Hydrology* **2015**, 193–209. [CrossRef]
33. Herb, W.; Janke, B.; Mohseni, O.; Stefan, H. Thermal pollution of streams by runoff from paved surfaces. *Hydrol. Process.* **2008**, *22*, 987–999. [CrossRef]
34. Rice, J.S.; Anderson, W.P., Jr.; Thaxton, C.S. Urbanization influences on stream temperature behavior within low-discharge headwater streams. *Hydrol. Res. Lett.* **2011**, *5*, 27–31. [CrossRef]
35. Paul, M.J.; Meyer, J.L. Stream in The Urban Landscape. *Annu. Rev. Ecol. Syst.* **2001**, *32*, 333–365. [CrossRef]
36. Menberg, K.; Blum, P.; Schaffitel, A.; Bayer, P. Long-Term Evolution of Anthropogenic Heat Fluxes into a Subsurface Urban Heat Island. *Environ. Sci. Technol.* **2013**, *47*, 9747–9755. [CrossRef]
37. Sinokrot, B.A.; Stefan, H.G. Stream temperature dynamics: Measurements and modeling. *Water Resour. Res.* **1993**, *29*, 2299–2312. [CrossRef]
38. O'Driscoll, M.A.; DeWalle, D.R. Stream-air temperature relations to classify stream-ground water interactions in a karst setting, central Pennsylvania, USA. *J. Hydrol.* **2006**, *329*, 140–153. [CrossRef]
39. Webb, B.W.; Nobilis, F. Long-term perspective on the nature of the water-air temperature relationship—A case study. *Hydrol. Process.* **1997**, *11*, 137–147. [CrossRef]
40. Gu, R.; McCutcheon, S.; Chen, C.J. Development of weather-dependent flow requirements for river temperature control. *Environ. Manag.* **1999**, *24*, 529–540. [CrossRef] [PubMed]
41. Raitz, K. *Appalachia: A Regional Geography: Land, People, and Development*; Routledge: Abingdon, UK, 2019; ISBN 978-0-429-72421-3.
42. Zajíček, A.; Kvítek, T.; Kaplická, M.; Doležal, F.; Kulhavý, Z.; Bystřický, V.; Žlábek, P. Drainage water temperature as a basis for verifying drainage runoff composition on slopes. *Hydrol. Process.* **2011**, *25*, 3204–3215. [CrossRef]
43. Houston, D.; Werritty, A.; Bassett, D.; Geddes, A.; Hoolachan, A.; McMillan, M. *Pluvial (Rain-Related) Flooding in Urban Areas: The Invisible Hazard*; Joseph Rowntree Foundation: London, UK, 2011.
44. LeBlanc, R.T.; Brown, R.D.; FitzGibbon, J.E. Modeling the effects of land use change on the water temperature in unregulated urban streams. *J. Environ. Manag.* **1997**, *49*, 445–469. [CrossRef]
45. Petersen, F.; Hubbart, J.A. Quantifying Escherichia coli and Suspended Particulate Matter Concentrations in a Mixed-Land Use Appalachian Watershed. *Water* **2020**, *12*, 532. [CrossRef]
46. Petersen, F.; Hubbart, J.A. Advancing Understanding of Land Use and Physicochemical Impacts on Fecal Contamination in Mixed-Land-Use Watersheds. *Water* **2020**, *12*, 1094. [CrossRef]
47. Zeiger, S.; Hubbart, J.A.; Anderson, S.H.; Stambaugh, M.C. Quantifying and modelling urban stream temperature : A central US watershed study. *Hydrol. Prochydrol. Process.* **2016**, *514*, 503–514. [CrossRef]
48. Hubbart, J.A.; Kellner, E.; Zeiger, S.J. A Case-Study Application of the Experimental Watershed Study Design to Advance Adaptive Management of Contemporary Watersheds. *Water* **2019**, *11*, 2355. [CrossRef]
49. Hewlett, J.D.; Lull, H.W.; Reinhart, K.G. In Defense of Experimental Watersheds. *Water Resour. Res.* **1969**, *5*, 306–316. [CrossRef]
50. Tetzlaff, D.; Carey, S.K.; McNamara, J.P.; Laudon, H.; Soulsby, C. The essential value of long-term experimental data for hydrology and water management: Long-term data in hydrology. *Water Resour. Res.* **2017**, *53*, 2598–2604. [CrossRef]
51. Kellner, E.; Hubbart, J.A. Application of the experimental watershed approach to advance urban watershed precipitation/discharge understanding. *Urban. Ecosyst.* **2017**, *20*, 799–810. [CrossRef]
52. The West Virginia Water Research Institute and The West Run Watershed Association. 2008; Watershed Based Plan for West Run. West Virginia Water Res. Inst. Available online: https://dep.wv.gov/WWE/Programs/nonptsource/WBP/Documents/WP/WestRun_WBP.pdf (accessed on 13 April 2020).
53. Arguez, A.; Durre, I.; Applequist, S.; Vose, R.S.; Squires, M.F.; Yin, X.; Heim, R.R.; Owen, T.W. NOAA's 1981–2010 U.S. Climate Normals: An Overview. *Bull. Am. Meteorol. Soc.* **2012**, *93*, 1687–1697. [CrossRef]

54. Kutta, E.; Hubbart, J. Climatic Trends of West Virginia: A Representative Appalachian Microcosm. *Water* **2019**, *11*, 1117. [CrossRef]
55. Kutta, E.; Hubbart, J. Observed Mesoscale Hydroclimate Variability of North America's Allegheny Mountains at 40.2° N. *Climate* **2019**, *7*, 1–24. [CrossRef]
56. Kutta, E.; Hubbart, J.A. Reconsidering meteorological seasons in a changing climate. *Clim. Change* **2016**, *137*, 511–524. [CrossRef]
57. United States Census Bureau. Available online: https://2020census.gov/?cid=23745:united%20states%20census%20bureau:sem.ga:p:dm:en:&utm_source=sem.ga&utm_medium=p&utm_campaign=dm:en&utm_content=23745&utm_term=united%20states%20census%20bureau (accessed on 12 April 2020).
58. Grime, J.P. Competitive Exclusion in Herbaceous Vegetation. *Nature* **1973**, *242*, 344–347. [CrossRef]
59. Grime, J.P. Benefits of plant diversity to ecosystems: Immediate, filter and founder effects. *J. Ecol.* **1998**, *86*, 902–910. [CrossRef]
60. Booth, D.B.; Roy, A.H.; Smith, B.; Capps, K.A. Global perspectives on the urban stream syndrome. *Freshw. Sci.* **2016**, *35*, 412–420. [CrossRef]
61. Walsh, C.J.; Roy, A.H.; Feminella, J.W.; Cottingham, P.D.; Groffman, P.M.; Morgan, R.P. The urban stream syndrome: Current knowledge and the search for a cure. *J. N. Am. Benthol. Soc.* **2005**, *24*, 706–723. [CrossRef]
62. Helsel, D.R.; Hirsch, R.M. Statistical methods in water resources. *Stat. Methods Water Resour.* **1992**. [CrossRef]
63. Bro, R.; Smilde, A.K. Principal component analysis. *Anal. Methods* **2014**, *6*, 2812–2831. [CrossRef]
64. Jay, A.; Avery, C.; Barrie, D.; Apurva, D.; DeAngelo Dzaugis, B.; Kolian, M.; Lewis, K.; Reeves, K.; Winner, D. *Fourth National Climate Assessment*; U.S. Government Publishing Office: Washington, DC, USA, 2018; pp. 33–71.
65. Beschta, R.L.; Taylor, R.L. Stream Temperature Increases and Land Use in a Forested Oregon Watershed. *JAWRA J. Am. Water Resour. Assoc.* **1988**, *24*, 19–25. [CrossRef]
66. Stefan, H.G.; Preud'homme, E.B. Stream Temperature Estimation from Air Temperature. *JAWRA J. Am. Water Resour. Assoc.* **1993**, *29*, 27–45. [CrossRef]
67. Mohseni, O.; Stefan, H.G. Stream temperature/air temperature relationship: A physical interpretation. *J. Hydrol.* **1999**, *218*, 128–141. [CrossRef]
68. Macedo, M.N.; Coe, M.T.; DeFries, R.; Uriarte, M.; Brando, P.M.; Neill, C.; Walker, W.S. Land-use-driven stream warming in southeastern Amazonia. *Philos. Trans. R. Soc. B Biol. Sci.* **2013**, *368*. [CrossRef]
69. Ma, Q.; Wu, J.; He, C. A hierarchical analysis of the relationship between urban impervious surfaces and land surface temperatures: Spatial scale dependence, temporal variations, and bioclimatic modulation. *Landsc. Ecol.* **2016**, *31*, 1139–1153. [CrossRef]
70. Story, A.; Moore, R.D.; Macdonald, J.S. Stream temperatures in two shaded reaches below cutblocks and logging roads: Downstream cooling linked to subsurface hydrology. *Can. J. For. Res.* **2003**, *33*, 1383–1396. [CrossRef]
71. Anderson, W.P.; Anderson, J.L.; Thaxton, C.S.; Babyak, C.M. Changes in stream temperatures in response to restoration of groundwater discharge and solar heating in a culverted, urban stream. *J. Hydrol.* **2010**, *393*, 309–320. [CrossRef]

Article

Resilient Urban Water Services for the 21th Century Society—Stakeholder Survey in Finland

Jyrki Laitinen [1,*], Johanna Kallio [2], Tapio S. Katko [3], Jarmo J. Hukka [3] and Petri Juuti [3]

1 Finnish Environment Institute SYKE, Latokartanonkaari 11, FI-00790 Helsinki, Finland
2 Centre for Economic Development, Transport and the Environment for Southeast Finland, FI-45100 Kouvola, Finland; johanna.kallio@ely-keskus.fi
3 Tapio Katko, Petri Juuti and Jarmo Hukka, Tampere University, Faculty of Built Environment, FI-33720 Tampere, Finland; tapio.katko@tuni.fi (T.S.K.); omraj@jarmohukka.fi (J.J.H.); petri.juuti@tuni.fi (P.J.)
* Correspondence: jyrki.laitinen@ymparisto.fi; Tel.: +358-295-251-346

Received: 21 November 2019; Accepted: 4 January 2020; Published: 9 January 2020

Abstract: Resilience has become a vital theme in the discussion concerning urban water services. Resilience in this context can be defined as both keeping up a good level of services, as well as rapid and fluent recovery from failures caused by natural disasters, unsound infrastructure or incorrect management. Although adequate water services resilience can be considered as sustainable, resilience is a wider concept than sustainability. In order to call water services resilient, all sections from policy and management to technical operation should be clear and coherent, and their operation in challenging situations also must be guaranteed. This study seeks a resilient approach to water services through a literature review, and a questionnaire to stakeholders; mainly water supply and sanitation experts. The results show that sufficient technology and good water quality are not sufficient for achieving resilient water services, but also education and institutional management are essential issues. These are accomplished by a methodical education system, capacity building, and good governance.

Keywords: good governance; sanitation; sustainability; water supply

---

1. Introduction

Water services—water supply and sanitation in this context—are essential services for human welfare. Yet, these services are not always organized and operated in an adequately planned and controlled mode of operation, even in many developed countries. Especially in urban areas, systems are vulnerable to internal or external disturbance, which might cause severe health, environmental and economic challenges for communities. These kinds of disturbance can include, for example, technical and economic problems (internal) or changes in environment or policy (external). Climate change is one example in external disturbances affecting raw water resources and wastewater management.

To maintain continuous and acceptable water services, decision makers, public servants, and experts have to be aware of the requirements of the community for adequate operation of the water systems utility. The requirements concern not only water utilities, but the whole process that affects the urban water cycle. Special challenges are faced due to disasters and climate change impacts. To prepare for recovering from these situations as soon as possible, and for providing uninterrupted water services, the water utility must become more resilient. There are several different issues and sectors that must be robust as prerequisites for considering a community's comprehensive water services resilient. Resilience in water services is in this study considered to include (a) ability to operate continuously and resist disturbances and (b) ability to recover after failures.

Applying integrated water resources management (IWRM) Koop and Leeuwen (2015) analyzed 45 municipalities in 27 countries using the improved city blueprint framework (CBF) [1]. They categorized five different levels of sustainability of urban IWRM, (1) cities lacking basic water services, (2) wasteful cities, (3) water-efficient cities, (4) resource-efficient and adaptive cities, and (5) water-wise cities. They emphasized the importance of effective governance, environmental awareness and community involvement for sustainable IWRM.

Closely related to IWRM and very much applicable when studying sustainability and resilience of urban water services is the concept of integrated urban water management (IUWM) [2]. In this approach drinking water, sanitation and storm water management are not developed, planned and implemented separately, their cross-scale interdependences must be acknowledged. This is a growing aspect especially in large urban centers.

Urban water services include water intake, treatment and distribution, wastewater collection, treatment and discharge back to natural waters. Storm water management also affects urban water services and that cannot be neglected, especially when considering impacts of climate change on urban water management [2]. The abovementioned issues are the main issues of urban water management globally, but their importance is different depending of the meteorological, hydrological, political, environmental and economic conditions of the country and the area [3].

Some areas may have severe problems because of water scarcity, while other areas have enough water sources, but not adequate water policy. In recent years there have been alarming news about water crisis in large urban areas all over the world [4]. Some studies have been carried out to compare resilience of urban water services in developed and developing economies [5], and impact of population and lifestyle changes [6]. The results show that there are parallels between water, human rights and reproductive justice crises in communities, and that e.g., UK may face a supply-demand gap by the 2080s [6]. In the UK, the Water Services Regulation Authority (Ofwat) has prepared a document, which gives proposals on water regulation in future [7]. The US Environmental Protection Agency has defined systems measures of water distribution system resilience for securing drinking water services in future [8]. One remarkable issue is that, when improving resilience, it is most effective when implemented at a local level [9].

The objective of this study was to determine what kind of institutional aspects should be developed to strengthen the resilience and sustainability of Finnish water services. This study concentrated on Finnish urban water services and their resilience by surveying stakeholders' points of view regarding the value and best practices of community water services. This survey was carried out via a questionnaire to find answers to the following research questions:

1. What is resilience in urban water management?
2. What are crucial aspects for securing adequate water services?

Sustainability and resilience in water services have been studied and applying these principles has been attempted in several countries, e.g., [7,8]. The studies often concentrate on some special subjects or themes, like urban water technology and infrastructure [10], operational management [11], governance [12], or they try to define a method for analyzing the system [13]. The research gap addressed by this study is to develop a wider perspective in technical, institutional and socioeconomic aspects of water resilience. Water resilience and water governance is still poorly understood, especially institutional and governance dimensions of building water resilience [14]. More research should be directed to factors, practices and governance principles that help increase the resilience of people, communities or the environment to water-related risks [15].

2. Materials and Methods

This study was carried out via a literature survey and a questionnaire sent to experts representing key water services stakeholders. Within the stakeholders, 67 individuals were selected from universities, ministries and institutions. In addition, the questionnaire was sent to 338 water utilities, in which there were 403 individual recipients. These 338 water utilities provide drinking water to more than 80.0% of Finland's population. Of these 470 recipients, 99 replied with response rate of 21%. This kind of scope of study concerning Finnish water services has been applied also in studies carried out in 2010 and

2011 by Technical University of Tampere [16,17]. The questionnaire is presented as an attachment in Supplementary Material S1.

The results of the questionnaire can be interpreted for forming a reasonably good and representative view of sustainable and resilient water services in Finland. The main stakeholders in Finland are ministries (responsible of legislation), regional authorities (one organization responsible of permitting and another of monitoring), municipalities (responsible of organizing water services) and water utilities (operation and maintenance). Other stakeholders to whom the questionnaire was sent were research organizations, universities, consultants, equipment and service providers, as well as some NGOs. The coverage of respondents was as follows: water utilities (44%), other water companies (19%), governmental organizations (14%), private companies (7%), municipalities (6%), universities (4%), and other miscellaneous bodies (6%). The questions are formulated especially to experts of urban water services. Normally the questionnaire was sent to the managing director of a water utility, or respective leaders in other sector bodies. It would be interesting to implement a survey also to customers; institutions and citizens, but then the questions should be different and sampling considerably larger.

The geographical distribution of the replies covered the whole country, largely representing the distribution of settlements: 23% of respondents were from the capital area or other big cities, 27% from Southern Finland (other than the capital area), 20% from Eastern Finland, 19% from Western Finland, and 11% from Northern Finland. This corresponds well to the distribution of the population. Of all respondents, 74% were men and 26% women, while 64% represented management positions and 22% experts.

The questions were formulated to gain an understanding of experts' and stakeholders' points of view regarding the following major aspects. The answers supported the results and conclusions when searching for answers to the research questions formulated in Introduction of this article.

i. Significance of water services failures and their significance in urban water services.
ii. Policy instruments and impact methods for ensuring continuous service in water supply and sanitation: a. pricing policy b. institutional strengthening c. service reliability, d. development planning.

The majority of the 15 questions were formulated, 10 dealt with substance and five with the background of the respondent. In three questions the respondents were asked to assess statements or arguments using a scale from 1 to 5, one question was completely open-ended. The majority of the questions included several alternative means or proposals for improving the current situation out of which from 1 to 3 were selected and ranked. In this way, it was possible to get a balanced overview on how the experts in water services prioritize the selected questions related to resilient water services in Finnish conditions.

This method was considered to suit well for finding answers in this kind of study and to the stated research questions [18]. Before sending the web-based questionnaire, it was pre-tested by five experts. In the beginning of the survey, a short description of all the questions was presented for giving an overview to the respondents before they replied to specific questions. The answers were compiled and analyzed in order to get a good impression and response to research questions that were set in the beginning of the study.

3. Results

3.1. Resilience in Urban Water Services—Literature Survey

Resilience in water services has no definite definitions in literature. The term has been increasingly used, especially during the last few years. However, it can be defined in several ways. Johannessen and Wansler (2017) stated that the resilience concept is generally not operationalized, and they investigated in their study how the resilience concept can be systematized, operationalized and applied better in urban water management [19]. United Nations International Strategy for Disaster Risk (UNISDR) defined the term resilience as follows: "the ability of a system, community or society exposed to hazards to resist, absorb, accommodate to and recover from the effects of a hazard in a timely and efficient manner" [20]. According to a thorough study by Folke (2016), resilience thinking is an integrative approach for dealing with the sustainability challenge [21]. It can be viewed as a subset of sustainability science with a focus on social-ecological systems of people, communities, economies, societies and cultures.

A major challenge is to make the now largely invisible infrastructure of water services more visible to decision-makers and citizens. From a historical context, water services are not only necessary but invaluable, and they are a key component of the national security of water supply. Only if water services fail, do they seem to get recognized. Resilient water services and systems are the foundation of well-being, and resiliency is the key for sustainable water services [22]. In considering urban water resilience, it is good to assess various scales in urban water systems depending on users (households, communities, cities), institutions (service providers and regulators), technologies and ecosystems [22]. According to Johannessen and Wansler (2017) [19] resilience in urban water services and defined three levels of disturbances as follows:

i. Socioeconomic disturbance, which is not associated with external hazards but within the urban water service infrastructure and the entities that manage and govern them.
ii. Hazard disturbance, i.e., external hazard, disaster, and crises-related disturbances that are outside the urban water service infrastructure.
iii. Long-term disturbances such as unsustainable resource extraction by the urban water services on the broader social-ecological system and vice versa.

Resilience has also been studied in other environment and infrastructure-related ensembles, such as housing. Miller (2015) concluded that sustainability, environmental performance and resilience are inter-related, and she used technical, social and economic approaches in her study [23]. This paper emphasizes the importance of cooperation and collaborative approach, which can be seen quite clearly also in resilience of water services. Bocchini et al. (2014) compared the terms 'sustainability' and 'resilience' in civil infrastructure and concluded that the proposed perspective and assessment technique is applicable to various types of civil infrastructure systems, although their case concentrated on transportation networks and bridge systems [24].

Linkov et al. (2013) formulated a resilience matrix for measuring overall system resilience, not only fragmented resilience in separate disciplines [25]. They defined four functions with respect to adverse events: (i) planning and preparation, (ii) adsorption, (iii) recovery and (iv) adaptation. In their resilience matrix these events are mapped to four functions:

(i) Physical: sensors, facilities, equipment, system states and capabilities
(ii) Information: creation, manipulation and storage of data
(iii) Cognitive: understanding, mental models, preconceptions, biases and values
(iv) Social: interaction, collaboration and self-synchronization between individuals and entities

This matrix was defined for supporting the decision-making process for perceiving the overall picture of possible disaster management.

It is challenging to combine different stakeholders' views in the same calculations or assessments. One way to navigate this problem is to visualize water supply systems graphically so that different views are illustrated. Using this scheme, Lehrman (2018) used so-called Sankey diagrams for engaging water policy makers on issues of social and environmental justice, ecological water use, sustainability, recreational access and urban/rural issues [26].

Cities generate more than 80% of the gross world product (GWP), so resilience of cities is important to maintain [27]. GWP is the combined gross national product (GNP) of all countries in the world including the total domestic and foreign output claimed by residents of a country. For this, sustainable water services are crucial and in the transition towards smarter cities, water issues play a significant role. Urban water security is strongly related to resilience [11]. Four issues can be pointed: welfare, equity, sustainability and water-related risks. While public administrations and political scientists are looking for mechanisms of good governance, they underestimate the quality and effectiveness of policy outcomes. Good governance is essential, but it does not guarantee outcomes that are effective in terms of solving the problems at hand. For the level of organizations, the Finnish Technical Research Centre VTT recommended the principle of "flexibility for change" for supporting organizational resilience [28].

In benchmarking water utilities, a wide variety of indicators are used [29,30]. It is important that, in addition to indicators for the performance of the physical infrastructure, there are also indicators illustrating management and financial performance [11]. These indicators point out that "We need to better understand the full potential of water-sensitive design, rainwater harvesting, recycling, reuse, pollution prevention and other innovative urban water approaches" [11].

Hordijk et al. (2014) explored water governance systems in four cities and assessed adaptation practices at three levels: resilience, transition and transformation [31]. They concluded that "the crucial question for the transformation of water governance systems in all cases will be whether, in the long run, participation and deliberative decision-making are extended to decisions about hard infrastructure and the provision of local water and sanitation services, and whether local powers are indeed empowered to hold the approach of water as an economic good to account". This complex problem can also be seen when comparing the combination of centralized and decentralized water systems approaches. In Melbourne, Australia, it was discovered that this kind of hybrid water system both reduced potable water demand and altered wastewater flow and contaminant concentration [32]. This improved the resilience of the water system to variable climate conditions.

In many countries, where water supply and sewer networks in cities are aging, resilience of urban water services is subject to risk of malfunction of deteriorated networks. Krueger et al. (2017) studied how to enhance water and sewer network resilience to external and internal threats [10]. They compared the functional topology of planned urban infrastructure networks to natural river networks draining natural landscapes. As implications, they emphasized the relevance of efficient planning of networks and observation of expected topological features.

Water supply and sewer networks are technically and financially remarkable parts of sustainability and resilience in water services, however these are not the only aspects of water and sewer network management. Sustainable water demand management (SWDM), was defined by Arfanuzzaman and Rahman (2017) in their research in Dhaka city, Bangladesh [33]. In their analyses, they covered the present condition of water demand, supply, system loss, pricing strategy, groundwater level and per capita water consumption. The main idea was to reduce the water footprint and pollution. To achieve SWDM political, financial, technical and legal control, a variety of methods are needed, e.g., 100% coverage of metering, pricing policy on water withdrawal, development of surface water sources and penalty or discount according to meeting the consumption goals.

Schifman et al. (2017) introduced a Framework for Adaptive Socio-Hydrology (FrASH) for planning of using green infrastructure in storm water management [34]. This approach requires cooperation between community organizations and increases stakeholder involvement. Thus, integrated urban water resources management can be a step towards sustainable city development. The authors see also that this concept can be applied to other environmental management plans and projects, and it can be considered suitable for planning sustainable urban water services. One important aspect in IUWM concept is separation of wastewater systems from rainwater drain systems. This was concluded in the Netherlands in a study of three case cities, namely Amsterdam, Rotterdam, and Utrecht [12].

The paradigms of sustainability and resilience in the built environment were the subject matter of research by Lizarralde et al. [35]. They found that there are different interpretations of these terms. This might explain tensions that occur when the paradigms of sustainability and resilience are translated into policy instruments. They name sustainability 'green' and resilience 'blue' and conclude that both academics and practitioners need more refined tools and conceptual frameworks to successfully achieve a turquoise agenda in the built environment.

3.2. Results of the Survey

In our survey, the respondents considered water services the most important part of municipal engineering. However, it must be stated that they represent particularly stakeholders of water services, like water utilities, consultants, researchers and authorities. In water use, water as a source of water supply was considered extremely important (see Figure 1). The exact question was "What is the most important use of water resources?" The respondents were given a scale from one to five, one meaning not important at all and five extremely important. So, the scale from all 99 answers ranges from 99 to 495. The same scale is used in Figure 2.

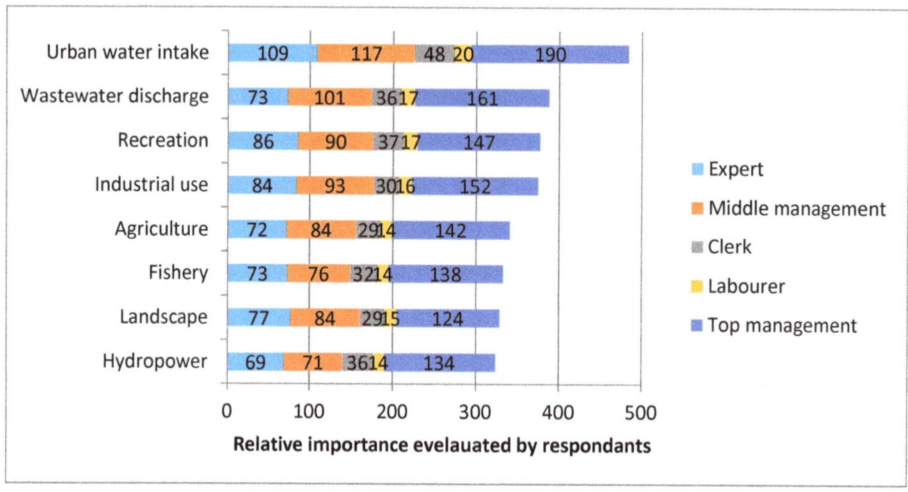

Figure 1. Ranking of water use priorities, illustrated by occupational groups.

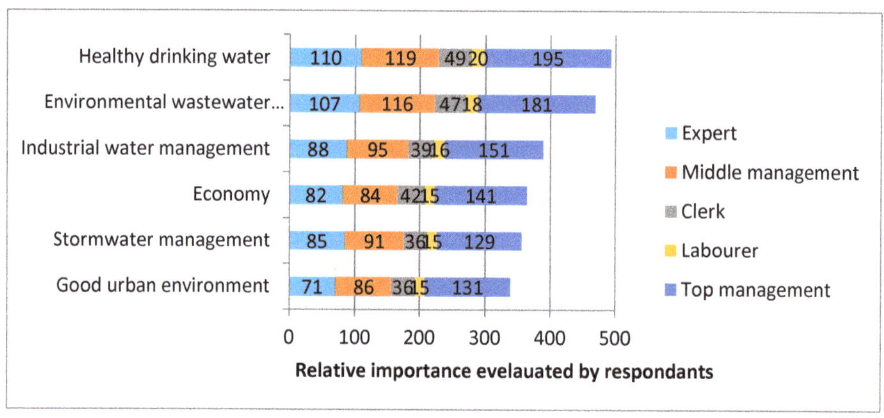

Figure 2. Relative importance of water services according to respondents, illustrated by occupational groups.

In the relative importance of water services, the most important aspect was healthy and secure water supply (scale from 99 to 495) followed by wastewater management that is secure for the environment (Figure 2). Participants responded to the statement 'Importance of functional water services'.

Pricing policy and institutional aspects were explored for getting views on policy instruments. Six statements were given, and respondents were asked to pick the most important one:

- Water fee must be decreased so that everybody can afford water services
- Water fee must be increased so that services can be improved
- Water fee is reasonable, services can be improved by more-efficient operational management
- Payments for owners of water utilities (usually municipalities) must be decreased
- Water supply network leakages should be decreased
- Sewer network leakages should be decreased

Most remarkable is that no-one thought that the water fee should be decreased and 31% thought that it should be increased for improving water services (Figure 3).

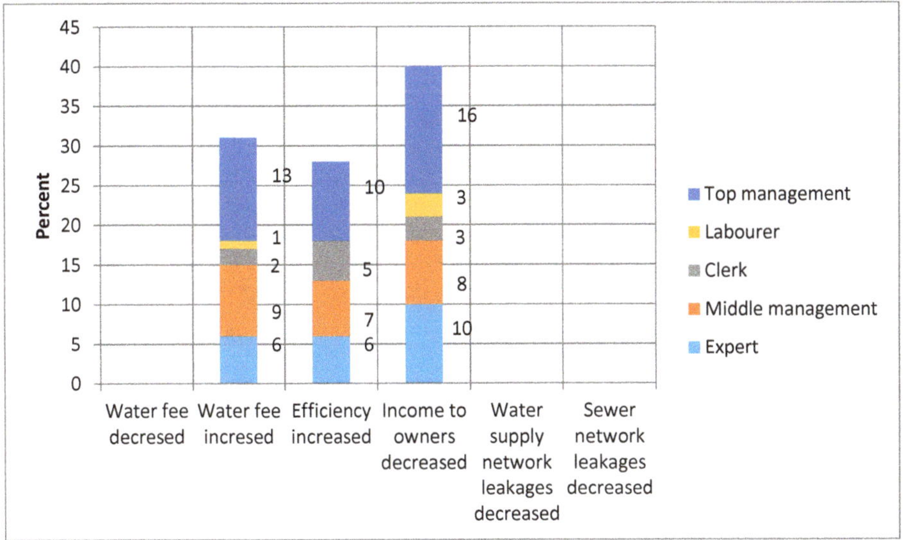

Figure 3. Policy instruments and pricing policy, illustrated by occupational groups.

Municipalities transfer part of water utilities' profits as income for other municipal costs (so-called reasonable rate of return). This might be a considerably large part of funds that could be used for development and renovation investments. Altogether, 41% of respondents had the view that the current rate of return is too high (this is too large a part of the water utilities budget) and it should be reduced.

One remarkable issue in Finnish water services for the time being is renovation of water pipe and sewer networks. Major parts of networks were constructed in the 1950–1960s and now it is time for major investments for maintaining a safe and acceptable level of water services. Considering the most important measures to ensure good water services, 64% of respondents thought that renovation of water pipe networks should be increased and 57% thought that renovation of sewer networks should be increased. Only 5% thought that the quality of drinking water should be improved, which indicates that the quality is good enough in most of Finnish communities.

The measures for ensuring continuous acceptable levels of water services, good data and information management, were considered the most important issues. Other issues identified as important were detailed planning for renovations and modelling as a tool for leakage monitoring. In ensuring the reliability of water services, skillful and sufficient personnel is important. Hence, national-level capacity building should be ensured. Network rehabilitation financing was also considered important for ensuring good-quality water services (see Figure 4).

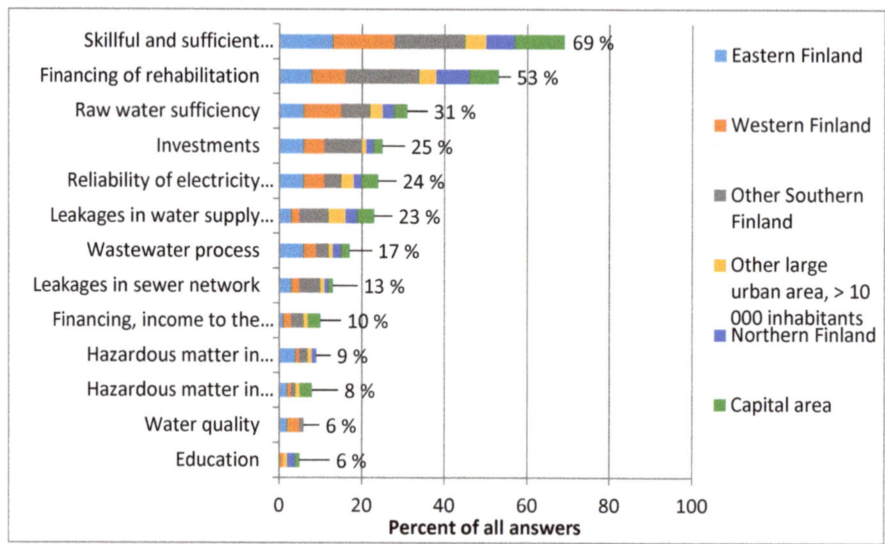

Figure 4. Topics for ensuring sufficient water services, illustrated by regions (in % of all answers).

The questionnaire included one open question: "What is the significance of adequate water services for your organization?" The 102 answers from 63 respondents were divided to seven groups; Operation of community, Health, Environment, Industry, Convenience, Economy, and Image of water services. Altogether, 29 (28%) of the answers emphasized operation of community, 24 (23%) the economy, and 19 (19%) the environment. This indicates that Finnish water experts consider water services as an eminent part of socioeconomic civil services.

The Finnish Act for Water Services says that the municipalities are in charge of water services, but they can outsource operation and purchase services from private or public organizations. This course of action is considered highly functional (89% of respondents) and flexible cooperation between public and private operators is important (70%). Only 1% thought that also private companies could be owners of a water utility.

4. Discussion and Conclusions

This study was carried out via a literature review and a questionnaire to professionals in water services. The coverage of the respondents was quite good when proportioned to the size of the country and the previous studies. This study dealt only with urban water services and its resilience. Issues in rural areas are different and the results could have been different if the questions had concerned also rural water supply and sanitation. In such a case, it would have been more difficult to find specific conclusions to the formulated research questions.

The first two questions in the questionnaire concerned the importance of water services in the field of other critical municipal services and water issues. These sectors are strongly connected, and as public services they must be planned and developed integrated. For example, very often the rehabilitation of water and sewer networks is sensible to implement together with street rehabilitation. Storm water management is also connected with water services, and especially due to climate change impacts, it must be taken into consideration together with water supply and sanitation as well as construction of streets and green infrastructure.

4.1. Results Reflected in the Survey

In the literature survey, the articles can be divided according to how they approached the concerned theme, or the subject they emphasized. According to the subject, most of them deal with water infrastructure and disturbances in service. In terms of approach, the most common concerns are water scarcity and sufficiency as well as water-smart cities. According to this study, the importance of these subjects can be confirmed, but there are also several other issues that cannot be neglected. The most important point of views in addition to subjects mentioned above are education and skilled personnel, good governance, institutional arrangement and financing. As a conclusion this literature survey gives an idea what resilience means in water services and how it can be reflected to the case of Finland.

Countries can learn from each other and by taking the different circumstances into account, they can improve the weaknesses of their own systems. Water scarcity is not a big problem in Finland, which can be seen also through the responses to our questionnaire survey. However, the methods to approach the strategy can be applied in Finnish water management.

One common issue in water services in western countries, reducing resiliency, is the deteriorating water infrastructure, especially water and wastewater networks, and the fact that the knowledge of networks and their real conditions is inadequate. In recent years, it can be seen that water utilities have been able to gain more knowledge and funding for systematic rehabilitation. Also, technical development provides better possibilities to implement thorough surveys regarding the state of networks. The respondents did not consider that water fees are too high in Finland. The average water fee, including drinking water supply and wastewater collection and treatment, is about 5 euros depending on the house type [36]. When average water consumption is about 130 L/person/day, this means that the cost of water services is about 2% of the household's income, assuming that two persons are working and receiving an average salary.

4.2. Results Reflected in Research Questions

The questionnaire was designed so that when analyzing the answers, the answers to research questions would also be gained. The results show that generally reliable water services were considered a very important part of municipal services. Within water services, safe drinking water was considered the most important issue, while environmentally adequate wastewater management was seen almost as important. Noteworthy is that the price of water was not considered a very important concern. This indicates that in Finland the price of supplied water is reasonable and in developing more sustainable and resilient urban water services, the stakeholders considered that willingness to pay is quite high. Concerning water utilities, the economy was still regarded an important part of their operational management. An open question on major concerns in urban water services revealed that 23% of the answers found the economy as one of the most important issues in practice. This is, however, not a concern of pricing, but of revenue sharing within the utility and its owner.

Aging infrastructure was still considered a big problem in urban water utilities, as discovered by Heino et al. already in 2011 [16]. The other topics that were raised as relevant in developing sustainable and resilient Finnish water services, were skilled personnel, financing of rehabilitation, and raw water sufficiency. The last issue, that is growing in significance in Finland too, is that due to climate change, seasonal water scarcities are expected in some parts of the country.

Good governance is a prerequisite for any society to have sustainable water services. Finland is often considered as one of the best countries in the world concerning low corruption, good public services and public private partnership. These institutional aspects are essential in water management, while continuous development and improvement are needed for avoiding regression. The results indicated that there is a clear commitment to this among water experts, and their knowledge should be integrated into decision-making for the good of society.

4.3. Resilient Finnish Water Services in the Future

Recovering from disaster or adversity requires proper technical and institutional preparedness. According to this study, technical resilience is considered strong in Finland. When water utilities are run in accordance to the full-cost recovery principle, it is easy to keep technical preparedness in good order. Institutional preparedness requires good consensus among water professionals, authorities and decision makers. This necessitates continuous discussion and mutual understanding in development and implementation of water services. This is not only a question of technical service, but a question of wider socio-institutional principles, how health and environment issues are dealt with within the whole society.

In the future, to maintain the current standard of water services and to strengthen resilience, some topics can be listed:

(a) Asset management as an important part in water services
(b) Good governance ensured
(c) Economic aspects as full-cost recovery
(d) Continuous improvement
(e) Institutional framework maintained
(f) Capacity building and human resources policy cannot be neglected.

More research could be done, for example, in comparing water services concerning their resilience in different countries with various institutional frameworks. It would also be interesting to study opinions of the customers by a comprehensive questionnaire targeted to the public. Public private partnership is an essential part in resilient water services, and therefore a thorough review of the institutional framework is needed.

4.4. Conclusions

The main achievements of this study are knowledge about main disturbances, which threaten trouble-free water services, and how to develop or keep sustainability and resilience of a water utility. According to this knowledge, a water utility can prepare its strategy and strengthen its resilience in its future operation. These achievements are explained in following aspects and topics.

The main aspects in resilient and sustainable water services in Finland can be concluded as follows:

(a) Maintaining in high-quality education and training at all levels; high school, college and university
(b) Skilled and motivated personnel is considered very important in water services performance
(c) Integrated knowledge in technical development, institutional aspects and socioeconomic needs
(d) Awareness of effects of external disturbances, e.g., climate change (droughts and floods), political situations and the changing urban environment.

This can be secured by regular training and education needs assessment and development, open discussion and cooperation between all operators and stakeholders, and realistic objectives that can be agreed and accepted among all parties. Need for open cooperation between public and private parties is obvious. Answers showed, however, quite clearly that water services should be owned by a public organization, which can purchase services from private companies. This is organized differently in some European countries, but for keeping this basic and necessary function accessible for all people, Finnish public–private partnership or cooperation with public ownership has worked socio-economically fluently, confidently, and equally. This can be ensured by keeping the core functions strictly controlled by the owner of the utility, and outsourcing only support functions, e.g., repairs, accounting, and cleaning work.

Supplementary Materials: The following are available online at http://www.mdpi.com/2073-4441/12/1/187/s1, Supplementary Materials S1: Questionnaire used in stakeholder survey.

Author Contributions: J.L. was the correspondent author and the principal researcher of this study. J.K. and J.J.H. gave their contribution in methodology and validation of the results. T.S.K. acted as a co-writer, and a supervisor together with P.J. All authors have read and agreed to the published version of the manuscript.

Funding: This research received no external funding.

Acknowledgments: This work was supported by the Ministry of Agriculture and Forestry and the Academy of Finland [number 288153]. The authors thank the peer reviewers and the editors for valuable comments and feedback.

Conflicts of Interest: The authors declare no conflict of interest.

References

1. Koop, S.H.A.; Leeuwen, C.J. Application of the improved City Blueprint Framework in 45 municipalities and regions. *Water Resour. Manag.* **2015**, *29*, 4629–4647. [CrossRef]
2. GWP. *Towards Integrated Urban Water Management*; Perspectives Paper of Global Water Partnership; Global Water Partnership: Stockholm, Sweden, 2011; p. 12.
3. Katko, T.S. *Finnish Water Services. Experiences in Global Perspective*; Finnish Water Utilities Association: Helsinki, Finland, 2016; p. 288.
4. Mitlin, D.; Beard, V.A.; Satterthwaite, D.; Du, J. *Unaffordable and Undrinkable: Rethinking Urban Access in the Global South*; World Resources Report; World Resources Institute: Washington, DC, USA, 2019; p. 60.
5. Mosley, E.A.; Bouse, C.K.; Hall, K.S. Water, human rights and reproductive justice: Implications for women in Detroit and Monrovia. *Environ. Justice* **2015**, *8*, 78–85. [CrossRef]
6. Manouseli, D.; Anderson, B.; Nagarajan, M. Domestic water demand during droughts in temperate climates: Syhthesising evidence for an integrated framework. *Water Resour. Manag.* **2018**, *32*, 433–447. [CrossRef]
7. Ofwat. *Towards Resilience: How We Will Embed Resilience in Our Work*; Ofwat: Birmingham, UK, 2015; p. 41.
8. United States Environmental Protection Agency. *Systems Measures of Water Distribution System Resilience*; EPA: Washington, DC, USA, 2015; p. 52.
9. Inha, L.M.; Hukka, J.J. Policies enabling resilience in Seattle's water services. *Eur. J. Creat. Pract. Cities Landsc.* **2019**, *2*, 93–120.
10. Krueger, E.; Klinkhamer, C.; Urich, C.; Zhan, X.; Rao, P.S.C. Generic patterns in the evolution of urban water networks: Evidence from a large Asian city. *Phys. Rev. E* **2017**, *95*, 032312. [CrossRef]
11. Hoekstra, A.Y.; Buurman, J.; van Ginkel, K.C.H. Urban water security: A review. *Environ. Res. Lett.* **2018**, *13*, 053002. [CrossRef]
12. Dai, L.; Wörner, R.; van Rijswick, H.F. Rainproof cities in the Netherlands: Approaches in Dutch water governance to climate-adaptive urban planning. *Int. J. Water Resour. Dev.* **2018**, *34*, 652–674. [CrossRef]
13. Nikolopoulos, D.; van Alphen, H.-J.; Vries, D.; Palmen, L.; Koop, S.; van Thienen, P.; Medema, G.; Makropoulos, C. Tackling the "New Normal": A resilience assessment method applied to real-world urban water systems. *Water* **2019**, *11*, 330. [CrossRef]
14. Rodina, L. Defining "water resilience": Debates, concepts, approaches, and gaps. *WIREs Water* **2019**, *6*, e1334. [CrossRef]
15. World Health Organization (WHO) and the United Nations Children's Fund (UNICEF). Progress on Drinking Water, Sanitation and Hygiene: 2017 Update and SDG Baselines. Available online: https://www.who.int/mediacentre/news/releases/2017/launch-version-report-jmp-water-sanitation-hygiene.pdf (accessed on 8 January 2020).
16. Heino, O.A.; Annina JTakala, A.J.; Katko, T.S. *Challenges to Finnish Water and Wastewater Services in the Next 20–30 Years*; E-Water, Official Publication of the European Water Association (EWA): Hennef, Germany, 2011.
17. Pietilä, P.E.; Katko, T.S.; Seppälä, O.T. Uniqueness of Water Services. *EWA* **2010**, 17. Available online: http://www.ewa-online.eu/tl_files/_media/content/documents_pdf/Publications/E-WAter/documents/6_Pietila_UNIQUENESSOFWATER.pdf (accessed on 8 January 2020).
18. Metsämuuronen, J. *Basics of Qualitative Research*; International Methelp Oy: Helsinki, Finland, 2008; p. 74. (In Finnish)
19. Johannessen, Å.; Wansler, C. What does resilience mean for urban water services? *Ecol. Soc.* **2017**, *22*, 1. [CrossRef]

20. UNISDR. *Terminology on Disaster Risk Reduction*; United Nations International Strategy for Disaster Risk: Geneva, Switzerland, 2009; Available online: http://www.unisdr.org/we/inform/terminology (accessed on 1 May 2009).
21. Folke, C. Resilience (Republished). *Ecol. Soc.* **2016**, *21*, 44. [CrossRef]
22. Howe, C.; Butterworth, J.; Smout, I.; Duffy, A.M.; Vairavamoorthy, K. *Sustainable Water Management in the City of the Future*; Findings from the SWITCH Project 2006–2011; UNESCO-IHE: Delft, The Netherlands, 2012; Available online: http://www.switchurbanwater.eu/outputs/pdfs/Switch_Final_Report.pdf (accessed on 1 January 2012).
23. Miller, W. What does built environment research have to do with risk mitigation, resilience and disaster recovery? *Sustain. Cities Soc.* **2015**, *19*, 91–97. [CrossRef]
24. Bocchini, P.; Frangopol, D.M.; Ummenhofer, T.; Zinke, T. Resilience and Sustainability of Civil Infrastructure: Toward a Unified Approach. *J. Infrastruct. Syst.* **2014**, *20*. [CrossRef]
25. Linkov, I.; Eisenberg, D.A.; Bates, M.E.; Chang, D.; Convertino, M.; Allen, J.H.; Flynn, S.E.; Seager, T.P. Measurable resilience for actionable policy. *Environ. Sci. Technol.* **2013**, *47*, 10108–10110. [CrossRef]
26. Lehrman, B. Visualizing water infrastructure with Sankey maps: A case study of mapping the Los Angeles Aqueduct, California. *J. Maps* **2018**, *14*, 52–64. [CrossRef]
27. Koop, S.H.A.; Leeuwen, C.J. The challenges of water, waste and climate change in cities. *Environ. Dev. Sustain.* **2017**, *19*, 385–418. [CrossRef]
28. Nieminen, M.; Talja, H.; Airola, M.; Viitanen, K.; Tuovinen, J. *Flexibility of Change*; VTT Technology: Espoo, Finland, 2017; p. 86. (In Finnish)
29. Berg, S.; Marques, R.C. Quantitative studies of water and sanitation utilities: A benchmarking literature survey. *Water Policy* **2011**, *13*, 591–606. [CrossRef]
30. Seppälä, O.T. Performance Benchmarking in Nordic Water Utilities. *Procedia Econ. Financ.* **2015**, *21*, 399–406. [CrossRef]
31. Hordijk, M.; Miranda, S.L.; Sutherland, C. Resilience, transition or transformation? A comparative analysis of changing water governance systems in four southern cities. *Environ. Urban.* **2014**, *26*, 130–146. [CrossRef]
32. Sapkota, M.; Arora, M.; Malano, H.; Moglia, M.; Sharma, A.; George, B.; Pamminger, F. An integrated framework for assessment of hybrid water supply systems. *Water* **2016**, *8*, 4. [CrossRef]
33. Arfanuzzaman, M.; Rahman, A.A. Sustainable water demand management in the face of rapid urbanization and ground water depletion for social-ecological resilience building. *Glob. Ecol. Conserv.* **2017**, *10*, 9–22. [CrossRef]
34. Schifman, L.A.; Herrmann, D.L.; Shuster, W.D.; Ossola, A.; Garmestani, A.; Hopton, M.E. Situating green infrastructure in context: A framework for adaptive socio-hydrology in cities. *Water Resour. Res.* **2017**, *53*, 10139–10154. [CrossRef]
35. Lizarralde, G.; Chmutina, K.; Bosher, L.; Dainty, A. Sustainability and resilience in the built environment: The challenges of establishing a turquoise agenda in the UK. *Sustain. Cities Soc.* **2015**, *15*, 96–104. [CrossRef]
36. Finnish Water Utilities Association. *Water Services Fees*; Vesilaitosyhdistyksen julkaisusarja nro 68; Finnish Water Utilities Association: Helsinki, Finland, 2017. (In Finnish)

Article

The Influence of Location on Water Quality Perceptions across a Geographic and Socioeconomic Gradient in Appalachia

Ross G. Andrew *, Robert C. Burns and Mary E. Allen

School of Natural Resources, West Virginia University, Morgantown, WV 26506, USA; robert.burns@mail.wvu.edu (R.C.B.); mary.allen1@mail.wvu.edu (M.E.A.)

* Correspondence: ross.andrew@mail.wvu.edu; Tel.: +1-304-293-9538

Received: 27 September 2019; Accepted: 22 October 2019; Published: 25 October 2019

Abstract: Understanding perceptions of water quality held by residents is critical to address gaps in public awareness and knowledge and may provide insight into what defines communities that are more/less resilient to changing water conditions locally. We sought to identify spatial patterns of water quality perceptions gathered in a survey of Southern West Virginia (WV) residents during spring/summer 2018. Using over 500 survey responses across 15 counties, we calculated spatial autocorrelation metrics and modeled the relationships between overall water quality perceptions and county-level socioeconomic endpoints, such as poverty rate, per capita income, and education level. We identified significant differences across counties labeled as socioeconomically "transitional", "at-risk", and "distressed", as it pertained to responses for water quality perceptions, education level, and income level. We also found significant positive relationships between overall water quality perceptions, elevation, and income level. We calculated an empirical semivariogram and fit an exponential model to explain a significant autocorrelation pattern within a range of 104.2 km. Using that semivariance function, we created a kriging interpolation surface across the study area to identify significant clusters of water quality perceptions. This work highlights the influence of location on water quality perceptions within Southern West Virginia, but the analytical framework should be considered in further research, when samples are spread across large areas with varying socioeconomics.

Keywords: water quality; environmental perceptions; human dimensions; spatial models; socioeconomics

---

1. Introduction

Water quality may be both broadly and specifically defined across chemical and biological continuums [1]. Endpoints such as pH, conductivity, total dissolved solids, and temperature are often cited as criteria for evaluation of the quality of water and aquatic environments [2,3]. Collection and measurement of water quality information is also a major component used for legal protection of waters of the United States under provisions of the Clean Water Act (CWA) as amended in 1972. Since the passage of the amended CWA, public awareness of water quality has been increasing, with concerns mostly about the physical, chemical, and biological conditions following dramatic crises, such as the Cuyahoga River fire of 1969 [4] or the Exxon Valdez spill in 1989 [5]. Understanding what water quality conditions exist, how they are changing, and how they are perceived or understood is a critical challenge that exists as much today as it has in the past throughout the world.

While the physiochemical endpoints are important for defining water quality as it pertains to the functionality of ecosystems and the services they provide [6], it is also important to define the level of awareness/perception of water quality by people living in a given area. Public awareness and knowledge of local conditions is valuable, as it may help create a sense of place or place attachment which may be linked to social, economic, and environmental benefits [7,8]. Sense of place is a multidimensional concept (person, place, and process) which refers to the way in which a person relates to and perceives the natural environment. People develop a sense of place as they get to know an area, depend on its natural resources, and assign the place meaning and value [9,10]. As a result of this development, a strong sense of place can lead to benefits such as an increase in visitation and economy, social bonding, support for conservation, and the promotion of sustainable uses of natural resources [7,10].

Over time, this can influence a community's ability to perceive risks and adapt to changes in the quality of their natural resources [11,12]. Communities which are more aware and/or knowledgeable of environmental conditions, such as relative water quality, may be more resilient to changes in resource use, climate, and policy [13]. In contrast, a community largely unaware of resource conditions related to water quality may be more vulnerable to deterioration of water quality and its associated negative health effects [14,15]. This contrast may occur across space, both locally and regionally, which interacts with other important factors, such as income and education, to influence quality of life for residents.

Previous studies have suggested the relevance of socioeconomic and demographic variables in explaining peoples' perceptions of environmental conditions and how they may respond to environmental issues at both the individual and community level [16,17]. For example, age, income, and length of residency have been significant factors predicting perceptions of water quality and health risks [11,16]. However, a general conclusion about these variables is difficult, as the direction and intensity of their relationship differ from place to place [18]. Research suggests that location of residence and proximity are important in explaining correlations between water quality perceptions and socioeconomic factors [19].

Defining the spatial patterns of public perceptions of water quality requires a baseline understanding of the setting across space and time. For example, within the Appalachian region of the US, a legacy of resource extraction [20], poverty [21], and pollution issues [22] all interact to define the current conditions that residents experience. Within Appalachia, small cities and towns exist, and such areas juxtapose relatively intact forests, rivers, and streams, within some of the oldest mountains on the planet. In particular, Southern West Virginia (WV) has experienced the ebbs and flows of extractive industries from timber harvesting and coal mining, bringing with them both prosperity and fallout [21]. This applies both economically and environmentally, as these industries left behind issues with sedimentation, acid mine drainage, and others which impact water quality. One of the more recent events occurred in January of 2014, when approximately 10,000 gallons of chemicals used to process coal spilled from a storage tank into the Elk River [23]. The Elk River is a primary municipal water source, serving about 300,000 people in the Charleston, WV area. Incidents such as this can have long-lasting impacts on the environment, economy, public health, and well-being. Thus, an understanding of public perceptions is important when it comes to water resource treatment, management, and policy-making. While economic and environmental impacts may be identified and illustrated in distinct units, human perceptions of water quality conditions may operate across more diffuse boundaries which are not strictly defined by census blocks or watershed boundaries [16,17]. Therefore, the main objective of this study was to identify spatial patterns of water quality perceptions in Southern WV as they relate to location and socioeconomic endpoints.

2. Materials and Methods

2.1. Data Collection

In the spring (February–May) and summer (June–August) of 2018, a water quality perceptions survey was created and distributed by West Virginia University (WVU) researchers to 8772 randomly generated addresses within the state of West Virginia, with an emphasis on the southern half of the state. The randomization of addresses was created by random draws from a third-party contractor database, following the methods outlined in a previous study focused on the northern part of West

Virginia and water quality perceptions [24]. Surveys have long been used in social science to collect quantitative data from large samples, and they are commonly administered through mail, online, or mixed survey modes [25]. Overall, when mail and online surveys are identical in design and administration method, research has shown no significant differences in response rates or the nature of the data resulting from survey mode [26]. In the present study, both a mail and email version of the survey were distributed to potential respondents, following the methodology outlined by the tailored design method [25]. This method uses personalization and repeated contacts to increase the likelihood that an individual will complete and return the survey. Each study participant was sent a hand-addressed packet of survey materials, which included a cover letter, a survey, and a US postage-paid business-reply envelope [25]. The online survey followed the same schedule. The surveys contained questions related to water quality perceptions, as well as information about the respondents themselves, such as their education and income levels. Location and demographic questions were also included in the survey.

Survey respondents were asked to rate the overall water quality of rivers, streams, and lakes near their home in West Virginia. Responses to overall water quality perceptions were rated on 5-point Likert scale, ranging from "very poor" (1) to "excellent" (5) water quality. Level of education and income were measured as ordinal variables and coded numerically. Location data was collected via coordinates assigned to the IP address on the email/online survey and the self-reported home zip code in the mail-back survey. Locations derived from IP address were also validated by comparing the automatically generated coordinates to the self-reported zip code information in the survey. If the IP address coordinates were not accurate (e.g., different state, far from home zip code, etc.), they were removed from analyses. All completed survey responses were then compiled into a shapefile, within ArcGIS, to be used for spatial analyses. Data used in the analyses are available online and noted as supplementary material to this manuscript.

Data were collected across 15 counties (Table 1) that span across a range of ~150 km in a north-to-south direction and ~230 km in an east-to-west direction. The entire study area represents an area of ~21,000 km^2. Mean elevation ranges from 218 m in the western counties to 777 m in the eastern counties. Furthermore, the climate across the study area varies only slightly, with mean annual temperatures in the east ~9.5 °C and the west ~13 °C, and mean annual precipitation ~105 cm in the east and ~109 cm in the west. The eastern counties do receive a much larger mean annual amount of snow (~152 cm) than the western counties (~35 cm), likely due to their higher elevation position in the mountains.

Table 1. Study area elevation and socioeconomic statistics summarized by county. The random intercept for each county-level water quality perception score is also provided for reference to the mixed-model framework.

County	Overall Mean WQ Perception Score	WQ Perception Score Random Effect Intercept	Mean Elevation (m)	ARC Economic Status	Unemployment Rate 2014–2016 (%)	Per Capita Income 2016 ($ USD)	Poverty Rate 2012–2016 (%)	High School Diploma 2012–2016 (%)	Bachelor's Degree 2012–2016 (%)
Boone	2.64	2.68	268.3	Distressed	9.3	19,098	26.0	79.2	8.7
Cabell	2.73	2.78	218.7	Transitional	5.2	27,563	21.8	87.0	26.1
Fayette	3.05	2.82	563.1	Distressed	8.3	18,149	19.1	81.7	13.2
Greenbrier	3.54	3.22	660.4	Transitional	6.3	22,761	18.4	83.8	19.6
Kanawha	2.80	2.79	260.5	Transitional	5.8	31,941	16.0	88.0	25.2
Lincoln	3.19	3.04	239.4	Distressed	9.3	17,379	25.2	79.1	9.0
Logan	2.84	2.88	247.9	Distressed	10.7	18,737	20.2	77.7	8.9
McDowell	1.77	2.16	475.1	Distressed	12.7	13,282	37.6	64.9	5.2
Mercer	3.02	2.78	772.3	At-Risk	7.1	20,397	21.1	83.0	19.5
Mingo	2.69	2.79	302.4	Distressed	12.6	14,743	26.0	73.9	9.9
Monroe	3.44	3.17	610.2	At-Risk	5.3	18,284	17.1	81.4	13.4
Raleigh	3.34	3.00	686.1	At-Risk	7.2	23,114	17.7	84.3	18.1
Summers	3.07	2.86	627.0	Distressed	6.8	15,025	18.4	82.8	14.8
Wayne	2.50	2.66	220.2	At-Risk	6.8	21,313	20.9	79.4	12.9
Wyoming	2.88	2.81	475.1	Distressed	9.8	15,743	20.1	77.4	8.2
Average	3.00	2.83	479.7	-	8.21	19,835	21.71	80.24	14.18

2.2. Data Analysis

Completed surveys with accurate spatial information were also assigned to their respective counties to allow for a county-level comparison of water quality perceptions across those grouped as "transitional", "at-risk", and "distressed". The Appalachian Regional Commission (ARC; https://www.arc.gov/) uses these categories within a multi-metric socioeconomic classification scheme to identify counties which may be vulnerable to high poverty rates, low education level, etc. Counties labeled as "distressed" rank in the bottom 10 percent of all United States counties with respect to index values of unemployment rate, per capita market income, and poverty rate. Counties labeled as "at-risk" and "transitional" rank in the bottom 10–25 percent and the middle 50 percent of all United States counties, respectively. These categories were compared using one-way analysis of variance to identify potential differences in water quality perception scores, and self-reported education and income level scores. In order to evaluate the influence of county on water quality perception scores, we constructed a linear mixed-effects model, using county as a random effect and income, education, distance to 2014 Elk River chemical spill, and elevation as fixed explanatory effects. This model was constructed in package "lme4" [27], within the R statistical environment [28]. The significance of the fixed effects elevation and reported income level was assessed using model comparisons of the full model and a reduced model with each significant variable removed. A likelihood ratio test was then conducted between the full and reduced model to evaluate the significance of the effect of each variable.

Spatial analyses were conducted in both ArcGIS (ESRI, Inc., Redlands, CA, USA), and package "synchrony" [29] within the R environment. We calculated overall spatial autocorrelation in water quality perception scores using Global Moran's I and used a Getis-Ord General G hotspot analysis to identify areas of high and low water-quality perception score clusters. We then calculated an empirical semivariogram, using the location coordinates. We calculated the significance of the empirical semivariogram, using 999 Monte Carlo randomizations, and compared different semivariance models (i.e., spherical, exponential, etc.), using AIC and root mean square error to select the best fit to the data. Using the best-fitting semivariance model, we created a kriging interpolation surface across the 15-county area to help identify and visualize any significant hot and cold spots of water quality perceptions.

3. Results

3.1. Survey Response and Socioeconomic Analysis

A total of 734 surveys were completed and returned (8.4% response rate overall). The mail-back surveys had a higher response rate (14.1%) than the email surveys (6.7%). Of all the surveys which were completed and returned via either method, a total of 508 (69.2%) surveys were completed with respondents answering the question about the location of their home zip code. Responses were obtained from 15 counties in Southern WV, with a mean of 33.9 ± 1.1 responses per county. The mean (3.0 ± 0.04) overall water quality perception score was scored on a scale of 1–5, with five representing the highest quality and one the lowest quality. The average response came from a resident with a self-reported education level of "some college" and household income of $50,000–$74,999.

West Virginia is the only state entirely within the "Appalachian Region" as defined by the ARC. The ARC criteria for the multi-metric index to classify counties into categories includes three-year mean unemployment rate, per capita market income, and poverty rate (Table 1). These values are then compared for each county to the US national average, to determine what percentile a particular county falls within across the index values. Distressed counties are classified as being the bottom 10% of all US counties. At-risk counties are classified as the lower 10%–25% of all US counties, and transitional counties are classified within 25%–75% of all US counties. Within our 15-county study area, eight counties are classified as distressed, four are at-risk, and three are transitional (Table 1). No counties were classified by the ARC as being competitive or attainment (the two highest performance categories). Across all 15 study counties, mean three-year unemployment rate (8.2% ± 0.6%) is higher than the mean value for WV, the Appalachian Region, and the US (6.5%, 6%, and 5.4%, respectively). The same pattern emerges with respect to study area poverty rate (21.7% ± 1.4%), when compared to the same regional and national averages (17.7%, 16.7%, and 15.1%, respectively). Within the study area, mean per capita income ($19,835 ± $1291) is lower than WV, Appalachian Region, and US means ($25,987, $29,765, and $40,679, respectively).

When comparing the survey response data across county classifications (i.e., distressed, at-risk, and transitional), significant differences exist for overall water quality perception scores (F = 5.67; $p < 0.01$), with the distressed county group having the lowest mean score (2.84 ± 0.07). Self-reported education and income level differed significantly among county status groups (F = 5.75; $p < 0.01$; F = 5.49, $p < 0.01$, respectively), with the distressed county group having the lowest mean score in both response variables (4.5 ± 0.13 and 2.78 ± 0.13, respectively). The linear mixed effects model to describe overall water quality perception scores returned elevation ($\chi^2 = 5.62$; $p < 0.05$) and reported income level ($\chi^2 = 7.92$; $p < 0.01$) as the only significant fixed effects. Both elevation and reported income showed a positive effect on water quality perception score, with elevation showing a slightly stronger effect and income having an effect closer to zero (Table 2). Reported education level and distance from the 2014 Elk River chemical spill were nonsignificant fixed effects. The random effect (intercept) of county only explained 10.2% of the variance in the model following the fixed effects.

Table 2. Linear mixed model results for overall water quality perception score as the dependent variable of interest. Note: * $p < 0.05$; ** $p < 0.01$.

Parameter	Estimate	Lower 95% CI	Upper 95% CI	Random Effect—County (SD)
Intercept	2.828	2.685	2.972	0.30
Elevation *	0.199	0.117	0.282	0.30
Education	−0.015	−0.038	0.007	0.30
Income **	0.068	0.044	0.092	0.30
Distance from Elk River Spill Site	−0.065	−0.136	0.007	0.30

3.2. Spatial Analysis

The locational data showed a significantly clustered pattern (Global Moran's I = 0.04; $p < 0.001$) with respect to the autocorrelation of water quality perception scores. Furthermore, the high values of these water quality perception scores were more significantly clustered (Getis-Ord General G = 0.52; $p = 0.04$) than would be expected if the underlying spatial arrangement were random. The empirical semivariogram was fit with four models, each having a somewhat similar AIC and RMSE value. The best-performing model used an exponential semivariance function (RMSE = 0.12; AIC = −95), with maximum likelihood estimates of a range of 104.2 km, nugget = 0.74, and sill = 1.08 (Figure 1). The kriging function produced a prediction surface (Figure 2), with root mean square error = 0.913, mean standardized error = 0.032, root mean square standardized error = 0.986, and average standard error = 0.928. A prediction standard error surface map also indicates relatively low prediction standard error across the majority of the study area (Figure 3).

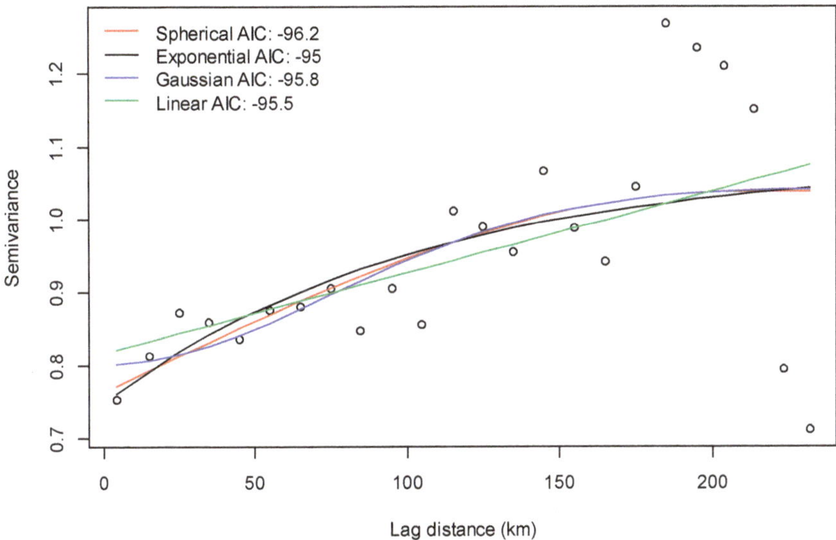

Figure 1. Empirical semivariogram plotted as points, with four fitted semivariance models shown as different colored lines. The best-fitting model (RMSE = 0.12) is represented by the exponential curve, with a range of 104.2 km, nugget = 0.74, and sill = 1.08.

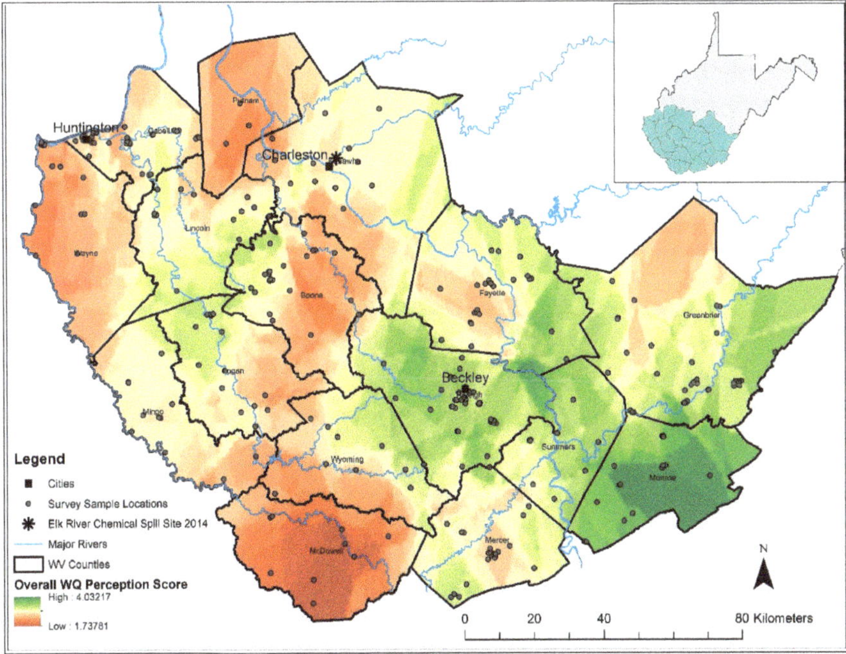

Figure 2. Kriging interpolated prediction surface using exponential semivariance model to predict overall water quality perception scores across study area in Southern West Virginia. Values range from 1 (lowest water quality perception; red) to 5 (highest water quality perception; green).

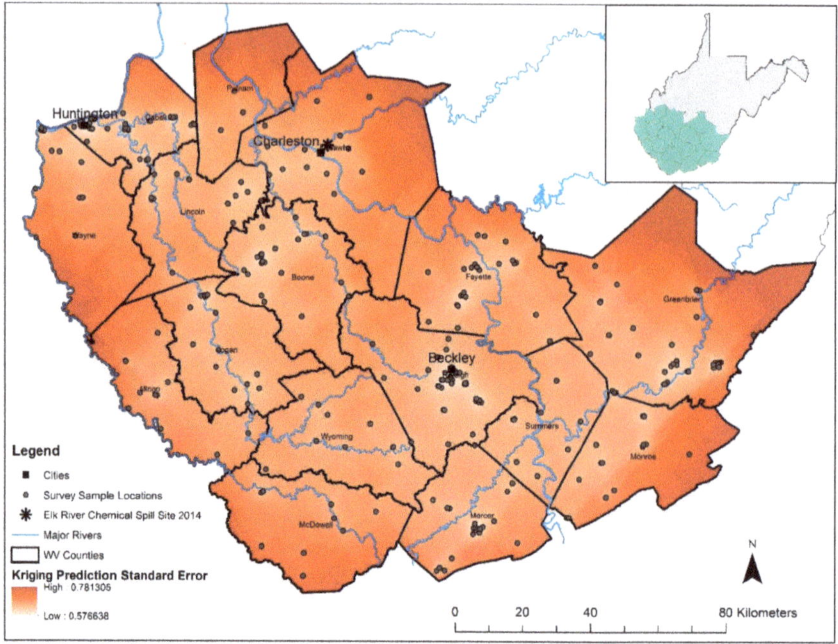

Figure 3. Kriging interpolated prediction surface standard error for overall water quality perception scores across study area in Southern West Virginia. Values range from 0.57 (lighter colors) to 0.78 (dark red) across the study area.

4. Discussion

Differences in water quality perceptions were identified across socioeconomic status categories for the counties represented in this study. Generally, lower socioeconomic status indicated lower water quality perception scores. This aligns with the socioeconomic status–health gradient, presented originally by Adler et al. to help define how socioeconomic status may interact with and influence disease and mortality in humans [30]. Some meaningful explanatory variables were uncovered in this study with respect to water quality perceptions across space. Specifically, we found income and elevation to have significant relationships with water quality perceptions within our study area. Elevated income may associate positively with environmental concern [31]. Also, income is frequently related to environmental quality experienced, which leads to adverse health conditions and outcomes at the lower levels of income [32]. In central Appalachia, elevated poverty and mortality rates are associated with areas connected to mountaintop coal mining [33], which may also be an underlying factor which influences water quality perceptions and awareness.

Changes in water quality perception across an elevation gradient follows a logical pattern of watershed mechanics along the river continuum. Water quality metrics change as you move downstream, due to both natural and anthropogenic influences [34,35]. From this, water quality perceptions would likely change as you move downstream, as well. Our results indicate this relationship between elevation and water quality perception within West Virginia, which rises in the east to elevations of approximately 1500 m and drops in the west to elevations around 170 m near the Ohio River. Along this elevation gradient, influences from human development, such as mining, agriculture, urbanization, and industrial activities, build cumulatively in a downstream direction. Uncovering the significance of elevation on water quality perceptions is both logical and encouraging, in that residents are somewhat aware of the downstream processes which modify water quality characteristics within the study area.

Both income and elevation were significant in explaining water quality perceptions in this study. The most easily visible spatial gradient within the study area is elevation, moving from east to west in a generally downhill direction. However, in some areas, the influence of socioeconomic factors seems to override or modify the general trend of perception driven by elevation. For example, samples obtained from McDowell County contained an average elevation of 475.1 m. Using the statistical relationship within our model for elevation alone, we would predict an average water quality perception score of around 3.26 within McDowell County. This contrasts greatly with the actual average value of 1.77, indicating a strong influence of the local socioeconomic conditions within the county to lower the water quality perception. McDowell County is a prime example due to its outstandingly low socioeconomic status (<1st percentile nationally for ARC index statistics), but this influence may operate to varying degrees and in both positive and negative directions across the landscape. Inequality of environmental conditions and contamination adjusted by community characteristics of race and poverty level was previously described as a way to define environmental justice, or lack thereof, in some cases [36]. Our findings in McDowell County, in particular, mirror this foundational theory, as lower-income areas tended to have lower water quality perceptions. While the present study does not include data on water quality concerns, lower perception scores could lead to higher levels of environmental concern. For example, residents who have low/poor perceptions of their local water quality would likely view their overall local environmental conditions negatively and thus be more highly concerned about environmental conditions than those living in areas with high/positive perceptions of water quality and the associated environment. This would align with extensions of the environmental justice literature, which indicate higher environmental concern in low-income communities [37] and countries [38] due to the increased risk of exposure to poor conditions.

Surprisingly, education level was not clearly related to water quality perceptions within our study. Higher education level may increase awareness related to water quality and use dynamics [39]. This could be related to our study area, which contains relatively low values of residents with both high school and bachelor's diplomas. For example, within our study area, 80.2% of residents hold a high school diploma. Within the greater Appalachian region and the entire US, those numbers rise to 85.9% and 87%, respectively. For bachelor's degree holders, the study area is 14.2%, while the Appalachian region is 23.2%, and the US average is 30.3%. This relatively low level of education across our study area may not allow a wide enough range of educational levels within the surveyed residents to elucidate a relationship between this variable and water quality perceptions.

A notable finding of this study is the lack of a relationship between proximity to the 2014 chemical spill on the Elk River and residents' overall water quality perceptions. While the spill and the survey samples were separated by four years of time, we suspected lingering effects of public perception within a given distance to the location of the spill site. Time lags between environmental science developments and public perception and understanding have been noted before with more theoretical applications [40]. The chemical spill in January 2014 along the Elk River was declared a State of Emergency by the WV Governor within hours of its discovery, so it is very likely none of our survey respondents were unaware of the event, assuming they were residents of WV in 2014. The lack of a relationship between proximity to the spill site and overall water quality perceptions may indicate a lack of lag time beyond four years for the public perception of water quality in this type of point-source pollution event. Perceptions of environmental concern and water quality have been shown to be influenced by proximity to natural resource extraction activities (i.e., oil and gas wells, and mines) in West Virginia [41]. However, this study showed slight effect sizes across small distances (5 km or less), which contrasts with the spatial extent of the 2014 Elk River chemical spill used in this study.

Spatial analysis of environmental perception data is a useful way to explore and illustrate the potential for coupled relationships of human and environmental systems. Pairing perceptions of environmental quality metrics along with perceptions of human uses of the environment via activities like recreation holds great value for the management of complex settings [42]. Spatial clustering of high and low environmental quality perception values offers insight to locations which may contain tradeoffs for management of environmental conservation and human needs [42]. In the present study, we demonstrate significant spatial clustering of water quality perception values within a region that contains vast potential for tradeoffs between human and environmental coupled systems. For example, high elevation areas hold higher scores for water quality perception but also contain mountains, which produce coal and timber. Therefore, understanding the spatial patterns of these perceptions illustrated in this work can lead to a more holistic view of these settings. Furthermore, spatial analysis of perception score semivariance to elucidate the range of correlation in values using social data holds great promise to examine the range of social autocorrelation across space. While typically used in spatial analysis of environmental variables [29], the application of this technique to social perception data helps define the range of distance at which social perceptions operate. This information is valuable in defining the scale at which coupling of environmental and social systems occurs.

5. Conclusions

In this study, spatial patterns of public perceptions of water quality were illustrated within the context of autocorrelation, as clustering and hot spots of water quality perception scores. Clustering of environmental perceptions is logical, following theoretical constructs of a "sense of place" that emerge to help define people's relations to their region and environment via social and natural features of their daily lives [43]. Clustering of water quality perceptions may be related to community features that help drive issue-based activism or concern for the health and safety of a localized area [16]. We demonstrate this potential across large spatial areas in Appalachia, with an upper limit of spatial autocorrelation (statistically shown as semivariance range in the variogram) of 104 km. While this distance would not represent a local community as in [16], it may elucidate spatial dependence of environmental perceptions tied to larger regional processes, such as coal mining or agriculture. This finding represents an avenue for further investigation of resident environmental perceptions across space, using more layered and complex suites of covariates. These types of analyses may solidify connections and more closely approximate the reality of socially driven environmental concern and stewardship.

Supplementary Materials: Data used are available at https://github.com/randrew4/spatial-water-quality-perceptions. Please contact the authors for specific information about the survey design and questions used.

Author Contributions: Conceptualization, R.G.A.; methodology, R.G.A. and R.C.B.; formal analysis, R.G.A.; data curation, R.G.A. and R.C.B.; writing—original draft preparation, R.G.A. and M.E.A.; writing—review and editing, R.G.A., R.C.B., and M.E.A.; supervision, R.C.B.; project administration, R.C.B.; funding acquisition, R.C.B.

Funding: This research was funded by US National Science Foundation-Experimental Program to Stimulate Competitive Research (through WV-HEPC-Division of Science and Research) RII Grant: OIA1458952. The APC was funded by the West Virginia University Institute of Water Security and Science.

Acknowledgments: Free and informed consent was asked from participants or their legal representatives and was obtained. The study protocol was approved by the Committee for the Protection of Human Subjects (West Virginia University Institutional Review Board IRB), by West Virginia University, West Virginia, United States, protocol No.1510895135, November 2015. The authors thank two anonymous reviewers and the editorial staff for constructive feedback that resulted in an improved manuscript.

Conflicts of Interest: The authors declare no conflicts of interest. The funders had no role in the design of the study; in the collection, analyses, or interpretation of data; in the writing of the manuscript; or in the decision to publish the results.

References

1. Chapman, D. *Water Quality Assessments-A Guide to Use of Biota, Sediments and Water in Environmental Monitoring*, 2nd ed.; Taylor & Francis: New York, NY, USA, 1996.
2. Shelton, B.L.R. *Field Guide for Collecting and Processing Stream-Water Samples for the National Water-Quality Assessment Program*; U.S. Geological Survey Open-File Report 94–455; U.S. Department of the Interior: Sacramento, CA, USA, 1994.

3. Bolstad, P.V.; Swank, W.T. Cumulative impacts of landuse on water quality in a southern appalachian watershed. *J. Am. Water Resour. Assoc.* **1997**, *33*, 519–533. [CrossRef]
4. Stradling, D.; Stradling, R. Perceptions of the burning river: Deindustrialization and Cleveland's Cuyahoga River. *Environ. Hist.* **2008**, *13*, 515–535. [CrossRef]
5. Palinkas, L.A.; Downs, M.A.; Petterson, J.S.; Russell, J. Social, Cultural, and Psychological Impacts of the "Exxon Valdez" Oil Spill. *Hum. Organ.* **1993**, *52*, 1–13. [CrossRef]
6. Keeler, B.L.; Polasky, S.; Brauman, K.A.; Johnson, K.A.; Finlay, J.C.; Neill, A.O. Linking Water Quality and Well-Being for Improved Assessment and Valuation of Ecosystem Services. *Proc. Nat. Acad. Sci.* **2012**, *109*, 18619–18624. [CrossRef] [PubMed]
7. Van Liere, K.D.; Dunlap, R.E. Moral norms and environmental behavior: An application of Schwartz's norm-activation model to yard burning. *J. Appl. Psychol.* **1978**, *8*, 174–188.
8. Ramkissoon, H.; David, L.; Smith, G.; Weiler, B. Testing the Dimensionality of Place Attachment and Its Relationships with Place Satisfaction and pro-Environmental Behaviours: A Structural Equation Modelling Approach. *JTMA* **2013**, *36*, 552–566. [CrossRef]
9. Williams, D.R.; Patterson, M.E. Environmental psychology: Mapping landscape meanings for ecosystem management. In *Integrating Social Sciences and Ecosystem Management: Human Dimensions in Assessment, Policy and Management*; Cordell, H.K., Bergstrom, J.C., Eds.; Sagamore: Champaign, IL, USA, 1999; pp. 141–160.
10. Vaske, J.J.; Kobrin, K.C. Place attachment and environmentally responsible behavior. *J. Environ. Ed.* **2001**, *32*, 16–21. [CrossRef]
11. Boudet, H.; Clarke, C.; Bugden, D.; Maibach, E.; Roser-Renouf, C.; Leiserowitz, A. "Fracking" controversy and communication: Using national survey data to understand public perceptions of hydraulic fracturing. *Energy Policy* **2014**, *65*, 57–67. [CrossRef]
12. Marlon, J.R.; van der Linden, S.; Howe, P.D.; Leiserowitz, A.; Woo, S.L.; Broad, K. Detecting local environmental change: The role of experience in shaping risk judgments about global warming. *J. Risk Res.* **2018**, *22*, 936–950. [CrossRef]
13. Jalan, J.; Somanathan, E.; Chaudhuri, S. Environment and Development Development Economics: Awareness and the Demand for Environmental Quality: Survey Evidence on Drinking Water in Urban India. *Environ. Dev. Econ.* **2009**, *14*, 665–692. [CrossRef]
14. Stern, P.C.; Dietz, T.; Black, J.S. Support for Environmental Protection: The Role of Moral Norms. *Popul. Envrion.* **1986**, *8*, 204–222. [CrossRef]
15. Ekmekçi, M.; Günay, G. Role of Public Awareness in Groundwater Protection. *Environ. Geol.* **1997**, *30*, 81–87. [CrossRef]
16. Brody, S.D.; Highfield, W.; Peck, B.M. Exploring the Mosaic of Perceptions for Water Quality across Watersheds in San Antonio, Texas. *Landsc. Urban Plan.* **2005**, *73*, 200–214. [CrossRef]
17. Hu, Z.; Morton, L.W.; Mahler, R.L. Bottled water: United States consumers and their perceptions of water quality. *Int. J. Environ. Res. Public Health* **2011**, *8*, 565–578. [CrossRef]
18. Wakefield, S.; Elliot, S.J.; Cole, D.C.; Eyles, J.D. Environmental risk and (re)action: Air quality, health, and civic involvement in an urban industrial neighborhood. *Health Place* **2001**, *7*, 163–177. [CrossRef]
19. Syme, G.J.; Williams, K.D. The psychology of drinking water quality: An exploratory study. *Water Resour. Res.* **1993**, *29*, 4003–4010. [CrossRef]
20. Lewis, R.L. Appalachian Restructuring in Historical Perspective: Coal, Culture and Social Change in West Virginia. *Urban Stud.* **1993**, *30*, 299–308. [CrossRef]
21. Partridge, M.D.; Betz, M.R.; Lobao, L. Natural Resource Curse and Poverty in Appalachian America. *Am. J. Agric. Econ.* **2013**, *95*, 449–456. [CrossRef]
22. Bernhardt, E.S.; Lutz, B.D.; King, R.S.; Fay, J.P.; Carter, C.E.; Helton, A.M.; Campagna, D.; Amos, J. How Many Mountains Can We Mine? Assessing the Regional Degradation of Central Appalachian Rivers by Surface Coal Mining. *Environ. Sci. Technol.* **2012**, *46*, 8115–8122. [CrossRef]
23. National Toxicology Program. NTP Research Program on Chemicals Spilled into the Elk River in West Virginia, Final Update. 2016; pp. 1–7. Available online: https://ntp.niehs.nih.gov/ntp/research/areas/wvspill/wv_finalupdate_july2016_508.pdf (accessed on 22 August 2019).
24. Levêque, J.G.; Burns, R.C. Predicting water filter and bottled water use in Appalachia: A community-scale case study. *J. Water Health* **2017**, *15*, 451–461. [CrossRef]

25. Dillman, D.A.; Smyth, J.D.; Christian, L.M. *Internet, Phone, Mail, and Mixed-Mode Surveys: The Tailored Design Method*; John Wiley & Sons: Hoboken, NJ, USA, 2014.
26. Loomis, D.K.; Paterson, S. A comparison of data collection methods: Mail versus online surveys. *J. Leis. Res.* **2018**, *49*, 133–149. [CrossRef]
27. Bates, D.; Maechler, M.; Bolker, B.; Walker, S. Fitting linear mixed-effects models using lme4. *J. Stat. Softw.* **2015**, *67*, 1–48. [CrossRef]
28. R Core Team. *R: A Language and Environment for Statistical Computing*; R Foundation for Statistical Computing: Vienna, Austria, 2018; Available online: https://www.R-project.org/ (accessed on 19 March 2019).
29. Gouhier, T.C.; Guichard, F. Synchrony: Quantifying variability in space and time. *Methods Ecol. Evol.* **2014**, *5*, 524–533. [CrossRef]
30. Adler, N.E.; Boyce, W.T.; Chesney, M.A.; Folkman, S.; Syme, S.L. Socioeconomic Inequalities in Health: No Easy Solution. *JAMA* **1993**, *269*, 3140–3145. [CrossRef]
31. Scott, D.; Willits, F.K. Environmental Attitudes and Behavior: A Pennsylvania Survey. *Environ. Behav.* **1994**, *26*, 239–260. [CrossRef]
32. Evans, G.W.; Kantrowitz, E. Socioeconomic status and health: The Potential Role of Environmental Risk Exposure. *Annu. Rev. Public Health* **2002**, *23*, 303–331. [CrossRef]
33. Hendryx, M. Poverty and Mortality Disparities in Central Appalachia: Mountaintop Mining and Environmental Justice. *J. Health Disparit. Res. Pract.* **2011**, *4*, 44–53.
34. Chang, H. Spatial Analysis of Water Quality Trends in the Han River Basin, South Korea. *Water Res.* **2008**, *42*, 3285–3304. [CrossRef]
35. Neal, C.; Jarvie, H.P.; Love, A.; Neal, M.; Wickham, H.; Harman, S. Water Quality along a River Continuum Subject to Point and Diffuse Sources. *J. Hydrol.* **2008**, *350*, 154–165. [CrossRef]
36. Bullard, R.D. *Dumping in Dixie: Race, Class, and Environmental Quality*; Boulder, Color. Westview; Routledge: Abingdon, UK, 1990.
37. Kurtz, H.E. Scale Frames and Counter-Scale Frames: Constructing the Problem of Environmental Injustice. *Political Geogr.* **2003**, *22*, 887–916. [CrossRef]
38. Fairbrother, M. Rich People, Poor People, and Environmental Concern: Evidence across Nations and Time. *Eur. Sociol. Rev.* **2013**, *29*, 910–922. [CrossRef]
39. Robinson, K.G.; Robinson, C.H.; Hawkins, S.A. Assessment of Public Perception Regarding Wastewater Reuse. *Water Sci. Technol. Water Supply* **2005**, *5*, 59–66. [CrossRef]
40. Ladle, R.J.; Gillson, L. The (Im) balance of Nature: A Public Perception Time-Lag? *Public Underst. Sci.* **2009**, *18*, 229–242. [CrossRef] [PubMed]
41. Levêque, J.G.; Burns, R.C. Water Quality Perceptions and Natural Resources Extraction: A Matter of Geography? *J. Environ. Manag.* **2019**, *234*, 379–386. [CrossRef] [PubMed]
42. Alessa, L.; Kliskey, A.; Brown, G. Social–ecological hotspots mapping: A spatial approach for identifying coupled social–ecological space. *Landsc. Urban Plan.* **2008**, *85*, 27–39. [CrossRef]
43. Cantrill, J.G. The Environmental Self and a Sense of Place: Communication Foundations for Regional Ecosystem Management. *J. Appl. Commun. Res.* **1998**, *26*, 301–318. [CrossRef]

Article

Factors Influencing the Adoption of Water Conservation Technologies by Smallholder Farmer Households in Tanzania

Srijna Jha [1,2,*], Harald Kaechele [1,3] and Stefan Sieber [1,2]

1 SusLAND: Sustainable Land Use in Developing Countries, Leibniz Centre for Agricultural Landscape Research (ZALF), Eberswalder Straße 84, 15374 Müncheberg, Germany; harald.kaechele@zalf.de (H.K.); stefan.sieber@zalf.de (S.S.)
2 Department of Agricultural Economics, Humboldt University Berlin, Unter den Linden 6, 10117 Berlin, Germany
3 Department of Landscape Management and Nature Conservation, Eberswalde University for Sustainable Development (HNEE), Schicklerstraße 5, 16225 Eberswalde, Germany
* Correspondence: srijna.jha@zalf.de; Tel.: +49-1636-960-469

Received: 20 November 2019; Accepted: 4 December 2019; Published: 13 December 2019

Abstract: In Tanzania, the increasing population coupled with climate change amplifies issues of food insecurity and negatively impacts the livelihoods of smallholder farmer households. To address these issues a range of water conservation techniques (WCTs) have been useful. However, the adoption of these WCTs in Tanzania has been limited due to many reasons. With the objective to better understand and identify the factors that significantly influence the adoption of WCTs in Tanzania, the study uses survey data from 701 smallholder farmer households and a bivariate logistic regression, to provide, for the first time, a comprehensive model for the adoption of WCTs in Tanzania that includes a range of individual, household, socio-economic, and farmer perception related variables (factors). The evaluation shows that 120 farmers (17.12%) adopted WCTs and finds the farmer perceptions of rainfall instability, household wealth, and food security to be crucial. The results suggest that policy interventions should encourage conservation behavior (especially when the rainfall is perceived to be uncertain), emphasize the economic and food security-related benefits of adopting WCTs, include strategies that make adoption of WCTs attractive to female-led households, attempt to reach greater number of farmers via social networks and provide better access to public funds for farmers.

Keywords: decision-making; logit regression; farmer perceptions; social networks; public funds; water conservation adoption

---

1. Introduction

Owing to the topography and the changing climate, Sub-Saharan Africa (SSA) continues to face issues of food and water insecurity. Here chronic food insecurity, including the threat of famine, as well as malnourishment remains endemic. The most vulnerable are the small-holder farmers in rural areas where agrarian dependency and sensitivity to climate fluctuations are greater [1].

Tanzania, a predominantly agrarian economy, is one of the fastest-growing economies in SSA, but economic growth has not equally benefited all areas of the country [2]. Agricultural production accounts for nearly half of Tanzania's GDP [3], but rural areas, in particular, remain underdeveloped, productivity in agriculture lags and resources to improve the agricultural sector are needed [3–5]. Agriculture in Tanzania is predominantly rain-fed and directly dependent on annual rainy seasons [3]. A close relationship between variations in the amount of rainfall and economic growth is observed by various studies in Tanzania [3,6,7].

The high dependency on rain-fed agriculture is significantly impacted by the variability of rainfall (amount and distribution) in Tanzania [8]. These issues of high fluctuations in rainfall often manifest as droughts, famines, and floods, severely impacting the livelihoods and food security of the small-holder farmers [9–11]. For example, in Tanzania droughts and floods have been reported to cause failure and damage to crop and livestock leading to chronic food shortages [9–11]. The studies conducted by [7,12] revealed that changes in rainfall patterns and amounts have and are predicted to lead to loss of crops and reduced livestock production. The agricultural sector suffers an estimated $200 million in average annual losses because of weather-related incidents, particularly drought. In 2017, aggregate food prices increased by 12% due to drought-related food shortages. Another drought in 2009 resulted in the mortality of 80% of livestock in northern Tanzania [5]. Such extreme weather-related events and variable rainfall severely undermine the local and national development goals, and are predicted to continue to amplify issues of food security in the region [13].

Furthermore, in Tanzania the agriculture sector uses the vast majority of water resource available. Agriculture accounts for around 89% of total water used in Tanzania, which is high against a global average of 70% [5]. This makes water a critical input for Tanzania's economy, which heavily relies on the performance of the agricultural sector. In such a scenario of high dependence on agriculture, variability of rainfall and increasing demand for water, the adaptation measures towards ensuring water availability on farmland become critical. Water conservation methods seem to be effective for food security and reducing poverty in Tanzania [14]. A study by [15] found evidence of increased crop yields and long term financial profitability in the West Usambara highlands of Tanzania due to adoption of soil and water conservation technologies (WCTs). Another study by [16] in a semi-arid region of Tanzania showed that there is scope to improve grain yields, with the little available rainfall, through the adoption of techniques that promote water availability and retention within the field. In this context, WCTs are imperative for better agricultural production and food security in Tanzania.

Several efficient WCTs have been developed and utilized by farmers in the last decades. However, the adoption of these WCTs remains low in Tanzania [17,18]. The reasons for this are many and vary location to location. Relatively little work has been done to examine the adoption of WCTs at the farm and household levels in Tanzania. A study by [17] presents a model for farm-level adoption of soil and WCTs in the West Usambara highlands of Tanzania, with a focus on socio-economic factors. A study by [19] emphasizes the role of institutional and economic factors for the adoption of soil and WCTs in the semi-arid areas of Tanzania. Another study by [18] concludes that better agricultural water management can be achieved in Tanzania by matching adaptation measures to the farmers' local conditions. As study by [20] suggests a need for formal integration of sociologic, economic, and psychological variables in the adoption models. Whereas [21] argue that many factors affect the adoption decision, which may be determined by the historical, political, ecological, socio-cultural, and economic conditions. In this sense most adoption studies are limited and a comprehensive model to study adoption of WCTs at the household level in Tanzania does not exist and often focus on one or two dimensions (individual, household, social, economic, technological, environmental).

Furthermore, we found no adoption studies that included farmer perceptions, along with individual, household, and socio-economic factors, to explain the adoption process of WCTs by smallholder farmer households in Tanzania. The studies by [22,23] strongly argue to expand the range of independent variables used in the technology adoption models, and to include variables representing a farmer's subjective perceptions along with the more standard individual, household, and socio-economic variables. In [24], a strong case is made for a better and comprehensive understanding of farmer perceptions and associated constraints for the design and promotion of soil and WCTs. The authors suggest that the decision of a farmer to invest in soil and WCTs is influenced by the perceptions of the farmers, which, in turn, is driven by a range of institutional, socio-economic, biophysical, and attitude related factors.

Similar to [21,24], we argue that adoption studies that include farmer perceptions, along with other individual, household, and socio-economic conditions are useful for understanding the adoption of WCT and may guide management and policy interventions better than one-dimensional adoption models. Therefore, the study takes a broader approach and provides a framework for analysis that includes a wide range of factors that reflect the individual, household, and socio-economic characteristics of the farmers, along with factors that reflect farmer perceptions. The study poses two research questions:

1. What is the adoption level of WCTs?
2. What are the most significant variables (factors, drivers of adoption) associated with adoption of "water conservation technologies" in Tanzania?

The study contributes first by adding to the very limited literature on the adoption of WCTs by smallholder farmer households in Tanzania. Second, the analysis provides new evidence on the significance of factors related to farmer perceptions for the successful adoption of WCTs and its relevance for policy. Third, it includes a wide range of factors (individual, household, socio-economic) that may influence the adoption of WCTs, which allows the study to be adapted easily. Hence it provides a unique dataset, discusses implications for the adoption of WCTs from a unique multidimensional perspective, and provides a starting point for policy interventions to better manage adoption of WCTs amongst smallholder farmer households.

2. Literature Review

2.1. Definition of Adoption

Adoptions are the processes governing the utilization of innovations. Innovations are defined here as new methods, customs, or devices used to perform new tasks [25]. In defining "adoption", the study considers adoption to be a dichotomous variable [26]. A farmer is defined as an adopter if he or she has one or more WCTs implemented and functioning [27] at the time of the survey. The WCTs, as reported by the farmers, referred to here, include rooftop rainwater harvesting (RWH), micro-dam surface water runoff collection, on-farm runoff water harvesting (planting pits, furrows), and diversion of water (spate irrigation) (Appendix A). The farmers based on relevance for their agricultural systems selected these four WCTs. For the purpose of this study the adoption of any of these four WCTs was considered as adoption.

2.2. Adoption Models

The study considers the three adoption models—innovation diffusion model, economic constraints model, and user context model—proposed by [28]. The innovation diffusion model follows the work of [29]. The economic constraints model emphasizes that the resources the potential adopter have, often determines their adoption behavior. The user context model assumes that the potential adopter's agro-ecological, institutional, and socio-economic factors drive adoption behavior [28]. In line with the objectives, the economic constraints model and the user-context adoption model are of particular interest. The models allow for inclusion of the variables related to resources available and the seldom-studied perception of an individual farmer regarding their socio-economic status and environment. The models signify the role of farmers in the adoption process.

2.3. Variables

In general, literature shows adoption of WCTs to be a function of a multitude of variables (factors), which differ study to study. Based on the economic constraint and user-context adoption models, for the study, it is assumed that adoption behavior (action) is influenced by individual, household, socio-economic characteristics, and perceptions of the farmers. Each of these factors is assumed to influence the adoption process.

2.3.1. Individual and Household

Age, health, gender of the head of the household, education, ability to read and write, and risk-taking attitude have been shown to be associated with adoption in numerous studies [30–35]. Most studies that have examined age as a determinant of water conservation have found that older people are more likely to be water conservers [36–38]. Poor health acts as a constraint and may result in low adoption rates of WCTs, since households may lose important labor due to illness [39]. Most studies show the gender of the head of the household to be positivity related (i.e., male-headed households are more likely to adopt new agricultural technologies compared to female-headed households) [40,41]. Inconsistencies emerge from the research investigating the impact of education on water conservation behavior. Some researchers report a positive relationship between education and water conservation [42–44]. Other researchers show an inverse relationship. In particular, they found that it is less educated individuals that show both more water conservation behavior and higher water conservation intentions [36,38]. Generally the ability to read and write allows the adopter access, comprehension, and deliberates the consequences of adoption of WCTs. As argued by [45], risk, uncertainty, and learning play a number of distinct roles in the process of adopting new technologies. Studies show risk aversion to be an important factor affecting adoption [46]. A few studies have found adoption to be strongly related to geographic location [39,47]. Household size facilitates the division of labor, investment, knowledge sharing, awareness, and experience required for the adoption process [48–51]. Household water consumption is inversely related to adoption [52] and there is a positive relationship between the number of residents and water use [36].

2.3.2. Socio-Economic

A wide range of social and economic parameters impact adoption. Social networks on the household level are important to adoption. They can be seen as a process of imitation of behavior, wherein contacts with others led to the spread of technology [53,54]. Social ties within and outside an organization provide extensiveness, quality, and diversity that may drive adoption behavior [55]. Several studies have found social networks to positively influence the adoption process [39,56–61]. Wealth and differential access to capital are often cited to explain different rates of adoption of WCTs. Credit constraints tend to negatively impact the adoption of WCTs, especially if a capital investment is required. Adoption behavior may not be affected by credit constraints if the adopter has other sources of finance, such as access to microcredits, public funds, and/or own savings [39,52,62–64]. Pieces of land owned or used by small-scale farmers (land fragmentation) could be a determinant of the adoption of land, soil, and water conservation measures [30]. Ownership of more pieces of land is associated with greater wealth and the increased availability of capital resources, which increases the likelihood of farmers making investments in land, soil, and water conservation measures [65]. Farm size is often cited as a crucial variable that impacts the adoption of agricultural technologies [66]. While some studies found a significant relationship between farm size and tractor adoption [67], others found no relation at all [68]. Literature shows a negative and significant effect of reliance on off-farm employment by poorer population segments. Higher opportunity costs of household labor are realized with the increased availability of off-farm work [52,69]. Results from studies that examine income stability as a determinant of water conservation behavior are consistent [63,70]. Research generally shows that individuals with a stable and/or high income conserve more water [71].

2.3.3. Farmer Perceptions

The diversity of activities, experiences, and observations from the past shape perceptions and influence actions. People's perception of climate change, the environment, and change in rainfall influences their level of concern, which affects their motivation to act [72]. In line with this assumption, some studies indicate that concern about climate change increases consumers' willingness to modify their behaviors [73]. The study of [74] sought to identify the relationship between specific knowledge of environmental problems and water conservation behavior, finding that individuals who reported greater awareness of environmental problems also reported greater conservation actions. Environmental knowledge has predictive power in terms of pro-environmental behavior [75]. A study by [76] found that when farmers perceive the rainfall to be getting less and unstably low they tend to adopt WCTs more. The study of [77] states that if rainfall instability is a continuing concern then farmers may adopt WCTs less. The perception of wealth, personal and household, is included to be an important variable for adoption decisions in developing countries [39]. Wealthy farmers may focus on other income-generating activities and they may give less attention to WCTs [78]. No studies have explored the impact of the farmers perception of his or her household food security on the adoption of WCTs. The studies of [30,79] discuss the role of household food security on adoption, but do not relate it to farmer perceptions.

3. Materials and Methods

3.1. Study Area and Data Collection

Two districts in Tanzania, Morogoro and Dodoma (Figure 1), were selected because of having the highest rates of food insecurity in Tanzania, 38% for Dodoma and 34% for Morogoro [80]. The two regions also vary in their topography and environmental conditions. Morogoro has a semi-humid climate with an annual rainfall of 600–800 mm. The region is diverse with flat plains, highlands, and dry alluvial valleys. The main crops are maize, sorghum, legumes, and rice, with partial livestock integration. Dodoma is the semi-arid region with an annual rainfall of 350–500 mm and mostly small hills and flat plains. The main crops are sorghum and millet, with extensive livestock integration [81]. Low yields and frequent crop failures due to climate change and limited use of agricultural technologies are reoccurring issues [82].

Figure 1. Map of Tanzania, showing the case study regions of Morogoro and Dodoma.

Data was collected as part of a larger survey carried out by the TRANSEC project [83] and analyzed using Stata IC 15.1. The questionnaire covered a wide range of issues including information on individual, household, and socio-economic background. Agricultural systems and farmer perceptions of climate change, their own household, and adaptation measures were also included. A total of 900 households were randomly selected in the Dodoma and Morogoro regions in rural Tanzania in 2016 [83]. Household heads were interviewed face to face and were all smallholder farms (under 2 ha).

Some respondents did not answer all of the questions; therefore, only 701 datasets, where full information for all relevant variables for this study was available, were used. The group consisted of 329 farmers from Morogoro and 372 farmers from Dodoma, of which 542 were males and 159 were females. The average household size was 5 people. A comparison of the data analyzed to the excluded data showed no significant difference between the included and excluded datasets with respect to these key variables (age, gender, health, education). Thus, it was appropriately assumed that no systematic bias exists and that the included datasets remain representative of the larger sample.

3.2. Model Selection and Adequacy

Opting for the best fit, a logit model (regression) was used to analyze the adoption of WCTs in Tanzania. The logit model allows for the outcome to have a binary value (i.e., adoption happens or it does not (1 or 0)) [84]. In this study, farmers were classified as adopters and non-adopters of WCTs. A value of 1 was assigned to farmers that adopted WCTs and 0 to non-adopters. Hence, for our case the logit model had the obvious advantage and suited the research objectives. Two automated variable selections were done using the backward and forward method. Independent variables were added and removed one by one until all the independent variables included in the model were found to be significant. Both forward and backward selection methods resulted in the same set of variables to be included in the model. Hence the model is well specified. To estimate the goodness-of-fit we used the Hosmer–Lemeshow test. The *p*-value for our model is 0.6216 and; therefore, adequately fits the data. The rate of correct classification is estimated to be 85.73%

3.3. Empirical Model Specification

The probability that a WCT will be adopted is defined as,

$$\mathrm{Logit}(Y) = \alpha + \Sigma\beta_1 X_1 + \Sigma\beta_2 X_2 \ldots + \Sigma\beta_n X_n + \varepsilon_i.$$

Here adoption, "Y", is a dummy and dependent variable indicating the decision to adopt or not. Y = dependent variable (adoption of water conservation measures), with 1 = adopters and 0 = non-adopter; α = intercept; $\beta_1, \ldots, \beta_n$ = coefficients of the independent variables indicating the influence of these variables on the likelihood of adoption; $X_1, \ldots, X_{19}$ = the independent variables. Appendix B gives the definition and summary statistics of all the independent variables (X_1 to X_{19}).

3.4. Variable Specification and Expected Outcomes

A range of individual, household, socio-economic, and farmer perception related variables were selected from literature to be included in the analysis (Appendix B). Variables such as education, loans (taken and given), and perception of personal wealth were found to be collinear to the ability to read and write, savings, and perception of household wealth, respectively, hence they were excluded from the analysis. The land owned and farm size was found to be homogenous (average was 0.97 ha), hence excluded from the analysis.

The expected effects of the independent variables included in the logit model for the adoption of WCTs by households are discussed and summarized in Table 1.

Table 1. The expected relationship between adoption of water conservation technologies (WCTs) and all independent variables.

Symbol	Variables	Hypothesis	Expected Outcome	References
		Individual and Household		
X_1	Adopter's age	Older farmers have traditional knowledge, experience, and a better understanding of their farming systems, and are better prepared for the adoption of WCTs. Hence, a positive relationship between age and the adoption of WCTs is expected.	Positive	[36–38]
X_2	Adopter's health	A healthy adopter is more likely to have time, energy, and money to invest in new innovations. Hence positively related to the adoption of WCTs.	Positive	[39]
X_3	Gender of head of household	Tanzania is primarily a patriarchal society, where the decision-making power lies with the heads of households who are mostly males. Female-led households face labor and time constraints. Hence gender (female) is anticipated to be negatively related to the adoption of WCTs.	Negative	[40,41]
X_4	Adopter's ability to read and write	The ability to read and write indicates a cognition level capable of efficiently receiving, processing, and analyzing information from various channels (media, internet). Hence, we would expect a positive relationship between the ability to read and write and the adoption of WCTs.	Positive	[42–44]
X_5	Adopter's attitude towards risk	Perceptibility towards risk would indicate curiosity and interest in the unknown. Hence, the attitude towards risk would be positively related to the adoption of WCTs.	Positive	[46]
X_6	Adopter's region	Since Morogoro is a wet region, whereas Dodoma is a dry region, the need for water conservation was expected to be much higher in Dodoma due to scarcity.	Positive	[39,47]
X_7	Household size	Assumed that a larger household size will facilitate the division of labor and better time management; hence positively related to adoption of WCTs.	Positive	[48–51]
X_8	Household water usage	A high water consumption pattern would indicate more demand for water and may result in scarcity; hence the likelihood of adoption of WCTs is negative.	Negative	[52]
		Socio-Economic		
X_9	Membership in social networks	In rural areas, social organizations have proved to benefit farmers from any kinds of support, such as access to credit and training from government institutions and NGOs. It allows for exchange of knowledge, ideas, resource sharing, and implementation strategies. Hence, being a member of a farmers' organization should increase the likelihood of the adoption.	Positive	[39,56–58, 60,61]
X_{10}	Access to micro-credits	Access to micro-credits would allow the adopter to access money required for initial setups and adoption of new innovation; hence positively related to the adoption of WTCs.	Positive	[39]
X_{11}	Access to public funds	Access to public funds and programs provides the adopter with knowledge, network, training, and economic resources for the adoption of WCTs.	Positive	[39,52,62–64]
X_{12}	Household savings	Having savings provide the adopter the security, the economic resource needed for adoption of WCTs.	Positive	[39]
X_{13}	Off-farm employment	Off-farm employment would indicate less labor for the farm; hence negatively related to adoption of WCTs.	Negative	[52,69]
X_{14}	Household income fluctuation	Stability and low to nil fluctuation of household income would allow the adopter more time and planning options; hence positively related to the adoption of WCTs.	Positive	[63,70]

Table 1. *Cont.*

Symbol	Variables	Hypothesis	Expected Outcome	References
		Farmer Perceptions		
X_{15}	Perception of change in rainfall	"Some" rainfall although erratic maybe preferred to less or no rainfall and that "consistent" rainfall may be preferred to less rainfall and erratic rainfall; hence positively related to adoption of WTCs.	Positive	[76]
X_{16}	Perception of climate change	An adopter's observation and recognition of a changing climate would indicate an awareness of the negative consequences of CC on poverty and food security. It affects the intention to behave and the attitude towards the behavior, in our case water conservation practices (theory of reasoned action); hence a perception of climate change and environment both would be positively related to adoption of WTCs.	Positive	[73]
X_{17}	Perception of change in environment	An adopter's observation and recognition of a changing environment would indicate an awareness of the negative consequences of degrading environment on poverty and food security. It affects the intention to behave and the attitude towards the behavior, in our case water conservation practices (theory of reasoned action); hence a perception of climate change and environment both would be positively related to adoption of WTCs.	Positive	[74,75]
X_{18}	Perception of household wealth	Better wealth status of the household would reflect a better capacity to access, utilize and diversify input resources needed for water conservation and positively related to adoption of WCTs.	Positive	[39,78]
X_{19}	Perception of household food security	Having a food secure household would allow the farmer to think and invest in practices to enhance his food-value chain system; hence positively related to the adoption of WCTs.	Positive	[30,79]

4. Results

Although 98% of the respondents in the study believed that the climate is changing, the rate of adoption of WCTs was at a low of 17.12%. Amongst the adopters of WCTs, 39% implemented on-farm runoff water harvesting, 35% implemented rooftop rainwater harvesting, 18% implemented micro-dams/surface water runoff collection, and 9% implemented diversion of water (spate irrigation). On-farm runoff water harvesting and rooftop rainwater harvesting were the most adopted WCTs. Whereas, the adoption of spate irrigation was found to be the most limited WCT.

Out of all the variables analyzed, the study finds six variables to have a significant impact on the adoption of WCTs in Tanzania. These are one individual and household variable—the gender of the head of household; two socio-economic variables—membership in social organizations and access to public funds; and three farmer perceptions related variables—farmer perception of change in rainfall, household wealth, and food security. This study finds that using a wide dataset to comprehensively study the adoption process of WCTs by smallholder farmers is useful. The results show that all, individual, household, socio-economic, and farmer perceptions factors are important to explain the adoption process of WCTs in the study region. Interestingly, of the six significant variables, we find three variables (50%) to be related to farmer perceptions. This strongly suggests that the perceptions of the farmers' largely shape their adoption decisions and that a failure to consider the characteristics of the household and environment as perceived by the adopters (farmers) themselves may significantly hinder the adoption of WCTs.

The results of the logit model (Table 2) further show that the individual, household, socio-economic, and farmer perceptions related variables affect the adoption of WCTs differently. In line with the expected outcomes (Table 1), the study finds that women-led households have a lower likelihood of adoption of WCTs and those farmers who have access to social networks and public funds have a higher likelihood of adopting WCTs. In contrast to the expected outcome (Table 1), the study finds that a farmer's perception of rainfall instability has a significant negative influence on the adoption of WCTs. Whereas, a positive perception of household wealth and food security by the farmer has a significant positive influence on the adoption of WCTs, as expected.

Table 2. Results of the logit model for the adoption of WCTs in Tanzania. Only the significant variables are shown.

Significant Independent Variables	Coefficient	p-Value	Significance
Individual and Household			
Gender of head of household—female	−0.570	0.071	*
Socio-Economic			
Membership in social network—yes	2.065	0.000	***
Access to public funds—yes	1.298	0.000	***
Farmer Perceptions			
Perception of change in rainfall—more Erratic	−0.789	0.002	***
Perception of change in rainfall—no change	−2.317	0.003	***
Perception of household wealth—better off	0.474	0.081	*
Perception of household food security—secure	0.626	0.012	**
Hosmer–Lemeshow test ^	0.6216		
Pseudo R-squared	0.2500		
Prediction statistics (correctly classified)	85.73%		

^ if the value is less than 0.05, the model is a poor fit; *** Significant at $p < 0.01$ (99%); ** Significant at $p < 0.05$ (95%); * Significant at $p < 0.1$ (90%) level.

The study finds no statistically significance for the individual and household variables related to adopter's age, health, ability to read and write, attitude towards risk, region, household size, household water usage, and adoption of WCTs in the study region. The same was found for socio-economic variables, such as access to microcredits, savings, off-farm employment, and household income fluctuations. Similarly, the farmers perception and recognition of the changing climate and environment has no statistical significance for the adoption of WCTs. Based on the literature review and general characteristics of the case study region we were expecting a positive (except household water usage and off-farm employment) relationship between these variables and the adoption of WCTs. However this was not observed in the sample analyzed.

In the next section, the six variables (factors) found to be significant for the adoption of WCTs in the case study region are discussed.

4.1. *Individual and Household*

Gender of Head of Household

Households with a female head are less likely to adopt WCTs as compared to households with male heads. We find that the WCTs in the case study region are labor intensive and need a sufficient input of time, which largely differs for males and females—the decision to spend time on activities related to conserving water vs. household activities. The labor and time constraint, hence, effects the relative advantage and compatibility of the WCTs with existing family and farming structures [22,85–87]. Several studies have investigated the association of the gender of the head of household and adoption. Most studies suggest that male-headed households are more likely to adopt new agricultural technologies compared to female-headed households [40,41]. In the study of [40] about improved maize technology in Ghana, they distinguish between the gender of the farmer and the gender of the head of the household. They find that the gender variable does not have any explanatory power regarding the decision to adopt, but the females living in female-headed households adopt at a lower rate than individuals in male-headed households. A study by [88], regarding peanut production in Eastern Uganda, finds that females living in female-headed households are less likely to adopt new varieties than females or males living in male-headed households. Their decision to adopt is affected by the available labor and time constraints.

4.2. Socio-Economic

4.2.1. Membership in Social Network

The study finds that farmers who are part of a social network are more likely to adopt WCTs than farmers who are not part of social networks. Social networks enable the sharing of information, thus making the complexity of the WCTs more manageable and highlighting the benefits of conservation behavior [85,86]. Consistent with previous studies [39,56–58], membership in social networks is useful in explaining the adoption decision of WCTs. The study by [17] finds that farmers who are part of a group are more likely to adopt soil and WCTs in Tanzania. The authors suggest that membership in farmer groups and contact with extension agents positively influences the adoption of WCTs in Tanzania. The study of [60] finds that being part of a group enhances social networking, which in turn facilitates the sharing of experiences and building confidence in those farmers interested in the agricultural technologies [61].

4.2.2. Access to Public Funds

Farmers who have access to public programs and funds are more likely to adopt WCTs. We find that access to public funds and programs facilitate the initial adoption, provide for an information exchange network and institutional support. The public programs and funds provide economic support and may be used to shape attitudes towards adoption of WCTs. For example, public programs and funds, which support smallholder financing, can be an important adoption driver to overcome wealth constraints to investment in new technologies [62]. Once the initial adoption process is carried out, the understanding of the expected outcomes and benefits [22,85–87], of a WCT may become evident. Access to public programs or funds also indicates the motive of the government towards the farmers [89]. Similar to the findings of this study, [52,63,64] find that public programs can play a positive role in creating incentives for adoption. Studies by [90] in Ghana and [78] highlight the significance of financial inputs for the adoption of WCTs by smallholder farmers.

4.3. Farmer Perceptions

4.3.1. Perception of Change in Rainfall

Contrary to our hypothesis, when a farmer perceives that the rainfall is stagnant and/or getting more erratic (unclear onset and ending of rains), meaning that the regularity of rainfall is uncertain, they are less likely to adopt WCTs. It was assumed that an element of uncertainty would encourage the farmer to prepare for potential shocks; however, this was not the case. We find that if the farmers perceive that rainfall (water) is getting scarce then the efforts to conserve water decrease. We find that conservation behavior is hampered when the resource available is limited, that is, there is no water to conserve. Uncertainty in the future leads to a diversion from conservation actions due to a low perceived benefits [85,86]. In contrast a study by [76] in Ethiopia found a positive association between farmer perception of low and erratic rainfall and adoption of WCTs. In a study in Tanzania, [77] outlines that when farmers experience unreliable rainfall as a constant, they may become habituated and; therefore, not perceive the risk as urgent or immediate.

4.3.2. Perception of Household Wealth Status

Farmers whose household wealth status is better off than last year, and as compared to the rest of the village, are more likely to adopt WCTs. Perception of a better household wealth status reflects a better capacity to access, utilize, and diversify input resources needed for the adoption of WCTs. It leads to a sense of economic security which facilitates the investment and learning about the benefits of WCTs [22,85–87]. Similar to [39], we find that a perception of household wealth positively influences the adoption process of WCTs. While we found no other studies investigating adoption and the perceptions related to wealth, we did find two studies by [56,91] who reported a positive influence of wealth on adoption in Tanzania and Ethiopia, respectively.

Other studies in Chile [52] and Ethiopia [92] have found that wealthier farmers were able to take on greater levels of risk, which gave them an advantage as adopters. In contrast, [77] finds that the wealthier farmers in Tanzania have more options to deal with water scarcity and low yields, which may make the issues of water conservation less important for them.

4.3.3. Perception of Household Food Security

Farmers who perceive their household to be food secure are more likely to adopt WCTs as compared to farmers who think they are food insecure. This is perhaps because enough food for the household members reflects a sufficient livelihood, which allows the farmer time and resources to test and understand the benefits of adopting WCTs. This, in turn, would form the behavior without the presence of significant stressors like food insecurity [22,85–87]. The impact of adoption of WCTs on food security has been studied extensively; however, no studies have explored the impact of farmer perception of household food security on the adoption of WCTs. The studies of [71,79] argue that stressors, such as food scarcity, have a negative impact on the adoption decision. Food insecure households must focus their efforts on coping strategies, rather than on conservation [93].

5. Conclusions

The study presents a model for the adoption of WCTs by smallholder farmers in Tanzania, based on data collected from 701 smallholder farmer households, and discusses the most significant factors that influence the adoption decisions of the farmers.

The adoption of WCTs in the case study area was at a low of 17% and emphasizes the urgency to comprehensively study adoption processes at the smallholder farmer household level.

The adoption of WCTs by smallholder farmer households in Tanzania can be effectively explained by integrating individual, household, socio-economic, and farmer perceptions related variables. The study finds that an integrated approach to study the adoption of WCTs better explains the adoption decisions, opportunities, and constraints that farmers face at the household level and allows for targeted agricultural management at the household level.

In order to ensure food and livelihood security for the most vulnerable, agriculture and policy interventions should better address gender-based disparities, such as labor and time constraints associated with adoption of WCTs. Furthermore, agriculture and policy interventions should aim to better reach, include, and integrate those farmers that are left behind (e.g., farmer's with no ties to social networks and farmer's with no access to public funds).

How farmers perceive the changes in rainfall, their household economy, and food security situation significantly influences their adoption decisions. Therefore, agriculture and policy interventions should emphasize the importance of adoption of WCTs in times when the farmers perceive the rainfall to be uncertain or scarce. Furthermore, agriculture and policy interventions should highlight the economic and food security related benefits that come from adoption of WCTs. The study provides evidence and determines that farmer perceptions are fundamental to studying and developing a well-targeted agricultural strategy at the household level; therefore, they should be included in adoption studies and further investigated. The study is limited by its scope, which focuses only on the determinants and not the effects of adoption of WCTs.

Author Contributions: Conceptualization, S.J.,H.K. and S.S.; Data Curation, S.J.; Formal Analysis, S.J.; Investigation, S.J.; Methodology, S.J.,H.K. and S.S.; Software, S.J.; Supervision, H.K. and S.S.; Validation, S.J., H.K. and S.S.; Visualization, S.J.; Writing—Original Draft, S.J.; Writing—Review and Editing, S.J., H.K. and S.S.

Funding: This research received no external funding.

Acknowledgments: We are thankful to the whole TRANSEC project team and SusLAND team for their administrative support and for providing necessary data and technical guidance. We are also grateful to the team at Sokoine University of Agriculture (SUA); for the provision of expertise; and technical support in the larger survey and data collection. We would also like to express our sincere thanks to the reviewers and peers who contributed their time to provide feedback and comments to revise the study.

Conflicts of Interest: The authors declare no conflicts of interest.

Appendix A

Types of WCTs used by the farmers and definitions.

1. Rooftop RWH involves three primary components: catchment, conveyance, and a collection device. The catchment is the rooftop where rainwater falls naturally. Rainwater drains down the slanted rooftop to the conveyance instruments, or gutters, at the base of the roof. These gutters transport the water from the rooftop to the collection device or storage tank [94].
2. Micro-dam surface water runoff harvesting is the collection of runoff (signifies the water running off surfaces on which rain has directly fallen) on small (~1–1000 m^2) treated catchments to channel it to a storage pit, adjacent cropping areas, or individual plants. The catchments are either modified by some kind of special tillage technique, earthen embankments, or masonry walls [94].
3. On-farm runoff water harvesting is the collection of runoff on a small catchment area to channel it using furrows to a planting pit for on-farm irrigation use. Furrows refer to a micro-catchment harvesting technique in which the rainfall water is harvested through the mulched ridges and the crop is planted in the furrows between the ridges [95]. Planting pits refer to ditches dug on farm with hand-hoe, to a depth exceeding 15 cm, with no soil disturbance in between [96].
4. Spate irrigation has been defined as a system diverting flash floods from the riverbed via canals to fields that may be located some distance from the water source. The word "spate" refers to flood water originating from episodic rainfall in the upper part of river catchments, which in the lower part is diverted from ephemeral rivers and spread over agricultural land [97].

Appendix B

Definition and statistical summary of all the independent variables used in the logit model for household adoption of WCTs in Tanzania using 701 observations

Symbol	Variables	Survey Question	Variable Type	Mean or Proportion
		Individual and Household		
X_1	Adopter's age	Age?	Continuous, 22 to 99	50.81
X_2	Adopter's health status	How healthy is adopter?	Categorical, 1 = Sick, 2 = Can manage, 3 = Healthy	Sick = 0.05, Can manage = 0.15, Health = 0.80
X_3	Gender of head of household	Gender?	Categorical, 0 = Male, 1 = Female	Male = 0.77, Female = 0.23
X_4	Adopter's Ability to read and write	Can the primary adopter read and write?	Binary, 0 = Cannot read and write, 1 = Can read and write	No = 0.33, Yes = 0.67
X_5	Adopter's risk taking attitude	Risk attitude (0: unwilling to take risk, 3: fully prepared to take risks)	Ordinal, 1 = Low, 2 = Medium, 3 = High	Low = 0.22, Medium = 0.50, High = 0.28
X_6	Adopter's Region	Which region of Tanzania does the respondent live and farm in?	Categorical, 1 = Morogoro, 2 = Dodoma	Morogoro = 0.47, Dodoma = 0.53
X_7	Household size	Household nucleus size?	Continuous, 0 to 10	4.59
X_8	Household water usage	Daily water use per household in liter?	Continuous, 4 to 200	70.06

Symbol	Variables	Survey Question	Variable Type	Mean or Proportion
		Socio-Economic		
X_9	Membership in social and or political organizations	Member in organizations/committees/and political parties?	Binary, 0 = Not a member, 1 = Yes, a member	No = 0.56, Yes = 0.44
X_{10}	Access to microcredit	Is anyone in the household member of a micro-credit group?	Binary, 0 = Has no access to microcredit, 1 = Has access to microcredit	No = 0.91, Yes = 0.09
X_{11}	Access to public funds	Has any of household received public transfers (cash/in-kind) in ref. period?	Binary, 0 = Has no access to public funds, 1 = Has access to public funds	No = 0.72, Yes = 0.28
X_{12}	Household savings	Do you have any savings?	Binary, 0 = Has no savings, 1 = Has savings	No = 0.23, Yes = 0.77
X_{13}	Off farm employment	Anyone in household engaged in non-farm self-employment in ref. period?	Binary, 0 = No, 1 = Yes	No = 0.64, Yes = 0.36
X_{14}	Household income fluctuation	How much did your household income fluctuate in 3 years?	Ordinal, 1 = Yes a bit, 2 = Not at all, 3 = Yes a lot	Yes a bit = 0.34, Not at all = 0.11, Yes a lot = 0.55
		Farmer Perceptions		
X_{15}	Perception of change in rainfall	Change in rainfall pattern last 3 years?	Continuous, 1 = Less rain, 2 = More erratic, 3 = No change, 4 = More rain	Less = 0.38, Erratic = 0.40, No change = 0.13, More = 0.10
X_{16}	Perception of climate change	Do you think the climate (weather) in general has been changing in the past 20 years?	Binary, 0 = Climate is not changing, 1 = Yes, climate is changing	No = 0.02, Yes = 0.98
X_{17}	Perception of environmental change	Do you think, your environment has changed in past 20 years?	Binary, 0 = Environment is not changing, 1 = Yes, environment is changing	No = 0.08, Yes = 0.92
X_{18}	Perception of household wealth status	Do you think your household is better off than last year?	Ordinal, 1 = Worse off, 2 = Same, 3 = Better off	Worse = 0.51, Same = 0.21, Better = 0.28
X_{19}	Perception of household food security	Do you consider your household being "food secure"?	Binary, 0 = Not food secure, 1 = Yes, food secure	No = 0.52, Yes = 0.48

References

1. Williams, P.A.; Crespo, O.; Abu, M.; Simpson, N.P. A systematic review of how vulnerability of smallholder agricultural systems to changing climate is assessed in Africa. *Environ. Res. Lett.* **2018**, *13*, 103004. [CrossRef]
2. Pelizzo, R.; Kinyondo, A. Growth, Employment, Poverty and Inequality in Tanzania. *JPAS* **2018**, *13*, 3. [CrossRef]
3. Noel, S. *The Economics of Climate Change in Tanzania Water Resources*; Institute of Resource Assessment, University of Dar es Salaam: Dar es Salaam, Tanzania, 2012.
4. Drakenberg, O.; Ek, G.; Fernqvist, K.W. *Environmental and Climate Change Policy Brief—Tanzania*; Sidas Technical report for Sida Helpdesk for Environment and Climate Change; Sida–the Swedish International Development Cooperation Agency: Stockholm, Sweden, 30 May 2016.
5. The World Bank Water Stress Could Hurt Tanzania's Growth and Poverty Reduction Efforts—New World Bank Report. Available online: https://www.worldbank.org/en/news/press-release/2017/11/06/water-stress-could-hurt-tanzanias-growth-and-poverty-reduction-efforts---new-world-bank-report (accessed on 2 May 2019).
6. Magehema, A.; Chang, L.; Mkoma, S. Implication of rainfall variability on maize production in Morogoro, Tanzania. *Int. J. Environ. Sci.* **2014**, *4*, 1077–1086.
7. Afifi, T.; Liwenga, E.; Kwezi, L. Rainfall-induced crop failure, food insecurity and out-migration in Same-Kilimanjaro, Tanzania. *Clim. Dev.* **2014**, *6*, 53–60. [CrossRef]

8. Kijazi, A.L.; Reason, C.J.C. Relationships between intraseasonal rainfall variability of coastal Tanzania and ENSO. *Theor. Appl. Climatol.* **2005**, *82*, 153–176. [CrossRef]
9. Jackson, S.; Claude, G.M.; Godfrey, F.K. Impacts of Climate Change, Variability and Adaptation Strategies on Agriculture in Semi Arid Areas of Tanzania: The Case of Manyoni District in Singida Region, Tanzania. *Afr. J. Environ. Sci. Technol.* **2018**, *12*, 323–334. [CrossRef]
10. Levira, P.W. Climate change impact in agriculture sector in Tanzania and its mitigation measure. *IOP Conf. Ser. Earth Environ. Sci.* **2009**, *6*, 372049. [CrossRef]
11. Majule, A.E.; Rioux, J.; Mpanda, M.; Karttunen, K. *Review of Climate Change Mitigation in Agriculture in Tanzania*; Food and Agricultural Organization of United Nations: Rome, Italy, 2014.
12. Rowhani, P.; Lobell, D.B.; Linderman, M.; Ramankutty, N. Climate variability and crop production in Tanzania. *Agric. For. Meteorol.* **2011**, *151*, 449–460. [CrossRef]
13. Shayo, C.M. Forests, Rangelands and Climate Change Adaptation in Tanzania. In Proceedings of the Workshop on Forests, Rangelands and Climate Change Adaptation, Johannesburg, South Africa, 17–19 June 2013; pp. 1–14.
14. *FAO Water for Sustainable Food and Agriculture A Report Produced for the G20 Presidency of Germany*; FAO: Rome, Italy, 2017.
15. Tenge, A.J.; De Graaff, J.; Hella, J.P. Financial efficiency of major soil and water conservation measures in West Usambara highlands, Tanzania. *Appl. Geogr.* **2005**, *25*, 348–366. [CrossRef]
16. Makurira, H.; Savenije, H.H.G.; Uhlenbrook, S.; Rockström, J.; Senzanje, A. The effect of system innovations on water productivity in subsistence rainfed agricultural systems in semi-arid Tanzania. *Agric. Water Manag.* **2011**, *98*, 1696–1703. [CrossRef]
17. Tenge, A.J.; De Graaff, J.; Hella, J.P. Social and economic factors affecting the adoption of soil and water conservation in West Usambara highlands, Tanzania. *L. Degrad. Dev.* **2004**, *25*, 348–366. [CrossRef]
18. Kimaro, J. A Review on Managing Agroecosystems for Improved Water Use Efficiency in the Face of Changing Climate in Tanzania. *Adv. Meteorol.* **2019**, *2019*, 1–12. [CrossRef]
19. Boyd, C.; Turton, C.; Hatibu, N.; Mahoo, H.F.; Lazaro, E.; Rwehumbiza, F.B.; Okubal, P.; Makumbi, M. The contribution of soil and water conservation to sustainable livelihoods in semi-arid areas of Sub-Saharan Africa. *Agric. Res. Ext. Netw. Pap.* **2000**, *102*, 1–16.
20. Edwards-Jones, G. Modelling farmer decision-making: Concepts, progress and challenges. *Anim. Sci.* **2006**, *82*, 783–790. [CrossRef]
21. Kideghesho, J.R.; Røskaft, E.; Kaltenborn, B.P. Factors influencing conservation attitudes of local people in Western Serengeti, Tanzania. *Biodivers. Conserv.* **2007**, *16*, 2213–2230. [CrossRef]
22. Adesina, A.A.; Baidu-Forson, J. Farmers' perceptions and adoption of new agricultural technology: evidence from analysis in Burkina Faso and Guinea, West Africa. *Agric. Econ.* **1995**, *13*, 1–9. [CrossRef]
23. Rahm, M.R.; Huffman, W.E. The Adoption of Reduced Tillage: The Role of Human Capital and Other Variables: Reply. *Am. J. Agric. Econ.* **1986**, *66*, 405–413. [CrossRef]
24. Moges, D.M.; Taye, A.A. Determinants of farmers' perception to invest in soil and water conservation technologies in the North-Western Highlands of Ethiopia. *Int. Soil Water Conserv. Res.* **2017**, *5*, 56–61. [CrossRef]
25. Sunding, D.; Zilberman, D. The agricultural innovation process: Research and technology adoption in a changing agricultural sector. *Handb. Agric. Econ.* **2001**, *1*, 207–261.
26. Lin, J.Y. Education and Innovation Adoption in Agriculture: Evidence from Hybrid Rice in China. *Am. J. Agric. Econ.* **1991**, *73*, 713–723. [CrossRef]
27. Doss, C.R. Analyzing technology adoption using microstudies: Limitations, challenges, and opportunities for improvement. *Agric. Econ.* **2006**, *34*, 207–219. [CrossRef]
28. Negatu, W.; Parikh, A. The impact of perception and other factors on the adoption of agricultural technology in the Moret and Jiru Woreda (district) of Ethiopia. *Agric. Econ.* **1999**, *21*, 205–216. [CrossRef]
29. Rogers, E.M. *Diffusion of Innovations*, 4th ed; Simon and Schuster: New York, NY, USA, 2010; ISBN 0-02-926671-8.
30. Mango, N.; Makate, C.; Tamene, L.; Mponela, P.; Ndengu, G. Awareness and adoption of land, soil and water conservation practices in the Chinyanja Triangle, Southern Africa. *Int. Soil Water Conserv. Res.* **2017**, *5*, 122–129. [CrossRef]

31. Willis, R.M.; Stewart, R.A.; Panuwatwanich, K.; Williams, P.R.; Hollingsworth, A.L. Quantifying the influence of environmental and water conservation attitudes on household end use water consumption. *J. Environ. Manag.* **2011**, *92*, 1996–2009. [CrossRef]
32. Sidibé, A. Farm-level adoption of soil and water conservation techniques in northern Burkina Faso. *Agric. Water Manag.* **2005**, *71*, 211–224. [CrossRef]
33. Campbell, H.E.; Johnson, R.M.; Larson, E.H. Prices, devices, people, or rules: The relative effectiveness of policy instruments in water conservation. *Rev. Policy Res.* **2004**, *21*, 637–662. [CrossRef]
34. Berk, R.A.; Schulman, D.; McKeever, M.; Freeman, H.E. Measuring the impact of water conservation campaigns in California. *Clim. Change* **1993**, *24*, 233–248. [CrossRef]
35. Hamilton, L.C. Saving water: A causal model of household conservation. *Sociol. Perspect.* **1983**, *26*, 355–374. [CrossRef]
36. Gregory, G.D.; Di Leo, M. Repeated Behavior and Environmental Psychology: The Role of Personal Involvement and Habit Formation in Explaining Water Consumption. *J. Appl. Soc. Psychol.* **2003**, *33*, 1261–1296. [CrossRef]
37. Russell, S.; Fielding, K. Water demand management research: A psychological perspective. *Water Resour. Res.* **2010**, *46*, 1–12. [CrossRef]
38. Clark, C.F.; Kotchen, M.J.; Moore, M.R. Internal and external influences on pro-environmental behavior: Participation in a green electricity program. *J. Environ. Psychol.* **2003**, *23*, 237–246. [CrossRef]
39. Abdulai, A.; Huffman, W. The Adoption and Impact of Soil and Water Conservation Technology: An Endogenous Switching Regression Application. *Land Econ.* **2015**, *90*, 26–43. [CrossRef]
40. Doss, C.R. Designing agricultural technology for African women farmers: Lessons from 25 years of experience. *World Dev.* **2001**, *29*, 2075–2092. [CrossRef]
41. Kumar, S.K. Adoption of hybrid maize in Zambia: effects on gender roles, food consumption, and nutrition. *Food Nutr. Bull.* **1995**, *16*, 1–3. [CrossRef]
42. De Oliver, M. Attitudes and inaction. A case study of the manifest demographics of urban water conservation. *Environ. Behav.* **1999**, *31*, 372–394. [CrossRef]
43. Gilg, A.; Barr, S. Behavioural attitudes towards water saving? Evidence from a study of environmental actions. *Ecol. Econ.* **2006**, *57*, 400–414. [CrossRef]
44. Lam, S.P. Predicting intention to save water: Theory of planned behavior, response efficacy, vulnerability, and perceived efficiency of alternative solutions. *J. Appl. Soc. Psychol.* **2006**, *36*, 2803–2824. [CrossRef]
45. Marra, M.; Pannell, D.J.; Abadi Ghadim, A. The economics of risk, uncertainty and learning in the adoption of new agricultural technologies: Where are we on the learning curve? *Agric. Syst.* **2003**, *75*, 215–234. [CrossRef]
46. Canales, E.; Bergtold, J.S.; Williams, J.; Peterson, J. Estimating farmers ' risk attitudes and risk premiums for the adoption of conservation practices under different contractual arrangements: A stated choice experiment. In Proceedings of the AAEA & WAEA Joint Annual Meeting; Agricultural and Applied Economics Association, Western Agricultural Economics Association, San Francisco, CA, USA, 26–28 July 2015.
47. Lesch, W.C.; Wachenheim, C.J. Factors Influencing Conservation Practice Adoption in Agriculture: A Review of the Literature. *Agribus. Appl. Econ. Rep.* **2014**. [CrossRef]
48. Boyer, T.A.; Kanza, P.; Ghimire, M.; Moss, J.Q. Household Adoption of Water Conservation and Resilience Under Drought: The Case of Oklahoma City. *Water Econ. Policy* **2015**, *1*, 1550005. [CrossRef]
49. Asfaw, D.; Neka, M. Factors affecting adoption of soil and water conservation practices: The case of Wereillu Woreda (District), South Wollo Zone, Amhara Region, Ethiopia. *Int. Soil Water Conserv. Res.* **2017**, *5*, 273–279. [CrossRef]
50. Millock, K.; Nauges, C. Household adoption of water-efficient equipment: The role of socio-economic factors, environmental attitudes and policy. *Environ. Resour. Econ.* **2010**, *46*, 539–565. [CrossRef]
51. Syme, G.J.; Shao, Q.; Po, M.; Campbell, E. Predicting and understanding home garden water use. *Landsc. Urban Plan.* **2004**, *68*, 121–128. [CrossRef]
52. Jara-Rojas, R.; Bravo-Ureta, B.E.; Díaz, J. Adoption of water conservation practices: A socioeconomic analysis of small-scale farmers in Central Chile. *Agric. Syst.* **2012**, *110*, 54–62. [CrossRef]
53. Nyangena, W. Social determinants of soil and water conservation in rural Kenya. *Environ. Dev. Sustain.* **2008**, *10*, 745–767. [CrossRef]

54. Mazzucato, V.; Niemeijer, D. The cultural economy of soil and water conservation: Market principles and social networks in Eastern Burkina Faso. *Dev. Change* **2000**, *31*, 831–855. [CrossRef]
55. Greenhalgh, T.; Robert, G.; Macfarlane, F.; Bate, P.; Kyriakidou, O. Diffusion of innovations in service organizations: Systematic review and recommendations. *Milbank Q.* **2004**, *82*, 581–629. [CrossRef]
56. Kassie, M.; Jaleta, M.; Shiferaw, B.; Mmbando, F.; Mekuria, M. Adoption of interrelated sustainable agricultural practices in smallholder systems: Evidence from rural Tanzania. *Technol. Forecast. Soc. Change* **2013**, *80*, 525–540. [CrossRef]
57. Di Falco, S.; Veronesi, M. Managing Environmental Risk in Presence of Climate Change: The Role of Adaptation in the Nile Basin of Ethiopia. *Environ. Resour. Econ.* **2014**, *57*, 553–577. [CrossRef]
58. Abdulai, A.N. Impact of conservation agriculture technology on household welfare in Zambia. *Agric. Econ.* **2016**, *47*, 729–741. [CrossRef]
59. De Graaff, J.; Amsalu, A.; Bodnár, F.; Kessler, A.; Posthumus, H.; Tenge, A. Factors influencing adoption and continued use of long-term soil and water conservation measures in five developing countries. *Appl. Geogr.* **2008**, *28*, 271–280. [CrossRef]
60. Bandiera, O.; Rasul, I. Social networks and technology adoption in Northern Mozambique. *Econ. J.* **2006**, *116*, 869–902. [CrossRef]
61. Kassie, M.; Teklewold, H.; Jaleta, M.; Marenya, P.; Erenstein, O. Understanding the adoption of a portfolio of sustainable intensification practices in eastern and southern Africa. *Land Use Policy* **2015**, *42*, 400–411. [CrossRef]
62. Drechsel, P.; Olaleye, A.; Adeoti, A.; Thiombiano, L.; Barry, B.; Vohland, K. Adoption driver and constraints of resource conservation technologies in sub-saharan Africa. *Berlin: FAO, IWMI, Humbold Universitaet*, 2005; Unpublished work.
63. Traore, N.; Landry, R.; Amara, N. On-Farm Adoption of Conservation Practices: The Role of Farm and Farmer Characteristics, Perceptions, and Health Hazards. *Land Econ.* **1998**, *74*, 114–127. [CrossRef]
64. Kassam, A.; Friedrich, T.; Shaxson, F.; Bartz, H.; Mello, I.; Kienzle, J.; Pretty, J. The spread of Conservation Agriculture: policy and institutional support for adoption and uptake. *Field Act. Sci. Rep. J. Field Act.* **2014**, *7*, 1–12.
65. Tadesse, M.; Belay, K. Factors influencing adoption of soil conservation measures in Southern Ethiopia: The case of Gununo area. *J. Agric. Rural Dev. Trop. Subtrop.* **2004**, *105*, 49–62.
66. Feder, G.; Just, R.E.; Zilberman, D. Adoption of agricultural innovations in developing countries: a survey. *Econ. Dev. Cult. Chang.* **1985**, *33*, 255–298. [CrossRef]
67. Tripathi, R.S.; Binswanger, H.P. The economics of tractors in South Asia; an analytical review. *Am. J. Agric. Econ.* **1979**, *61*, 581. [CrossRef]
68. Doss, C.; Mwangi, W.; Verkuijl, H.; De Groote, H. *Adoption of Maize and Wheat Technologies in Eastern Africa: A Synthesis of The Findings of 22 Case Studies*; CIMMYT: Mexico City, Mexico, 2003.
69. Giller, K.E.; Andersson, J.A.; Corbeels, M.; Kirkegaard, J.; Mortensen, D.; Erenstein, O.; Vanlauwe, B. Beyond conservation agriculture. *Front. Plant Sci.* **2015**, *6*, 870. [CrossRef]
70. Liu, T.; Bruins, R.J.F.; Heberling, M.T. Factors influencing farmers' adoption of best management practices: A review and synthesis. *Sustainability* **2018**, *10*, 432. [CrossRef]
71. Rasoulkhani, K.; Logasa, B.; Reyes, M.P.; Mostafavi, A. Understanding fundamental phenomena affecting the water conservation technology adoption of residential consumers using agent-based modeling. *Water* **2018**, *10*, 993. [CrossRef]
72. Swim, J.; Clayton, S.; Doherty, T.; Gifford, R.; Howard, G.; Reser, J.P.; Stern, P.C.; Weber, E.U. Psychology and Global Climate Change: Addressing a Multi-faceted Phenomenon and Set of Challenges A. *Am. Psychol.* **2009**, *66*, 241–250. [CrossRef]
73. Semenza, J.C.; Hall, D.E.; Wilson, D.J.; Bontempo, B.D.; Sailor, D.J.; George, L.A. Public perception of climate change voluntary mitigation and barriers to behavior change. *Am. J. Prev. Med.* **2008**, *35*, 479–487. [CrossRef]
74. Clark, W.A.; Finley, J.C. Determinants of water conservation intention in Blagoevgrad, Bulgaria. *Soc. Nat. Resour.* **2007**, *20*, 613–627. [CrossRef]
75. Kaiser, F.G.; Wölfing, S.; Fuhrer, U. Environmental attitude and ecological behaviour. *J. Environ. Psychol.* **1999**, *19*, 1–19. [CrossRef]

76. Kahsay, H.T.; Guta, D.D.; Birhanu, B.S.; Gidey, T.G. Farmers' Perceptions of Climate Change Trends and Adaptation Strategies in Semiarid Highlands of Eastern Tigray, Northern Ethiopia. *Adv. Meteorol.* **2019**, *2019*, 1–13. [CrossRef]
77. Slegers, M.F.W. "If only it would rain": Farmers' perceptions of rainfall and drought in semi-arid central Tanzania. *J. Arid Environ.* **2008**, *72*, 2106–2123. [CrossRef]
78. Teshome, A.; de Graaff, J.; Kassie, M. Household-Level Determinants of Soil and Water Conservation Adoption Phases: Evidence from North-Western Ethiopian Highlands. *Environ. Manag.* **2016**, *57*, 620–636. [CrossRef]
79. Karki, L.B.; Bauer, S. Assessment of impact of project intervention and factors determining technology adoption at small farm households level. In Proceedings of the The Deutscher Tropentag, Berlin, Germany, 5–7 October 2004.
80. URT. *Tanzania National Sample Census of Agriculture 2007/2008 Small Holder Agriculture: Regional Report—Singida Region*; United Republic of Tanzania, Ministry of Agriculture, Food Security and Cooperatives, Ministry of Livestock Development and Fisheries, Ministry of Water and Irrigation, Ministry of Agriculture, Livestock and Environment: Zanzibar City, Tanzania, April 2012.
81. Graef, F.; Sieber, S.; Mutabazi, K.; Asch, F.; Biesalski, H.K.; Bitegeko, J.; Bokelmann, W.; Bruentrup, M.; Dietrich, O.; Elly, N.; et al. Framework for participatory food security research in rural food value chains. *Glob. Food Sec.* **2014**, *3*, 8–15. [CrossRef]
82. Habtemariam, L.T.; Mgeni, C.P.; Mutabazi, K.D.; Sieber, S. The farm income and food security implications of adopting fertilizer micro-dosing and tied-ridge technologies under semi-arid environments in central Tanzania. *J. Arid Environ.* **2019**, *166*, 60–67. [CrossRef]
83. Trans-SEC—Innovating Strategies to safeguard Food Security using Technology and Knowledge Transfer: A people-centred Approach. Available online: http://www.trans-sec.org/ (accessed on 29 April 2019).
84. Shakya, P.B.; Flinn, J.C. Adoption of modern varieties and fertilizer use on rice in the eastern tarai of Nepal. *J. Agric. Econ.* **1985**, *36*, 409–419. [CrossRef]
85. Fishbein, M.; Ajzen, I. *Belief, Attitude, Intention and Behaviour*; Addison-Wesley: Boston, MA, USA, 1975; ISBN 0201020890.
86. Hill, R.J.; Fishbein, M.; Ajzen, I. Belief, Attitude, Intention and Behavior: An Introduction to Theory and Research. *Contemp. Sociol.* **1977**, *6*, 244. [CrossRef]
87. Lewin, K. Defining the "field at a given time" . *Psychol. Rev.* **1943**, *50*, 292–310. [CrossRef]
88. Tanellari, E.; Kostandini, G.; Bonabana-Wabbi, J.; Murray, A. Gender impacts on adoption of new technologies: the case of improved groundnut varieties in Uganda. *Afr. J. Agric. Resour. Econ.* **2014**, *9*, 1–9.
89. Wolka, K. Effect of soil and water conservation measures and challenges for its adoption: Ethiopia in focus. *J. Environ. Sci. Technol.* **2014**, *7*, 185–199. [CrossRef]
90. Darkwah, K.A.; Kwawu, J.D.; Agyire-Tettey, F.; Sarpong, D.B. Assessment of the determinants that influence the adoption of sustainable soil and water conservation practices in Techiman Municipality of Ghana. *Int. Soil Water Conserv. Res.* **2019**, *7*, 248–257. [CrossRef]
91. Teklewold, H.; Kassie, M.; Shiferaw, B. Adoption of multiple sustainable agricultural practices in rural Ethiopia. *J. Agric. Econ.* **2013**, *64*, 597–623. [CrossRef]
92. Anley, Y.; Bogale, A.; Haile-Gabriel, A. Adoption decision and use intensity of soil and water conservation measures by smallholder subsistence farmers in Dedo district, Western Ethiopia. *L. Degrad. Dev.* **2007**, *18*, 289–302. [CrossRef]
93. Dil Farzana, F.; Rahman, A.S.; Sultana, S.; Raihan, M.J.; Haque, M.A.; Waid, J.L.; Choudhury, N.; Ahmed, T. Coping strategies related to food insecurity at the household level in Bangladesh. *PLoS ONE* **2017**, *12*, 0171411. [CrossRef]
94. Beckers, B.; Berking, J.; Schütt, B. Ancient Water Harvesting Methods in the Drylands of the Mediterranean and Western Asia. *eTopoi. J. Anc. Stud.* **2013**, *2*, 145–164.
95. Gan, Y.; Siddique, K.H.M.; Turner, N.C.; Li, X.G.; Niu, J.Y.; Yang, C.; Liu, L.; Chai, Q. Ridge-Furrow Mulching Systems-An Innovative Technique for Boosting Crop Productivity in Semiarid Rain-Fed Environments. In *Advances in Agronomy*; Elsevier: Amsterdam, The Netherlands, 2013; pp. 429–476. ISBN 9780124059429.

96. Rockström, J.; Kaumbutho, P.; Mwalley, J.; Nzabi, A.W.; Temesgen, M.; Mawenya, L.; Barron, J.; Mutua, J.; Damgaard-Larsen, S. Conservation farming strategies in East and Southern Africa: Yields and rain water productivity from on-farm action research. *Soil Tillage Res.* **2009**, *103*, 23–32. [CrossRef]
97. Komakech, H.C.; Mul, M.L.; Van Der Zaag, P.; Rwehumbiza, F.B.R. Water allocation and management in an emerging spate irrigation system in Makanya catchment, Tanzania. *Agric. Water Manag.* **2011**, *98*, 1719–1726. [CrossRef]

MDPI
St. Alban-Anlage 66
4052 Basel
Switzerland
Tel. +41 61 683 77 34
Fax +41 61 302 89 18
www.mdpi.com

Water Editorial Office
E-mail: water@mdpi.com
www.mdpi.com/journal/water

www.ingramcontent.com/pod-product-compliance
Lightning Source LLC
LaVergne TN
LVHW071554170726
843515LV00008B/2288

9783036502281